Frontiers in Theoretical and Applied Physics/UAE 2017 (FTAPS 2017)

Journal of Physics: Conference Series Volume 869

Sharjah, United Arab Emirates
22 - 25 February 2017

Editors:

Ali S. Alnaser
Shahin Abdel-Naby
Raza Syed
Said Sakhi
Yousef Salamin
Mehmet Egilmez
Sami El-Khatib
Nasser Hamdan
Nidhal Guessoum
Tariq Majeed

ISBN: 978-1-5108-4443-8
ISSN: 1742-6588

Printed from e-media with permission by:

Curran Associates, Inc.
57 Morehouse Lane
Red Hook, NY 12571

CURRAN ASSOCIATES INC. proceedings.com

Printed by Curran Associates, Inc. (2017)

For permission requests, please contact the Institute of Physics at the address below.

Institute of Physics
Dirac House, Temple Back
Bristol BS1 6BE UK

Phone: 44 1 17 929 7481
Fax: 44 1 17 920 0979

techtracking@iop.org

Additional copies of this publication are available from:

Curran Associates, Inc.
57 Morehouse Lane
Red Hook, NY 12571 USA
Phone: 845-758-0400
Fax: 845-758-2633
Email: curran@proceedings.com
Web: www.proceedings.com

Table of contents

Volume 869

Frontiers in Theoretical and Applied Physics/UAE 2017 (FTAPS 2017)
22–25 February 2017, American University of Sharjah (AUS), UAE

Accepted papers received: 27 June 2017
Published online: 11 July 2017

Preface

Papers

IOP Conf. Series: Journal of Physics: Conf. Series **869** (2017) 011001 doi:10.1088/1742-6596/869/1/011001

Preface:

The International Conference Frontiers in Theoretical and Applied Physics |UAE 2017 was organized by the Department of Physics at the American University of Sharjah (AUS) in partnership with the Emirates Mars Mission and Mohammed Bin Rashid Space Center (MBRSC) in Dubai-United Arab Emirates. The conference was endorsed by the American Physical Society (APS), Materials Research Society (MRS), the Institute of Physics (IOP) in the UK, and the International Astronomical Union (IAU)-Office of Astronomy for Development. It was held on the beautiful campus of the American University of Sharjah (AUS) in the United Arab Emirates between 22-25February,2017. The conference was the first international and interdisciplinary physics conference in the region that attracted such large number of world-leading scientists and experts in the fields of atomic, molecular and optical physics, condensed matter and materials science, astrophysics and space science, nuclear and high energy physics, and mathematical physics. It had a broad international participation from more than 45 countries, with registered delegates numbering over 330. The scientific program included the presentation and discussion of 300 papers, classified as 5 keynote lectures (each lasted 1 hour), 60 invited talks (each lasted thirty minutes), 145 contributed talks (12 minutes each), and 90 posters. The keynote lectures were attended by all conference attendees in addition to undergraduate and graduate students from AUS, and other guests from neighboring institutions. The invited talks were scheduled into parallel sessions, followed by contributed talks, corresponding to the different research areas that constitute the themes of the conference. Those themes included: Atomic, Molecular, and Laser Physics; Materials Science; Astrophysics and Planetary Science; Mathematical Physics; the Solar System and Mars; and High Energy and Particle Physics.

Submission of papers for the conference proceedings was voluntary. 125 manuscripts were received for consideration in the proceedings and 96 were accepted for publication after a peer-review conducted by the conference's scientific committee in consultation with experts in the corresponding fields. The keynote and invited speakers, in addition to some of the international advisory board members, were invited to submit their papers for publication in a dedicated special issue: Focus issue on Theoretical and Applied Physics in Physica Scripta, the broad scope journal of the Royal Swedish Academy of Sciences. Those invited papers are to be reviewed and handled by Physica Scripta itself. The editors of this focus issue are Ali S. Alnaser, Roland Allen, and Suzy Lidström, Editor-in-chief of Physica Scripta.

Based on an assessment conducted by Roland Allen and Suzy Lidström, with help from the conference scientific committee, the best poster prize presented by Physica Scripta was awarded to two separate teams from the Beirut Arab University and from Mohammed Bin Rashid Space Center.

We are very pleased and proud to have held this truly international conference for it provided a unique forum for physicists, astronomers, and space and material scientists from around the world to present the latest developments in their various dynamic fields of physics and meet at the American University of Sharjah. We believe that the conference had provided ample opportunity for intensified exchange and debate on contemporary research issues, and allowed for exchanges of ideas, and exploring different means of regional and international collaborations. We hope that hosting such international and interdisciplinary conference at the American University of Sharjah (AUS) would add to its visibility in the contemporary research arena, where it came as a timely activity that is consistent with the university's current growth and research directions that are geared to meet the UAE and the region's science and technology interests.

Besides the exciting and rich scientific program, a group of space scientists and astrophysicists enjoyed a visit to the Mohammed Bin Rashid Space Center (MBRSC) in Dubai, where they got to know about the center's activities and future plans. Additionally, the conference attendees and their companions enjoyed the social program that was arranged by the organizing committee in the touristic and historical places around the cities of Sharjah and Dubai. The social program included trips to Sharjah Cornish and its historic museums; Dhow Cruises looking over the old and modern parts of Dubai; Safari with BBQ dinner and entertainment in the Arabian Desert; visit to the top of Burj Khalifa (124 floor) and the famous Dubai Mall; in addition to experiencing indoor skiing at the Mall of the Emirates; and tours in Dubai Aquarium and Underwater Zoo. The delegations also enjoyed the Gala dinner, along with shows of cultural performances, shows and dances, which took place at the Radisson Blue Resort facing the Arabian Gulf on the night of February 23.

We would like to thank the administration at the American University of Sharjah for their support and for providing the convenient and beautiful venue of the conference; and for facilitating all the necessary logistics. We are thankful to Mohammed Bin Rashid Space Center (MBRSC) for providing a generous financial support. We would also like to thank our keynote and invited speakers, and all participants who contributed with oral or poster presentations, and other attendees for their valuable contributions to the success of the conference.

Given the huge success of the first edition and in response to the expressed wish by most of the conference attendees, it is certain that we will be hosting the second International Conference Frontiers on Theoretical and Applied Physics again at the American University of Sharjah in the winter of 2019 or 2020. The exact date will be announced before the end of this year and further details will be made clear and circulated soon after that. We will be delighted to host the next conference and we look forward to welcoming you again.

More information about the conference can be found at:

www.aus.edu/ftaps2017

The list of people who were involved in the organization and contributed to the success of the conference is given below.

Conference Chair: Ali S. Alnaser- American University of Sharjah (AUS)

Keynote Speakers:

Andrew Wee (National University of Singapore (NUS), Singapore)
David Villeneuve (National Research Council of Canada)
Joachim Hermann Ullrich (Physikalisch-Technische Bundesanstalt, Germany)
Mark Blamire (University of Cambridge)
Sushil Atreya (University of Michigan)

Invited Speakers:

Alaa Ibrahim (American University of Cairo, Egypt)
Alex Vazdekis (Instituto de Astrofsica de Canarias (IAC), Spain)
Anil Bhardwaj (Space Physics Laboratory, Vikram Sarabhai Space Centre, India)
Chihiro Tao (National Institute of Information and Communications Technology, Japan)
Francesco Arneodo (NYUAD, UAE)
Ilias Fernini (University of Sharjah, UAE)
Ingo Mueller-Wodarg (Imperial College, London, UK)
Jean-Yves Chaufray (Laboratoire de Météorologie Dynamique, France)

IOP Conf. Series: Journal of Physics: Conf. Series **869** (2017) 011001 doi:10.1088/1742-6596/869/1/011001

Margaret Hanson (University of Cincinnati, USA)
Maurizio Salaris (Liverpool John Moores University, UK)
Mike Beasely (Instituto de Astrofsica de Canarias (IAC), Spain)
Ricardo Schiavon (Liverpool John Moores University, UK)
Richard de Grijs (Kavli Institute for Astronomy and Astrophysics, China)
Roger V. Yelle (University of Arizona, USA)
Santi Cassisi (INAF, Italy)
Steve Bougher (University of Michigan, USA)
Syed A. Haider (Space & Atmospheric Science Division, Physical Research Laboratory, India)
Tahar Amari (Centre de Physique Théorique, France)
Walid Azzam (University of Bahrain, Bahrain)
Wing-Hue Ip (National Central University, Taiwan)
Yasumasa Kasaba (Tohoku University, Japan)
Zouhair Benkhaldoun (Cadi Ayyad University, Morrocco)
Amine Cassimi (Centre de Recherche sur les Ions, les Matériaux et la Photonique, France)
Chunlei Guo (University of Rochester, USA)
David Villeneuve (National Research Council of Canada)
Deepak Mathur (Tata Institute of Fundamental Research, India)
Dieter Bauer (University of Rostock, Germany)
Goong Chen (Texas A & M University, USA)
Igor Litvinyuk (Griffiths University, Australia)
Jiansheng Liu (Shanghai Institute for Optics and Fine Mechanics, China)
Kiyoshi Ueda (Tohoku University, Japan)
Koji Sugioka (RIKEN Center for Advanced Photonics, Japan)
Matthias Kling (Max Planck Institute for Quantum Optics, Germany)
Ravi Bhurdwaj (University of Ottawa, Canada)
Reinhard Dörner (Goethe Universität, Germany)
Robert Moshammer (Max-Planck-Institut für Kernphysik Heidelberg, Germany)
Suhail Zubairy (Texas A&M University, USA)
Yunpei Deng (Paul Scherrer Institute, Switzerland)
Abdel Khaleq Alsmadi (Kuwait University, Kuwait)
Chris Leighton (University of Minnesota, USA)
Erick Fullerton (Center for Memory and Recording Research, UCSD, USA)
Heinrich Nakotte (New Mexico State University, USA)
Husam N. AlShareef (KAUST, Saudi Arabia)
Jason Robinson (University of Cambridge, U.K.)
Khalid Eid (Miami University, USA)
Lutfi Ozyuzer (Izmir institute of Technology, Turkey)
Mihai Irimia-Vladu (Institute for Surface Technologies and Photonics, Austria)
Mukhles Sowwan (Okinawa Institute of Science and Technology Graduate University, Japan)
Nouar Tabet (Qatar Environment and Energy Research Institute, Qatar)
Omar F. Mohammed (KAUST, Saudi Arabia)
Osman Bakr (KAUST, Saudi Arabia)
Parinda Vasa (Indian Institute of Technology Bombay, India)
Roland Allen (Texas A&M University, USA)
Sebastian Bergeret (Material Physics Center (MPC/CFM), Spain)
Suzy Lidstrom (Physica Scripta, Sweden)
Udo Schwingeschlögl (KAUST, Saudi Arabia)

IOP Conf. Series: Journal of Physics: Conf. Series **869** (2017) 011001 doi:10.1088/1742-6596/869/1/011001

Editors of the Proceedings:

Ali S. Alnaser, Shahin Abdel-Naby, Raza Syed, Said Sakhi, Yousef Salamin, Mehmet Egilmez, Sami El-Khatib, Nasser Hamdan, Nidhal Guessoum, Tariq Majeed - American University of Sharjah

Scientific Committee:

Ali S. Alnaser- American University of Sharjah (AUS)
Mahmoud Rasras- Masdar Institute of Science and Technology, UAE
Maryam Al-Shamsi- Mohammad Bin Rashid Space Center, UAE
Mehmet Egilmez- American University of Sharjah, UAE
Nidhal Guessoum- American University of Sharjah, UAE
Omar F. Mohammed- King Abdullah University of Science and Technology, Saudi Arabia
Tariq Majeed- American University of Sharjah, UAE
Usama Al-Khawaja- UAE University, Al Ain, UAE
Yousef Salamin- American University of Sharjah. Sharjah, UAE

International Advisory Board:

Pran Nath (Northeastern University, USA)
Deepak Mathur (TIFR, India)
Andrew Szentgyorgyi (CFA, USA)
Vitaly I. Konov (NSI, Moscow, Russia)
Sushil Atreya (University of Michigan, USA)
Suzanne Lidström (Physica Scripta, RSAS, Sweden)
Jan Alexander Jung (University of Alberta, Canada)
Reinhard Dörner (Goethe Universität, Germany)
Paul Corkum (University of Ottawa, Canada)
Ruxin Li (Chinese Academy of Science, China)
Udo Schwingeschlögl (KAUST, Saudi Arabia)
Suhail Zubairy (Texas A&M University, USA)
James Colgan (Los Alamos National Laboratory, USA
Ferenc Krausz (MPQ, Germany)
Husam N. Alshareef (KAUST, KSA)
Mostafa El-Syed (GIT, SA)
Dan Neumann (NIST, USA)
Metin Arik (Bogazici University, Turkey)
Dieter Hartmann (Clemson University, USA)
Ata Sarajedini (University of Florida, USA)
Alexandre Vazdekis (IA, Canary Islands, Spain)
Anil Bhardwaj (Vikram Sarabhai Space Centre, India)
MooYoung Choi (Seoul National University, Korea)
John D. Barrow (University of Cambridge, UK)
Jason Robinson (University of Cambridge, UK)
Andrew Wee (NUS, Singapore)

Local Organizing Committee:

Asad Hasan, AUS
Isra Siry, AUS
Nasser Hamdan, AUS
Said Sakhi, AUS

Shahin AbdelNaby, AUS
Randa Asa'd, AUS
Raza Syed, AUS
Sami El-Khatib, AUS
Sarah Amiri, MBRSC-Dubai

Local Logistics:
Noor AbdulMaati, AUS
Mazhar Iqbal , AUS
Naveed Ali, AUS
Faisal Masood, AUS
Mohammad Idrees, AUS.

IT-Support:
Nabeel Amireh
Huzefa Pedhiwala
Aamenah Bashir
Farouk Achoui
Rizwan Bakhsh

IOP Conf. Series: Journal of Physics: Conf. Series **869** (2017) 011001 doi:10.1088/1742-6596/869/1/011001

Some photos from the conference:

IOP Conf. Series: Journal of Physics: Conf. Series **869** (2017) 011001 doi:10.1088/1742-6596/869/1/011001

IOP Conf. Series: Journal of Physics: Conf. Series **869** (2017) 011001 doi:10.1088/1742-6596/869/1/011001

IOP Conf. Series: Journal of Physics: Conf. Series **869** (2017) 011001 doi:10.1088/1742-6596/869/1/011001

IOP Conf. Series: Journal of Physics: Conf. Series **869** (2017) 011001 doi:10.1088/1742-6596/869/1/011001

IOP Conf. Series: Journal of Physics: Conf. Series **869** (2017) 011001 doi:10.1088/1742-6596/869/1/011001

APS physics
MRS
IAU OAD
WILEY
Frontiers in Theoretical and Applied Physics UAE
FTAPS2017
February 22-25, 2017 | Main Building, AUS
AUS | American University of Sharjah

IOP Conf. Series: Journal of Physics: Conf. Series **869** (2017) 011001 doi:10.1088/1742-6596/869/1/011001

IOP Conf. Series: Journal of Physics: Conf. Series **869** (2017) 011001 doi:10.1088/1742-6596/869/1/011001

IOP Conf. Series: Journal of Physics: Conf. Series **869** (2017) 011001 doi:10.1088/1742-6596/869/1/011001

IOP Conf. Series: Journal of Physics: Conf. Series **869** (2017) 011001 doi:10.1088/1742-6596/869/1/011001

IOP Conf. Series: Journal of Physics: Conf. Series **869** (2017) 011001 doi:10.1088/1742-6596/869/1/011001

IOP Conf. Series: Journal of Physics: Conf. Series **869** (2017) 011001 doi:10.1088/1742-6596/869/1/011001

IOP Conf. Series: Journal of Physics: Conf. Series **869** (2017) 011001 doi:10.1088/1742-6596/869/1/011001

IOP Conf. Series: Journal of Physics: Conf. Series **869** (2017) 011001 doi:10.1088/1742-6596/869/1/011001

IOP Conf. Series: Journal of Physics: Conf. Series **869** (2017) 011002 doi:10.1088/1742-6596/869/1/011002

Peer review statement

All papers published in this volume of *Journal of Physics: Conference Series* have been peer reviewed through processes administered by the proceedings Editors. Reviews were conducted by expert referees to the professional and scientific standards expected of a proceedings journal published by IOP Publishing.

IOP Conf. Series: Journal of Physics: Conf. Series **869** (2017) 012001 doi:10.1088/1742-6596/869/1/012001

Theoretical electronic structure of the molecule SrCl for the feasibility of laser cooling

B Rafei, G Younes and M Korek

Faculty of Science, Beirut Arab University, P.O. Box 11-5020 Riad El Solh, Beirut 1107 2809, Lebanon

E-mail: Mahmoud.korek@bau.edu.lb

Abstract. Electronic structure of the molecule SrCl has been investigated by employing the *ab initio* methods CASSCF/MRCI+Q with Davidson correction in the representation $^{2s+1}\Lambda^{\pm}$, neglecting spin-orbit effects, and using the computational program Molpro. The potential energy curves of the low-lying doublet and quartet electronic states have been investigated. The harmonic vibrational wave number ω_e, the relative electronic energy T_e, referred to the ground state, the rotational constant B_e and the equilibrium internuclear distance R_e, have been calculated for the investigated electronic states. Moreover, the static dipolar moments have been determined in term of the internuclear distance R_e.

1. Introduction

Because of their wide-ranging applications in many areas, the strontium monohalides SrX have received considerable attention from both experimentalists and theoreticians. The Strontium monohalide molecules are present in sunspots, SC-stars, cool M-giant stars, stellar red-giant stars, and in S-type stars [1]. The abundances of these species are important for understanding the evolutionary phases, activities and interactions of the stars with cometary material [2]. Since the SrX molecules are produced in the gaseous phase at high temperature, it is very difficult to realize such experimental measurement at high temperature to produce these molecules and observe their spectra [3, 4]. Therefore, theoretical studies are quite useful for the experimentalists to identify the electronic structures and their corresponding spectroscopic constants.

The alkaline earth mono-halides all have diagonal Franck–Condon Factors (FCFs) and are, in principle, amenable to laser cooling. While many of these molecules are studied experimentally [5-7], there is absence of the study of the electronic structure of the molecules SrX, particularly for the higher excited electronic states. Shuman *et al.* [8] introduced the first direct cooling of a polar molecule (SrF) by using an optical cycling scheme with Doppler and Sisyphus cooling forces. This significant innovation needs deep knowledge of the electronic structure and its ro-vibrational energy levels of the ultracold candidate molecules. Moreover, highly diagonally distributed FCFs and the short radiative lifetimes (between 10^{-6} and10^{-9} s) are needed for these ultracold molecules. While the FCF calculation is based on the accurate potential energy curve of two electronic states, the radiative lifetimes need the calculation of the dipole moment.

DeMill [9] has detailed a prototype design for quantum computation using ultracold polar molecules, trapped in a one-dimensional optical lattice, partially oriented in an external electric field, and coupled by the dipole-dipole interaction. This offers a promising platform for quantum computing because scale-up appears feasible to obtain large networks of coupled qubits.

The spectra of alkaline earth monohalides have been frequently investigated in the literature [10-16] by using various theoretical and experimental spectroscopic techniques. For the molecule SrCl, the $A^2\Sigma^+$ - $X^2\Sigma^+$ transition has been observed and analysed by Brinkmann *et al.* [17], Singh *et al.* [18], and Schröder *et al.* [19], while the transition $B^2\Sigma^+$ - $X^2\Sigma^+$ has been reported by Ernest and Schröder [20].

In the present work, the theoretical electronic structure of SrCl molecule is investigated, using *ab-initio* methods. The potential energy curves (PECs) are plotted, the electronic energy with respect to the ground state T_e, the harmonic frequency ω_e, the rotational constant B_e, the internuclear distances at equilibrium R_e and the static dipole moments, have all been calculated.

2. Computational methods

Electronic structure of the molecule SrCl has studied by using Multi-Configuration Self-Consistent Field (MCSCF) methods [21]. This generation is followed by the Multi Reference Configuration Interaction (MRCI+Q) technique, with Davidson correction +Q [22]. These calculations have been performed employing the computational chemistry program MOLPRO [23], using the graphical user interface GABEDIT [24]. For the strontium atom, Sr, the basis set is an Effective Core Potential; ECP28MWB [25], choosing 28 inner electrons in the core. The chlorine atom is taken as an all electrons scheme with the basis aug-cc-pVQZ;c (augmented correlation consistent polarized valence quintet zeta). Therefore, 27 electrons will remain in the valence shells for an SrCl molecule. The active space contains 18 active orbitals corresponding to 6σ[Sr: 5s, $4d_0$, $5p_0$, 6s; Cl: $3p_0$, $4s_0$], 3π[Sr: $4d_{\pm1}$, $5p_{\pm1}$;Cl: $3p_{\pm1}$] and 1δ[Sr: $4d_{\pm2}$] distributed as 6a1, 3b1, 3b2, and 1a2 noted [6,3,31] in the irreducible representation of the C2v point group. We choose to freeze 20 electrons among the 27 free electrons for this molecule. This implies that 7 valence electrons are considered in the MRCI calculation and used for the excitation to the different active orbitals.

3. Results and discussion

The SrCl molecule is assumed to be ionic Sr^+Cl^-. Our computational calculations have been done for the first lowest 5 electronic states of the SrCl molecule. The theoretical energies as functions of the internuclear distance R_e, varying from 1.88 A to 3.08 A, have been calculated. The ground state is $X^2\Sigma^+$, and the lowest excited electronic states are all found located below 35100 cm^{-1}. The potential energy curves of these electronic states are given in Figure 1.

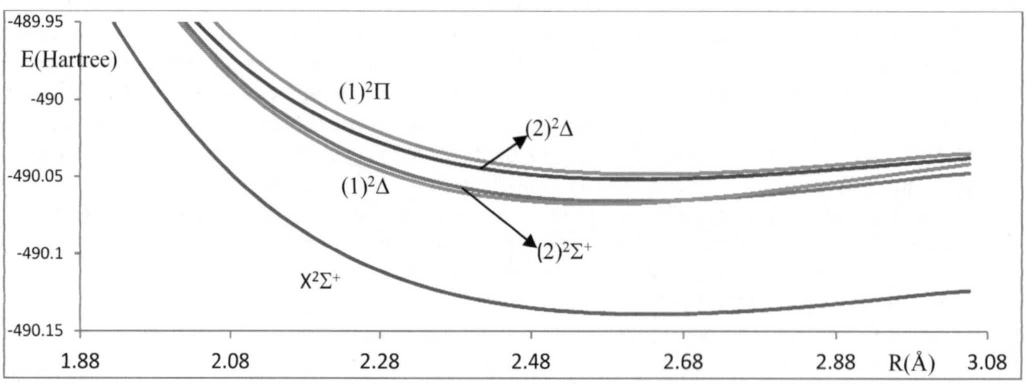

Figure 1. Potential energy curves of the molecule SrCl.

Theoretical energy values, T_e, the spectroscopic constants ω_e and B_e, as well as the internuclear distance at equilibrium, R_e, are represented in the Table 1, along with different experimental and theoretical results given in the literature.

IOP Conf. Series: Journal of Physics: Conf. Series **869** (2017) 012001 doi:10.1088/1742-6596/869/1/012001

Table 1: Spectroscopic constants of the molecule SrCl

state	T_e (cm^{-1})	R_e (Å)	B_e (cm^{-1})	ω_e (cm^{-1})
	0.0	2.63[a]	0.10370[a]	289.6[a]
$X^2\Sigma^+$		2.57[b]	0.10156[b]	302[b]
		2.57[c]		
$(1)^2\Pi$	15700[a]	2.58[a]	0.10546[a]	323[a]
	14818[b]	2.55[e]	0.103989[c]	309[b]
	15616[c]			
	16073[a]	2.60[a]	0.105449a	299[a]
$(2)^2\Sigma^+$	15719[b]	2.55[e]	0.103067[d]	306[b]
	16605[f]			
$(1)^2\Delta$	19198[a]	2.63[a]	0.0963[a]	280[a]
	18607[f]			
$(2)^2\Delta$	35072[a]	2.47[a]	0.11053[a]	364[a]
	34085[b]			356[b]

[(a)]Our theoretical values (MRCI); [(b)]Experimental values Ref. [12]; [(c)]Experimental values Ref. [18]; [(d)]Experimental values Ref. [15], [(e)]Deduced values from (c) and (d); [(f)] Theoretical values Ref. [10].

Comparison of our calculated values of T_e with those obtained experimentally in the literature [9, 13, 19], shows very good agreement with the relative difference $0.54\% \leq \Delta T_e/T_e \leq 5.6\%$. For the internuclear distance, R_e, our theoretical values are very significant, since the relative difference from those obtained experimentally in the literature [9, 13, 19] shows excellent agreement with the value $1.2\% \leq \Delta R_e/R_e \leq 2.3\%$. While for the ω_e and B_e constants, the relative differences show very good agreement with the experimental data [9, 13, 19] with relative differences $1.2\% \leq \Delta R_e/R_e \leq 2.3\%$ and $1.2\% \leq \Delta R_e/R_e \leq 2.3\%$, respectively. One can notice that for the majority of the investigated electronic states, good agreement with the published data in the literature. This confirms accuracy of the new data given in Table 1.

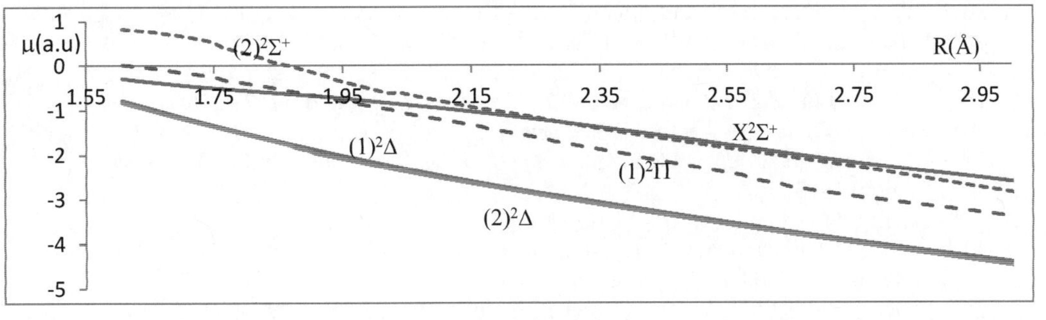

Figure 2. Dipole moment curves of the molecule SrCl.

For the study of the bond nature and the polarity of the molecule SrCl, we show in Figure 2, the variation of the static dipole moment of this molecule in terms of the internuclear distance R by taking the Sr atom at the origin. Note that the dipole moment curves are in the negative region and that the sign convention corresponds to the polarity of $Sr^{\delta+}Li^{\delta-}$. Moreover, these dipole moment curves increase in the negative region, which points out polarized states that dissociate into ionic fragments. Note also the linear trend of the static dipole moment curves in the negative region. This behaviour of a bond from such curve can be understood by using the Pacchioni *et al.* method [26]. For two electric charges +q and –q separated by the internuclear distance R_e, the dipole moment curve is a straight line if the compound formed by the two charges is ionic, and curved if the compound has a dominating covalent character.

4. Conclusion

As implemented in the use of *ab-initio*/MRCI+Q calculations, the adiabatic potential energy and the dipole moment curves for the low-lying electronic states of SrCl molecule have been calculated. From the plots of these curves the spectroscopic constants R_e, ω_e, T_e, and B_e, are determined for the bound electronic states. Comparing the obtained values with the available data, both theoretical and experimental, shows very good agreement, which confirms the validity and the accuracy of the newly investigated constants in the present work. These data can lead to the calculation of the Franck-Condon factor, which demonstrate feasibility of laser cooling of the molecule SrCl.

References

[1] Sri Ramachandran P, Rajamanickam N, and Bagare S P 2006 *Serbian Astronomical Journal* **172** 13
[2] Short C I, Hauschildt H P, Ernest W E and Schröder J O 2006 *ApJ* **641** 494
[3] Langhoff S R, Bauschlicher C W and Patridge H J 1988 *Chem. Phys.* **89** 396
[4] Carlson K D and Claydon C R 1967 *Adv. High Temp. Chem.* **1** 43
[5] Ernest W E and Schröder J O 1984 *J. Chem. Phys.* **81** 136
[6] Schröder J O and Ernest W E 1985 *J. Mol. Spectrosc.* **112** 413
[7] Ernest W E and Schröder J O 1986 *J. Mol. Spectrosc.* **117** 444
[8] Shuman E S, Barry J F and DeMille D 2010 *Nature* **467** 822.
[9] DeMille D 2002 *Phys. Rev. Lett.* **88** 067901
[10] Allouche A R, Wannous G and Aubert-Frecon M 1993 *Chem. Phys.* **170** 11
[11] Dagdigian P J, Cruse H W and Zare R N J. 1974 *Chem. Phys.* **60** 2330
[12] Huber K P and Herzberg G, *Molecular spectra and molecular structure*, 4 (Van Nostrand Princeton 1978)
[13] Berg L E, Royen P and Weijnitz P 1990 *Mol. Phys.* **69** 385
[14] Torring T, Ernst W E and Kandler J 1989 *J. Chem. Phys.* **90** 4927
[15] Novikov M M and Gurvich L V 1965 *Opt. Spectrosc.* **19** 76
[16] Schröder J O, Nitsch C and Ernst W E 1988 *J. Mol. Spectrosc.* **132** 166
[17] Brinkmann U, Schmidt V H, and Telle H1982 *Chem. Phys.* **64** 19
[18] Singh J, Nai K P R, Upadhya K N and Rai D K 1970 *Opt. Pura Apl.* **3** 76
[19] Schröder J O, Zeller B and Ernest W E, 1988 *J. Mol. Spectrosc.* **127** 255
[20] Ernest W E and Schröder J O 1983 *Chem. Phys.* **78** 363
[21] Atkins P W and Friedman R S, *Molecular Quantum Mechanics* (Oxford University Press, 3thd Edition) 1997
[22] Davidson E R and Silver D W 1977 *Chem. Phys. Lett.* **52** 403
[23] Werner H J *et al a package of ab initio programs* 2010.1, http://www.molpro.net/info/users.
[24] Allouche A R 2011 *J. Comput. Chem.* **32** 174
[25] Dolg M, Stoll H, Preuss H and Pitzer R M 1993 *J. Phys. Chem.* **97** 5852
[26] Pacchioni P and Illas F 1995 *Chem. phys.* **199** 155

IOP Conf. Series: Journal of Physics: Conf. Series **869** (2017) 012002 doi:10.1088/1742-6596/869/1/012002

Energy levels and radiative rates for transitions in Mo XV

F El-Sayed[1,2] and S M Attia[1,2]

[1] Physics Department, Faculty of science, Kafrelsheikh University, Kafr El-Sheikh, Egypt

[2] Physics Department, Faculty of Applied Science, Umm Al-Qura University, Makkah, Saudi Arabia

E-mail: fatma_mahrous@yahoo.com

Abstract. Energy levels, wavelengths, transition probabilities, oscillator strengths, line strengths, and lifetimes have been calculated for transitions among the fine-structure levels belonging to the $(1s^22s^22p^6)3s^23p^63d^{10}$, $3s^23p^63d^94l$, $3s^23p^53d^{10}4l$, and $3s3p^63d^{10}4l$ (l= s, p, d, f) configurations of the Ni-like Molybdenum, Mo XV. The results for electric dipole (E1), electric quadrupole (E2), magnetic dipole (M1), and magnetic quadrupole (M2) transitions among the lowest levels of Mo XV have been reported and compared with available NIST results.

1. Introduction

Coherent and saturated wavelengths toward the water window are important objectives in the development of X-ray lasers [1]. The are required for holography [2], microscopy of biological specimens [3], for deflectometry [4] and radiography [5] of dense plasmas relevant to inertial confinement fusion and laboratory astrophysics.

The atomic data for Ni-like ions are needed in the development of X-ray lasers, there are few studies on energy levels, oscillator strengths, and spontaneous radiative decay rates [6-16]. Hagelstein [6] calculated electron excitation cross sections and rate coefficients between the $3s^23p^63d^{10}$ Ni-like Gd XXXVII ground state and the singly excited states with an N-shell electron. Brage *et al.* [7] identified lines of the $3d^94s$-$3d^94p$ transitions in Ni-like Zr XIII, Nb XIV, and Mo XV in the 340-500 Å region in spectra emitted from laser produced plasmas.

Zhang *et al.* [8], using the Dirac-Fock-Slater (DFS) code, obtained radiative rates and collision strengths for transitions from the ground state, and performed their calculations for all ions with Z in the range Z = 60 – 92 among the 249 energetically lowest levels. Aggarwal *et al.* [9], using the fully relativistic GRASP code, calculated energy levels and radiative rates for allowed transitions among the 107 fine-structure levels belonging to the $(1s^22s^22p^6)3s^23p^63d^{10}$, $3s^23p^63d^94l$, $3s^23p^53d^{10}4l$, and $3s3p^63d^{10}4l$ configurations of the Ni-like ions Nd XXXIII, Sm XXXV, Eu XXXVI, Ta XXXXVI, and W XXXXVII. Dong *et al.* [10] calculated the 45 low-lying energy levels and the transition probabilities among the levels for Ni-like ions with $74 \leq Z \leq 84$, and considered the main correlation, relaxation and relativistic effects. The energy levels, transition probabilities, and electron impact excitations for possible X-ray line emission have been calculated using the flexible atomic code (FAC), for Ni-like Ta XXXXVI [11], La XXX [12], Au XXXXXII [13], Xe XXVII [14], and Sn XXIII [15]. Xie Lu-You *et al.* [16], using the multiconfiguration Dirac-Fock (MCDF) method, calculated wavelengths and transition probabilities of the two-electron and one-photon (TEOP) transitions for Ni-like ions with $47 \leq Z \leq 92$.

IOP Conf. Series: Journal of Physics: Conf. Series **869** (2017) 012002 doi:10.1088/1742-6596/869/1/012002

2. Calculations

The General Purpose Relativistic Atomic Structure Package (GRASP) [17–19] and the fully relativistic MCDF approach with the multiconfiguration Dirac-Fock and General Matrix Elements (MCDFGME) program [20] have been used in energy levels, transition probabilities, oscillator strengths, and line strengths calculations.

The GRASP is a fully relativistic code and the relativistic corrections arising from the Breit interaction and QED effects have been included. The standard scheme of orbital optimization is based on an iterative procedure involving the radial functions of the orbitals and the expansion coefficients until the required degree of self-consistency is achieved.

The MCDF is the fully relativistic program of Desclaux and Indelicato, which is available at http://dirac.spectro.jussieu.fr/mcdf, and is used with the inclusion of the relativistic two-body Breit interaction and of the quantum electrodynamic corrections that are due to self-energy and vacuum polarization.

In this paper, we present fine-structure energy levels as well as wavelengths, oscillator strengths, and transition probabilities for all fine-structure transitions between the levels of Mo XV. The data refer to 107 fine-structure levels belonging to the $(1s^22s^22p^6)3s^23p^63d^{10}$, $3s^23p^63d^94l$, $3s^23p^53d^{10}4l$, and $3s3p^63d^{10}4l$ (l = s, p, d, f) configurations.

3. Results and Discussion

The three resonance lines $3d^{10}\ {}^1S_0$-$3d^94p$ (${}^3P_1$, ${}^1P_1$, and ${}^3D_1$) were first measured with an uncertainty of ±0.02 Å by Alexander *et al.* [21]. Two electric quadrupole lines $3d^{10}\ {}^1S_0$-$3d^94s\ {}^{1,3}D_2$ were observed with an uncertainty ±0.005 Å by Klapisch et al. [22]. These lines were also observed by Mansfield *et al.* [23]. The energy levels, wavelengths, and intensities for Mo ions have been compiled [24, 25].

Table 1 presents GRASP and MCDF energy levels for Mo XV. The GRASP with and without the inclusion of Breit and QED effects (GRASP^a and GRASP^b) and MCDF results are in comparison with the available NIST results. Our Breit- and QED energies are lower than the corresponding Coulomb energies by up to 0.07 Ryd (≈ 0.19%), and our GRASP and MCDF energies are lower than NIST values by up to 0.15 Ryd and 0.07 Ryd, respectively, except $3d^94f\ {}^3P_2^o$ level.

Table 1. Energy Levels of Mo XV, their threshold energies (in Ryd), and lifetimes τ (in s).

Index	Configuration	LSJ	GRASP^a	GRASP^b	MCDF	NIST	τ (s)
1	$3s^23p^63d^{10}$	${}^1S_0^e$	0.000	0.000	0.000	0.000	---
2	$3s^23p^63d^94s$	${}^3D_3^e$	15.323	15.319	15.390	15.445	---
3	$3s^23p^63d^94s$	${}^3D_2^e$	15.370	15.366	15.436	15.490	1.144E-07
4	$3s^23p^63d^94s$	${}^3D_1^e$	15.575	15.562	15.633	15.689	4.202E-03
5	$3s^23p^63d^94s$	${}^1D_2^e$	15.622	15.608	15.679	15.732	7.599E-08
6	$3s^23p^63d^94p$	${}^3P_2^o$	17.482	17.479	17.566	17.614	1.251E-10
7	$3s^23p^63d^94p$	${}^3F_3^o$	17.552	17.548	17.623	17.674	1.174E-10
8	$3s^23p^63d^94p$	${}^3F_2^o$	17.764	17.751	17.828	17.881	1.138E-10
9	$3s^23p^63d^94p$	${}^3P_1^o$	17.768	17.757	17.843	17.894	1.022E-10
10	$3s^23p^63d^94p$	${}^3F_4^o$	17.812	17.804	17.880	17.936	8.003E-11
11	$3s^23p^63d^94p$	${}^1D_2^o$	17.911	17.903	17.972	18.025	7.879E-11
12	$3s^23p^63d^94p$	${}^3D_3^o$	17.953	17.946	18.015	18.069	7.094E-11
13	$3s^23p^63d^94p$	${}^1P_1^o$	17.955	17.945	18.010	18.064	2.374E-12
14	$3s^23p^63d^94p$	${}^3P_0^o$	17.962	17.945	18.036	18.088	9.096E-11

Tables 2 and 3 present GRASP wavelengths (λ_{ij}), length form of oscillator strengths (f_{ij}), transition probabilities (A_{ji}), and line strengths (S_{ji}) for E1 and M2 transitions, and for E2 and M1 transitions, respectively. The two forms of oscillator strengths (f_L and f_V) agree to within 20% for a majority of strong transitions ($f \geq 0.01$). Table 4 shows a comparison between our results and NIST results in Mo XV. Generally, there are no discrepancies between our results and available NIST results, where the difference is within 0.1-1%, except the $3d^{10}\,{}^1S_0^e - 3d^9 4f\,{}^1P_1^o$ transition, the difference is 1.7%. GRASP lifetimes of excited levels in Mo XV have been presented in Table 1. In the lifetime calculations, we include the contributions of all four types of transitions (E1, E2, M1 and M2).

Table 2. GRASP wavelengths (Å), transition probabilities (in s^{-1}), oscillator strengths and line strengths (in a. u.) for E1 and M2 transitions in Mo XV.

Transition	λ_{ij}	E1 transitions			M2 transitions		
i - j		A_{ji}	f_{ij}	S_{ji}	A_{ji}	f_{ij}	S_{ji}
			$3d^{10}$ -$3d^9 4p$				
${}^1S_0^e - {}^3P_2^o$	52.13	0.000E+00	0.000E+00	0.000E+00	4.830E+03	9.841E-09	6.238E-01
${}^1S_0^e - {}^3F_2^o$	51.34	0.000E+00	0.000E+00	0.000E+00	1.293E+02	2.554E-10	1.546E-02
${}^1S_0^e - {}^3P_1^o$	51.32	1.555E+07	1.842E-05	3.112E-06	0.000E+00	0.000E+00	0.000E+00
${}^1S_0^e - {}^1D_2^o$	50.90	0.000E+00	0.000E+00	0.000E+00	6.182E+03	1.200E-08	7.081E-01
${}^1S_0^e - {}^1P_1^o$	50.78	4.099E+11	4.755E-01	7.949E-02	0.000E+00	0.000E+00	0.000E+00
${}^1S_0^e - {}^3D_1^o$	50.26	8.855E+10	1.006E-01	1.664E-02	0.000E+00	0.000E+00	0.000E+00
${}^1S_0^e - {}^3D_2^o$	50.14	0.000E+00	0.000E+00	0.000E+00	2.403E+02	4.527E-10	2.552E-02
			$3d^{10}$ -$3d^9 4f$				
${}^1S_0^e - {}^3P_1^o$	36.60	1.734E+10	1.044E-02	1.258E-03	0.000E+00	0.000E+00	0.000E+00
${}^1S_0^e - {}^3P_2^o$	36.53	0.000E+00	0.000E+00	0.000E+00	3.622E+05	3.623E-07	7.901E+00
${}^1S_0^e - {}^1D_2^o$	36.39	0.000E+00	0.000E+00	0.000E+00	1.090E+05	1.082E-07	2.333E+00
${}^1S_0^e - {}^3D_1^o$	36.26	2.849E+11	1.685E-01	2.012E-02	0.000E+00	0.000E+00	0.000E+00
${}^1S_0^e - {}^3D_2^o$	36.15	0.000E+00	0.000E+00	0.000E+00	3.004E-01	2.942E-13	6.217E-06
${}^1S_0^e - {}^3F_2^o$	36.03	0.000E+00	0.000E+00	0.000E+00	5.832E+04	5.675E-08	1.188E+00
${}^1S_0^e - {}^1P_1^o$	35.41	8.175E+12	4.609E+00	5.372E-01	0.000E+00	0.000E+00	0.000E+00

Table 3. GRASP wavelengths (Å), transition probabilities (in s^{-1}), oscillator strengths and line strengths (in a. u.) for E2 and M1 transitions in Mo XV.

Transition	λ_{ij}	E2 transitions			M1 transitions		
i - j		A_{ji}	f_{ij}	S_{ji}	A_{ji}	f_{ij}	S_{ji}
			$3d^{10}$ -$3d^9 4s$				
${}^1S_0^e - {}^3D_2^e$	59.30	8.739E+06	2.304E-05	2.862E-02	0.000E+00	0.000E+00	0.000E+00
${}^1S_0^e - {}^3D_1^e$	58.56	0.000E+00	0.000E+00	0.000E+00	8.532E-01	1.316E-12	1.905E-08
${}^1S_0^e - {}^1D_2^e$	58.38	1.316E+07	3.362E-05	3.986E-02	0.000E+00	0.000E+00	0.000E+00
			$3d^{10}$ -$3d^9 4d$				
${}^1S_0^e - {}^3S_1^e$	42.98	0.000E+00	0.000E+00	0.000E+00	8.702E+03	7.228E-09	7.680E-05
${}^1S_0^e - {}^3P_2^e$	42.57	1.771E+05	2.406E-07	1.106E-04	0.000E+00	0.000E+00	0.000E+00
${}^1S_0^e - {}^1P_1^e$	42.55	0.000E+00	0.000E+00	0.000E+00	4.072E+02	3.316E-10	3.489E-06
${}^1S_0^e - {}^1D_2^e$	42.32	4.899E+07	6.577E-05	2.970E-02	0.000E+00	0.000E+00	0.000E+00
${}^1S_0^e - {}^3P_1^e$	42.20	0.000E+00	0.000E+00	0.000E+00	1.494E+04	1.196E-08	1.248E-04
${}^1S_0^e - {}^3D_1^e$	42.06	0.000E+00	0.000E+00	0.000E+00	2.397E+03	1.907E-09	1.983E-05
${}^1S_0^e - {}^3D_2^e$	41.90	3.561E+07	4.686E-05	2.053E-02	0.000E+00	0.000E+00	0.000E+00

Table 4. Comparison of the present wavelengths with NIST results for transitions in Mo XV.

i	j	Type	GRASP	MCDF	NIST
$3d^{10}\ ^1S_0^e$	$3d^94s\ ^3D_2^e$	E2	59.303	58.989	58.832
$3d^{10}\ ^1S_0^e$	$3d^94s\ ^1D_2^e$	E2	58.384	58.083	57.927
$3d^{10}\ ^1S_0^e$	$3d^94p\ ^3F_2^o$	E1	51.319	51.067	50.928
$3d^{10}\ ^1S_0^e$	$3d^94p\ ^3D_3^o$	E1	50.782	50.593	50.448
$3d^{10}\ ^1S_0^e$	$3d^94p\ ^3D_1^o$	E1	50.259	50.051	49.914
$3d^{10}\ ^1S_0^e$	$3d^94f\ ^3P_1^o$	E1	36.595	36.479	36.376
$3d^{10}\ ^1S_0^e$	$3d^94f\ ^1P_1^o$	E1	35.405	35.439	36.060

4. Conclusion

In conclusion, we report energy levels, transition probabilities, oscillator strengths, and line strengths for E1, E2, M1, and M2 transitions of Mo XV. Two calculations have been performed, using a GRASP and MCDF codes. A comparison made between GRASP and MCDF energy levels, as well as with the available NIST values, shows satisfactory agreement in general. The present paper reports a wide range of energy levels and transitions in Mo XV. Hopefully, they will help in the development of Ni-like X-ray lasers.

Acknowledgement: We thank J. Desclaux and P. Indelicato for the MCDFGME code.

References

[1] Zhang J et al 1997 *Phys. Rev. lett.* **78** 3856
[2] Trebes J E et al 1987 *Science* **238** 517
[3] Da Silva L B et al 1992 *Science* **258** 269
[4] Ress D et al 1994 *Science* **265** 514
[5] Kalantar D H et al 1996 *Phys. Rev. Lett.* **76** 3574
[6] Hagelstein P L 1986 *Phys. Rev. A* **34** 874
[7] Brage T and Litzén U 1987 *Phys. Sr.* **35** 662
[8] Zhang H L, Sampson D H and Fontes C J 1991 *At. Data Nucl. Data Tables* **48** 91
[9] Aggarwal K M et al 2000 *At. Data Nucl. Data Tables* **74** 157
[10] Dong C Z, Fritzsche S and Xie L Y 2003 *J. Quant. Spectrosc. Radiat. Transfer* **76** 447
[11] Zhong J Y, Zhang J, Zeng J L, Zhao G and Gu M F 2005 *At. Data Nucl. Data Tables* **89** 101
[12] Zhong J Y, Zhao G and Zhang J 2006 *At. Data Nucl. Data Tables* **92** 481
[13] Zeng J, Zhao G and Yuan J 2007 *At. Data Nucl. Data Tables* **93** 199
[14] Huang J, Zhao Q and Jiang G 2007 *At. Data Nucl. Data Tables* **93** 864
[15] Liang Y Q and Zhong J Y 2008 *At. Data Nucl. Data Tables* **94** 807
[16] Lu-You X, Zhong D C-, Jun J, Jian-Jie W and Jun Y 2008 *Chin. Phys. B* **17** 3294
[17] Grant I P, McKenzie B J, Norrington P H, Mayers D F and Pyper N C 1980 *Comput. Phys. Commun* **21** 207
[18] Dyall K G, Grant I P, Johnson C T, Parpia F A and Plymmer E P 1989 *Comput. Phys. Commun.* **55** 425
[19] Norrington P H, http://www.am.qub.ac.uk/DARC/
[20] Desclaux J P and Indelicato P, MCDFGME, 2005, http://dirac.spectro.jussieu.fr/mcdf
[21] Alexander E, Even-Zohar M, Fraenkel B S and Goldsmith S 1971 *J. Opt. Soc. Am.* **61** 508
[22] Klapisch M *et al* 1978 *Phys. Rev Lett.* **41** 403
[23] Mansfield M W D, Peacock N J, Smith C C, Hobby M G and Cowan R D 1978 *J. Phys. B* **11** 1521
[24] Shirai T, Nakai Y, Ozawa K, Ishii K, Sugar J and Mori K 1987 *J Phys. Chem. Ref. Data* **16** 327
[25] Sugar J and Musgrove A 1988 *J. Phys. Chem. Ref. Data* **17** 155

Coulomb potential effects in strong-field photoelectron holography with a midinfrared laser pulse

G L Shi[1], J W Xu[1], H T Zhang[1], C Lin[1], X H Song[1], J Chen[2,3] and W F Yang[1]

[1]Department of Physics, College of Science, Shantou University, Shantou, Guangdong 515063, China
[2]Center for Applied Physics and Technology, Peking University, Beijing 100084, China
[3] Institute of Applied Physics and Computational Mathematics, P. O. Box 8009, Beijing 100088, China

E-mail: wfyang@stu.edu.cn

Abstract. A profound ringlike pattern coming from the interplay between the intra- and the intercycle interference of electron trajectories can be observed in the deep tunneling ionization regime, and an appropriate experimental condition for the observation of the photoelectron holography was provided.

1. Introduction

Coulomb potential effects in photoelectron holography (PH) with midinfrared wavelength laser pulses are investigated by solving numerically the time-dependent Schrödinger equation (TDSE) and using a generalized quantum-trajectory Monte Carlo (GQTMC) method. It is found that the Coulomb potential plays a key role in the interference of PH, especially in the parameter region $\gamma \sim 1$ ($\gamma = \sqrt{\frac{I_p}{2U_p}}$ is the Keldysh parameter, U_p is the ponderomotive energy or the averaged quiver energy of a free electron in the laser field, $U_p = \frac{I}{4\omega^2}$, where I is the laser intensity and ω is the angular frequency, and I_p is the ionization potential) in which the potential barrier combining with the laser and Coulomb field changes drastically. The interference structure of PH can be observed in the photoelectron momentum distribution (PMD) in the tunneling regime, where $\gamma < 1$ either with or without considering the long-range Coulomb potential. The interference structure of PH cannot be observed without the long-range Coulomb potential in the parameter region, where $\gamma \sim 1$. With the help of the GQTMC simulation, we show that the wavepacket dynamics of rescattered electrons can be influenced strongly by the Coulomb field. The rescattered electrons miss the first return to the parent ion without the long-range Coulomb potential. Therefore, the interference between the direct and rescattered electron wavepackets can hardly occur without the driving of the Coulomb force, which agrees well with the TDSE simulation. Moreover, the interference structures in PMD for different laser parameters provide the information of electron dynamics with sub-optical-cycle resolution.

IOP Conf. Series: Journal of Physics: Conf. Series **869** (2017) 012003 doi:10.1088/1742-6596/869/1/012003

2. Discussion and Results

Atomic photoionization under intense laser irradiation is a fundamental process in strong-field light-matter interactions. Since above-threshold ionization (ATI) was firstly observed more than thirty years ago [1], a series of experimental discoveries together with subsequent theoretical efforts have greatly advanced our understanding of the underlying physics of the interaction between atoms and intense laser fields [2, 3]. Recently, holographic *fork* structures have been observed in the photoelectron momentum distribution of ionized metastable xenon atoms [4]. This photoelectron holography, which records underlying electron dynamics on a sub-cycle time scale, offers opportunities to extract information of core and electron dynamics encoded in the interference patterns in the PMD [5–11].

Calculations based on numerical solution of the time-dependent Schrödinger equation can quantitatively reproduce well the experimental observations of PH [4–11]. However, it is difficult to explore the underlying physics. Very recently, a quantum-trajectory Monte Carlo (QTMC) method was developed from the classical-trajectory Monte Carlo method [12–14] by including quantum interference effect after the tunneling [15]. However, the QTMC method, which treats ionization rate with the quasistatic approximation, is only rigorously valid in the limit $\gamma \ll 1$ [16]. Here we develop a generalized QTMC (GQTMC) method [17, 18] which could be applied to relatively large region of the Keldysh parameter γ. This method provides an efficient way to probe the nonadiabatic effects, Coulomb potential effect, and unravel the underlying ultrafast electron dynamics encoded in the interference structures in the PMDs.

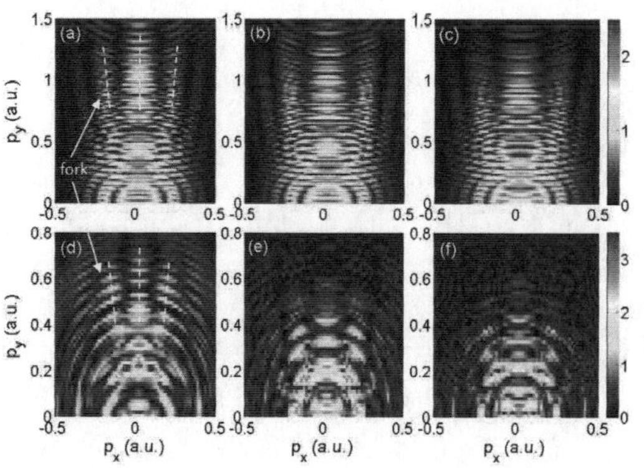

Figure 1. Photoelectron momentum distribution for ionization of Xe by a laser pulse with 1700 nm, (a), (b), and (c) 7.0×10^{13} W/cm^2, and $\gamma = 0.55$, and (d), (e) and (f) 1.2×10^{13} W/cm^2, and $\gamma = 1.33$, calculated by the TDSE with (a) and (d) the full long-range Coulomb potential, (b) and (e) the modified potential with $r_c = 5$ a.u. and (c) and (f) the modified potential with $r_c = 2$ a.u.

We have carried on the numerical calculation by solving the TDSE [19–21]. Firstly, we compare the 2D PMDs in two different parameter regions by solving the TDSE with different Coulomb potential models: figures 1 (a) and (d) using the long-range Coulomb potential of Xe atom; (b) and (e) using the modified potential with cutoff radius $r_c = 5$ a.u. and (c) and (f) $r_c = 2$ a.u. For the calculation in the upper panel of figure (1), the laser pulse has a wavelength of 1700 nm, and peak intensity of 7.0×10^{13} W/cm^2 with Keldysh parameter of 0.55, which would be considered to be in the tunneling ionization regime. For the calculation in the lower panel of figure (1), the laser has same wavelength and a large peak intensity of 1.2×10^{13} W/cm^2. The Keldysh parameter is 1.33 which is slightly larger than 1.

It can be clearly seen that the influence of the Coulomb potential on the PMDs is different in different parameter regions. In the tunneling regime ($\gamma = 0.55$), the *fork* structure which has been identified as the typical interference structure of the PH in previous theoretical and experimental results [22, 23] can be clearly observed both in the long-range Coulomb potential and the short-range potential results. The interference between the rescattered and direct electron wavepackets in the tunneling regime cannot be erased by reducing the cutoff radius of the potential, which suggests that: (a) when the electron appears in the continuum it is under the strong influence of the laser electric field, (b) the electron has a large kinetic energy when it is driven back to the nucleus by the strong laser field, and (c) the atomic potential force can be neglected comparing with the force of the laser electric field. These agree with the physical mechanism of the tunneling ionization and recollision processes in strong laser fields [24]. For the parameter region where $\gamma \sim 1$, the *fork* interference structure can be identified in PMD with the full long-range Coulomb potential (see figure 1(d)). However, the cutoff energy of the *fork* interference structure becomes smaller with decreasing cutoff radius of the Coulomb potential in figures 1(e) and (f). When the cutoff radius is reduced to $r_c = 2$ a.u., the PH interference structure can hardly be observed (see figure 1(f)). As a result, we see a substantial change of the interference structure in PMDs by modifying the Coulomb potential, which indicates that the Coulomb potential plays an essential role in the electron dynamics in the parameter region where $\gamma \sim 1$. In order to understand this critical effect of Coulomb potential on the interference

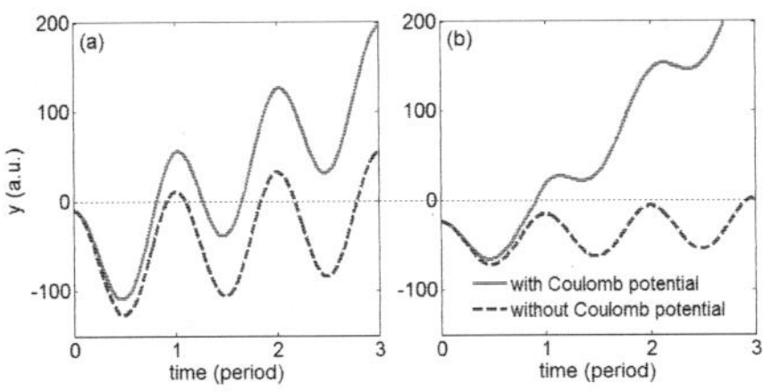

Figure 2. Typical trajectories of electrons along the laser polarization direction calculated by the semiclassical model with the same laser parameters (a) in figure 2(a); and (b) in figure 2(d).

of rescattered and direct electron wavepackets, we plot in figures 2(a) and (b) typical electron trajectories calculated by the GQTMC with and without considering the Coulomb potential for $\gamma = 0.55$ and 1.33. In the case of tunneling ionization regime $\gamma = 0.55$, the electron has a large excursion amplitude, and when it goes far away from the parent ion, the Coulomb force becomes much smaller than the force of laser electric field. Thus, the force of laser electric field dominates the motion of electron in this case. The electron is ionized near the peak of the laser field and can be driven past their parent ion more than once. It can be seen in figure 2(a) that the two electron trajectories with and without considering the Coulomb potential have similar trends. The first and subsequent revisits of the ionized electron to the core take place in both of the electron trajectories. These typical recollision trajectories combining with similar statistic distributions of rescattered electrons result in the similar interference structure in the momentum spectra in tunneling ionization regime (see figures 1(a)-(c)).However, the situation is quite different in the parameter region where $\gamma \sim 1$. The electron driven by the force of electric field and the Coulomb force will revisit the core after one optical period (the red line in figure 2(b)). The ionized electron has a relatively small excursion amplitude comparing with that in tunneling ionization regime and is affected strongly by the Coulomb potential in this small region around the parent ion. When the effect of Coulomb potential is not considered (the blue dashed line in figure 2(b)), the electron will miss the important first revisit to the parent ion. Therefore, GQTMC results are consistent with those of TDSE, and both of them demonstrated that the effect of the Coulomb potential is essential to generate the interference between the direct and rescattered electron wavepackets in the parameter region where $\gamma \sim 1$, and the force of the laser electric field dominates the rescattering dynamics when $\gamma < 1$. From figure 1, one can see that the interference structures are also different in different parameter regions. In conclusion, we have investigated 2D PMDs with midinfrared wavelength laser pulses by solving the TDSE for different laser intensities and wavelengths. By comparing the PMDs for the long-range and short-range Coulomb potentials, the dynamic mechanism of the PH interference, viz., the interference between the direct electron and rescattered electron wavepackets was clearly revealed. This interference can always occur in the long-range Coulomb interaction for different laser parameter regions. In tunneling regime where $\gamma < 1$, this interference and the interference structure which originates from the interference between the two electron wave packets appear on the opposite sides of the ion but have the same drift momentum, can be observed in high energy region of PMD both for the long-range and short-range Coulomb potentials. In the parameter region where $\gamma \sim 1$, the interference structures of PH and ATI can be observed in PMDs for the long-range Coulomb potential. However, only the interference structure of ATI can be distinguished in PMDs for the short-range Coulomb potential.

To explore the physical reason behind the interference structures, we have also used the GQTMC to calculate typical electron trajectories and the distributions of the rescattered electrons with and without considering the Coulomb potential. The results obtained are consistent with the TDSE simulations. It is shown that in the tunneling regime where $\gamma < 1$, the force of the laser electric field plays a dominant role in the rescattering dynamics, which leads to similar interference structures in PMDs for the long-range and short-range Coulomb potentials. When the tunneling picture is not valid ($\gamma \sim 1$), the Coulomb force is relatively strong and the Coulomb potential has significant effects on the motion of electrons and the rescattering processes.

References

[1] Agostini P, Fabre F, Mainfray G, Petite G and Rahman N K 1979 *Phys. Rev.Lett.* **42** 1127
[2] Becker W, Grasbon F, Kopold R, Milošević D B, Paulus G G and Walther H 2002 Adv. At. *Mol. Opt. Phys.* **48** 35
[3] Corkum P B 1993 *Phys. Rev. Lett.* **71** 1994
[4] Huismans Y *et al* 2011 *Science* **331** 61
[5] Hickstein D D *et al* 2012 *Phys. Rev. Lett.* **109** 073004
[6] Meckel M *et al* 2014 *Nature Phys.* **10** 594
[7] Huismans Y *et al* 2012 *Phys. Rev. Lett.* **109** 013002
[8] Marchenko T, Huismans Y, Schafer K J and Vrakking M J J 2011 *Phys. Rev.* A **84** 053427
[9] Yang W F, Song X H and Chen Z J 2012 *Opt. Express* **20** 12067
[10] Yang W F, Sheng Z H, Feng X, Wu M L, Chen Z J and Song X H 2014 *Opt. Express* **22** 2519
[11] Bian X B, Huismans Y, Smirnova O, Yuan K J, Vrakking M J J and Bandrauk A D 2011 *Phys. Rev.* A **84** 043420
[12] Brabec T, Ivanov M Y and Corkum P B 1996 *Phys. Rev.* A **54** R2551
[13] Hu B, Liu J and Chen S G 1997 *Phys. Lett.* A **236** 533
[14] Chen J, Liu J and Chen S G 2000 *Phys. Rev.* A. **61** 033402
[15] Li M *et al* 2014 *Phys. Rev. Lett.* **112** 113002
[16] Keldysh L V 1965 *Sov. Phys. JETP.* **20** 1307
[17] Yang W F, Zhang H T, Lin C, Xu J W, Sheng Z H, Song X H, Hu S L and Chen J 2016 *Phys. Rev.* A **94** 043419
[18] Song X H, Lin C, Sheng Z H, Liu P, Chen Z J, Yang W F, Hu S L, Lin C D and Chen J 2016 *Sci. Rep.* **6** 28392
[19] Yang W F, Song X H, Zeng Z, Li R, and Xu Z 2010 *Opt. Express* **18** 2558
[20] Lin C, Zhang H T, Sheng Z H, Yu X H, Liu P, Xu J W, Song X H, Hu S L, Chen J and Yang W F 2016 *Acta Phys. Sin.* **65** 223207
[21] Song X H, Xu J W, Lin C, Sheng Z H, Liu P, Yu X H, Zhang H T, Hu S L, Chen J, Xu S P, Chen Y J, Quan W, Liu X J, Yang W F 2017 *Phys. Rev.* A **95** 033426
[22] Gong X C *et al* 2017 *Phys. Rev. Lett.* **118** 143203
[23] Huismans Y *et al* 2011 *Science* **331** 61
[24] Lewenstein M, Balcou P, Ivanov M Y, L'Huillier A and Corkum P B 1994 *Phys. Rev.* A **49** 2117

IOP Conf. Series: Journal of Physics: Conf. Series **869** (2017) 012004 doi:10.1088/1742-6596/869/1/012004

Breakdown of odd-harmonic rule in the high harmonic generation spectra of the hydrogen molecular ion

H Iravani and M Vafaee

Department of Chemistry, Tarbiat Modares University, P. O. Box 14115-175, Tehran, I. R. Iran

E-mail: hossein.iravani@modares.ac.ir

Abstract. We numerically solve the fully-dimensional electronic time-dependent Schrödinger equation for an H_2^+ molecular ion. The occurrence of electron localization is investigated to better understand the complex patterns appearing in the high-order harmonic generation (HHG) spectrum. Our studies show that changing trends of electronic acceleration are affected by electron localization at large enough internuclear distances. This effect leads to the complex patterns in the HHG spectrum.

1. Introduction

According to the odd-harmonic rule, the harmonic angular frequencies, nω_o, of photons produced by recollision of the electron with the parent ion in an intense laser field are odd multiples of the laser angular frequency, ω_o, but some irregularities have been seen in high harmonic generation (HHG) spectra.

Different studies have been carried out to explain the origin of these complex patterns. Martin's group reported that these patterns result from electron localization, which induces a permanent dipole during dissociation, even for pulses containing tens of laser cycles. They claimed that the permanent dipole associated with the linear Stark effect leads to a relative phase between the two consecutive electron round trips initiated in the adjacent laser half-cycles. As this phase approaches π, the odd harmonics are replaced by the even ones [1-2].

Ahmadi *et al.* discussed elaborately in Ref. [3] that these complexities (non-adiabatic frequency redshift of the harmonics and spatially asymmetric emission along the z direction) are related to the falling edge of the laser pulse. According to Ahmadi *et al.*, the H_2^+ system responds non-adiabatically to the rapidly changing laser field, and the emission along the polarization direction is spatially asymmetric in this region.

In this work, we investigate the HHG of the H_2^+ molecular ion exposed to an intense laser field with a 2-8-2 trapezoidal envelope, I =3 $\times 10^{14}$ Wcm^{-2} intensity, and 800 nm wavelength, by solving the time-dependent Schrödinger equation numerically for the fully-dimensional electronic dynamics without nuclear dynamics, i.e., with a fixed R_0 inter-nuclear distance. We choose three different R_0 and study the effect of increasing R_0 on electronic population, electronic acceleration and HHG in the simulation box. Throughout this paper, atomic units (a.u.) with $e = \hbar = m_e = 1$, will be used, unless otherwise stated.

IOP Conf. Series: Journal of Physics: Conf. Series **869** (2017) 012004 doi:10.1088/1742-6596/869/1/012004

2. Computational method

After separating the center-of-mass motion and ignoring the nuclear dynamics, the time-dependent Schrödinger equation (TDSE) for a hydrogen molecular ion can be written, in terms of the electron's coordinates z and ρ, as [4-5]

$$i\frac{\partial\psi(z,\rho;R_0,t)}{\partial t}=\widehat{H}(z,\rho;R_0,t)\psi(z,\rho;R_0,t), \tag{1}$$

where R_0, to be treated as a parameter, denotes the internuclear distance (parallel to both electronic coordinate z and the laser polarization direction) and $\hat{\mathrm{H}}$ is the Hamiltonian, given by

$$\widehat{H}(z,\rho;R_0,t)=-\frac{2m_N+m_e}{4m_Nm_e}\left[\frac{\partial^2}{\partial z^2}+\frac{\partial^2}{\partial\rho^2}+\frac{1}{\rho}\frac{\partial}{\partial\rho}\right]+V_C(z,\rho;R_0,t), \tag{2}$$

with

$$V_C(z,\rho;R_0,t)=-\frac{1}{\sqrt{(z+\frac{R_0}{2})^2+\rho^2}}-\frac{1}{\sqrt{(z-\frac{R_0}{2})^2+\rho^2}}+\left(\frac{2m_N+2m_e}{2m_N+me}\right)zE_0f(t)\cos(\omega t). \tag{3}$$

In the above equations, E_0 is the laser peak amplitude, m_e and m_N are masses of the electron and single nucleus, respectively, ω is the laser angular frequency, and $f(t)$ is the laser pulse envelope. The envelope rises linearly during the first two cycles, then remains constant for eight cycles and decreases during the last two cycles.

The TDSE is numerically integrated using unitary split-operator methods [6-7]. Details of the numerical procedure are described in our previous work [8-9]. The finest grid size values in our numerical integration are, respectively, 0.13 and 0.2 for z and ρ coordinates and the time steps are 0.02. The size of the simulation box is chosen as $z_{max}=150$ and $\rho_{max}=125$.

An HHG spectrum is calculated as the square of the windowed Fourier transform of the dipole acceleration $a_z(t)$ in the electric field direction (z), or

$$S(\omega)=\left|\int_0^T<\psi(z,\rho;R_0,t)|a_z(t)\mid\psi(z,\rho;R_0,t)>_{z,\rho}\times H(t)\,exp[-i\omega t]\,dt\right|^2,$$

where

$$H(t)=\frac{1}{2}\left[1-\cos(2\pi\frac{t}{T})\right], \tag{4}$$

is the Hanning function and T is the total pulse duration. The Hanning filter reduces the effect of nondecaying components in the dipole acceleration that last after the laser pulse is switched off [10].

3. Results and discusions

Ahmadi *et al.* reported that for the 14-cycle trapezoidal laser pulses with $I=3\times10^{14}$ Wcm^{-2} intensity and 800 nm wavelength, no considerable final electron localization was achieved and the HHG spectrum was dominated by the odd harmonic orders [3]. As is shown in figure 1 for $R_0=2.08$, the odd harmonic orders are distinct. However, by increasing the distance between the hydrogen nuclei, the HHG spectra become complicated. For $R_0=3.66$, the peaks become multiple and a red-shift to even harmonic orders is seen.

In order to investigate the origin of these complexities, we consider total acceleration in the z direction and also the acceleration in the left and right sides of the simulation box. According to figure 2, the variation of total acceleration at $R_0=2.08$ follows the pattern of laser field. However, the pattern of total acceleration changes becomes complicated by increasing R_0.

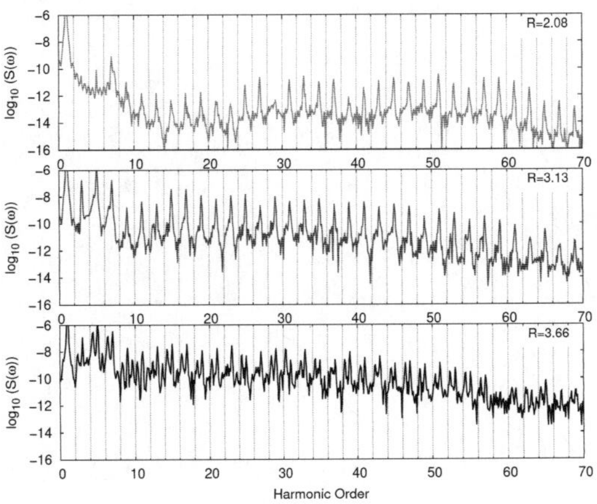

Figure 1. High harmonic generation spectra of H_2^+ system with different internuclear distances ($R_0 = 2.08, 3.13$ and 3.66) under 12-cycle trapezoidal laser pulse of 800 nm wavelength and $I = 3\times10^{14}$ Wcm^{-2} intensity.

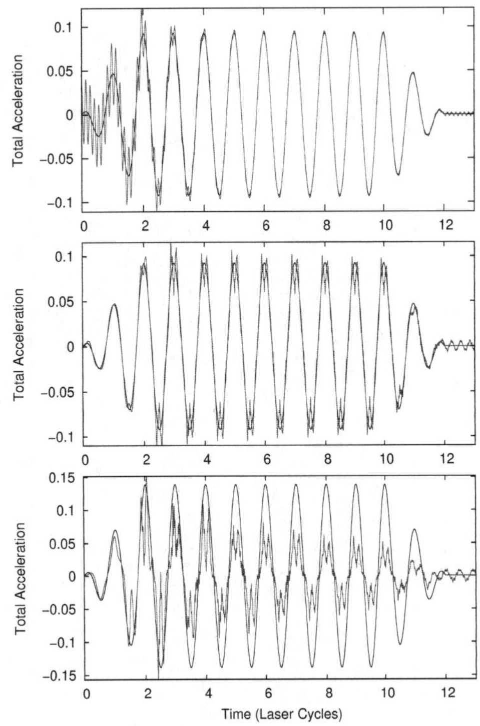

Figure 2. Total acceleration of H_2^+ system with different internuclear distances (2.08, 3.13 and 3.66 respectively). Laser parameters are the same as figure 1.

Interestingly, the total acceleration intensity decreases gradually at $R_0 = 3.66$ due to the decline of electronic population in the simulation box. It seems that the doublet peaks appearing in the HHG spectrum of H_2^+ at $R_0 = 3.13$ and $R_0 = 3.66$ (see figure 1) are related to doublet peaks of the total acceleration at similar internuclear distances.

Left- and right-accelerations (Acceleration(L) and Acceleration(R)) are obtained, respectively, by integration over the left and right parts of the simulation box. The changes of Acceleration(R) and Acceleration(L) are shown in Fig. 3. At $R_0 = 2.08$, these changes follow the laser field pattern. But at $R_0 = 3.13$, two periodic trends could be detected in them. One is dominated by the laser field and the other is controlled by electron localization (this trend could also be found in figure 2 for $R_0 = 3.66$). The changing trends at $R_0 = 3.66$ are similar to $R_0 = 3.13$, but the structure of the left and right accelerations get more complex.

In figure 4, Acceleration(R) and population of the right side of the simulation box (Population(R)) are displayed with the laser field. In order to compare these graphs more easily, we raise the intercept of Acceleration(R) and laser field graphs. According to this figure, when the laser field increases in the positive direction, the population in the right side of the simulation box decreases but the acceleration of electrons increases in that side of the simulation box.

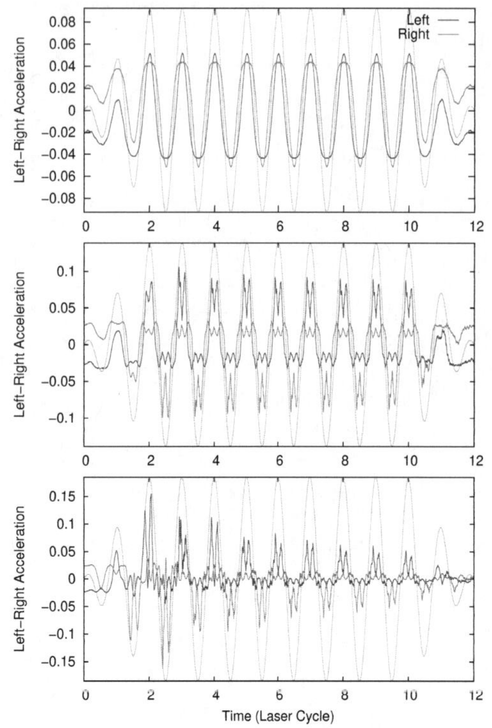

Figure 3. Acceleration of H_2^+ system in the left (blue) and right (red) side of simulation box. From top to bottom, the graphs are related to internuclear distances of 2.08, 3.13 and 3.66, respectively. Laser parameters are the same as in figure 1.

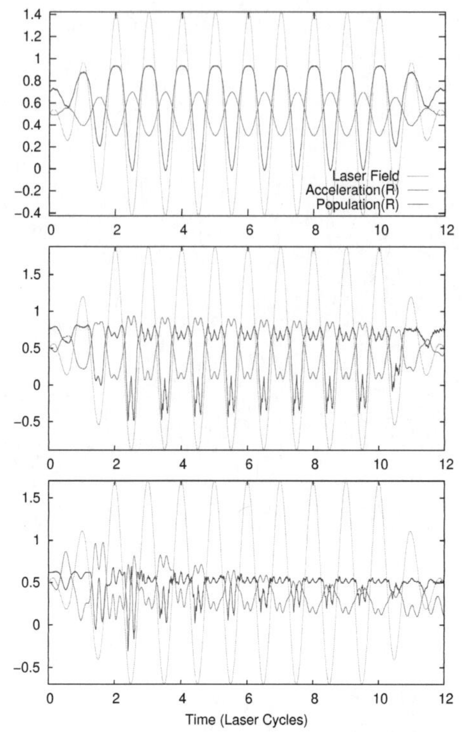

Figure 4. Changing of Acceleration(R) with respect to Population(R). From top to bottom, the graphs are related to internuclear distances of 2.08, 3.13 and 3.66, respectively. Laser parameters are the same as in figure 1.

4. Conclusions

At the lowest internuclear distance ($R_0 = 2.08$), the pattern of total acceleration and left and right accelerations follows the laser field. By increasing R_0, these patterns become more complex and another periodic trend appears which is governed by electron localization.

References

[1] Silva R E F, Riviere P, Morales F, Smirnova O, Ivanov M and Martin F 2016 *Sci. Rep* **6** 32653
[2] Morales F, Riviere P, Richter M, Gubaydullin A, Ivanov M, Smirnova O and Martin F 2014 *J. Phys.* B **47** 204015
[3] Ahmadi H, Vafaee M and Maghari A 2016 *Phys. Rev.* A **94** 033415
[4] Hiskes J R 1961 *Phys. Rev.* **122** 1207
[5] Vafaee M 2008 *Phys. Rev.* A **78** 023410
[6] Bandrauk A D and Shen H 1993 *J. Chem. Phys.* **99** 1185
[7] Feit M D, Fleck J A and Steiger A 1982 *J. Comput. Phys.* **47** 412
[8] Vafaee M and Sabzyan H 2004 *J. Phys.* B **37** 4143
[9] Vafaee M, Sabzyan H, Vafaee Z and Katanforoush A 2006 *Phys. Rev.* A **74** 043416
[10] Camp S, Schafer K J and Gaarde M B 2015 *Phys. Rev.* A **92** 013404

IOP Conf. Series: Journal of Physics: Conf. Series **869** (2017) 012005 doi:10.1088/1742-6596/869/1/012005

Interferometer-free Fourier-synthesized laser field generator estimated by molecular tunnelling ionization

H Ohmura and N Saito

National Institute of Advanced Industrial Science and Technology (AIST), Higashi, Tsukuba, Ibaraki 305-8565, Japan
E-mail: hideki-ohmura@aist.go.jp

Abstract. Intense nanosecond four-colour Fourier-synthesized laser fields induce orientation-selective ionization based on directionally asymmetric molecular tunneling ionization. The interferometer-free laser field generator ensures high stability and high reproducibility. Phase-sensitive, orientation-selective molecular tunneling ionization provides a simple way to estimate the relative phase differences between the fundamental light and each harmonic by data-fitting analysis.

1. Introduction

Optical excitation using a two-color phase-controlled laser field consisting of the fundamental and its harmonic, has been theoretically explored by Brumer and Shapiro [1]. For weak laser fields ($I < 10^{12}$ W/cm^2), a scheme involving quantum interference between two photo-transitions can be used to control the population of a targeted state by constructive or destructive interference [1]. In contrast, for intense laser fields ($I > 10^{12}$ W/cm^2), a scheme involving phase-controlled electric fields that steer the motions of charges or dipoles has been studied. Typical examples of such schemes are found in tunneling ionization (TI) [2-4]. Phase-controlled two-color laser fields have been used to control such coherent electron motions in TI [5-16].

An extension of this methodology is the well-known synthesis by Fourier transformation theory of arbitrary light waveforms with a series consisting of a fundamental frequency of light and its harmonics. We have reported the use of Fourier-synthesized laser fields to achieve quantum control of molecular TI [17]. Directionally asymmetric molecular TI induced by intense ($I = 5 \times 10^{12}$ W/cm^2) nanosecond Fourier-synthesized four-color laser fields consisting of fundamental, second-, third-, and fourth-harmonic light can make possible orientation-selective molecular TI (OSM-TI) and has been applied to the measurement of the three relative phase differences between the fundamental light and each of its harmonics [17]. In this study, we extend this technology to an interferometer-free Fourier-synthesized laser field generator. This methodology has enabled us to observe the phase-sensitive OSM-TI signals induced by Fourier-synthesized laser fields with high stability and reproducibility. This generator, however, does not separate each beam spatially; therefore, the three relative phase differences cannot be controlled and estimated independently. We have developed a procedure for analyzing the data that enables estimation of the three relative phase differences between the fundamental and each of its harmonics.

2. Experiment

The experimental setup consists of a pulsed Nd:YAG laser, an interferometer-free Fourier-synthesized laser field generator, and a time-of-flight mass spectrometer (TOF-MS) equipped with a supersonic molecular beam source. The Q-switched Nd:YAG laser (Spectra-Physics, LAB150) generated horizontally polarized laser pulses with a 10-ns duration and a bandwidth of 1.0 cm^{-1} at a wavelength of 1064 nm and was operated at a repetition rate of 10 Hz without injection seeding. Fourier-synthesized laser fields consisting of a fundamental light and its harmonics—hereafter, $\omega + 2\omega + 3\omega + 4\omega$ (four-color) laser fields—were generated by an interferometer-free Fourier-synthesized laser field generator [Fig. 1]. All of the nonlinear optical crystals used in this study were β-barium borate, type I phase matching, with thicknesses of 10 mm for the 2ω pulses, 8 mm for the 3ω pulses, and 6 mm for the 4ω pulses. Then the vertical polarization component of ω + 2ω + 3ω + 4ω pulses was selectively transmitted by a polarizer (air-spaced Glan–Thompson prism). The electric field of the linearly polarized Fourier-synthesized laser field is given by

$$E(t) = E_1 \cos(\omega t) + \Sigma^{4}_{n=2} E_n \cos(n\omega t + \phi_{1n}) , \qquad (1)$$

where $E_i (i = 1, 2, 3, 4)$ is the amplitude of the electric field of each component and ϕ_{1n} ($n = 2, 3, 4$) is the relative phase difference between ω and its harmonic field. The Fourier-synthesized laser fields were passed through a phase-shifting plate (synthesized quartz, thickness of 10 mm) that could be rotated around the incident angle of 45° to control the relative phase differences, ϕ_{1n}. The value of ϕ_{1n} could be changed by the optical length, which was determined by the difference in the value of the refractive index due to dispersion. We note that because all the beams were transmitted through the phase-shifting plate, each ϕ_{1n} could not be controlled independently. A procedure was required to estimate each ϕ_{1n}. The Fourier-synthesized $\omega + 2\omega + 3\omega + 4\omega$ beams were directed toward the TOF-MS and were focused on the supersonic molecular beam by a concave mirror with a focal length of 120 mm. The total intensity $I = I_1 + I_2 + I_3 + I_4$ was 5×10^{12} W/cm^2 at the focus, and the I_2/I_1, I_3/I_1, and I_4/I_1 ratios were about 0.56, 0.25, and 0.063, respectively, where I_1, I_2, I_3, and I_4 are the intensities of the ω, 2ω, 3ω, and 4ω pulses, respectively.

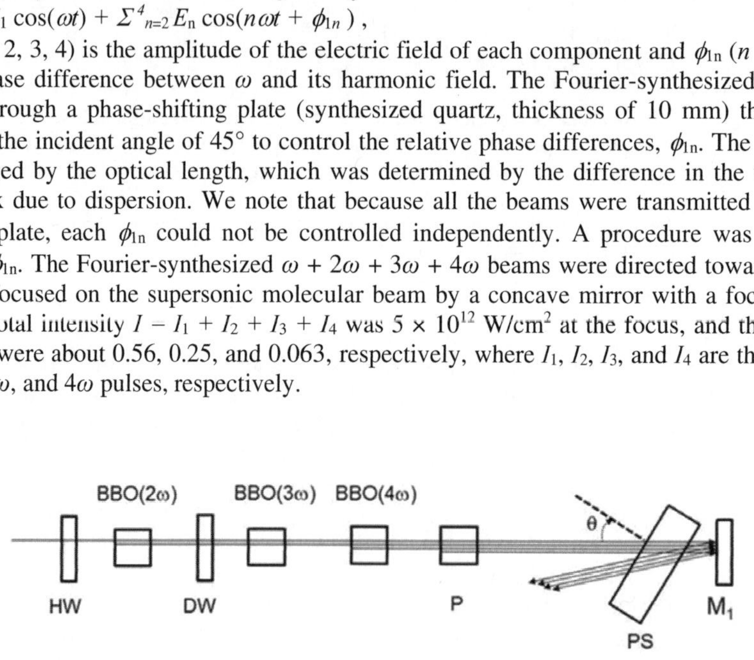

Figure 1. Schematics of the optics of a robust interferometer-free Fourier-synthesized laser field generator. The components are labeled as follows: HW, half-wave plate for the ω laser beam; BBO(nω), nonlinear optical crystal for generation of the nth harmonic (n = 2, 3, 4); DW, dual-wavelength wave plate (λ for the ω laser beam and λ/2 for the 2ω laser beam); P, polarizer; PS, phase-shifting plate; and M, mirror. The offset between the four beams is shown for clarity only, as the four beams overlapped completely in the experiment.

The principle of orientation-selective molecular tunneling ionization (OSM-TI) has been described previously [12,15]. Briefly, according to the most fundamental theory of molecular TI, the molecular Ammosov–Delone–Krainov model [18,19], photoelectrons are preferentially removed via tunneling from the large-amplitude lobe of the highest occupied molecular orbital in the opposite direction of the electric field vector. Because of the angular dependence of the TI rate between the electric field vector and the molecular axis, molecules oriented in a certain direction are selectively ionized in a randomly oriented gas-phase molecular ensemble. The amplitude of the electric field is larger in the positive direction than in the negative direction when (ϕ_{12} ϕ_{13} ϕ_{14}) = (0, 0, 0) [17]. The directional asymmetry is reversed when (ϕ_{12} ϕ_{13} ϕ_{14}) = (π, 0, π) (not shown). In the case of molecules with asymmetric structure, the TI rates of those with positive molecular orientations are expected to differ from those with negative molecular orientations when TI occurs at the positive field maxima. As a result, the asymmetric waveform of the Fourier-synthesized laser field can selectively ionize molecules while discriminating the head-tail order of molecules with asymmetric structures, a distinction that cannot be achieved by single-frequency laser fields with symmetric waveforms [17]. The phase dependences of OSM-TI are applicable to the estimation of ϕ_{1n} (n = 2, 3, 4) [17].

A TOF-MS equipped with a pulsed supersonic molecular beam source was used to detect the ionized molecules and their photo-fragments. TOF spectra were recorded with a digital oscilloscope. The polarization direction of the Fourier-synthesized laser fields was set to be vertical and parallel to the detection axis. We defined ϕ_{1n} (n = 2, 3, 4) = 0 as the indication that the electric field maxima were pointing toward the ion detector.

3. Results and Discussion

We selected carbonyl sulphide (OCS) as the focus of this study because we have already investigated its behavior by Fourier-synthesized laser fields generated by a multicolor Mach–Zehnder interferometer [17]. The OC^+ photo-fragments exhibited a pair of peaks in the TOF spectra. The first peak resulted from ions ejected directly toward the detector, and the second from ions that were first ejected in the backward direction before being accelerated toward the detector by the extraction field. The breaking of the forward-backward symmetry was clearly observed in the TOF spectra. The backward peak of the OC^+ ions predominated at $\theta = 45°$. The forward-backward asymmetries of the S^+ photo-fragment ions were less pronounced when the OCS molecules were irradiated with Fourier-synthesized $\omega + 2\omega + 3\omega + 4\omega$ pulses because the photo-fragments induced by irradiation of only the 4ω pulse, which is not phase-dependent with a symmetric signal shape, overlapped to cover up the forward-backward asymmetry.

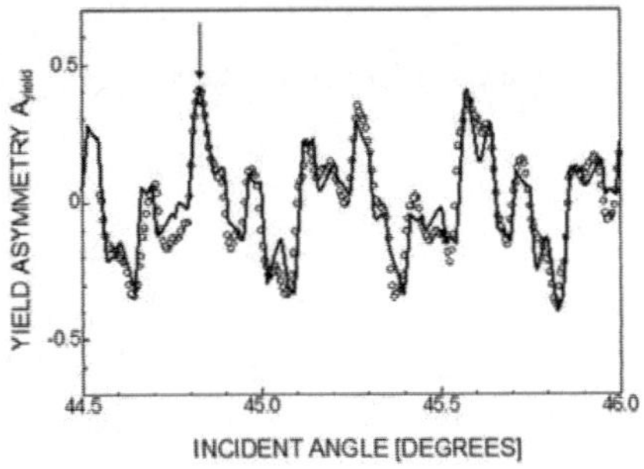

Figure 2. Yield asymmetries of OC^+ ions $A_{yield} = (I_F - I_B)/(I_F + I_B)$ as a function of angle incident to the phase-shifting plate θ for the ω + 2ω + 3ω + 4ω field. The solid curve is the result of fitting analysis. The arrow shows the angle at forward asymmetry. (ϕ_{12} ϕ_{13} ϕ_{14}) = (−0.05π, −0.14π, 0.15π) at θ = 44.83° .

We define the forward-backward yield asymmetry of the photo-fragment ions by

$$A_{yield} = (I_F - I_B)/(I_F + I_B), \quad (2)$$

where I_F (I_B) is the signal intensity of the forward (backward) photo-fragment emission. Figure 2 depicts A_{yield} of the OC^+ photo-fragment as a function of the incident angle θ for the $\omega + 2\omega + 3\omega + 4\omega$ laser field. Clearly phase-dependent behaviors of A_{yield} were observed for both OC^+ and S^+. The behavior of the A_{yield} as a function of incident angle θ for the $\omega + 2\omega + 3\omega + 4\omega$ laser field was complicated because A_{yield} was scanned through θ as a function of ϕ_{12}, ϕ_{13}, and ϕ_{14} for the $\omega + 2\omega + 3\omega + 4\omega$ laser field. The yield asymmetries of the OC^+ and S^+ ions were completely out of phase with each other (not shown). These results show that Fourier-synthesized $\omega + 2\omega + 3\omega + 4\omega$ pulses can distinguish the molecular orientation of the head-tail order [17]. Despite the fact that A_{yield} behaved in a complicated way as a function of the incident angle θ for the $\omega + 2\omega + 3\omega + 4\omega$ laser field, we have found the phase-sensitive signals A_{yield} induced by Fourier-synthesized laser fields to be highly stable and reproducible.

We now discuss the data-fitting procedure for estimating all of the ϕ_{1n} values from the phase-dependent OSM-TI. To estimate all of the ϕ_{1n} values, we tried to fit the phase-dependent signal A_{yield} to a calculated curve considering the difference of the optical lengths of the fundamental and its harmonics induced by the phase-shifting plate [20]. The solid curve in figure 2 shows the results of the fit. We could reproduce the complicated phase-dependent behaviour of A_{yield} with respect to the incident angle θ for many periods. Having obtained the fitted curve, we could extract all the ϕ_{1n} values at any incident angle θ. The characteristic set of relative phase differences (ϕ_{12} ϕ_{13} ϕ_{14}) was (-0.05π, -0.14π, 0.15π) at $\theta = 44.83°$ with an uncertainty of $\pm 0.05\pi$. This procedure enabled us to estimate all the ϕ_{1n} values through phase-dependent OSM-TI in an intuitive manner by using the molecular orientation of the head-tail order [20].

4. Conclusions

We have demonstrated robust generation of Fourier-synthesized laser fields and their estimation using the phase-sensitive behavior of OSM-TI induced by nanosecond four-color Fourier-synthesized laser fields. The interferometer-free laser field generator that ensures high stability and high reproducibility is significant in light of the previously reported work [17] where we used a multicolor Mach–Zehnder interferometer that is vulnerable to a variety of fluctuations. We have developed a procedure for analyzing the data that enables estimation of the three relative phase differences between the fundamental and each of its harmonics.

Acknowledgments

This work was supported by Japan Society for the Promotion of Science (KAKENHI) grant nos. 24340097, 16H04103, and by New Energy and Industrial Technology Development Organization.

References

[1] Shapiro M and Brumer P 2003 *Principles of the Quantum Control of Molecular Processes* (New York: Wily)
[2] Keldysh L V 1965 *Sov. Phys. JETP* **20** 1307
[3] Corkum P B, Burnett N H and Brunel F 1989 *Phys. Rev. Lett.* **62** 1259
[4] Uiberacker M *et al* 2007 *Nature* **446** 627
[5] Schafer K J and Kulander K 1992 *Phys. Rev.* A **45** 8026
[6] Schumacher D W, Weihe F, Muller H G and Bucksbaum P H 1994 *Phys. Rev. Lett.* **73** 1344
[7] Bandrauk A D and Chelkowski S 2000 *Phys. Rev. Lett.* **84** 3562
[8] Ray D *et al* 2009 *Phys. Rev. Lett.* **103** 223201
[9] Betsch K J, Pinkham D W and Jones R R 2010 *Phys. Rev. Lett.* **105** 223002
[10] Ray D *et al* 2011 *Phys. Rev.* A **83** 013410
[11] Shafir D, Soifer H, Bruner B D, Dagan M, Mairesse Y, Patchkovskii S, Ivanov M Y, Smirnova O and Dudovich N 2012 *Nature* **485** 343
[12] Ohmura H, Saito N and Tachiya M 2006 *Phys. Rev. Lett.* **96**, 173001
[13] Ohmura H and Tachiya M 2008 *Phys. Rev.* A **77** 023408
[14] Ohmura H, Saito N, Nonaka H and Ichimura S 2008 *Phys. Rev.* A 77, 053405
[15] Ohmura H, Saito N and Morishita T 2011 *Phys. Rev.* A **83** 063407
[16] Ohmura H, Saito N and Morishita T 2014 *Phys. Rev.* A **89** 013405
[17] Ohmura H and Saito N 2015 *Phys. Rev.* A **92** 053408
[18] Tong X M, Zhao Z X and Lin C D 2002 *Phys. Rev.* A **66** 033402
[19] Alnaser A S *et al* 2004 *Phys. Rev. Lett.* **93** 113003
[20] Ohmura H and Saito N in preparation

On the exact rotational and internal Hamiltonian for a non-relativistic closed many-body system

I Barth and K Renziehausen

Max Planck Institute of Microstructure Physics, Weinberg 2, 06120 Halle (Saale), Germany

E-mail: barth@mpi-halle.mpg.de

Abstract. Without applying Born-Oppenheimer approximation, the non-relativistic Hamiltonian can be separated into Hamiltonians for the translation of the center of mass and for the rotational and internal motions of the closed many-body system. This exact rotational and internal Hamiltonian can be expressed in terms of three Euler angles for three independent rotations of the system and the rotated Jacobi coordinates for the internal motions.

1. Introduction

In the realm of non-relativistic quantum mechanics of molecules, the original form of the total Hamiltonian can be separated into Hamiltonians for translational and non-translational motions. One usually first applies Born-Oppenheimer approximation (BOA) to separate the non-translational motion into the motions of electrons and nuclei and then separate the nuclear motions into the rotational and vibrational motions [1]. But for a closed many-body system consisting of arbitrary interacting particles (including muons for example), where BOA cannot be applied, the Hamiltonian cannot be separated into Hamiltonians of specific particle sorts. In this case, one can still rewrite the non-translational Hamiltonian specifically for the rotational and internal motions of the system, that can be described separately by three Euler angles and by the rotated Jacobi coordinates. In this work, we derive this form of the rotational and internal Hamiltonian for an arbitrary non-relativistic closed many-body system (for H_2^+ and its isotopomers, see Ref. [2]). To the best of our knowledge, such derivation has not been performed since the first publication of the Schrödinger equation [3]. In addition, we derive the general expression for exact stationary wave functions for the mentioned arbitrary many-body system.

2. Model system

We consider a general non-relativistic closed many-body system without external and spin-dependent interactions. It contains $N \geq 3$ interacting particles with corresponding position vectors $\mathbf{R}_i$ in the laboratory frame (X, Y, Z) and masses m_i. The time-dependent wave function $\Psi(\{\mathbf{R}_i\}, t)$ of this system is the solution of the time-dependent Schrödinger equation $i\hbar\dot{\Psi} = \hat{H}\Psi$, where the time-independent Hamiltonian for two-body interactions (e.g. Coulomb interactions) is equal to the sum of kinetic operators of the particles and two-body potentials $V_{ij} = V_{ji}$ [4]:

$$\hat{H} = -\frac{\hbar^2}{2}\sum_{i=1}^{N}\frac{1}{m_i}\nabla^2_{\mathbf{R}_i} + \sum_{\substack{i,j=1\\ i\neq j}}^{N} V_{ij}(|\mathbf{R}_i - \mathbf{R}_j|). \tag{1}$$

3. Translation of the center of mass

The motion of the system can be separated into the translational motion of the center of mass and the rotational and internal motions of the system. Using Jacobi coordinates (cf. Ref. [4]),

$$\mathbf{r}_j = \mathbf{R}_{j+1} - \frac{1}{M_j}\sum_{i=1}^{j} m_i \mathbf{R}_i \qquad (j = 1,\ldots,N), \tag{2}$$

with $M_j = \sum_{i=1}^{j} m_i$ $(i = 1,\ldots,N)$ and $\mathbf{R}_{N+1} = \mathbf{0}$, the total Hamiltonian (1) is rewritten as

$$\hat{H} = \hat{H}_{\text{trans}} + \hat{H}_{\text{rot,int}}, \tag{3}$$

where the Hamiltonians for the translational and the rotational and internal motions are

$$\hat{H}_{\text{trans}} = -\frac{\hbar^2}{2M}\nabla^2_{\mathbf{r}_{\text{com}}}, \tag{4}$$

$$\hat{H}_{\text{rot,int}} = -\frac{\hbar^2}{2}\sum_{i=1}^{N-1}\frac{1}{\mu_i}\nabla^2_{\mathbf{r}_i} + 2\sum_{i=1}^{N}\sum_{j=i+1}^{N} V_{ij}\left(\left|\frac{M_{i-1}}{M_i}\mathbf{r}_{i-1} - \frac{M_{j-1}}{M_j}\mathbf{r}_{j-1} - \sum_{k=i}^{j-1}\frac{m_{k+1}}{M_{k+1}}\mathbf{r}_k\right|\right), \tag{5}$$

and the reduced mass μ_i is defined as

$$\mu_i = \left(\frac{1}{M_i} + \frac{1}{m_{i+1}}\right)^{-1} \qquad (i = 1,\ldots,N-1). \tag{6}$$

Furthermore, the total mass of the system and the position vector of the center of mass are defined as $M = M_N$ and $\mathbf{r}_{\text{com}} = -\mathbf{r}_N$, respectively. We note that there is no mass-polarization term in the Hamiltonian for the rotational and internal motions, as reported in Refs. [2, 5], because we do not consider the nuclear center of mass but the total center of mass.

4. Rotation of the many-body system

The non-translational motion of the system can in turn be separated into the rotational motion and the internal motion of the many-body system. We use Euler angles (Ψ,Θ,Φ) to describe three independent rotations of the system in the standard-y-convention, i.e., the coordinate system is rotated by the angle Ψ around the initial Z-axis, then by the angle Θ around the new Y-axis, and finally by the angle Φ around the final Z-axis. The corresponding rotation matrices are

$$R_Z(\Psi) = \begin{pmatrix} \cos\Psi & \sin\Psi & 0 \\ -\sin\Psi & \cos\Psi & 0 \\ 0 & 0 & 1 \end{pmatrix}, \tag{7}$$

$$R_{Y'}(\Theta) = \begin{pmatrix} \cos\Theta & 0 & -\sin\Theta \\ 0 & 1 & 0 \\ \sin\Theta & 0 & \cos\Theta \end{pmatrix}, \tag{8}$$

$$R_{Z''}(\Phi) = \begin{pmatrix} \cos\Phi & \sin\Phi & 0 \\ -\sin\Phi & \cos\Phi & 0 \\ 0 & 0 & 1 \end{pmatrix}. \tag{9}$$

The coordinates of N particles in the rotated system $(x,y,z) = (X''',Y''',Z''')$ with reference to the center of mass are calculated as

$$\mathbf{Q}_i = R(\Psi,\Theta,\Phi)\,(\mathbf{R}_i - \mathbf{r}_{\text{com}}) \qquad (i = 1,\ldots,N), \tag{10}$$

where the total rotation matrix is

$$R(\Psi,\Theta,\Phi) \;=\; R_{Z''}(\Phi)R_{Y'}(\Theta)R_Z(\Psi). \tag{11}$$

Furthermore, we define the rotated Jacobi coordinates in the rotated system (x,y,z) as

$$\mathbf{q}_j \;=\; \mathbf{Q}_{j+1}-\frac{1}{M_j}\sum_{i=1}^{j} m_i\mathbf{Q}_i \;=\; R(\Psi,\Theta,\Phi)\mathbf{r}_j \qquad (j=1,\dots,N-1). \tag{12}$$

In order to fix the spatial orientation of the coordinates $\mathbf{q}_j$, we choose Euler angles $(\Psi,\Theta,\ \Phi)$ such that $\mathbf{q}_{N-2}=(0,0,R)^T$ with $R\geq 0$, and $\mathbf{q}_{N-1}=(\rho,0,\zeta)^T$ with $\rho\geq 0$. These coordinates for $N=3$ are in accordance with the Jacobi coordinates of the nuclei and the electron of the H_2^+ molecule. The inverse transformation of Eq. (12) is

$$\mathbf{r}_j \;=\; R^T(\Psi,\Theta,\Phi)\mathbf{q}_j \qquad (j=1,\dots,N-1), \tag{13}$$

where $R^T(\Psi,\Theta,\Phi)$ is the transpose matrix of $R(\Psi,\Theta,\Phi)$. Both sides of Eq. (13) have $3(N-1)$ independent coordinates, containing $3(N-2)$ rotated Jacobi coordinates in $\mathbf{q}_i$ $(i=1,\dots,N-1)$ and 3 Euler angles (Ψ,Θ,Φ). Then, the Hamiltonian for the rotational and internal motions (5) is rewritten as

$$\begin{aligned}\hat{H}_{\mathrm{rot,int}} \;=\;& -\frac{\hbar^2}{2}\sum_{i=1}^{N-3}\frac{1}{\mu_i}\nabla^2_{\mathbf{q}_i}-\frac{\hbar^2}{2\mu_{N-1}}\left(\frac{\partial^2}{\partial\rho^2}+\frac{1}{\rho}\frac{\partial}{\partial\rho}+\frac{\partial^2}{\partial\zeta^2}\right)\\ &-\frac{\hbar^2}{2\mu_{N-2}}\frac{1}{R^2}\frac{\partial}{\partial R}\left(R^2\frac{\partial}{\partial R}\right)+\frac{1}{2\mu_{N-1}}\frac{1}{\rho^2}\hat{\mathcal{J}}^2_{N-1\,z}\\ &+\frac{1}{4\mu_{N-2}}\frac{1}{R^2}\left(\sum_{i,j=1}^{N-1}\left(\hat{\mathcal{J}}_{i+}\hat{\mathcal{J}}_{j-}+\hat{\mathcal{J}}_{i-}\hat{\mathcal{J}}_{j+}\right)+\hbar\frac{\zeta}{\rho}\sum_{i=1}^{N-1}\left(\hat{\mathcal{J}}_{i+}-\hat{\mathcal{J}}_{i-}\right)\right)\\ &+2\sum_{i=1}^{N}\sum_{j=i+1}^{N}V_{ij}\left(\left|\frac{M_{i-1}}{M_i}\mathbf{q}_{i-1}-\frac{M_{j-1}}{M_j}\mathbf{q}_{j-1}-\sum_{k=i}^{j-1}\frac{m_{k+1}}{M_{k+1}}\mathbf{q}_k\right|\right).\end{aligned} \tag{14}$$

For the H_2^+ molecule ($N=3$), this Hamiltonian is similar to that of Ref. [2]. In Eq. (14), the angular momentum operators are defined as

$$\hat{\mathcal{J}}_{i\pm} \;=\; \begin{cases} e^{\pm i\phi_i}\left(\frac{\zeta_i}{\rho_i}\hat{\mathcal{J}}_{iz}\pm\hbar\left(\rho_i\frac{\partial}{\partial\zeta_i}-\zeta_i\frac{\partial}{\partial\rho_i}\right)\right) & (i=1,\dots,N-3)\\ -i\hbar e^{\mp i\Phi}\left(-\frac{1}{\sin\Theta}\frac{\partial}{\partial\Psi}\pm i\frac{\partial}{\partial\Theta}+\cot\Theta\frac{\partial}{\partial\Phi}\right) & (i=N-2)\\ \frac{\zeta}{\rho}\hat{\mathcal{J}}_{N-1\,z}\pm\hbar\left(\rho\frac{\partial}{\partial\zeta}-\zeta\frac{\partial}{\partial\rho}\right) & (i=N-1)\end{cases}, \tag{15}$$

$$\hat{\mathcal{J}}_{iz} \;=\; \begin{cases} -i\hbar\frac{\partial}{\partial\phi_i} & (i=1,\dots,N-3)\\ -i\hbar\frac{\partial}{\partial\Phi} & (i=N-2)\\ -i\hbar\frac{\partial}{\partial\Phi}-\sum_{j=1}^{N-3}\hat{\mathcal{J}}_{jz} & (i=N-1)\end{cases}, \tag{16}$$

and ρ_i, ζ_i, ϕ_i are the cylindrical coordinates of $\mathbf{q}_i$ $(i=1,\dots,N-3)$. The Hamiltonian $\hat{H}_{\mathrm{rot,int}}$ commutes with $\hat{\mathbf{J}}^2$ and $\hat{J}_Z=-i\hbar\frac{\partial}{\partial\Psi}$ but not with $\hat{\mathcal{J}}_{iz}$ $(i=1,\dots,N-1)$. On the other hand, $\hat{\mathbf{J}}^2$, $\hat{J}_Z$, $\hat{\mathcal{J}}_{iz}$ $(i=1,\dots,N-1)$ commute among each other. The normalized angular eigenfunctions of these $N+1$ operators are written as

$$\mathcal{D}^{J*}_{M,T,T_1,\dots,T_{N-3}}(\Psi,\Theta,\Phi,\phi_1,\dots,\phi_{N-3}) \;=\; \sqrt{\frac{2J+1}{8\pi^2\,(2\pi)^{N-3}}}\,D^{J*}_{M,T}(\Psi,\Theta,\Phi)\prod_{i=1}^{N-3}e^{iT_i\phi_i}, \tag{17}$$

where $D^{J*}_{M,T}(\Psi,\Theta,\Phi)$ is the complex conjugate of the Wigner D-matrix [6]. The quantum numbers for rotation are $J\in\mathbb{N}_0$, $M,T\in\{-J,-J+1,\dots,J-1,J\}$, $T_i\in\mathbb{Z}$ $(i=1,\dots,N-3)$.

5. Stationary wave functions of the many-body system

The Hamiltonian $\hat{H}_{\text{rot,int}}$ commutes not only with $\hat{\mathbf{J}}^2$ and $\hat{J}_Z$ but also with the parity operator $\hat{\Pi}$ with respect to the center of mass. The solutions of the time-independent Schrödinger equation for the rotational and internal motions,

$$\begin{aligned} &\hat{H}_{\text{rot,int}}\Psi_{J,M,\Pi,n}(\Psi,\Theta,\Phi,R,\rho,\zeta,\mathbf{q}_1,\ldots,\mathbf{q}_{N-3}) \\ = \; & E_{J,M,\Pi,n}\Psi_{J,M,\Pi,n}(\Psi,\Theta,\Phi,R,\rho,\zeta,\mathbf{q}_1,\ldots,\mathbf{q}_{N-3}), \end{aligned} \tag{18}$$

are the stationary wave functions $\Psi_{J,M,\Pi,n}$ and the corresponding eigenenergies $E_{J,M,\Pi,n}$ that depend on quantum numbers $J \in \mathbb{N}_0$, $M \in \{-J,-J+1,\ldots,J-1,J\}$ for the rotational motion, $n \in \mathbb{N}_0$ for the internal motion, and $\Pi \in \{-1,+1\}$ for the parity of the many-body system. Since the Hamiltonian $H_{\text{rot,int}}$ does not commute with $\hat{\mathcal{J}}_{iz}$ $(i = 1,\ldots,N-1)$, the stationary wave functions $\Psi_{J,M,\Pi,n}$ can be expanded as

$$\begin{aligned} &\Psi_{J,M,\Pi,n}(\Psi,\Theta,\Phi,R,\rho,\zeta,\mathbf{q}_1,\ldots,\mathbf{q}_{N-3}) \\ = \; & \sum_{T=-J}^{J}\sum_{\{T_i\}}\mathcal{D}^{J*}_{M,T,\{T_i\}}(\Psi,\Theta,\Phi,\{\phi_i\})\psi_{J,M,T,\{T_i\},n}(R,\rho,\zeta,\{\rho_i\},\{\zeta_i\}), \end{aligned} \tag{19}$$

where $\psi_{J,M,T,\{T_i\},n}(R,\rho,\zeta,\{\rho_i\},\{\zeta_i\})$ are the real and even wavefunctions. Using the parity relation for the Wigner D-matrix,

$$\hat{\Pi}D^{J*}_{M,T}(\Psi,\Theta,\Phi) = (-1)^{J+T}D^{J*}_{M,-T}(\Psi,\Theta,\Phi), \tag{20}$$

we are able to show that the following relation must be satisfied:

$$\psi_{J,M,-T,\{-T_i\},n}(R,\rho,\zeta,\{\rho_i\},\{\zeta_i\}) = (-1)^{J+T}\Pi\psi_{J,M,T,\{T_i\},n}(R,\rho,\zeta,\{\rho_i\},\{\zeta_i\}). \tag{21}$$

Finally, with

$$\tilde{\psi}_{J,M,T,\{T_i\},n}(R,\rho,\zeta,\{\rho_i\},\{\zeta_i\}) = \begin{cases} \frac{1}{2}\psi_{J,M,0,\{T_i\},n}(R,\rho,\zeta,\{\rho_i\},\{\zeta_i\}) & T=0 \\ \psi_{J,M,T,\{T_i\},n}(R,\rho,\zeta,\{\rho_i\},\{\zeta_i\}) & T\neq 0 \end{cases}, \tag{22}$$

we obtain the general expression for the stationary wave functions as

$$\begin{aligned} &\Psi_{J,M,\Pi,n}(\Psi,\Theta,\Phi,R,\rho,\zeta,\{\mathbf{q}_i\}) \\ = \; & \sum_{T=0}^{J}\sum_{\{T_i\}}\left(\mathcal{D}^{J*}_{M,T,\{T_i\}}(\Psi,\Theta,\Phi,\{\phi_i\}) + (-1)^{J+T}\Pi\mathcal{D}^{J*}_{M,-T,\{-T_i\}}(\Psi,\Theta,\Phi,\{\phi_i\})\right) \\ & \tilde{\psi}_{J,M,T,\{T_i\},n}(R,\rho,\zeta,\{\rho_i\},\{\zeta_i\}). \end{aligned} \tag{23}$$

6. Conclusion

We have derived the exact rotational and internal Hamiltonian for an arbitrary non-relativistic closed many-body system and the corresponding stationary wave functions, where the rotational and internal motions of the system can be described separately by three Euler angles and by the rotated Jacobi coordinates.

References

[1] Born M and Oppenheimer J R 1927 *Ann. Phys.* **389** 457–484
[2] Hilico L, Billy N, Grémaud B and Delande D 2000 *Eur. Phys. J.* D **12** 449–466
[3] Schrödinger E 1926 *Phys. Rev.* **28** 1049–1070
[4] Greiner W 2001 *Quantum Mechanics: An Introduction* 4th ed (Berlin: Springer) pp 373–377
[5] Bredtmann T, Diestler D J, Li S, Manz J, Pérez-Torres J F, Tian W, Wu Y, Yang Y and Zhai H 2015 *Phys. Chem. Chem. Phys.* **17** 29421–29464
[6] Wigner E P 1959 *Group Theory and its Application to the Quantum Mechanics of Atomic Spectra* (New York: Academic Press)

About the proof that the s-wave functions of the hydrogen atom obey the Ehrenfest equation of motion

K Renziehausen and I Barth

Max Planck Institute of Microstructure Physics, Weinberg 2, 06120 Halle (Saale), Germany

E-mail: ksrenzie@mpi-halle.mpg.de

Abstract. Applying the concept of quantum hydrodynamics on a hydrogen atom, one can derive an equation of motion which describes the time derivative of the mass current density for the relative motion of the electron and the proton. Since this derivation can be performed with the Ehrenfest theorem, we call this equation the Ehrenfest equation of motion. As quantum hydrodynamics is consistent with Schrödinger's quantum mechanics, the stationary s-wave functions satisfy both the Schrödinger equation and the Ehrenfest equation of motion. It is the purpose of this paper to prove that the stationary s-wave functions satisfy the Ehrenfest equation of motion. From the quantum hydrodynamical point of view, this result for s-wave functions can be interpreted that the Coulomb force density that attracts the electron to the proton is compensated by a quantum force density that is related to the dispersion of the probability density.

1. Introduction

The concept of quantum hydrodynamics (QHD) for one-electron systems was introduced by Madelung and Bohm [1–3]. Besides other authors, QHD was developed further by Epstein and Wyatt [4, 5]. In Refs. [4, 5], an equation is discussed that is related to the momentum balance of the system. Since the derivation of this momentum balance equation can be performed by calculating the time derivative of the mass current density with the Ehrenfest theorem [6], we call this equation the Ehrenfest equation of motion (EEM); it is given by

$$\frac{\partial \vec{j}_m}{\partial t} + \nabla \underline{\underline{\Pi}} \quad = \quad \vec{f}. \tag{1}$$

2. System

We analyse below a hydrogen atom in a spherical coordinate system where its origin is located at the center of mass of the atom. For this system, the quantities appearing in equation (1) – these are the Coulomb force density $\vec{f}$, the divergence of the momentum flow density tensor $\nabla\underline{\underline{\Pi}}$, and the mass current density $\vec{j}_m$ – refer to the relative motion of the electron and the proton.

In the following, the proof is shown that all s-wave functions Ψ_n of the hydrogen atom with the principal quantum number $n = 1, 2, 3, \dots$ satisfy the EEM (1). This is a generalization of the calculations in [7], where this proof was performed explicitly for the 1s and the 2s wave functions. Since the s-wave functions are real-valued, $\vec{j}_m$ vanishes

$$\vec{j}_m = \vec{0}. \tag{2}$$

The Coulomb force density is given by

$$\vec{f} = -D\frac{e^2}{4\pi\varepsilon_0}\frac{1}{r^2}\vec{e}_r. \tag{3}$$

Here, e is the elementary charge, ε_0 is the electric constant, and $D = |\Psi_n|^2$ is the probability density of finding the electron and the proton at the relative position $\vec{r} = \vec{r}_e - \vec{r}_p$, where $\vec{r}_e$ and $\vec{r}_p$ are the electron's and proton's positions. Then, $r = |\vec{r}|$ is both the electron-proton distance and the radius in the mentioned spherical coordinate system, and the vector $\vec{e}_r$ is the radial unit vector in this system. As the next step, we calculate $\nabla\underline{\underline{\Pi}}$.

We can split up the momentum flow density tensor $\underline{\underline{\Pi}}$ into the classical tensor $\underline{\underline{\Pi}}^{\mathrm{Cl}}$, the osmotical tensor $\underline{\underline{\Pi}}^{\mathrm{Os}}$, and the product of the unit matrix $\underline{\underline{1}}$ and the scalar pressure P [5]

$$\underline{\underline{\Pi}} = \underline{\underline{\Pi}}^{\mathrm{Cl}} + \underline{\underline{\Pi}}^{\mathrm{Os}} + \underline{\underline{1}}P. \tag{4}$$

The motivation for the naming of the classical tensor $\underline{\underline{\Pi}}^{\mathrm{Cl}}$ is that if the physical system is described by pure classical physics then equation (4) becomes $\underline{\underline{\Pi}} = \underline{\underline{\Pi}}^{\mathrm{Cl}}$. In contrast, both the osmotical tensor $\underline{\underline{\Pi}}^{\mathrm{Os}}$ and the scalar pressure P are related to the dispersion of the quantum wave packet and they are pure quantum quantities.
The classical tensor $\underline{\underline{\Pi}}^{\mathrm{Cl}}$ is given by

$$\underline{\underline{\Pi}}^{\mathrm{Cl}} = m_\mu D(\vec{v}\otimes\vec{v}), \tag{5}$$

where $m_\mu = m_\mathrm{e}m_\mathrm{p}/(m_\mathrm{e}+m_\mathrm{p})$ is the reduced mass of the electron (with the mass m_e) and the proton (with the mass m_p), and the term $\vec{v}\otimes\vec{v}$ is a dyadic product of the relative velocity $\vec{v}$ of the two particles. As the mass current density $\vec{j}_m$ is related to this velocity $\vec{v}$ by $\vec{j}_m = m_\mu D\vec{v}$, both of them vanish for the analyzed system – so, the classical tensor $\underline{\underline{\Pi}}^{\mathrm{Cl}}$ vanishes, too.
Moreover, the osmotical tensor $\underline{\underline{\Pi}}^{\mathrm{Os}}$ is given by

$$\underline{\underline{\Pi}}^{\mathrm{Os}} = m_\mu D(\vec{d}\otimes\vec{d}). \tag{6}$$

In the equation above, a dyadic product of the osmotic velocity $\vec{d}$, defined by

$$\vec{d} = -\frac{\hbar}{2m_\mu}\frac{\nabla D}{D}, \tag{7}$$

appears. In our spherical coordinate system, the s-wave functions Ψ_n depend only on the radius r, and not on the angles ϑ and φ. So, it holds for the probability density D, too. Thus, only the radial component d_r of the osmotical velocity does not vanish, and therefore only the Π^{Os}_{rr}-component of the tensor $\underline{\underline{\Pi}}^{\mathrm{Os}}$ is not zero. The result for this tensor component Π^{Os}_{rr} is

$$\Pi^{\mathrm{Os}}_{rr} = \frac{\hbar^2}{4m_\mu}\frac{1}{D}\left(\frac{\partial D}{\partial r}\right)^2. \tag{8}$$

The scalar pressure P is given by

$$P = -\frac{\hbar^2}{4m_\mu}\triangle D = -\frac{\hbar^2}{4m_\mu}\left(\frac{2}{r}\frac{\partial D}{\partial r} + \frac{\partial^2 D}{\partial r^2}\right). \tag{9}$$

Due to equation (4) and the disappearance of the classical tensor $\underline{\underline{\Pi}}^{\rm Cl}$, we have to sum up the two tensor divergence terms $\nabla\underline{\underline{\Pi}}^{\rm Os}$ and $\nabla(\underline{\underline{1}}P)$ for the calculation of $\nabla\underline{\underline{\Pi}}$. By regarding how to calculate tensor divergences in spherical coordinates [8] and by taking into account equations (8) and (9), finally, we get:

$$\nabla\underline{\underline{\Pi}} = -\frac{\hbar^2}{4m_\mu}\left\{\frac{1}{D}\frac{\partial D}{\partial r}\left[\frac{1}{D}\left(\frac{\partial D}{\partial r}\right)^2 - \frac{2}{r}\frac{\partial D}{\partial r} - 2\frac{\partial^2 D}{\partial r^2}\right] - \frac{2}{r^2}\frac{\partial D}{\partial r} + \frac{2}{r}\frac{\partial^2 D}{\partial r^2} + \frac{\partial^3 D}{\partial r^3}\right\}\vec{e}_r. \tag{10}$$

As the next step, we multiply the EEM (1) with the vector $-\frac{4m_\mu}{\hbar^2}\vec{e}_r$ and use the results (equations (2), (3) and (10)) of the mass current density $\vec{j}_m$, the Coulomb force density $\vec{f}$ and the divergence of the momentum flow density tensor $\nabla\underline{\underline{\Pi}}$. In addition, we regard that the modified Bohr radius for a finite nuclear mass is given by $a_\mu = \frac{4\pi\varepsilon_0\hbar^2}{m_\mu e^2}$. Then, we get this differential equation:

$$\frac{1}{D}\frac{\partial D}{\partial r}\left[\frac{1}{D}\left(\frac{\partial D}{\partial r}\right)^2 - \frac{2}{r}\frac{\partial D}{\partial r} - 2\frac{\partial^2 D}{\partial r^2}\right] - \frac{2}{r^2}\frac{\partial D}{\partial r} + \frac{2}{r}\frac{\partial^2 D}{\partial r^2} + \frac{\partial^3 D}{\partial r^3} = \frac{4}{a_\mu r^2}D. \tag{11}$$

3. Proof

Now, we show that this differential equation is solved by the quantum textbook result for stationary s-wave functions Ψ_n with the principal quantum number $n = 1, 2, 3, \ldots$ (see e.g. [9]). Therefore, we introduce a parameter $\lambda = \frac{2}{na_\mu}$ and a dimensionless coordinate $\rho = \lambda r$. The mentioned quantum textbook result for Ψ_n is

$$\Psi_n(\rho) \;=\; C\, g(\rho)\, \exp\left(-\frac{\rho}{2}\right), \tag{12}$$

where C is a real prefactor independent of the coordinate ρ. In addition, $g(\rho)$ is equal to the associated Laguerre polynomial $L_n^1(\rho)$ that satisfies this differential equation [9]:

$$\rho\frac{\partial^2 L_n^1(\rho)}{\partial\rho^2} + (2-\rho)\frac{\partial L_n^1(\rho)}{\partial\rho} + (n-1)L_n^1(\rho) \;=\; 0. \tag{13}$$

Therefore, the probability density D is given by

$$D(\rho) \;=\; C^2\,[g(\rho)]^2\, \exp(-\rho)\,. \tag{14}$$

As the next step, we handle equation (14) as an ansatz for $D(\rho)$ and insert it into the differential equation (11). After some straightforward transformations, we find the differential equation for $g(\rho)$,

$$g\frac{\partial^3 g}{\partial\rho^3} + \left(1-\frac{2}{\rho}\right)\left(\frac{\partial g}{\partial\rho}\right)^2 - \frac{\partial g}{\partial\rho}\frac{\partial^2 g}{\partial\rho^2} + \left(\frac{2}{\rho}-1\right)g\frac{\partial^2 g}{\partial\rho^2} - \frac{2g}{\rho^2}\frac{\partial g}{\partial\rho} - \frac{n-1}{\rho^2}g^2 = 0, \tag{15}$$

and define

$$y(\rho) := \rho\frac{\partial^2 g}{\partial\rho^2} + (2-\rho)\frac{\partial g}{\partial\rho} + (n-1)g, \tag{16}$$

where, according to equation (13), the solution of $y(\rho) = 0$ is $g(\rho) = L_n^1(\rho)$, and

$$h(\rho) := \frac{y}{\rho}. \tag{17}$$

Now, we evaluate

$$\begin{aligned} g\frac{\partial h}{\partial \rho} &= g\frac{\partial}{\partial \rho}\left[\frac{\partial^2 g}{\partial \rho^2} + \left(\frac{2}{\rho} - 1\right)\frac{\partial g}{\partial \rho} + \frac{n-1}{\rho} g\right], \\ &= g\frac{\partial^3 g}{\partial \rho^3} + \left(\frac{2}{\rho} - 1\right) g\frac{\partial^2 g}{\partial \rho^2} - \frac{2g}{\rho^2}\frac{\partial g}{\partial \rho} + \frac{n-1}{\rho} g\frac{\partial g}{\partial \rho} - \frac{n-1}{\rho^2} g^2, \end{aligned} \tag{18}$$

and

$$\begin{aligned} -\frac{\partial g}{\partial \rho} h &= -\frac{\partial g}{\partial \rho}\left[\frac{\partial^2 g}{\partial \rho^2} + \left(\frac{2}{\rho} - 1\right)\frac{\partial g}{\partial \rho} + \frac{n-1}{\rho} g\right], \\ &= -\frac{\partial g}{\partial \rho}\frac{\partial^2 g}{\partial \rho^2} + \left(1 - \frac{2}{\rho}\right)\left(\frac{\partial g}{\partial \rho}\right)^2 - \frac{n-1}{\rho} g\frac{\partial g}{\partial \rho}. \end{aligned} \tag{19}$$

Combining equations (18) and (19), the differential equation (15) is rewritten as

$$g\frac{\partial h}{\partial \rho} - \frac{\partial g}{\partial \rho} h = 0, \tag{20}$$

or simply

$$g\frac{\partial h}{\partial \rho} = \frac{\partial g}{\partial \rho} h. \tag{21}$$

Now, we can use the new version (21) of the differential equation (15) for $g(\rho)$ to prove that $g(\rho) = L_n^1(\rho)$ solves this differential equation (15). It is trivial that $h(\rho) = 0$ solves equation (21), and then $y(\rho) = \rho h(\rho) = 0$ is true. So, we have shown that $g(\rho) = L_n^1(\rho)$ solves (15) and, as a consequence, we have shown that the quantum textbook result for Ψ_n is a solution of the EEM (1), too.

4. Discussion

The fact that the s-wave functions Ψ_n satisfy the EEM means that these wave functions have such a form that the Coulomb force density, $\vec{f}$, that attracts the electron to the proton, and the divergence of the momentum flow density tensor, $\nabla\underline{\underline{\Pi}}$, that is related to the dispersion of the probability density, are consistent. Hereby, since only the quantum quantities $\underline{\underline{\Pi}}^{\mathrm{Os}}$ and $\underline{\underline{P}}$ contribute to the tensor $\underline{\underline{\Pi}}$, the term $\nabla\underline{\underline{\Pi}}$ can be interpreted as a pure quantum force density.

The arguments presented in this paper advance a new view of the properties of the s-wave functions Ψ_n that are obtained by the analysis of the EEM.

References

[1] Madelung E 1926 *Z. Phys.* **40** 322–326
[2] Bohm D 1952 *Phys. Rev.* **85** 166–179
[3] Bohm D 1952 *Phys. Rev.* **85** 180–193
[4] Epstein S T 1975 *J. Chem. Phys.* **63** 3573–3574
[5] Wyatt R E 2005 *Quantum Dynamics with Trajectories: Introduction to Quantum Hydrodynamics* (New York: Springer) pp 326–328
[6] Schwabl F 2005 *Quantum Mechanics* softcover reprint of hardcover 3rd ed. (Berlin: Springer) pp 28–30
[7] Deb B M and Bamzai A S 1978 *Mol. Phys.* **35** 1349–1367
[8] Lai M, Krempl E and Ruben D 2009 *Introduction to Continuum Mechanics* 4th ed. (Burlington, MA: Butterworth-Heinemann) p 65
[9] Bransden B H and C J Joachain 2003 *Physics of Atoms and Molecules* 2nd ed. (Harlow: Pearson Education) pp 155–161

IOP Conf. Series: Journal of Physics: Conf. Series **869** (2017) 012008 doi:10.1088/1742-6596/869/1/012008

Lasing due to the excited state in quantum dot lasers

M Abusaa[1], J Danckaert[2] and E A Viktorov[3]

[1]Physics Department, Arab American University, Palestine
[2]Applied Physics Research Group (APHY), Vrije Universiteit Brussel, Belgium
[3]Institute for Laser Physics of Vavilov SOI Corporation, Russia

E-mail: muayad.abusaa@aauj.edu

Abstract. Quantum Dot Lasers (QDLs) are promising sources of light because of their favorable properties compared to other light sources. Emission in QDLs can access transitions in ground state (GS) and excited state (ES). Lasing due to the ES extends the spectral range and enables the laser to generate high output powers. Thus, lasing action due to the ES or to the dual lasing regime (GS and ES simultaneously) is expected to increase the applicability of QDLs in many future applications. We present a partially microscopic rate equation model that takes into account lasing action due to both the GS and the ES and distinguishes between both types of carriers (electrons and holes). Also, we present all possible steady-state solutions and we apply a stability analysis to the solutions to determine all stable lasing regimes (lasing due to the GS, lasing due to the ES and the dual lasing regime) to highlight the role of ES transitions. Specifically, we address the appearance of lasing due to the ES to the larger population of the ES and hence to the larger gain in higher injected current regimes.

1. Introduction

Quantum dot lasers (QDLs) are expected to be the new generation of lasers in many future applications such as telecommunications and biomedical applications because they possess many favourable properties compared to other sources of light. QDL is a semiconductor laser (SCL) whose active medium is a quantum dot (QD) double hetero structure. A QD hetero structure forms the ultimate case of quantization of carriers whose motion is restricted in all three spatial dimensions and, hence, the energy levels of QDLs are truly discrete. Thus, electrically pumped QDLs have lower threshold current density, higher differential gain, higher temperature stability and broader modulation bandwidth, compared to other SCLs. For many applications, QDLs should be able to generate high output powers. Emission in QDLs can access transitions in the ground state (GS), the excited state (ES), and both states. Lasing action due to the ES is expected to exhibit higher output power that is not achievable with other SCLs. Also, lasing action due to the ES extends the spectral range, which can enlarge the field of applications. Thus, it is of vital importance to highlight the role of lasing due to the ES in controlling the dynamical behaviour of QDLs [1, 2].

The two-state operation (GS and ES transitions) in QDLS has been reported experimentally and theoretically in many dynamical regimes [3-7]. In epitaxial growth techniques, QDs are grown on a substrate that is known as the wetting layer (WL). Thus, in electrically pumped QDLs we pump the WL and carriers relax from the WL to the energy levels of the QDL. In this work, we use a partially microscopic rate equation model that we presented in [3] to highlight the lasing predominantly via the

ES in QDLs. The model takes care of both the emission from the GS and the ES, and distinguishes between both types of carriers (electrons and holes).

2. Partially microscopic rate equation model

The following model consists of rate equations for the electromagnetic field GS- intensity (I_g), ES - intensity (I_e) and for the GS and ES occupational probabilities for electrons and holes ($n^g_{e,h}, n^e_{e,h}$) in the dot where the lower indices, *e* and *h,* stand for electron and hole, respectively, while the upper indices, *g* and *e,* stand for ground and excited states respectively.

$$I'_g = \left(2g\left(n^g_e + n^g_h - 1\right) - 1\right)I_g, \tag{1}$$

$$I'_e = \left(4g\left(n^e_e + n^e_h - 1\right) - 1\right)I_e, \tag{2}$$

$$n'^g_e = \eta\left(2Bn^e_e\left(1 - n^g_e\right) - \left(n^g_e + n^g_h - 1\right)I_g\right), \tag{3}$$

$$n'^e_e = \eta\left(J\left(1 - n^e_e\right) - R_e n^e_e - Bn^e_e\left(1 - n^g_e\right) - \left(n^e_e + n^e_h - 1\right)I_e\right), \tag{4}$$

$$n'^g_h = \eta\left(2B\left(n^e_h - n^g_h\right) - \left(n^g_e + n^g_h - 1\right)I_g\right), \tag{5}$$

$$n'^e_h = \eta\left(J\left(1 - n^e_h\right) - R_h n^e_h - B\left(n^e_h - n^g_e\right) - \left(n^e_e + n^e_h - 1\right)I_e\right), \tag{6}$$

where the prime means differentiation with respect to time. Here, $\eta = \tau_p \tau^{-1}$ describes slow relaxation of the carriers with respect to the photon lifetime, where τ_p is the photon lifetime and τ is the carrier recombination time. The factors 2 and 4 account for the degeneracy in the GS and the ES respectively. The higher degeneracy of the ES is due to the existence of two closely spaced p-levels, which has been verified experimentally [8]. Also, J stands for the pump current per dot. The terms $2g\left(n^g_e + n^g_h - 1\right)$ and $4g\left(n^e_e + n^e_h - 1\right)$ are the gains defined by the populations of both types of carriers in both states (GS and ES) and the effective gain factor g, which is scaled to the cavity losses, and is assumed to be identical for both states. The term $\left(1 - n^{g,e}_{e,h}\right)$ is the Pauli blocking factor, written in this form because the relaxation of carriers is limited by the filling of the states according to the Pauli exclusion principle. When the state is empty, the relaxation rate of carriers has a maximum value, and as the population increases, the rate of relaxation of carriers decreases. B is the ratio of the carrier recombination time to the capture time and it is an important intradot timescale that distinguishes QDLs form other SCLs. R_e and R_h represent the escape coefficients of electrons and holes from the ES to the WL, respectively. In this model, we assume direct pumping of the ES from the wetting layer. Also, we assume that electrons and holes are captured at the same rates B, while escape of coefficients from the GS to the ES is ignored because it is very small compared to the capture rates [9, 10].

3. Results and discussion

The three nonzero intensity steady state solutions are: lasing due to the GS only ($I_g \neq 0$ and $I_e = 0$), lasing due to the ES only ($I_g = 0$ and $I_e \neq 0$) and the dual lasing regime ($I_g \neq 0$ and $I_e \neq 0$). We examine the stability of these solutions by applying Routh-Hurwitz conditions [11] at different values of *g*. For $g = 0.75$, we find that lasing starts at a low threshold injected current (first threshold) due to the GS transitions and the lasing action remains stable up to the appearance of a second threshold at which a dual lasing stable regime emerges and the single state (GS) regime loses stability. For $g = 0.65$, we find that the only stable regime is lasing due to the ES only and for $g = 0.7056$, we find that the dual lasing regime is the only stable regime. From the intensity expressions and occupational probabilities for both stable regimes, and by checking the solutions for different capture rates, we can define a critical effective gain factor (g_c) at which lasing action takes place first due to both GS and ES simultaneously. By plotting the second threshold current density versus g for a large range of capture rates, we find that all curves emerge from the same point g_c. At large capture rates, g_c depends on the escape rates of electrons and holes from the dot to the WL, through

$$g_c = \left(3R_e + \sqrt{9R_e^2 + 8R_e\left(R_h - R_e\right)}\right)/8R_e, \tag{7}$$

which yields the value $g_c = 0.7056$ for InAs/GaAs QDL operating at 1.3 μm, and where $R_e = 1$ and $R_h = 0.75$.

Depending on the value of g, we can achieve stability for all three mentioned solutions. For $g > g_c$, lasing takes place first due to the GS and this regime tracks in stability up to the appearance of the second threshold at which lasing takes place due to the dual lasing regime. In this case, lasing due to the ES only is absent (figure 1 (a)). For $g = g_c$, lasing takes place due to the dual lasing regime which is the only stable regime in this case (figure 1 (b)). For $g < g_c$, lasing takes place due to the ES only and lasing due to the GS is absent (figure 1 (c)).

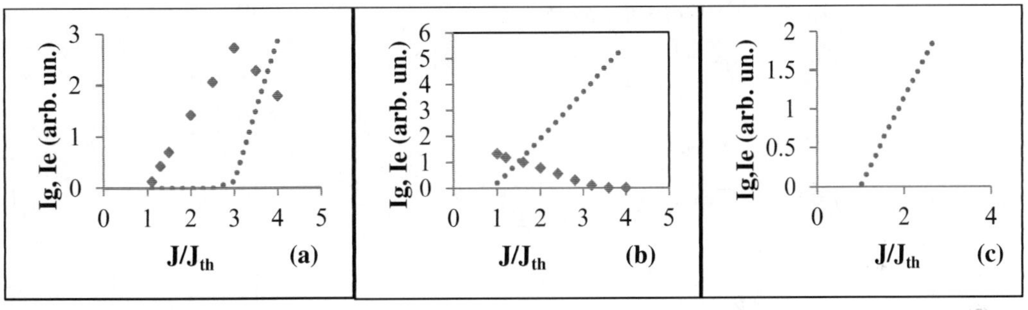

Figure 1: Bifurcation diagrams of the stable GS and ES steady-state intensities. The blue dots stand for the intensity of the GS and the red dashed line stands for the intensity of the ES. $\eta = 0.01$, $\tau = 1$ ns, $\tau_p = 10$ ps, $R_h = 0.75$, $R_e = 1$ and $B = 100$. (a) $g = 0.75$ (b) $g = 0.7056$ (c) $g = 0.65$.

In QDLs, lasing action takes place as a result of the efficient energy relaxation into the lowest available energy states. Thus, for $g > g_c$, lasing due to the GS takes place first for QDLs operating at low level of injection and increasing of the injected current leads to a larger population of the ES, which, in turn, leads to the appearance of the ES threshold, and to dual lasing in both the GS and the ES. GS lasing takes place first due to the higher gain and further increasing the injected current results in lasing predominantly via the ES. In lasing due to ES and in dual lasing regime the intensity of ES increases with injected current because the hole levels are closely spaced in both states and as lasing due to ES takes place, ES is preferred because of the higher degeneracy of ES compared to GS.

Either GS threshold or ES threshold may appear first, or both GS and ES may start lasing at the same injected current. From the steady-state solutions, we find that lasing threshold of the ES is directly proportional to g while lasing threshold of the GS is inversely proportional to g. Thus, lasing action takes place due to the ES only for g less than g_c.

4. Conclusion

QDLs operate at both GS and ES transitions. For higher injected currents, lasing due to ES is dominant because of the higher degeneracy of the ES. ES emission has its origin in the slow band relaxation of carriers and the discrete-nature of density of states for QD Structures. Also, the effective gain factor plays an important role in determining the possible stable lasing regimes.

References

[1] Malic E, Ahn K J, Bormann M J P, Hovel P, Scholl E, Knorr A, Kuntz M and Bimberg D 2006 *Appl. Phys. Lett.* **89** 101107
[2] Rafailov E U, Cataluna M A and Sibbett W 2007 *Nature Photonics* **1** 395
[3] Abusaa M, Danckaert J, Viktorov E A and Erneux T 2013 *Phys. Rev.* A **87** 063827
[4] Markus A, Chen J X, Paranthoen C, Fiore A, Platz C and Gauthier-Lafaye O 2003 *Appl. Phys. Lett.* **82** 1818-20
[5] Viktorov E A, Mandel P, Tanguy Y, Houlihan J and Huyet G 2005 *Appl. Phys. Lett.* **87** 053113
[6] Viktorov E A , Cataluna M A, O'Faolain L, Krauss T F, Sibbett W, Rafailov E U and Mandel P 2007 *Appl. Phys. Lett.* **90** 121113
[7] Cataluna M A, Sibbett W, Livshits D A, Weimert J, Kovsh A R and Rafailov E U 2006 *Appl. Phys. Lett.* **89** 081124
[8] Williamson A J, Wang L W and Zunger A 2000 *Phys. Rev.* B **62** 12963
[9] Sugawara M, Mohki K and Shoji H 1997 *Appl. Phys. Lett.* **71** 2791
[10] Ishida M, Hatori N, Akiyama T, Otsubo K, Nakata Y, Ebe H, Sugawara M and Arakawa Y 2004 *Appl. Phys. Lett.* **85** 4145
[11] Uspensky J V 1969 *Theory of Equations* (New York: McGraw-Hill)

IOP Conf. Series: Journal of Physics: Conf. Series **869** (2017) 012009 doi:10.1088/1742-6596/869/1/012009

Interplay between pulse parameters and alignment of H_2^+ in strong laser fields

M Abu-samha[1] and L B Madsen[2]

[1] College of Sciences and Humanities, Fahad bin Sultan University, Tabuk 71454, KSA
[2] Department of Physics and Astronomy, Aarhus University, 8000 Aarhus C, Denmark

E-mail: mabusamha@fbsu.edu.sa

Abstract. We present an *ab initio* study of strong field ionization of aligned H_2^+ by general elliptically polarized laser pulses containing few optical cycles. In particular, we address the effects of laser pulse parameters (light ellipticity and carrier-envelope phase) and molecular alignment on photoelectron momentum distributions (PMDs) of H_2^+.

1. Introduction

Strong field experiments and theory exploit observables such as photoelectron momentum distributions (PMDs), in investigations of atomic-scale electron dynamics, including attosecond streaking [1], strong-field circular dichroism [2,3] and attosecond ionization dynamics or attoclock experiments [4–9].

In strong-field ionization of molecules [10–13], the nature of the probed targets has effects on the PMDs; for example, interferences due to the multicenter character, coupling of electronic and nuclear motions, and multielectron effects. Strong-field studies of H_2^+ using general elliptically polarized laser [14–17] addressed role of the Coulomb potential, coupling of the nuclear and electronic motions, and light-induced subcycle dynamics and how these factors impact above-threshold ionization and PMDs. In Ref. [16], PMDs for H_2^+ and D_2^+ suggest that vibronic coupling is negligible, which validates solution of the time-dependent Schrödinger equation (TDSE) at frozen geometric structure [17]. In Ref. [17], the TDSE calculations were performed in two dimensions and the resulting PMDs revealed an offset angle which is a function of both intensity and wavelength.

In a recent three-dimensional TDSE study of aligned Ar $3p_x$ and H_2^+ in circularly polarized fields [18], the PMDs corresponding to molecular hydrogen showed a substantial alignment effect due to the anisotropy of the molecular potential and possible coupling to low-lying excited states. In the present work, we expand on the results published in Ref. [18]. Namely, we study interplay between the laser pulse parameters and the molecular alignment, and how it affects PMDs. In particular, we address effects of light ellipticity and carrier-envelope phase on PMDs and lateral PMD (LPMDs) of aligned H_2^+.

2. Computational details

The laser pulse is described by $\vec{F}(t) = -\partial_t \vec{A}(t)$, where $\vec{A}(t)$ is the vector potential, given as $\vec{A}(t) = A_0 f(t) \left(\cos(\omega t + \varphi)\cos(\epsilon/2)\hat{i} + \sin(\omega t + \varphi)\sin(\epsilon/2)\hat{j}\right)$, with A_0 the amplitude, ω the

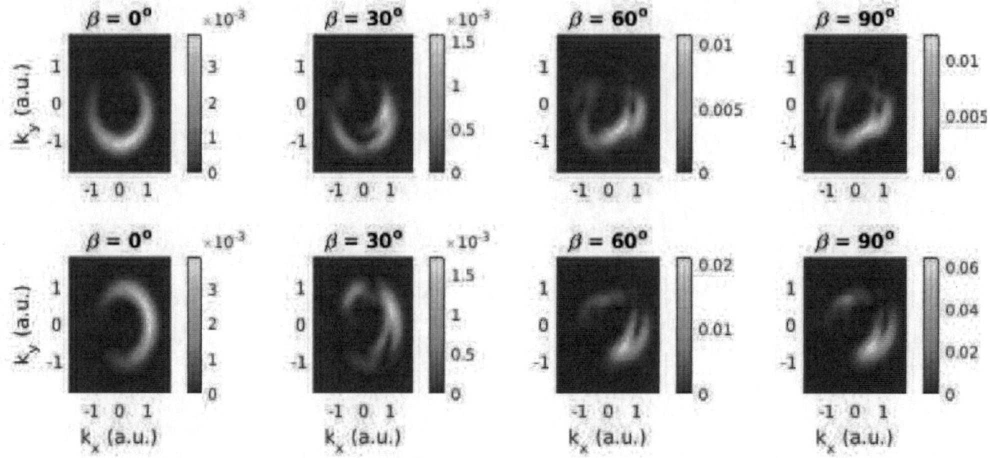

Figure 1. PMDs for aligned H_2^+ at CEP values of -$\pi/2$ (top panels) and 0 rad (bottom panels). The laser pulse is circular, contains 2 cycles, and has peak intensity of 0.2 PW/cm^2. The results in top panels are adapted from Ref. [18].

carrier angular frequency, φ the carrier-envelope phase (CEP) value, ϵ the light ellipticity, and $f(t) = \sin^2(\omega t/2N)$ the envelope for an N-cycle pulse. In the present work, $\omega = 0.057$ a.u. corresponding to 800 nm wavelength, and $N = 2$ optical cycles. Atomic units (a.u.) are used.

The TDSE calculations were carried out for $\epsilon = \pi/2$ (corresponding to circular polarization) and for $\epsilon = 1.35$ rad, following the experiment of Ref. [7]. To address the effect of CEP, the calculations were performed for CEP values $\varphi = -\pi/2$ and 0 in the circular polarization limit.

In TDSE calculations, the time-dependent wavefunction is expanded in spherical harmonics for the angular degrees of freedom and the TDSE is solved in the velocity gauge with a grid representation for the reduced radial wave functions [19]. We use an equidistant grid with 4096 points that extends up to 400 a.u. and a time-step of $\Delta t = 0.005$ a.u. in our split-step Crank-Nicholson propagator. PMDs are computed by projecting the wavepacket at the end of the laser pulse on Coulomb scattering states [18, 20].

3. Results and discussion

We explore the PMDs originating from the σ_g ground state of the H_2^+ molecule at internuclear separation of 5 a.u., and at alignment angles $\beta = 0°$, 30°, 60°, and 90°. The alignment angle β is defined as the angle between the molecular frame and the lab frame z-axes. In our TDSE calculations, the molecular axis is defined along the molecular frame z-axis, whereas the laser polarization is defined in the xy-plane of the lab frame. For $\beta = 0°$, the molecular axis is perpendicular to the plane of laser polarization. For $\beta = 90°$, on the other hand, the molecular axis and the lab frame x-axis are aligned.

In figure 1(top panels), we present PMDs for H_2^+ in circularly polarized laser pulses containing 2 optical cycles, with a laser peak intensity 0.2 PW/cm^2 and CEP value of $\varphi = -\pi/2$. At $\beta = 0°$, the PMDs is similar to those obtained for atomic targets, in particular, the PMDs show an offset angle relative to the predictions of strong field approximation, in which the molecular potential is neglected [21]. At $\beta = 30°$, new features can be seen in the PMDs. These features evolve with molecular alignment as can be seen for results at $\beta = 60°$ and 90°. These results make an interesting comparison to the atomic case, for which the angular distribution of photoelectrons is insensitive to orbital alignment [18].

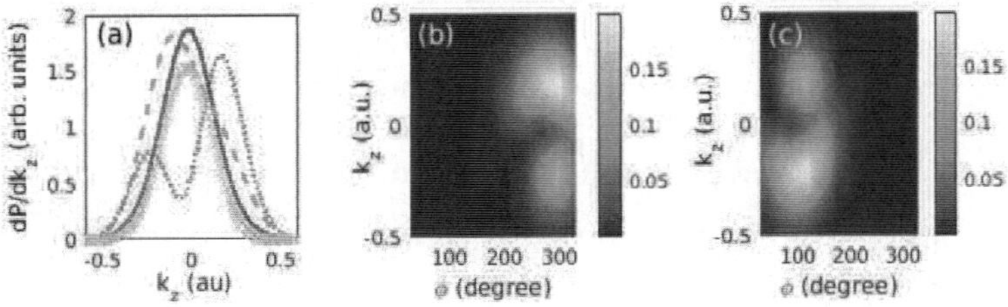

Figure 2. (a) Lateral PMDs for H_2^+ at alignment angle $\beta = 0°$ (solid line), 30° (dots), 60° (dashed), and 90° (crosses). In (b) and (c), we present LPMD for H_2^+ at alignment angle $\beta = 30°$ and φ values of (b)-$\pi/2$ and (c)$\pi/2$.

Next, we investigate effects of CEP on PMDs of H_2^+. In figure 1(bottom panels), we present PMDs for H_2^+ for $\varphi = 0$, and otherwise same laser parameters as results in figure 1 (top panels). At $\beta = 0°$, changing the CEP results in a simple overall clockwise rotation of the PMD. At $\beta = 30°$, 60°, and 90°, changing the CEP leads to different angular distribution of photoelectrons. For alignment angles different from $\beta = 0°$, the molecular potential is strongly anisotropic in the frame of laser polarization. Hence, the PMDs reflect the anisotropic molecular potential felt by the ionized electron.

The results in figure 1 are highly relevant in view of the ongoing discussion on time-delay in tunneling ionization, see e.g., Refs. [4, 5, 22, 23]. In the attoclock scheme, the shift in the peak of the photoelectron momentum distribution compared to that expected from semi-classical modelling has been used to address the question of a possible time-delay [4, 5, 22, 23]. The results in figure 1 show that the momentum distribution is highly affected by the molecular orientation, and hence the short-range part of the molecular potential. For some alignments, we do not even see a clear peak (figure 1, $\beta = 60°$). This shows that not only Coulomb effects and polarization effects as in [7], but also the short-ranged part of the molecular potential, need to be considered in the interpretation of attoclock experiments.

In figure 2(a), we present lateral PMDs (LPMD) for H_2^+ for alignment angles $\beta = 0$, 30°, 60° and 90°. The lateral PMDs show strong dependence on molecular alignment. For $\beta = 0°$ and 90°, the lateral PMDs peak about $k_z = 0$, and the LPMD are reflection symmetric across $k_z = 0$. We interpret this feature in light of coupling to excited states, in particular the σ_u state. The population of σ_u is found to be zero for alignment angles $\beta = 0°$ and 90°, analysis performed at the end of the laser pulse. At alignment angles of 30° and 60°, however, the lateral PMDs are asymmetric across $k_z = 0$.

From figure 2, we find the LPMD to be particularly interesting for H_2^+ at alignment angle $\beta = 30°$. In this case, the lateral PMD shows a bimodal structure along the k_z axis, with larger ionization probability in the positive k_z direction. The asymmetry of ionization probability along k_z is due to interplay between laser carrier envelope phase and anisotropic molecular potential. We have checked that a change in the CEP by π shifts the distribution by 180° in φ as can be seen in figure 2, which leads to a reflection of the distribution with respect to $k_z = 0$, in accordance with the change in the effective molecular potential seen by the outgoing electron upon that CEP change. Simultaneously, at $\beta = 30°$, there is a relatively stronger coupling to the σ_u state, and the resulting double-peak structure in the lateral PMD may be consistent with ionization out of this state.

To address the effect of light ellipticity on PMDs, TDSE calculations were carried out for H_2^+ in an elliptically polarized laser with light ellipticity $\epsilon = 1.35$ rad. The results are shown in figure 3 for different molecular alignment. One can clearly see that the PMDs show strong dependency on alignment angle, regardless of light ellipticity. This is due to the anisotropic

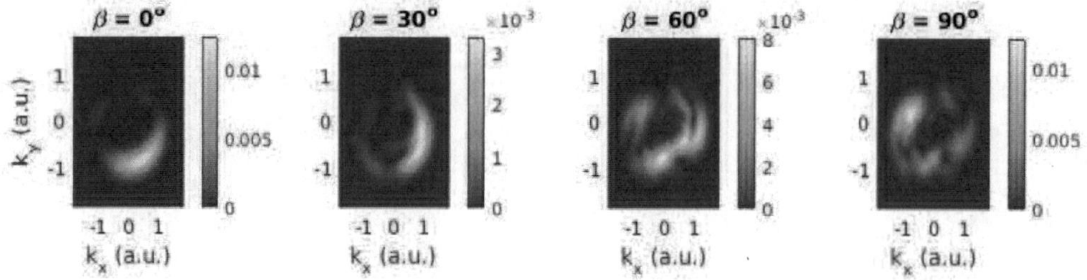

Figure 3. Same as figure 1 (top panels) for an ellipticity $\epsilon = 1.35$ rad.

nature of the H_2^+ potential. For $\beta = 0°$, $30°$, and $60°$, changing the ellipticity have minor effects on PMDs. However, at $\beta = 90°$, changing the ellipticity has substantial effect on the PMDs. In this case, the peak maximum shifts by a considerable angle.

4. Conclusions

The present study indicates that it is not straightforward to interpret the effects of laser pulse parameters on PMDs of aligned H_2^+, due to the anisotropic nature of the short-range molecular potential and possible coupling to low-lying excited states.

Acknowledgments

The computations were performed on the cluster of the Centre of Scientific Computing Aarhus (CSCAA).

References

[1] Krausz F and Ivanov M 2009 Attosecond physics. *Rev. Mod. Phys.* **81** 163–234
[2] Lux C, Wollenhaupt M, Bolze T, Liang Q, Köhler J, Sarpe C and Baumert T 2012 *Angew. Chem. Int. Ed.* **51** 5001–5005
[3] Lux C, Wollenhaupt M, Sarpe C and Baumert T 2015 *Chem. Phys. Chem.* **16** 115–137
[4] Torlina L *et al* 2015 *Nat. Phys.* **11** 503–508
[5] Eckle P, Pfeiffer A N, Cirelli C, Staudte A, Dorner R, Muller H G, Buttiker M, and Keller U 2008 *Science* **322** 1525–1529
[6] Zipp L J, Natan A and Bucksbaum P H 2014 *Optica* **1** 361–364
[7] Pfeiffer A N, Cirelli C, Smolarski M, Dimitrovski D, Abu-samha M, Madsen L B and Keller U 2012 *Nat. Phys.* **8** 76–80
[8] Xie X 2015 *Phys. Rev. Lett.* **114** 173003
[9] Staudte A *et al* 2009 *Phys. Rev. Lett.* **102** 033004
[10] Maurer J, Dimitrovski D, Christensen L, Madsen L B and Stapelfeldt H 2012 *Phys. Rev. Lett.* **109** 123001
[11] Akagi H, Otobe T, Staudte A, Shiner A, Turner F, Dorner R, Villeneuve D M and Corkum P B 2009 *Science* **325** 1364–1367
[12] Holmegaard L et al 2010 *Nat. Phys.* **6** 428–432
[13] Dimitrovski D, Maurer J, Stapelfeldt H and Madsen L B 2014 *Phys. Rev. Lett.* **113** 103005
[14] Odenweller M *et al* 2014 *Phys. Rev.* A **89** 013424
[15] Odenweller M *et al* 2011 *Phys. Rev. Lett.* **107** 143004
[16] Spanner M *et al* 2012 *J Phys.* B **45** 194011
[17] He P-L, Takemoto N and He F 2015 *Phys. Rev.* A **91** 063413
[18] Abu-samha M and Madsen L B 2016 *Phys. Rev.* A **94** 023414
[19] Kjeldsen T K, Nikolopoulos L A A and Madsen L B 2007 *Phys. Rev.* A **75** 063427
[20] Madsen L B, Nikolopoulos L A A, Kjeldsen T K and Fernandez J 2007 *Phys. Rev.* A **76** 063407
[21] Martiny C P J, Abu-samha M and Madsen L B 2009 *J Phys* B **42** 161001
[22] Torlina L, Kaushal J and Smirnova O 2013 *Phys. Rev.* A **88** 053403
[23] Landsman A L, Weger M, Maurer J, Boge R, Ludwig A, Heuser S, Cirelli C, Gallmann L and Keller U 2014 *Optica* **1** 343

IOP Conf. Series: Journal of Physics: Conf. Series **869** (2017) 012010 doi:10.1088/1742-6596/869/1/012010

A new formula for the statistical weight for a sequence of Rydberg levels in an atom or ion

M Eriksson and H Lennerstad

Blekinge Institute of Technology, Department of mathematics and natural sciences, 371 79 Karlskrona, Sweden

E-mail: meo@bth.se

Abstract. We present a new formula for the total statistical weight of all Rydberg levels $G^{ion}(n_l, n_h)$ for which the principal quantum number n is between higher, n_h, and a lower, n_l, limits. This formula can be used for all atoms in the periodic table and for all corresponding ions.

1. Introduction

The statistical weight g of a set of quantum states defined by a set of quantum numbers describing a system is the total number of states possible with this set of quantum numbers. If all the states corresponding to the same set have a negligible energy difference, this set of states is often called one state and its statistical weight is called its degeneracy. In atomic physics the energy difference between different m_j states is negligible, as long as the states are not in a strong magnetic or electric field. Atomic states are mostly described by terms that describe how the spin and angular momenta of the valence electrons have coupled and the resulting final angular momentum J. Those states have the statistical weight $2J + 1$, which correspond to the number of possible m_j states for a given J.

A Rydberg level is a level for which the remaining ion after removing the valence electron is in the ground level. This means that the number of Rydberg levels in the theoretical limit of no perturbation is unlimited. But in all real cases there is a non-zero perturbation, that is why the number of Rydberg levels are limited by a highest possible principal quantum number n_h. In this work we have obtained a formula for the total statistical weight $G^{ion}(n_l, n_h)$ for all Rydberg levels for which $n_l \le n \le n_h$. It turns out that the statistical weight for all $n_l \le n \le n_h$ Rydberg levels is a function of the orbital and spin angular momenta of the ground state of the mother ion, L_p and S_p, respectively. That is why we write $G^{ion}(n_l, n_h) = G(L_p, S_p, n_l, n_h)$.

When calculating the population distribution between states, the statistical weights are crucial. For atoms and ions, the population of different ionization stages depends on the partition function of each corresponding ion, which in turn is a weighted sum of the statistical weight of each bound J state. A major part of those partition functions are from the Rydberg series, that is why those are usually given extra attention when calculating partition functions [2]. In the case of low perturbation the number of Rydberg levels contributing to the partition function are very high, in which case our formula can be very useful.

We present the formula in Section 2 followed by the proof of the formula in Section 3. In Section 4 we discuss how our formula could be used for atomic and ionic partition functions.

IOP Conf. Series: Journal of Physics: Conf. Series **869** (2017) 012010 doi:10.1088/1742-6596/869/1/012010

2. The formula for statistical weight of a Rydberg sequence

For a specific ion there is a total statistical weight, $G^{ion}(n_l, n_h)$ for all $n_l \leq n \leq n_h$ Rydberg levels. This number is a function of the chosen n_l and n_h and of the orbital and angular momenta L_p and S_p of the ground state of the corresponding mother ion. In atomic physics this is called the parent term of the mother ion. The formula reads

$$G^{ion}(n_l, n_h) = G(L_p, S_p, n_l, n_h) = \frac{1}{3}(2L_p + 1)(2S_p + 1)(N_f - N_{break}), \tag{1}$$

in which $N_f = n_f(n_f + 1)(2n_f + 1)$ and $N_{break} = n_{break}(n_{break} + 1)(2n_{break} + 1)$.

3. Derivation of a formula for the statistical weight of a Rydberg sequence

Let $G(L_p, S_p, n)$ be the statistical weight of all levels for which the principal quantum number is n. We now prove the following theorem.

Theorem: The total statistical weight G of the levels for which the parent term spin and orbital angular momentum quantum numbers are S_p and L_p and the principal quantum number of the valence electron is n. Using the selection rules in atomic physics, we show that

$$G(L_p, S_p, n) = \sum_{l=0}^{n-1} \sum_{L_2=|L_p-l|}^{L_p+l} \sum_{S_2=|S_p-\frac{1}{2}|}^{S_p+\frac{1}{2}} \sum_{J=|L_2-S_2|}^{L_2+S_2} (2J+1) = 2\,(2S_p+1)\,(2L_p+1)\,n^2. \tag{2}$$

In some of these sums, indicies may be half-integers. The summation indicies however run in unit steps.

Proof: The proof uses the elementary summation rule

$$\sum_{k=m}^{n} k = \frac{1}{2}(n(n+1) - m(m-1)), \tag{3}$$

many times. First, we evaluate the two innermost sums

$$\sum_{S_2=|S_p-\frac{1}{2}|}^{S_p+\frac{1}{2}} \sum_{J=|L_2-S_2|}^{L_2+S_2} (2J+1). \tag{4}$$

The substution $J' = J + \frac{1}{2}$ and $S_2' = S_2 + \frac{1}{2}$ provides integer arguments in the sums, namely

$$\sum_{S_2=|S_p-\frac{1}{2}|}^{S_p+\frac{1}{2}} \sum_{J=|L_2-S_2|}^{L_2+S_2} (2J+1) = 2 \sum_{S_2'=S_p}^{S_p+1} \sum_{J'=|L_2-S_2'+\frac{1}{2}|+\frac{1}{2}}^{L_2+S_2'} J'. \tag{5}$$

Furthermore we can observe that the sum over S_2' has only two terms, namely

$$\sum_{S_2'=S_p}^{S_p+1} \sum_{J'=|L_2-S_2'+\frac{1}{2}|+\frac{1}{2}}^{L_2+S_2'} J' = \sum_{J'=|L_2-S_p+\frac{1}{2}|+\frac{1}{2}}^{L_2+S_p} J' + \sum_{J'=|L_2-S_p-\frac{1}{2}|+\frac{1}{2}}^{L_2+S_p+1} J'. \tag{6}$$

Now considering the first term, if $L_2 \geq S_p$ it follows that $|L_2 - S_p + \frac{1}{2}| + \frac{1}{2} = L_2 - S_p + 1$, while from $L_2 \leq S_p$ we get $|L_2 - S_p + \frac{1}{2}| + \frac{1}{2} = S_p - L_2$.

IOP Conf. Series: Journal of Physics: Conf. Series **869** (2017) 012010 doi:10.1088/1742-6596/869/1/012010

For the second term, $L_2 \geq S_p + 1$ implies $\left|L_2 - S_p - \frac{1}{2}\right| + \frac{1}{2} = L_2 - S_p$, while $L_2 \leq S_p + 1$ gives $\left|L_2 - S_p - \frac{1}{2}\right| + \frac{1}{2} = S_p - L_2 + 1$. Remarkably, all three cases give the same expression by the elementary summation rule, or

$$L_2 \geq S_p + 1 \Rightarrow \sum_{J'=L_2-S_p+1}^{L_2+S_p} J' + \sum_{J'=L_2-S_p}^{L_2+S_p+1} J' = 2L_2 + 2S_p + 4L_2S_p + 1, \tag{7}$$

$$L_2 = S_p \Rightarrow \sum_{J'=L_2-S_p+1}^{L_2+S_p} J' + \sum_{J'=S_1-L_2+1}^{L_2+S_p+1} J' = 2L_2 + 2S_p + 4L_2S_p + 1, \tag{8}$$

$$L_2 \leq S_p - 1 \Rightarrow \sum_{J'=S_p-L_2}^{L_2+S_p} J' + \sum_{J'=S_p-L_2+1}^{L_2+S_p+1} J' = 2L_2 + 2S_p + 4L_2S_p + 1. \tag{9}$$

Thus we obtain

$$\sum_{S_2=\left|S_p-\frac{1}{2}\right|}^{S_p+\frac{1}{2}} \sum_{J=|L_2-S_2|}^{L_2+S_2} (2J+1) = 2\,(2S_p+1)\,(2L_2+1)\,. \tag{10}$$

Hence, if $S_p > L_p + n - 1$, it remains to calculate

$$G(L_p, S_p, n) = 2\,(2S_p+1) \sum_{l=0}^{n-1} \sum_{L_2=|L_p-l|}^{L_p+l} (2L_2+1)\,. \tag{11}$$

Dividing the summation over l, to avoid the absolute value, provides

$$\sum_{l=0}^{n-1} \sum_{L_2=|L_p-l|}^{L_p+l} (2L_2+1) = \sum_{l=0}^{L_p} \sum_{L_2=L_p-l}^{L_p+l} (2L_2+1) + \sum_{l=L_p+1}^{n-1} \sum_{L_2=l-L_p}^{L_p+l} (2L_2+1)\,, \tag{12}$$

from which the summation rule gives

$$\begin{aligned} G(L_p, S_p, n) &= (L_p+1)^2\,(2L_p+1) + (n^2\,(2L_p+1) - (L_p+1)^2\,(2L_p+1)), \\ &= 2\,(2S_p+1)\,(2L_p+1)\,n^2, \end{aligned} \tag{13}$$

which is the theorem. Now it follows from the sum of quadratic numbers that

$$\begin{aligned} G(L_p, S_p, n_l, n_h) = \sum_{n=n_l}^{n_h} G(L_p, S_p, n) &= 2\,(2S_p+1)\,(2L_p+1) \sum_{n=n_l}^{n_h} n^2, \\ &= \frac{1}{3}(2L_p+1)(2S_p+1)(N_f - N_{break}). \end{aligned} \tag{14}$$

4. Statistical weights and partition functions

For using either the Saha equation or free energy minimization methods [1, 4] to calculate the ionization balance of a plasma, the partition functions of all the involved ions in the plasma are needed. A partition function Q is defined as

$$Q = \sum_{i=0}^{\rho} e^{-\frac{E_i}{k_B T}}, \tag{15}$$

where E_i is the energy of the state i. The state i is characterized by its configuration, term, quantum number for its total orbital momentum j and quantum number of the z-projection of its orbital momentum m_j. The number of bound m_j states for the ion is ρ. For almost all practical cases the energy splitting due to external electric or magnetic fields is very small compared to the $k_B T$ factor, that is why the energy is considered degenerate with respect to orientation of the angular momentum. Because of this degeneracy, the following simplified form for the partition function is used in astrophysics

$$Q = \sum_{i=0}^{\mu} g_i e^{-\frac{E_i}{k_B T}} = \sum_{i=0}^{\mu} (2J_i + 1) e^{-\frac{E_i}{k_B T}}, \tag{16}$$

where the states i now are characterized only by configuration, term and the quantum number J. The number of bound i states are now μ and g_i is the number of possible projections of the total orbital momentum and is called the statistical weight of the state i. The observed levels are mostly tiny fractions of all bound levels when calculating the partition functions, that is why computational atomic codes, such as the Cowan code [5], are needed to obtain the energies to calculate the partition function. Following is a four-step description of how the formula for the statistical weights can be used to estimate partition functions. (1) Determine the highest value of n for the bound Rydberg states, n_h. There are some methods for determining n_h in the literature [3]. (2) Determine the lowest value of n, n_l, for which the excitation energy of the corresponding levels differ negligibly from the excitation energies of the n_h levels. (3) Make a list of the set of bound levels which are not Rydberg levels or Rydberg levels with $n < n_l$, ψ_i^{oth}. (4) Calculate the partition function from

$$Q = \sum_{i=1}^{|\Psi_{oth}|} g_i^{oth} \cdot e^{-\frac{E_i^{oth}}{k_B T}} + G(L_p, S_p, n_l, n_h) \cdot e^{-\frac{E_R}{k_B T}}, \tag{17}$$

where $|\Psi_{oth}|$ is the cardinality of the set of bound states ψ_i^{oth}, and g_i^{oth} and E_i^{oth} are the corresponding statistical weights and energies. The energy E_R is the energy of the highest excited bound Rydberg levels.

5. Summary

This paper presented a formula for the total statistical weight of all Rydberg levels defined by $n_l \le n \le n_h$, see equation (1). It is shown to be a function of the ground parent term of the mother ion. This formula can be used in calculations of population distributions in atomic physics. One example of future possible utility is for calculation of partition functions.

References

[1] Däppen W 1980 *A&A* **91** 212
[2] Halenka J and Gabrowski B 1977 *A&A* **54** 757
[3] Mihalas D 1978 *Stellar Atmospheres* 2nd ed. (San Francisco: Freeman)
[4] Stix M and Skaley D 1990 *A&A* **232** 234
[5] Robert D C 1981 *The Theory of atomic structure and spectra* (Berkeley: University of California Press)

IOP Conf. Series: Journal of Physics: Conf. Series **869** (2017) 012011 doi:10.1088/1742-6596/869/1/012011

Comparative study of the scaling behavior of the Rényi entropy for He-like atoms

M Farid[1], A Abdel-Hady[2] and I Nasser[3]

[1]Department of Electronics, Computing and Mathematics, University of Derby, Derby, DE22 1GB, United Kingdom

[2]Department of Basic Sciences, Faculty of Engineering, Egyptian Chinese University, Cairo, Egypt

[3]Department of Physics, King Fahd University of Petroleum & Minerals, Dhahran 31261, Saudi Arabia

E-mail: m.farid@derby.ac.uk

Abstract. We solve the Schrödinger equation in the spherical or Hylleraas-coordinate systems, and within the framework of the Ritz's variational principle. The eigenvalues, and the eigenfunctions $\psi(\mathbf{r})$ in $\mathbf{r}$- or Hylleraas-space for the 1s^2-state of the He-like atoms as a function of two variational parameters are calculated. Using a simple scaling procedure, we calculate the scaled wavefunction as a function of the nuclear charge Z. Given the density of states, $\rho(\mathbf{r})$, the scaling behavior of the information entropies, e.g., Fisher, Shannon and Rényi's entropies, with their powers and products, as functions of Z are calculated. Scaled wavefunctions for the 1s^2-state of the He-like atoms, with exchange, have been used to study the scaling behavior. Our results agree with the published results. Furthermore, we present a simple logarithmic equation that shows the dependence of information entropies on Z for He-like atoms. The formulation enhances the computational efficiency of the entropies and other related quantities.

1. Introduction

In atomic and molecular physics [1], the electronic density $\rho(\mathbf{r})$ plays a crucial role in calculating many measurable physical quantities that are related to the expectation values. Lately, $\rho(\mathbf{r})$ has been introduced to calculate many information entropies [2], such as the Shannon [3], the Fisher [4] and the Rényi [5] entropies.

Recently, Rényi entropy and its power have been investigated in different areas of science and engineering [6]. Fisher information, I_ρ, Shannon entropy, S_ρ, and Rényi entropy, R_β, are defined, respectively, as follows

$$I_\rho \equiv \int \rho(\mathbf{r_1}, \mathbf{r_2})\ \nabla(ln[\mathbf{r_1}, \mathbf{r_2}])^2\ \mathbf{dr_1 dr_2}, \tag{1}$$

$$S_\rho \equiv -\int \rho(\mathbf{r_1}, \mathbf{r_2})\ ln[\rho(\mathbf{r_1}, \mathbf{r_2})]\ \mathbf{dr_1 dr_2}, \tag{2}$$

$$R_\beta \equiv \frac{1}{1-\beta}\ ln \int \rho(\mathbf{r_1}, \mathbf{r_2})^\beta\ \mathbf{dr_1 dr_2}, \tag{3}$$

where $\mathbf{dr} = r^2 dr\ d\Omega$ and $d\Omega = sin\theta\ d\theta\ d\phi$. In three dimensions, the Shannon entropy power, J_ρ, is defined as [7]

$$J_\rho \equiv \frac{1}{2\pi e} e^{2S_\rho/3}. \tag{4}$$

The Fisher-Shannon information product, P_ρ, is defined by

$$P_\rho \equiv I_\rho J_\rho/3, \tag{5}$$

which satisfies the uncertainty relation $P_\rho > 0$.

The quantity I_ρ is considered as a gauge of the sharpness or concentration of the density $\rho(\mathbf{r})$. Moreover, the Fisher information, I_ρ, constitutes a measure of the gradient content of $\rho(\mathbf{r})$, which makes it quite sensitive even to small localized perturbations of $\rho(\mathbf{r})$. The Shannon entropy S_ρ provides a measure of information about the density of states in the respective spaces. A more localized distribution $\rho(\mathbf{r})$, corresponds to the smaller value of S_ρ. The smaller S_ρ is, the more concentrated the wave function of the state becomes. S_ρ is a limit value of the Rényi entropy as $\beta \to 1$. In atomic systems, the Shannon entropy power, J_ρ, measures the spatial delocalization of the electronic cloud.

In this paper, we consider the simple uncorrelated two electron (i.e., He- Like structure) wavefunction [1]

$$\phi_1(r_1, r_2; a) = \frac{a^3}{\pi} e^{-a(r_1+r_2)}, \tag{6}$$

where $a = Z - z_0$ and z_0 is the screening charge which we set to 5/16 (and also to 0 when no screening is employed). We also consider the asymmetric and exchange, but uncorrelated wavefunction ϕ_2, given here as

$$\phi_2(r_1, r_2; a, b) = A\left[e^{-ar_1-br_2} + e^{-br_1-ar_2}\right], \tag{7}$$

where a and b are both variational parameters and A is a normalization constant. The density of states in this context is given by

$$\rho_n(r_1, r_2; a, b) = |\phi_n(r_1, r_2; a, b)|^2, \tag{8}$$

subject to the normalization condition

$$(4\pi)^2 \int r_1^2 r_2^2 \ \rho(r_1, r_2; a, b) \ dr_1 dr_2 = 1. \tag{9}$$

Alternatively, for a wavefunction, ϕ_2, the Hylleraas coordinate system may be used, in which $s = r_1 + r_2$, $t = r_1 - r_2$, and $u = r_{12}$, subject to the conditions $-u \leq u$, $0 \leq s \leq \infty$ and the volume element $d\tau = \pi^2(s^2 - t^2)u$.

On the other hand, for the wavefunctions ϕ_1, it is straightforward to show that the Shannon entropy takes the form

$$S_\rho = 6 + 2\ ln(\pi) - 6\ ln(a), \tag{10}$$

while the Rényi entropy takes the form

$$R_\beta = \frac{(2\beta - 2)\ ln(a^3/\pi) - 6\ ln(\beta)}{1 - \beta} \longrightarrow S_\rho, \tag{11}$$

which tends to S_ρ as $\beta \to 1$.

IOP Conf. Series: Journal of Physics: Conf. Series **869** (2017) 012011 doi:10.1088/1742-6596/869/1/012011

Table 1. The quantities for seven He-like atoms, using the wavefunction ϕ_2. The literature for energy is cited as Ref. [8]. All results are given in atomic units.

	He Z=2	Li^+ Z=3	Be^{2+} Z=4	B^{3+} Z=5	C^{4+} Z=6	B^{5+} Z=7	O^{6+} Z=8
a	1.1885	2.0789	2.9847	3.9012	4.8256	5.7561	6.6914
b	2.1831	3.2949	4.3897	5.4734	6.5491	7.6187	8.6834
$-E$	2.8744	7.247	13.621	21.996	32.371	44.746	59.121
$-E^{[8]}$	2.9037	7.279	13.655	22.030	32.406	44.781	59.156
I_ρ	11.502	28.995	54.491	87.990	129.48	178.98	236.48
S_ρ	5.4482	2.5234	0.5745	-0.8920	-2.0690	-3.0521	-3.8965
R_1	5.4452	2.5229	0.5742	-0.8920	-2.0680	-3.0527	-3.8968
J_ρ	2.2085	0.3147	0.0858	0.0322	0.0147	0.0076	0.0043
P_ρ	8.4678	3.0422	1.5595	0.9472	0.6366	0.4564	0.3435
R_2	3.5356	0.6415	-1.2950	-2.7550	-3.9278	-4.9082	-5.7505
R_3	2.6474	-0.2354	-2.1677	-3.6254	-4.7968	-5.7762	-6.6179

Table 2. The scaling properties found in Table 1.

$a = 0.9121$ Z-0.6754	$b = 1.0818$ Z-0.0297
$E = -0.996$ Z$^2 + 0.573$ Z	$R_1 = -6.715 \ln(Z)+9.9773$
$J_\rho = 45.32$ Z$^{-4.477}$	$I_\rho = 2.6126$ Z$^{2.176}$
$P_\rho = 39.464$ Z$^{-2.3}$	$R_2 = -6.676 \ln(Z)+8.0485$
$R_3 = -6.663 \ln(Z)+7.1536$	

Table 3. Selected quantities for the three wavefunctions of the He-atom.

	$\phi_1(z_0 = 0)$	$\phi_1(z_0 = 5/16)$	ϕ_2
$-E$	4.0	2.8476	2.8744
R_1	4.1305	5.1496	5.4452
R_2	2.2894	3.3088	3.5356
R_3	1.4264	2.4458	2.6474
I_ρ	16.0	11.390	11.502
$\langle r_1^2 \rangle$	0.75	1.0535	1.2378
$\langle p_1^2 \rangle$	4.00	2.8476	2.8756
$\langle r_1^2 \rangle \langle p_1^2 \rangle$	3.00	3.00	3.5596

2. Results and Discussion

A symmetric and exchange wavefunction, ϕ_2, is used to calculate the values for the Shannon and Rényi entropies and other related quantities such as the Shannon entropy power, Fisher-Shannon information product. The main results are displayed in table 1. From table 1, the following observations can be made:

- Given the variational parameters a and b, good agreement with the exact values for the energy, at least to 3 decimal places, is observed. To facilitate the comparison with Ref. [8], their data has been truncated to only five digits.
- R_1 changes its sign for $Z > 4$. Furthermore, R_2 and R_3 also change their sign for $Z > 2$.
- The inequality of the product $P_\rho > 0$, is satisfied.

- The product and the power decrease gradually.

Table 2 lists the scaling laws for each quantity sought which was computed using curve-fitting. Furthermore, in this comparative study, table 3 provides a few quantities that present the following observations:

- Neglecting the electron-electron interaction in the unscreened wavefunction $\phi_1(z_0 = 0)$ gives lower bounds on all quantities.
- Including the electron-electron interaction, as in the $\phi_1(z_0 = 5/16)$, improves the results.
- Including the electron-electron interaction and exchange further improves the results.
- For both the unscreened wavefunction $\phi_1(z_0 = 0)$ and the screened wavefunction $\phi_1(z_0 = 5/16)$, the product $\langle r_1^2\rangle \langle p_1^2\rangle = \langle r_2^2\rangle \langle p_2^2\rangle = 3$, where the expectation values $\langle r_1^2\rangle$ and $\langle p_1^2\rangle$ for a single electron, are defined as follows

$$\left\langle r_1^2 \right\rangle = (4\pi)^2 \int dr_1 dr_2 \; r_1^4 r_2^2 (|\phi_i|)^2, \tag{12}$$

and

$$\left\langle p_1^2 \right\rangle = (4\pi)^2 \int dr_1 dr_2 \; r_1^2 r_2^2 \phi_i \; \frac{1}{r_1} \partial_{r_1,r_1}(r_1\phi_i). \tag{13}$$

- The effect of the exchange of the two electrons for $\langle r_1^2\rangle \langle p_1^2\rangle$ or $\langle r_2^2\rangle \langle p_2^2\rangle$ for $Z < 5$ is vividly clear. However, it diminishes and may be neglected for values of Z greater than 5.

3. Conclusion

The scaled ground state wavefunction, in r-space, of two variational parameters is used to calculate the ground state energies for seven He-like atoms. In addition, the information entropies for Fisher, Shannon and Rényi have been calculated and their scaled expressions have been derived. The dependence of such quantities on the atomic number Z is clear. Furthermore, to a good degree of accuracy, we are able to deduce expressions for Rényi entropy and consequently Shannon entropy, and other related quantities employing simple equations. This helps in computing these quantities efficiently. To our knowledge, the results are new and encouraging to pursue further developments using the J-matrix method [9] and correlated wavefunctions with different potentials.

References

[1] Bransden B H and Joachain C J 2003 *Physics of atoms and molecules* 2nd ed (Prentice Hall)
[2] Sen K (ed) 2011 *Statistical Complexity, Applications in electronic structure* (Switzerland: Springer International Publishing) pp 215 – 236
[3] Shannon C E 1948 *Bell Syst. Tech. J.* **27** 379
[4] Fisher R A 1925 *Proc. Cambridge Philos. Soc.* **22** 700
[5] Rényi A 1961 *Proc. Fourth Berkeley Symp. Math. Stat. and Probability, Vol. 1* (Berkeley, CA., USA: University of California Press) p 547
[6] Sen K (ed) 2014 *Electronic Structure of quantum confined atoms and molecules* (Switzerland: Springer International Publishing Switzerland) pp 91 – 100
[7] Romera E and Dehesa J S 2004 *J. Chem. Phys.* **120** 8906
[8] Nakashima H and Nakatsuji H 2008 *J. of Chem. Phys.* **128** 154107
[9] Nasser I, Abdelmonem M S and Abdel-Hady A 2014 *Mol. Phys.* **112** 2608

IOP Conf. Series: Journal of Physics: Conf. Series **869** (2017) 012012 doi:10.1088/1742-6596/869/1/012012

Theoretical calculation of the electronic structure of ZnO molecule

N El-Kork[1], S Mahmoud[2], M Bechelani[2], P Miele[2] and M Korek[3]

[1]Khalifa University, P.O. Box 57, Abu-Dhabi, United Arab Emirates
[2]Institut Européen des Membranes, UMR 5635 ENSCM UM2 CNRS, Université Montpellier 2, Place Eugène Bataillon, 34095 Montpellier, France
[3]Faculty of Science, Beirut Arab University, P.O. Box 11-5020, Lebanon

Email: nayla.elkork@kustar.ac.ae,

Abstract. The lowest potential energy curves, for the electronic states of the molecule ZnO in the representation $^{2s+1}\Lambda^{(+/-)}$ have been performed via Complete Active Space Self Consistent Field (CASSCF) using the Multireference Configuration Interaction (MRCI) method with Davidson correction (+Q). An excitation and deexcitation model has been proposed, in analogy to the emission of ZnO nanoparticles. In addition, the minimum energy level with respect to the ground state Te, the internuclear distance at equilibrium R_e, the rotational constant B_e, the vibrational frequency ω_e and the static and transition dipole moment μ have been investigated for some considered electronic states. Ten new electronic states have been investigated here for the first time.

1. Introduction

Metal oxides are characterized by their use in a large versatility of areas, such as enzyme mimetics [1], pharmacology [2-3], fuel additives [4], sensors [5], solar cells [6-8], and antibacterial agents [9]. Zinc Oxide for example, has many applications in optoelectronics and biomedical sciences . Many of these applications rely on the fact that ZnO nanoparticles or quantum dots exhibit quantum confinement effects i.e. their emission properties depend on their size. Mainly the bandgap of such nanometric semi-conductors varies with the size of the nanoparticle. The main objective of this work is to investigate ZnO properties at the smallest possible unit scale, i.e the molecular state. More specifically, we try to understand the electronic structure of the lowest lying electronic states of the metal oxide molecule ZnO. To this end, we performed high level calculations of the electronic structure of ZnO molecule. Ab-initio calculations have been performed via the quantum chemical package MOLPRO [10], taking the advantage of the graphical user GABEDIT [11]. The Zinc atom is treated by using ECP10MDF for s, p, d functions, while the oxygen atom is treated in all electronic scheme by using the AV5Z basis set for s, p, d and f functions. The quality of the selected basis sets is checked by comparing our CI calculations for the ground and several excited electronic states of isolated Zn and O atoms to the experimental data in NIST Atomic Spectra Database. We were additionally interested by the spectroscopic properties of ZnO, and its corresponding dipole moment variation in function of the internuclear distance.

IOP Conf. Series: Journal of Physics: Conf. Series **869** (2017) 012012 doi:10.1088/1742-6596/869/1/012012

2. Results and discussion

The potential energy curves of ZnO molecule (singlet state) are plotted in figure 1. Singlet states present mostly low energy bound states, with double minima at higher levels (($2)^1\Pi$, $(1)^1\Delta$).There were no avoided crossings in the studied curves.

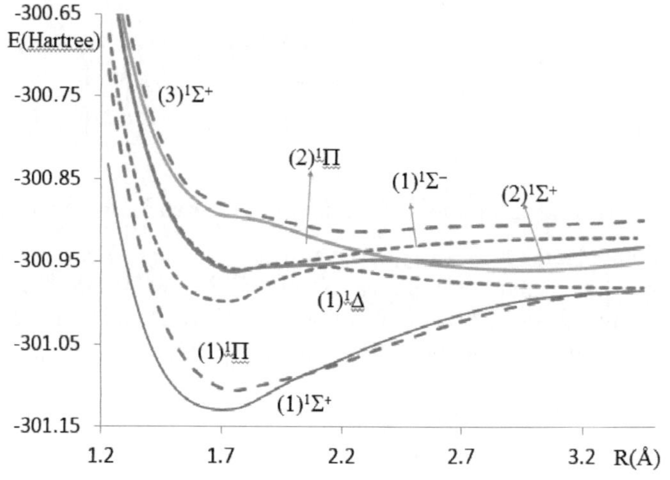

Figure 1. Singlet states potential energy curves.

2.1. Energy levels diagrams

An effective way to represent energy levels of a molecules, and understand possible emission mechanisms is to draw its energy level diagram. In this case one considers the lowest energy value for each potential energy curve, and represents it by a horizontal line. The corresponding Molecular energy level diagram for the molecule ZnO is shown in figure 2.

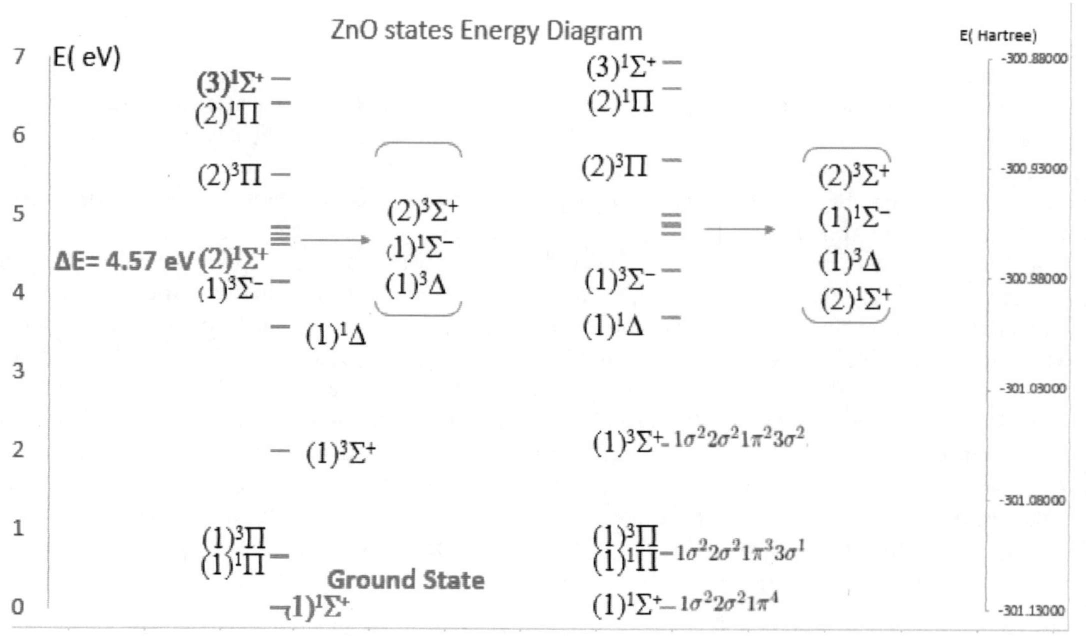

Figure 2. ZnO energy level diagram.

IOP Conf. Series: Journal of Physics: Conf. Series **869** (2017) 012012 doi:10.1088/1742-6596/869/1/012012

ZnO bulk solid is known to emit light with two distinct bands, in UV (~ 3.37 EV) and in the visible part of the spectrum (~2.34 eV) [12]. Quantum dots have been consequently used to understand the origin of the emission bands, in one of the studies for example, the green band was attributed to the excitation of electrons from the conduction band to a deep trap (originating from an oxygen vacancy defect), while the UV emission was attributed to the excitation of electrons from a shallow donor to the valence band [13]. In all cases, ZnO nanoparticles exhibited the behaviour of a band structure whose band gap increased in size, and the nanoparticles got smaller. If one imagines the smallest possible unit of ZnO particles to be the ZnO molecule, then according to the quantum confinement model, its bandgap would be in the orders of 6 eV. In terms of molecular states, the valence band would correspond to a HOMO state while the conduction band, a LUMO one. Up to our knowledge, one experimental study was able to reproduce the emission spectra of ZnO molecules through laser ablation of a ZnO target in vacuum and under oxygen pressure. Even though the spectra were not very clear, ZnO molecular bands were attributed in the green part of the spectrum (between 2.37 eV and 2.18 eV) and violet bands (370-390) were clearly discerned, however they were attributed to ZnO clusters. According to our energy levels diagram, and keeping in mind the molecular states transition rules, an analogy to the band model can be drawn at the molecular level: For the conduction and valence band in nanoparticles corresponds at the molecular level, the HOMO and LUMO levels. The HOMO state would be the electronic ground state $1^1\Sigma^+$, while the LUMO level could be the state $3^1\Sigma^+$. The difference between these two states being equal to about 6.5 eV, as predicted by quantum confinement. Concerning the deep traps, responsible for the green light emission and the shallow donor that caused the violet emission in nanoparticles, their analogous are the allowed states between the HOMO and LUMO states in molecules. More specifically, if an electron is excited to the state $3^1\Sigma^+$ (LUMO level), the energy level diagram and potential energy curves show that its deexcitation to the level $2^1\Sigma^+$ (including the vibrational levels) correspond to an energy of 2.18eV. Thus, the electronic state $2^1\Sigma^+$ plays in this case the role of a deep trap. In other words, this molecular transition would be the analogous of an electron passing from the valence band to a deep drap, as modelled for the green emission in nanoparticles. Similarly, an electron passing from some vibrational level in $2^1\Sigma^+$ to one in the ground state has enough energy to emit light in the violet regime. In this case, the vibrational level would be the shallow donor and the ground state would act as the valence band (or HOMO level).

2.2. *Dipole moment curves*

Dipole moment curves show the variation of the dipole moment with internuclear distance. In this case, we show the dipole moment curves of the singlet states of ZnO molecule.

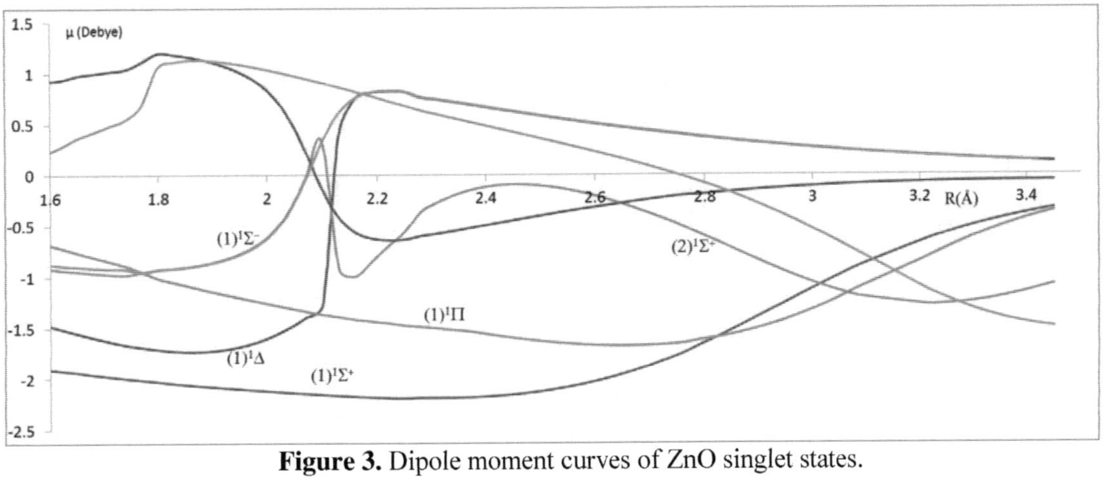

Figure 3. Dipole moment curves of ZnO singlet states.

IOP Conf. Series: Journal of Physics: Conf. Series **869** (2017) 012012 doi:10.1088/1742-6596/869/1/012012

2.3 Spectroscopic constants

Most of the work found in the literature about ZnO molecules is of theoretical nature, and mainly concerns the two lowest states $(1)^1\Sigma$, and $(1)^3\Pi$. In this paper, we were able to introduce ten additional states. The following table shows the spectroscopic constants of some of the low energy levels of ZnO molecule, in good agreement with previous data in the literature.

Table 1. Spectroscopic constants of ZnO molecule.

State	R_e (a. u.)	T_e (cm^{-1})	B_e (cm^{-1})	ω_e (cm^{-1})	
$(1)^1\Sigma^+$	3.22	0	0.45	770	This work
	3.19	0		766	[15]
	3.24	0		727	[16]
	3.22	0		738	[17]
$(1)^1\Delta$	3.25	28797	0.44	825	
$(2)\ ^1\Sigma^+$	3.305	36897	0.43	1095	
	5.07	39717	0.18	1967	

3. Conclusion

In this paper we have presented a study of the electronic structure of the diatomic molecule ZnO. From the potential energy curves we deduced an energy level diagram that shows the energy difference between the electronic levels. We consequently propose a model that shows an analogy between the molecular and nanometric scales of Zinc Oxide. We have also showed the permanent transition dipole moments and spectroscopic constants of ZnO molecule.

References

[1] Park B *et al* 2008 *Inhalation. Toxicol.* **20** 547
[2] Lin W, Huang Y, Zhou X and Ma Y 2006 *Int. J. Toxicol.* **25** 451
[3] Celardo I, Pedersen J.Z, Traversa E and Ghibelli L 2011 *Nanoscale* **3** 1411
[4] Kaur A and Gupta U 2009 *J. Mater. Chem.* **19** 8279
[5] Reddy S, Kumara Swamy B E and Jayadevappa H 2012 *Electrochim. Acta* **61** 78
[6] Beck W J E, Wienk M M and Janssen R A J 2004 *Adv. Mater.* **16** 1009
[7] Chen W, Chen J, Feng Y, Hong L, Chen Q, Wu L, Lin Xand Xia X 2012 *Analyst* **137** 1706
[8] Tong D T, Wu P, Su P,Wang H,Wang D and Tian H 2012 *Mater. Lett.* **70** 94
[9] Celardo I, Pedersen J Z, Traversa E and Ghibelli L 2011 *Nanoscale* **3** 1411
[10] MOLPRO, version 2010 1, a package of ab initio programs, H-J Werner et.al, see http://www.molpro.net
[11] Gabedit is a Graphical User Interface to Gamess- US, Gaussian, Molcas, Molpro and MPQC computational chemistry pakage http:/gabedit.sourceforge.net/
[12] Zeng H *et al* 2010 *Advanced Functional Materials* **20(4)** 561
[13] Xu X *et al* 2012 *Journal of Applied Physics* **111(8)** 083521
[14] Acquaviva S, D'Anna, E and De Giorgi M L 2007 *Journal of Applied Physics* **102(7)** 073109
[15] Boughdiri S *et al* 2008 *Chem. Phys. Lett.* **462** 18
[16] Bauschlicher Jr C W, and Harry P 1998 *The Journal of chemical physics* **109 (19)** 8430
[17] Zack L N, Pulliam R L, and Ziurys L M 2009 *Journal of Molecular Spectroscopy* **256(2)** 186

IOP Conf. Series: Journal of Physics: Conf. Series **869** (2017) 012013 doi:10.1088/1742-6596/869/1/012013

An approximate solution method for the quasipotential equations with local spin-orbital potentials

S I Fialka, V N Kapshai and Y A Grishechkin

Department of Physics, Gomel State University, 104 Sovetskaya str., 246019 Gomel, Belarus
E-mail: sergey.fialka@gmail.com

Abstract. An exact solution of partial quasipotential equations for spin-0 and spin-½ particles with superposition of N δ-potentials is obtained. Based on this, a method for an approximate solution using a superposition of δ-potentials instead of an arbitrary smooth potential is presented. A smooth analytic potential allowing existence of resonant states is considered. Scattering cross section dependences on various potential parameters are investigated.

1. Introduction

Quasipotential approach had proved highly successful as a transparent and sequential method for describing relativistic two-particle systems [1]. But the construction of the interaction quasipotential within the framework of quantum field theory is a complicated problem. For this reason phenomenological potentials are often used. The explicit analogy between the nonrelativistic (the Schrödinger and the Lippmann-Schwinger equations) and the relativistic quasipotential equations, thus, turns out to be very useful. In spite of this there are only a few potentials which allow exact analytical solution.

We focus on the δ-potentials which have been well investigated in quantum mechanics [2-5], but have not been widely used in the quasipotential approach up to the present time.

In this paper we present the solution of quasipotential equations for scattering states with nonzero orbital momentum. A system of two relativistic equal-mass particles with spin 0 and 1/2 is considered. We use a local spin-orbital quasipotential, which reduces the three-dimensional integral equation to partial equations – one-dimensional integral equations for a certain orbital momentum.

2. The local spin-orbital quasipotential case

In the case of a spin-orbital quasipotential, which is local in the Lobachevski momentum space $V_{\sigma\mu}(E_q;\vec{p},\vec{k})=V_{\sigma\rho}(E_q;(\vec{p}(-)\vec{k})^2)D^{1/2}_{\rho\mu}(\vec{n}_{kp})$, with D the Wigner rotation) [6], the quasipotential equation for scattering states of particles with spin 0 and 1/2 in the relativistic configuration representation takes the form

$$\psi_\sigma(\vec{q},\vec{r})=\xi(\vec{q},\vec{r})\chi^{1/2}_\sigma+\int G_0(E_q;r,r')V_{\sigma\rho}(E_q;r')\psi_\rho(\vec{q},\vec{r}')d\vec{r}'. \quad (1)$$

Here $\psi_\sigma(\vec{q},\vec{r})$ is two-component wave function of the relative motion of two particles, $\vec{p}$ and $\vec{k}$ are the initial and final relative momenta of the particles in the center-of-inertia system, $2E_q=2\sqrt{\vec{q}^2+m^2}$ is the two-particle system's energy, $\chi^{1/2}_\sigma$ are the normalized eigenvectors of the spin. In this paper we limit ourselves to the Green's function of Logunov-Tavkhelidze equation, which in momentum

IOP Conf. Series: Journal of Physics: Conf. Series **869** (2017) 012013 doi:10.1088/1742-6596/869/1/012013

representation is $G_0(E_q,E_p)=(E_q^2-E_p^2+i0)^{-1}$. Transformation between momentum representation and configuration representation is performed by functions $\xi(\vec{q},\vec{r})$, the "plane waves" in Lobachevsky space [7].

The local quasipotential allows to reduce three-dimensional integral equations to partial (one-dimensional) equations. Expansion of wave function, "plane wave", Green's function on the basis of the spherical spinors – eigenfunctions of $\hat{\vec{J}}^2$, $\hat{J}_Z$, $\hat{\vec{L}}^2$, $\hat{\vec{S}}^2$ operators [8, 9], yields:

$$\xi(\vec{q},\vec{r})\chi_\sigma^{1/2}=\frac{4\pi}{qr}\sum_{j\ell M} i^\ell s_\ell(w_q,r)\Omega_{jlM}^{1/2}(\vec{n}_r)\chi_\sigma^{1/2\dagger}\Omega_{jlM}^{1/2*}(\vec{n}_q);\tag{2}$$

$$\psi_\sigma(\vec{q},\vec{r})=\frac{4\pi}{qr}\sum_{j\ell M} i^\ell \psi_\ell(w_q,r)\Omega_{jlM}^{1/2}(\vec{n}_r)\chi_\sigma^{1/2\dagger}\Omega_{jlM}^{1/2*}(\vec{n}_q);\tag{3}$$

$$G_0(E_q;r,r')=\frac{1}{rr'}\sum_{j\ell M} G_0^{(\ell)}(E_q;r,r')\Omega_{jlM}^{1/2}(\vec{n}_r)\Omega_{jlM}^{1/2\dagger}(\vec{n}_{r'}).\tag{4}$$

Here we use the following parametrization: $E_q=m\cosh w_q$, $q=m\sinh w_q$; where w_q is the rapidity. Partial Green's functions of Logunov-Tavkhelidze equation [7] and their asymptotic behavior have the following form

$$G_0^{(\ell)}(E_q;r,r')=\frac{1}{im\sinh(2w_q)}\times$$
$$\times\left[\frac{e_\ell^{(1)}(w_q,r)e_\ell^{(1)*}(w_q,r')}{1-e^{-\pi m(r-r')}}+\frac{e_\ell^{(2)}(w_q,r)e_\ell^{(2)*}(w_q,r')}{1-e^{\pi m(r-r')}}-\frac{e_\ell^{(1)}(w_q,r)e_\ell^{(2)*}(w_q,r')}{1-e^{-\pi m(r+r')}}-\frac{e_\ell^{(2)}(w_q,r)e_\ell^{(1)*}(w_q,r')}{1-e^{\pi m(r+r')}}\right];\tag{5}$$

$$G_0^{(\ell)}(E_q;r,r')\Big|_{r\to\infty}=-\frac{2s_\ell^*(w_q,r')}{m\sinh(2w_q)}e^{i(mrw_q-\pi/2\ell)}.\tag{6}$$

Explicit form of functions $s_\ell(w_q,r)$ and $e_\ell^\pm(w_q,r)$ are given by Kadyshevsky *et al.* [7] in terms of Legendre functions of complex order. We have established that the partial functions can be expressed simply in terms of Legendre functions of integer order

$$s_\ell(w,r)=i^{-(\ell+1)}\frac{e^{mr\pi}}{\Gamma(imr)}Q_\ell^{imr}(\coth w);\quad e_\ell^{(1,2)}(w,r)=i^\ell\frac{\Gamma(imr-\ell)\Gamma(\mp imr+\ell+1)}{\Gamma(imr)}P_\ell^{\pm imr}(\coth w).$$

We will consider a simple spin-orbital potential without E_q dependence (I – identity matrix)

$$V_{\sigma\rho}(E_q;r)=V_{\sigma\rho}(r)=\left(I_{\sigma\rho}+g^2 2(\hat{\vec{\ell}}\hat{\vec{s}})_{\sigma\rho}\right)V(r).\tag{7}$$

Substituting Eqs. (2), (3), (4) and (7) into Eq. (1), we obtain the partial equations

$$\psi_\ell(w_q,r)=s_\ell(w_q,r)+\int_0^\infty G_0^{(\ell)}(w_q;r,r')\left(1-g^2(1+\kappa)\right)V(r')\psi_\ell(w_q,r')\,dr',\tag{8}$$

where the parameter κ depends on the spin orientation and takes on all integer values except zero: $-(\ell+1)$ if $j=\ell-1/2$ and ℓ if $j=\ell+1/2$.

3. Superposition of δ-potentials

Consider Eq. (8) with a finite, but varied, number of δ-potentials

$$V(r)=\sum_{n=1}^N A_n\delta(r-a_n).\tag{9}$$

Substituting Eq. (9) into Eq. (8), we obtain the following algebraic system of equations

$$\psi_\ell(w_q,r)=s_\ell(w_q,r)+\sum_{n=1}^{N}G_0^{(\ell)}(w_q;r,a_n)\big(1-g^2(1+\kappa)\big)A_n\,\psi_\ell(w_q,a_n)\,. \tag{10}$$

Using a_v $(v=1,2,...,N)$ instead of r, we obtain a system of linear equations with respect to $\psi_\ell(w_q,a_n)$:

$$\sum_{n=1}^{N}\Big[I_{vn}-G_0^{(\ell)}(w_q;a_v,a_n)\big(1-g^2(1+\kappa)\big)A_n\Big]\psi_\ell(w_q,a_n)=s_\ell(w_q,a_v)\,.$$

To obtain the quantities which characterize scattering states, we have to consider the wave function at large distance. Taking into account asymptotic behavior of the partial Green's functions (6), equations (10) as $r\to\infty$ yields

$$\psi_\ell(w_q,r)\big|_{r\to\infty}=s_\ell(w_q,r)+\left(\sum_{n=1}^{N}\frac{-2s_\ell^*(w_q,a_n)}{m\sinh(2w_q)}\big(1-g^2(1+\kappa)\big)A_n\,\psi_\ell(w_q,a_n)\right)e^{i(mr\,w_q-\pi/2\ell)}\,. \tag{11}$$

In quantum mechanics, the partial scattering amplitude is defined as the coefficient divided by momentum in front of the scattered wave. In our case, relativistic partial amplitudes have the form

$$f_{\ell,\kappa}(w_q)=\frac{1}{m\sinh w_q}\sum_{n=1}^{N}\frac{-2s_\ell^*(w_q,a_n)}{m\sinh(2w_q)}\big(1-g^2(1+\kappa)\big)A_n\,\psi_\ell(w_q,a_n)\,. \tag{12}$$

If the initial state is not polarized, the total scattering cross section can be represented as the sum of the partial cross sections and can be expressed via the relativistic scattering amplitude [10]

$$\sigma=\sum_{\ell=0}^{\infty}\sigma_\ell=4\pi\sum_{\ell=0}^{\infty}\Big[(\ell+1)\big|f_{\ell,-(\ell+1)}(w_q)\big|^2+\ell\big|f_{\ell,\ell}(w_q)\big|^2\Big]. \tag{13}$$

Substituting A_n and a_n parameters into expression for determining the scattering amplitude (12) and producing calculation of the partial cross sections (13), we obtain exact solution of the two-particle equation with superposition of δ-potentials. Based on these results, we can also find numerical approximation to the solution for a wide class of potential fields, with the condition that the potential function tends to zero fast enough as $r\to\infty$.

4. Numerical solution for smooth potential

As an example, consider the following analytic potential [11]

$$V(r)=U\,r^2\,\frac{\cosh((\pi-\beta)mr)}{\cosh(\pi mr)}\,. \tag{14}$$

Range of the variable r $(0\le r\le\infty)$ should first be divided into two parts: semi-infinite $r\ge r_N$, where potential field is assumed to be zero, and finite $0\le r<r_N$, where the superposition of δ-potentials (9) is used instead of the smooth potential (14). The last interval is divided into N sub-intervals of the form $[r_n;r_{n+1}]$ with δ-potential in the middle and the parameters take the following form

$$A_n=(r_n-r_{n-1})V(a_n);\quad a_n=(r_n+r_{n-1})/2.$$

Accuracy of the numerical method can be improved by increasing the density of δ-functions near sharp variations of the potential field. Fig. 1a shows the dependence of partial cross section σ_0 on energy E_q for different numbers of δ-functions N at $U=30,\beta=\pi/3$. In the case of s-states ($\ell=0$) the partial cross section is independent of g^2. Clearly, with increasing number of δ-functions the result converges quickly to the constant (exact) solution.

Figure 1b shows dependence of the total cross section σ on energy E_q for $g^2=0.015$ at $U=4$ and $\beta=\pi/3$. In order to obtain exact dependence, 7 first partial cross sections were summed. The partial cross sections with $\ell\ge7$ give no contribution to the total cross section in shown energy range for considered parameters. The superposition of $N=100$ δ-functions was used.

In contrast to the case of two spinless particles ($g^2 = 0$), resonance energy can be divided depending on strength of the spin-orbital interaction, as shown in Fig. 1b. Just as in the case of two spinless particles, increased height of the barrier (increase parameter U and decrease β) leads to an increased number of resonant states. Resonance energies can be calculated using the complex scaling method [11].

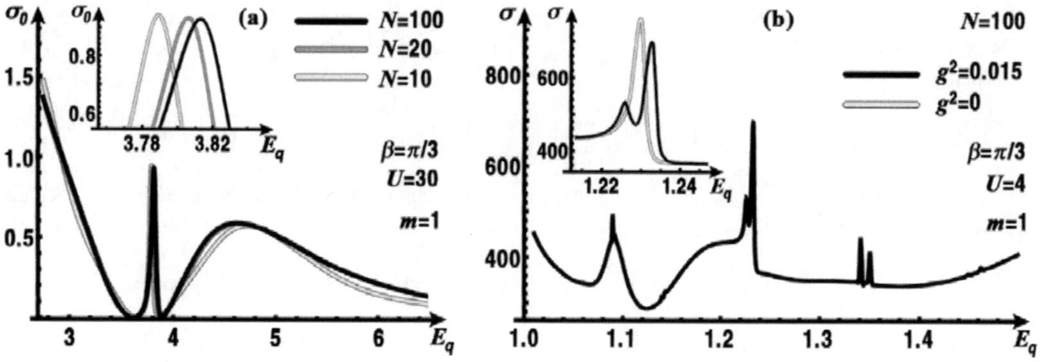

Figure 1. (a) The dependence of partial cross section σ_0 on energy E_q for different number of δ-functions. (b) The dependence of total cross section σ on two-particle system's energy E_q.

It should be mentioned that consideration of scattering states with $\ell > 0$ is important, as is clearly shown in Fig. 1b. In this case, when $\ell = 0, 1, 2,$ there is no resonance, with a small imaginary part and partial cross sections show no narrow peaks, in contrast to the states with $\ell > 2$.

5. Conclusion

Exact solution of partial quasipotential equations for spin-0 and spin-½ particles with superposition of N δ-potentials was presented. The compact general expression for determining the partial scattering amplitudes was obtained. By changing coe cients next to δ-functions, a numerical solution for a wide class of potential fields can be found out and more realistic systems can be considered.

References

[1] Matveev V A, Savrin V I, Sissakian A N and Tavkhelidze A N 2002 *Theor. Math. Phys.* **132** 1119
[2] Albeverio S, Gesztesy F, Hoegh-Krohn R and Holden H 2005 *Solvable models in quantum mechanics: second edition* (AMS Chelsea Publishing, Providence, RI)
[3] Kapshai V, Alferova T and Elander N 2003 *Int. J. Quantum chem.* **91** 533
[4] Kapshai V, Shilyaeva K and Elander N 2009 *J. Phys. B: At. Mol. Opt. Phys.* **42** 044001
[5] Fialka S and Kapshai V 2017 *AIP Conference Proceedings* **1796**, 020015
[6] Skachkov N B 1975 *Theor. Math. Phys.* **22** 213
[7] Kadyshevsky V G, Mir-Kasimov R M and Skachkov N B 1968 *Nuovo Cimento A* **55** 233; Kadyshevsky V G, Mir-Kasimov R M and Skachkov N B 1972 *Fiz. Elem. Chast. At. Yadra* **2** 635
[8] Newton R G 1982 *Scattering theory of waves and particles* (Springer-Verlag: New York)
[9] Varshalovich D A, Moskalev A N and Khersonskii V K 1988 *Quantum theory of angular momentum* (World Scientific Pub.)
[10] Landau L D and Lifshitz E M 1977 *Quantum Mechanics. Non-relativistic Theory* (Pergamon Press Ltd.)
[11] Kapshai V N, Shilyaeva K P and Grishechkin Y A 2011 *Problems of physics, mathematics and technics* **9** 33

IOP Conf. Series: Journal of Physics: Conf. Series **869** (2017) 012014 doi:10.1088/1742-6596/869/1/012014

Dissociation of soliton molecules under periodic perturbation in dipolar quantum gases

Sh R Otajonov[1], B Kh Turmanov[1], U Al Khawaja[2] and B B Baizakov[1]

[1]Physical-Technical Institute, Uzbek Academy of Sciences, 100084, Tashkent, Uzbekistan
[2]Physics Department, United Arab Emirates University, P.O. Box 15551, Al-Ain, UAE

E-mail: u.alkhawaja@uaeu.ac.ae

Abstract. In this paper we develop a collective coordinate approach for the dynamics of two solitons, governed by the nonlocal Gross-Pitaevskii equation (GPE). The model adequately describes both the small amplitude and large amplitude dynamics of soliton bound states (also known as soliton molecules), including the dissociation phenomena. Detailed comparison between theoretical predictions and numerical simulations of the governing GPE is presented.

1. Introduction

Solitons as localized nonlinear waves exhibit many interesting properties [1]. In particular, they can form stable bound states known as *soliton molecules* (SM). For the first time SM were experimentally observed in dispersion - managed optical fibers [2]. These soliton compounds may find important applications in fiber optic communication systems to enhance their data carrying capacity [3]. The existence of two-soliton bound states in Bose - Einstein condensates (BEC) with contact atomic interactions and some dynamic phenomena involving SM were reported in [4, 5].

As the solitons are neutral objects, interactions between them in homogeneous optical fibers and ordinary BEC occur indirectly, through modification of the medium properties. Essential point to be stressed here is that, solitons in media with uniform dispersion and local nonlinearity can either attract or repel each other, depending on the phase difference between individual solitons. The interaction between solitons in uniform media occurs only when their waveforms appreciably overlap. In contrast to this, in nonlocal media such as dipolar BEC, solitons can interact even though their waveforms do not overlap.

True molecular type of interaction, when two solitons attract each other at long distances and repel when they are close, was found to exist in dipolar quantum gases. Consequently, at some fixed separation between solitons attraction (due to dipolar forces) and repulsion (due to anti-phase configuration) can balance each other, and this corresponds to stationary state of the molecule. Small displacement of solitons from their stationary positions gives rise to oscillations near the stationary state. In previous works [6, 7] a variational approach (VA) with Gauss-Hermite trial function was employed to find the waveform of a two-soliton molecule and stationary separation between solitons. Despite very good agreement between the VA and numerical simulations of the Gross-Pitaevskii equation (GPE) for small amplitude dynamics, the model [6, 7] becomes inadequate for large amplitude dynamics of the molecule. In particular, it cannot describe the dissociation phenomena, when the molecule breaks into individual solitons under external forcing or parametric perturbations.

This paper is aimed at developing the collective coordinate approach (CCA) for the dynamics of two-soliton molecules, which can properly describe both the small amplitude and large amplitude oscillations of the molecule, including the dissociation phenomena. In similar setting the CCA for two solitons, governed by the standard nonlinear Schrödinger equation (NLSE), was introduced in Ref. [8]. However, as we mentioned above, standard NLSE does not support stable SM's with finite binding energy. In this work we extend the CCA [8] to soliton bound states in nonlocal GPE and find novel phenomena, not observed in the NLSE. In subsequent sections we present the mathematical model and corresponding numerical simulations.

2. Theory and numerical simulations

The dynamics of a quasi-one dimensional dipolar BEC is governed by the nonlocal GPE, which is written in normalized units as follows

$$i\frac{\partial\psi}{\partial t}+\frac{1}{2}\frac{\partial^2\psi}{\partial x^2}+g(t)\psi\int_{-\infty}^{+\infty}R(|x-z|)\,|\psi(z,t)|^2dz=0, \tag{1}$$

where $\psi(x,t)$ is the mean-field wave function of the condensate, g is the coefficient, responsible for the long - range dipolar atomic interactions. In experiments the strength of dipolar interactions can be made time-dependent $g\to g(t)$ by means of external polarizing magnetic fields [9]. $R(x)$ is the response function, which characterizes the degree of non-locality of the medium. Below we consider the simplest Gaussian form $R(x)=\exp(-x^2/2w^2)/(\sqrt{2\pi}w)$ which is sufficient for qualitative understanding of the properties of dipolar BEC. Also, for simplicity we excluded the contact interactions term $\sim q\,|\psi|^2\,\psi$ from the governing equation. This is justified for condensates of atomic species having large natural magnetic dipole moment, such as ^{52}Cr ($d=6\mu_B$), ^{164}Dy ($d=10\mu_B$) and ^{168}Er ($d=7\mu_B$), with μ_B being the Bohr magneton.

It can be verified, that equation (1) is obtained from the following Lagrangian density

$$\mathcal{L}=\frac{i}{2}(\psi\psi_t^*-\psi^*\psi_t)+\frac{1}{2}|\psi_x|^2-\frac{g(t)}{2}|\psi(x,t)|^2\int_{-\infty}^{\infty}R(x-z)|\psi(z,t)|^2dz. \tag{2}$$

To develop the collective coordinate approach, we consider two anti-phase Gaussian functions

$$\psi(x,t)=A\exp\left(-\frac{(x+\xi)^2}{2a^2}+iv(x+\xi)+i\varphi\right)-A\exp\left(-\frac{(x-\xi)^2}{2a^2}-iv(x-\xi)+i\varphi\right). \tag{3}$$

The time-dependent parameters of individual solitons $A(t)$, $a(t)$, $\xi(t)$, $v(t)$, $\varphi(t)$, meaning the amplitude, width, center-of-mass position, velocity and phase, respectively, are considered as collective coordinates. The averaged Lagrangian $L=\int_{-\infty}^{\infty}\mathcal{L}dx$ is evaluated using Eq. (3)

$$\begin{aligned}L &= A^2a\sqrt{\pi}\left[\frac{1}{2a^2}+v^2+2v\xi_t+2\varphi_t-\left(\frac{1}{2a^2}-\frac{\xi^2}{a^4}+2\xi v_t+2\varphi_t\right)\exp\left(-\frac{\xi^2+a^4v^2}{a^2}\right)\right]- \\ & g(t)\frac{A^4a^2\sqrt{\pi}}{\sqrt{2\,(a^2+w^2)}}\left[1-4\cos\left(\frac{a^2v\xi}{a^2+w^2}\right)\exp\left(-\frac{a^6v^2+2a^4v^2w^2+3a^2\xi^2+2\xi^2w^2}{2a^2\,(a^2+w^2)}\right)+\right. \\ & \left.\exp\left(-\frac{2\xi^2}{a^2+w^2}\right)+\exp\left(-\frac{2\xi^2+2a^4v^2}{a^2}\right)+\exp\left(-\frac{2\xi^2(a^2+w^2)+2a^4v^2w^2}{a^2\,(a^2+w^2)}\right)\right].\end{aligned} \tag{4}$$

IOP Conf. Series: Journal of Physics: Conf. Series **869** (2017) 012014 doi:10.1088/1742-6596/869/1/012014

In deriving the last expression for L we assumed the parameter a to be time-independent, as in Ref. [8]. The norm of the molecule is reduced due to destructive interference of two anti-phase solitons placed at a distance 2ξ from each other: $N_m = \int_{-\infty}^{\infty} |\psi|^2 dx = 2A^2 a\sqrt{\pi}[1-\exp(-\xi^2/a^2)]$. Obviously, when the solitons are far apart ($\xi \gg a$), the interference is negligible, and the norm of the molecule is twice of the single soliton's norm $N_m = 2N = 2A^2 a\sqrt{\pi}$.

As we have mentioned above, the stationary waveform and small amplitude dynamics of SM is well described by VA [6, 7]. For a given set of input parameters w, g_0, N_m, the width a_m of a SM can be evaluated from the algebraic equation [6]

$$a_m^4(a_m^4 + 4w^4)(a_m^2 + w^2)^{-7/2} - 4\sqrt{2\pi}/(g_0 N_m) = 0. \tag{5}$$

This equation describes the bond length of a two-soliton molecule $\Delta = 4a_m/\sqrt{\pi}$ as a function of its norm N_m. The wave function for SM $\psi_m = A_m x \exp(-x^2/2a_m^2)$ with a norm $N_m = A_m^2\, a_m^3\sqrt{\pi}/2$, used in [6], can be represented by two anti-phase Gaussians (3) with parameters

$$A = \left[2N_m/(a_m\sqrt{\pi}\, e)\right]^{1/2}, \quad a = a_m\, e/4, \quad \xi = a_m, \text{ where } e = 2.71828... \tag{6}$$

In the left panel of figure 1 we compare the shape of a two-soliton molecule $\psi_m(x)$ found from equation(5) and two anti-phase Gaussians with parameters according to equation(6). Numerically "exact" waveform constructed using the Nijhof's method [10], is also presented. Apparently, the results of VA and Nijhof's method almost coincide.

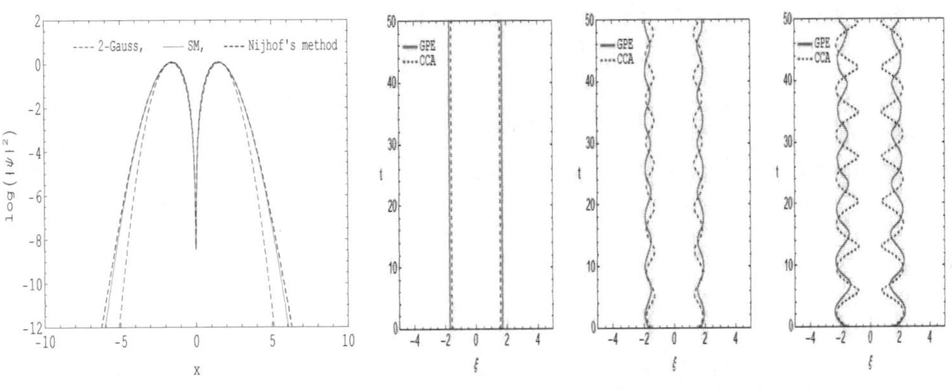

Figure 1. Left panel: The shape of a SM (green solid line) found from equation (5) using $w = 5$, $g_0 = 20$, $N_m = 4$, compared with two anti-phase Gaussians (3) with $a = 1.032$, $A = 1.046$, $\xi = 1.519$, $v = 0$, $\varphi = 0$ (red dashed line). Numerically "exact" waveform is constructed by Nijhof's method (blue dashed line). Right panels: Oscillations of solitons when they are set in motion in opposite directions with initial velocity (from left to right): $v = 0$, $v = 0.2$ and $v = 0.5$, according to GPE (red solid line) and CCA (blue dashed line).

Once the averaged Lagrangian (4) is available and the stationary state of the SM is found, its dynamics under parametric perturbations, such as time dependent strength of dipolar interactions $g(t)$, can be explored using the coupled system of Euler-Lagrange equations for collective coordinates $\eta \to A(t)$, $\xi(t)$, $v(t)$, $\varphi(t)$

$$\frac{d}{dt}\frac{\partial L}{\partial \eta_t} - \frac{\partial L}{\partial \eta} = 0. \tag{7}$$

IOP Conf. Series: Journal of Physics: Conf. Series **869** (2017) 012014 doi:10.1088/1742-6596/869/1/012014

In the right panels of figure 1 we show the dynamics of the molecule when the solitons are set in motion in opposite directions with some initial velocity. In the unperturbed molecule ($v = 0$) the solitons stay motionless at their equilibrium positions. If the molecule is excited due to initial velocity ($v > 0$), the solitons start to oscillate. This is the evidence of the molecular type of interaction between solitons in dipolar BEC. Another important aspect to be stressed here is that, collisions between dipolar solitons are strongly inelastic, which shows up as a phase shift after each collision.

In figure 2 we present typical examples of the dynamics of a two-soliton molecule under periodically varying strength of dipolar interactions. As can be seen from this figure, the developed model adequately describes the oscillations of solitons due to parametric perturbation, and quite accurately predicts the time instance, at which the molecule dissociates.

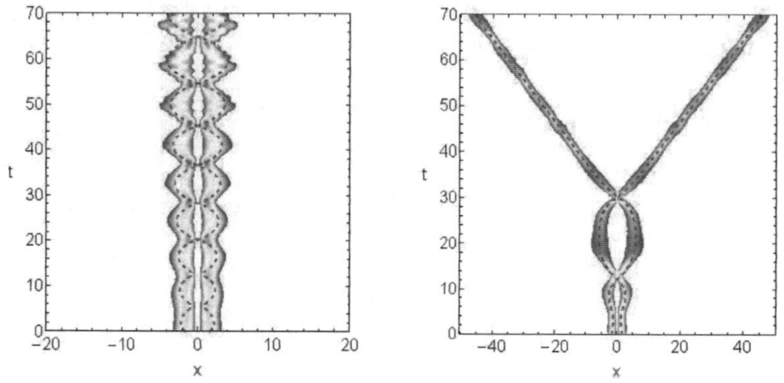

Figure 2. Dynamics of a two-soliton molecule under periodically varying strength of dipolar interactions $g(t) = g_0(1 + \varepsilon \sin(\omega t))$ according to CCA (red dashed line) and GPE (density plot $|\psi|^2$). Left panel: The solitons vibrate under weak perturbation $\varepsilon = 0.1$ with frequency $\omega = 0.756$. Right panel: At stronger perturbation $\varepsilon = 0.63$ with the same frequency the molecule decays into freely moving solitons at $t \sim 30$. The parameter values are the same as in figure 1.

3. Conclusions

We have developed a collective coordinate approach for the dynamics of two bright solitons, moving and colliding in dipolar quantum gases. The advantage of the present model over previously reported ones is that, it can adequately describe both the small amplitude and large amplitude dynamics of soliton molecules, including the dissociation phenomena, when the molecule decays into separate freely moving solitons. Theoretical predictions are compared with numerical simulations of the governing GPE, and good agreement between them is found.

References

[1] Remoissenet M 2003 *Waves Called Solitons. Concepts and Experiments* (Berlin: Springer)
[2] Stratmann M, Pagel T and Mitschke F 2005 *Phys. Rev. Lett.* **95** 143902
[3] Mitschke F, Hause A and Mahnke C 2016 *Eur. Phys. J. Special Topics* **225** 2453
[4] Al Khawaja U and Stoof H T C 2011 *New J. Phys.* **13** 085003
[5] Al-Marzoug S M, Al-Amoudi S M, Al Khawaja U, Bahlouli H, Baizakov B B 2011 *Phys. Rev.* E **83** 026603
[6] Turmanov B Kh, Baizakov B B, Umarov B A and Abdullaev F Kh 2015 *Phys. Lett.* A **379** 1828
[7] Baizakov B B, Al-Marzoug S M and Bahlouli H 2015 *Phys. Rev.* A **92** 033605
[8] Baron H E, Luchini G and Zakrzewski W J 2014 *J. Phys. A: Math. Gen.* **47** 265201
[9] Giovanazzi S, Görlitz A and Pfau T 2001 *Phys. Rev. Lett.* **89** 130401
[10] Nijhof J H B, Forysiak W and Doran N J 2000 *IEEE J. Select. Topics Quant. Electron.* **6** 330

IOP Conf. Series: Journal of Physics: Conf. Series **869** (2017) 012015 doi:10.1088/1742-6596/869/1/012015

Probing the quantum vacuum with petawatt lasers

W T Hill III[1] **and L Roso**[2]

[1]Joint Quantum Institute, Institute for Physical Science and Technology and Department of Physics, University of Maryland, College Park, MD 20742, USA

[2]Centro de Láseres Pulsados, CLPU, Salamanca, Spain

E-mail: wth@umd.edu

Abstract. Due to the bosonic nature of the photon, increasing the peak intensity through a combination of raising the pulse energy and decreasing the pulse duration will pile up more and more photons within the same finite region of space. In the absence of material, this continues until the vacuum is stressed to the point of breakdown and virtual particles become real. The critical intensity where this occurs for electrons and positrons – the so-called Schwinger limit – is predicted to be $\sim 10^{29}$ W/cm^2. At substantially lower intensities, however, nonlinear aspects of the *quantum vacuum* associated with polarization of the vacuum can be explored. These studies become viable at the petawatt level where 10^{23} W/cm^2 and above can be reached. This is an era into which we are just embarking that will provide critical tests of QED and theories beyond the *Standard Model* of particle physics.

1. Introduction

Shortly after Dirac's 1930 model of the negative energy sea, the physics community began thinking about how to coax an electron-positron pair out of the vacuum. The simplest idea was to convert a high-energy photon into matter, $q^\mu \to p^\mu + p'^\mu$ where p^μ, p'^μ and q^μ are the four momenta of the electron, positron and photon, respectively. While a single photon cannot conserve energy and momentum simultaneously, the first realistic proposal came from Oppenheimer and Plesset, who suggested a vacuum photoelectric effect [1], "if we allow gamma-rays of energy γ to fall upon a nucleus, we should expect pairs to appear; the kinetic energy of the pairs would be $\gamma - 2mc^2$ ... in the process the nucleus necessarily takes up a small recoil momentum." Bethe and Heitler studied this process more thoroughly and pair production via this mechanism often bears their name [2]. Breit and Wheeler [3] proposed a different approach that relied on additional photons (nk^μ) instead of a nucleus to conserve momentum, $q^\mu + nk^\mu \to p^\mu + p'^\mu$. This is the mechanism exploited by the SLAC E-144 experiment [4], which used nonlinear Compton scattering to transform a few laser photons (nk^μ) into the requisite gamma photons.

Devising all-optical schemes to reach the critical field and intensity,

$$E_{cr} = \frac{m^2c^3}{e^-h} \simeq 1.3 \times 10^{16} \text{ V/cm} \quad (1)$$

$$I_{cr} = (\varepsilon_0 c/2)\,|E_{cr}|^2 \simeq 2 \times 10^{29} \text{ W/cm}^2, \quad (2)$$

where pairs can be produced directly in a focused laser beam is a quest that began only ten years after the laser was invented [5]. In equations 1 and 2, m and e are the mass and charge of the

IOP Conf. Series: Journal of Physics: Conf. Series **869** (2017) 012015 doi:10.1088/1742-6596/869/1/012015

electron, c is the speed of light in vacuum, $\hbar$ is Planck's constant (h) divided by 2π and ε_0 is the vacuum permittivity. The critical intensity in equation 2 is associated with an effective field of the same magnitude as the critical static field predicted by Schwinger [6] (Eq. 1). An intensity of 10^{29} W/cm^2 is controversial and has been the subject of much discussion; some theoretical investigations even predict pair production at intensities as low as 10^{25} W/cm^2 [7–10]. While I_{cr} is out of reach with current technology, 10^{25} W/cm^2 is just around the corner.

2. Probing the quantum vacuum

We are very close, technologically, to being able to explore a different, but related proposal, suggested by Halpern in 1933 [11] – photon-photon scattering. He noted, "Two possible types of phenomena must be considered separately in connection with the foregoing: (1) All incident quanta have the same direction of propagation; (2) The incident quanta have different directions of propagation. Since we are only interested in purely radiation phenomena the frequencies in the second case should lie below mc^2/h so that no permanent formation of electron-positron pairs can occur."

Clearly, Halpern envisioned an experiment with two beams of low-energy photons where scattered photons are detected. The cross section for the process is calculated to be [12–14]

$$\frac{d\sigma}{d\Omega} = \frac{139\alpha^4\omega^6}{(180\pi)^2 m^8}\left(3+\cos^2\theta\right)^2, \tag{3}$$

$$\sigma = \frac{973\alpha^4\omega^6}{10125\pi m^8}. \tag{4}$$

Equation 3 (4) is the differential (total) cross section with ω being the photon angular frequency and α ($\sim 1/137$) is the fine-structure constant. With new multi-petawatt lasers and high-repetition rate, 1 PW lasers either in operation or coming online in the near future, we will soon be able to exceed the 10^{23} W/cm^2 threshold where these cross sections are predicted to be measureable. Scattering experiments require two synchronized beams but may be very difficult because of the physical vacuum requirements to ensure that photons do not scatter off real matter (electron, molecule, etc.). Nevertheless, viable designs for scattering experiments have been proposed [15, 16]. These will require clever ways to create extreme physical vacua with fewer than one real particle in the focal volume.

Photon-photon scattering can be viewed as vacuum polarization, a fundamentally quantum mechanical process caused by vacuum fluctuations – the appearance and disappearance of particle-antiparticle pairs. When these virtual pairs occur in the presence of intense fields, the vacuum will be polarizable, which will cause it to exhibit birefringence, among other things. Birefringence can be induced by either electric or magnetic fields and detected by the rotation of the plane of polarization or the change in ellipticity of a probe beam's polarization. While there have been no experiments with optical fields at suitable intensities, there have been measurements with quasi-DC magnetic fields that have set threshold magnetic field strengths below which vacuum polarization is not observed [17–19]. Recently, however, it is interesting to note that polarized light was observed by the Very Large Telescope at the European Southern Observatory on Cerro Paranal in Chile. The investigators have interpreted this result in terms of vacuum birefringence caused by a magnetic field of magnitude $\sim 10^{13}$ G generated by a neutron star [20]. This could be the first experimental evidence of vacuum polarization, giving more impetus to laser-based searches.

Vacuum polarization will provide critical tests of nonlinear aspects of quantum electrodynamics (QED). Due to relativistic considerations, the field-field coupling depends only on $E^2 - c^2B^2$ or $\vec{E}\cdot\vec{B}$. The lowest order coupling, $\mathcal{L}_0 = \varepsilon_0\left(E^2 - c^2B^2\right)/2$, simply accounts for the energy of the field. To study field-field coupling it is necessary to include extra terms (based

IOP Conf. Series: Journal of Physics: Conf. Series **869** (2017) 012015 doi:10.1088/1742-6596/869/1/012015

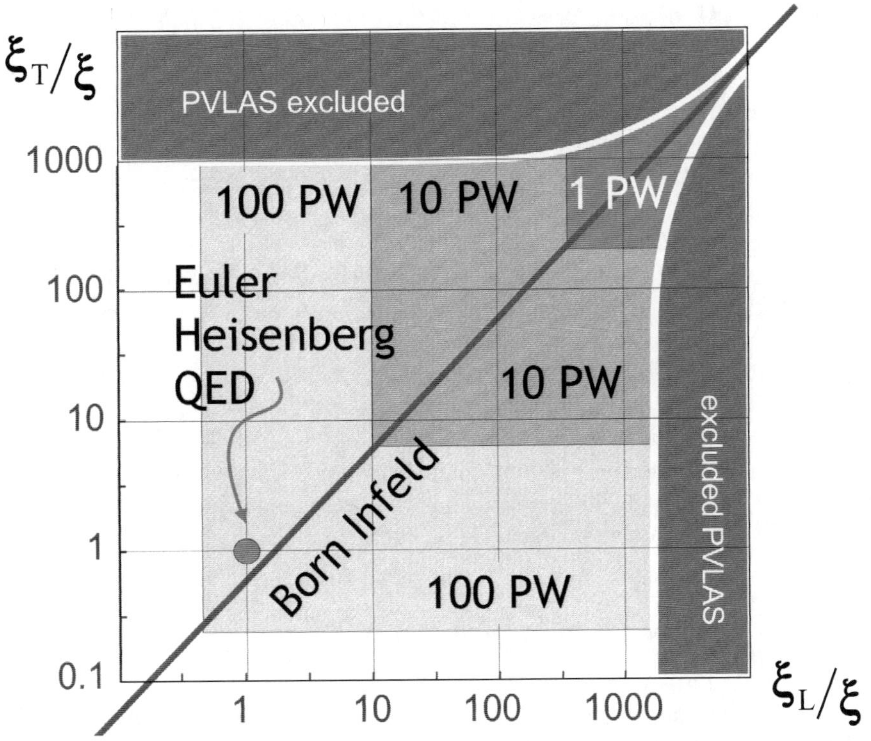

Figure 1. Single-shot sensitivity parameter space normalized to the QED prediction, $\xi_L/\xi_T \equiv 1$ [22] (blue point). The blue line corresponds to the Born-Infeld theory [23] and the red shaded regions is the parameter space excluded by the latest PVLAS data [24,25]. The sensitivity scales as the square root of the number of shots. Thus, 10^4 shots at 1 PW puts one in the 100 PW, single-shot range. This plot is adopted from Ref. [16].

on the only two allowed). Thus, the lowest order terms of the Lagrangian are [16, 21, 22]

$$\mathcal{L} = \mathcal{L}_0 + \xi_L \mathcal{L}_0^2 + (7/4)\xi_T \mathcal{G}^2, \tag{5}$$

where $\mathcal{G} = \varepsilon_0 c \vec{E} \cdot \vec{B}$. The coupling factors, $\xi_{L,T}$, are the longitudinal and transverse responses that are model dependent. In QED $\xi_L = \xi_T = \xi$ [22], where

$$\xi \equiv \frac{8\alpha^{2}\hbar^{3}}{45 m^4 c^5} \simeq 6.7 \times 10^{-30}\ \text{m}^3/\text{J}. \tag{6}$$

A ξ_L/ξ_T ratio different from 1 could indicate new physics, such as the existence of axion-like particles or mini-charged particles (with a charge less than the elementary electric charge, e) and perhaps a departure from the *Standard Model* [16]. It is important to note that the nonlinear terms in Eq. 5 generated at the focus of an intense laser field also give rise to an effective index of refraction different from 1, resulting in an anomalous phase shift of a probe beam traversing the region. About ten years ago, Ferrando *et al* [22] suggested measuring this phase shift to determine ξ, which to our knowledge has not been attempted.

IOP Conf. Series: Journal of Physics: Conf. Series **869** (2017) 012015 doi:10.1088/1742-6596/869/1/012015

3. Technological outlook

While the investigation of vacuum polarization will be difficult, new short-pulsed petawatt lasers make it possible to consider the all-optical approaches mentioned above. Figure 1 shows that it becomes possible to study the ξ_L/ξ_T ratio with single-shot petawatt lasers with powers above ~ 1 PW. More parameter space becomes available at higher powers. Verification of QED, for example, requires a single-shot 100 PW laser. The red shaded region in Fig. 1 is the region excluded by the ground-based magnetic measurements, leaving plenty of room to test QED. It is important to recognize that the sensitivity to new physics is really a problem of signal to noise. A high-rep rate 1 PW laser capturing thousands of shots a day, as will be possible with the VEGA laser facility in Salamanca, Spain, can explore the same parameter space as a single-shot 10 or 100 PW laser. Thus, direct-observation, all-optical experiments (i.e., capturing scattered photons or measuring phases or polarization changes) with both types of lasers should be possible in the near future. In addition, indirect experiments should also be considered. One example might be to exploit plasmas to generate megagauss magnetic fields [26] to generate magnetic birefringence, perhaps in conjunction with quasi-static DC magnetic fields. The future looks bright for new investigations of the *quantum vacuum* and the various proposals for quantitative measurements now need to be given serious consideration.

References

[1] Oppenheimer J R and Plesset M S 1933 *Phys. Rev.* **44** 53
[2] Bethe H and Heitler W 1934 *Proceedings of the rhe Royal Society A* **146** 83
[3] Breit G and Wheeler J A 1934 *Phys. Rev.* **46** 1087
[4] Bamber C, Boege S J, Koffas T, Kotseroglou T, Melissinos A C, Meyerhofer D D, Reis D A, Ragg W, Bula C, McDonald K T, Prebys E J, Burke D L, Field R C, Horton-Smith G, Spencer J E, Walz D, Berridge S C, Bugg W M, Shmakov K and Weidemann A W 1999 *Phys. Rev. D* **60** 092004
[5] Brezin E and Itzykson C 1970 *Phys. Rev. D* **3** 618
[6] Schwinger J 1951 *Phys. Rev.* **82** 664
[7] Bell A R and Kirk J G 2008 *Phys. Rev. Lett.* **101** 200403
[8] Kirk J G, Bell A R and Arka I 2009 *Plasma Phys. Control. Fusion* **51** 085008
[9] Ruf M, Mocken G R, Müller C, Hatsagortsyan K Z and Keitel C H 2009 *Phys. Rev. Lett.* **102** 080402
[10] Bulanov S S, Esirkepov T Z, Thomas A G R, Koga J K and Bulanov S V 2010 *Phys. Rev. Lett.* **105** 220407
[11] Halpern O 1933 *Phys. Rev.* **44** 855
[12] Euler H and Kockel B 1935 *Naturwissenschaften* **23** 246
[13] Euler H 1936 *Ann. Phys.* **418** 398
[14] Liang Y and Czarnecki A 2012 *Can. J. Phys.* **90** 11
[15] Heinzl T, Liesfeld B, Amthor K U, Schwoerer H, Sauerbrey R and Wipf A 2006 *Optics Communications* **267** 318
[16] Tommasini D, Novoa D and Roso L 2014 *Progress in Ultrafast Intense Laser Science X* (*Springer Series in Chemical Physics* vol 106) ed Yamanouchi K, Paulus G and Mathur D (Springer Science) p 137
[17] Cameron R, Cantatore G, Melissinos A C, Ruoso G, Semertzidis Y, Halama H J, Lazarus D M, Prodell A G, Nezrick F, Rizzo C and Zavattini E 1993 *Phys. Rev. D* **47** 3707
[18] Cadène A, Berceau P, Fouché M, Battesti R and Rizzo C 2014 *Eur. Phys. J. D* **68** 16
[19] Valle F D, Ejlli A, Gastaldi U, Messineo G, Milotti E, Pengo R, Ruoso G and Zavattini G 2016 *Eur. Phys. J. C* **76** 24
[20] Mignani R P, Testa V, Caniulef D G, Taverna R, Turolla R, Zane S and Wu K 2017 *MNRAS* **465** 492
[21] Heisenberg W and Euler H 1936 *Z. Phys.* **98** 714
[22] Ferrando A, Michinel H, Seco M and Tommasini D 2007 *Phys. Rev. Lett.* **99** 150404
[23] Born M and Infeld L 1934 *Proc. R. Soc. London* **144** 425
[24] Zavattini G, Gastaldi U, Pengo R, Ruoso G, Valle F D and Milotti E 2012 *Int. J. Mod. Phys. A* **27** 1260017
[25] Valle F D, Milotti E, Ejlli A, Messineo G, Piemontese L, Zavattini G, Gastaldi U, Pengo R and Ruoso G 2014 *Phys. Rev. D* **90** 092003
[26] Gopal A, Minardi S, Burza M, Genoud G, Tzianaki I, Karmakar A, Gibbon P, Tatarakis M, Persson A and Wahlström C G 2013 *Plasma Phys. and Control. Fusion* **55** 035002

IOP Conf. Series: Journal of Physics: Conf. Series **869** (2017) 012016 doi:10.1088/1742-6596/869/1/012016

Electronic polarizability of light crude oil from optical and dielectric studies

A K George and R N Singh

Department of Physics, College of Science, Sultan Qaboos University, P. O. Box 36, Al Khodh 123, Muscat, Sultanate of Oman.
Email: akgeorge@squ.edu.om

Abstract. In the present paper we report the temperature dependence of density, refractive indices and dielectric constant of three samples of crude oils. The API gravity number estimated from the temperature dependent density studies revealed that the three samples fall in the category of light oil. The measured data of refractive index and the density are used to evaluate the polarizability of these fluids. Molar refractive index and the molar volume are evaluated through Lorentz-Lorenz equation. The function of the refractive index, F_{RI}, divided by the mass densityρ, is a constant approximately equal to *one-third* and is invariant with temperature for all the samples. The measured values of the dielectric constant decrease linearly with increasing temperature for all the samples. The dielectric constant estimated from the refractive index measurements using Lorentz-Lorentz equation agrees well with the measured values. The results are promising since all the three measured properties complement each other and offer a simple and reliable method for estimating crude oil properties, in the absence of sufficient data.

1. Introduction

Crude oil is a complex combination of hydrocarbons consisting predominantly of aliphatic, alicyclic and aromatic hydrocarbons covering carbon numbers, which may vary from C1 to C60. The determination of the physical and thermodynamic properties of a particular crude oil is important to provide a basis for further investigations related to exploration and transportation. Both density and refractive index are important properties of crude oils that are routinely monitored, most of the time independently. Specific gravity (typically reported as American Petroleum Institute gravity – API gravity) [1] as determined by the density, not only determines if a crude oil is light or heavy but also is an important parameter for the experimental determination of the interfacial tension and other transport properties.

Along with density (ρ), refractive index (n) and electronic polarizability (α) are important physical properties of crude oil and are intimately connected to density [2-6]. Refractive index is easily measurable and because it correlates to density, it has been used as proxy measurement for density in some processes [7]. Measurements for both parameters are not always available, and it is desirable to find an accurate correlation between the two properties for crude oils. Electronic polarizability tells us how much the crude oil can respond to applied external electric fields. In a particular case of hydrocarbon and crude oil systems, where polar interactions are week, intermolecular attractions are determined by the polarizability that can be directly related to the refractive index through the Lorentz-Lorentz equation [8]. The dielectric property has been widely used [9-15] to characterize crude oil for its reformation under thermal changes, the precipitation of asphaltenes with the addition of hexane, dielectric properties based on infrared spectroscopy, etc. The dielectric constant is closely related to the refractive index through Lorentz-Lorentz equations. In this paper we are making a comparison of the theoretical value of the dielectric constant computed using the refractive index data to the directly measured values.

IOP Conf. Series: Journal of Physics: Conf. Series **869** (2017) 012016 doi:10.1088/1742-6596/869/1/012016

2. Experimental measurements

The density of the samples was measured as a function of temperature using an *Anton Paar* precision density meter. The details of the measurements are given in an earlier [9] publication. An *Abbé* refractometer was used to measure the refractive index of the samples as a function of temperature ranging from 20 to 50°C. The liquid sample is sandwiched between the illuminating prism and the refracting prism. Dielectric investigations were performed using a *Wayne Kerr* (WK6500B) Impedance Analyser along with a Test Fixture. The temperature variation of capacitance and dielectric loss (tan Δ) are studied using 1Vp-p oscillating level of 100 kHz frequency. The variation of dielectric constant with temperature was estimated from observed capacitance.

3. Results and discussion

All three samples of oil studied (*Sayyala, Mabrook and Daleel*) are procured from the oil fields which are located in the different regions of the Sultanate of Oman. Hereafter we term them as, Sample 1 (*Sayyala*), Sample 2 (*Mabrook*) and Sample 3 (*Daleel*).

3.1. Density and API gravity

The density of the three crude oil samples decreases linearly with temperature as can be seen in Fig 1. The API gravity was measured using the equation $API = 141.5/\rho - 131.5$, where ρ is the density at 15.6°C, is found by extrapolation. The API gravity of the oils are: 44.3°, 42.0° and 37.9° for samples 1, 2 & 3 respectively. According the API classification [1], all three samples are termed as *light oil.*

3.2. Refractive index as a function of temperature

We used an *Abbe* Refractometer to measure refractive index as a function of temperature from 20°C to 50°C in steps of 5° C. The refractive index is found to decrease linearly with increasing temperature for all samples (Fig 2). Samples with higher API have lower refractive index.

3.3. Correlation between refractive index and electronic polarizability.

Polarizability is the ability for a molecule to be polarized. Polarizabilities determine the dynamical response of a bound system to external fields, and provide insight into a molecule's internal structure. Large molecules, as in hydrocarbon liquids, get grossly affected by applied electric field.

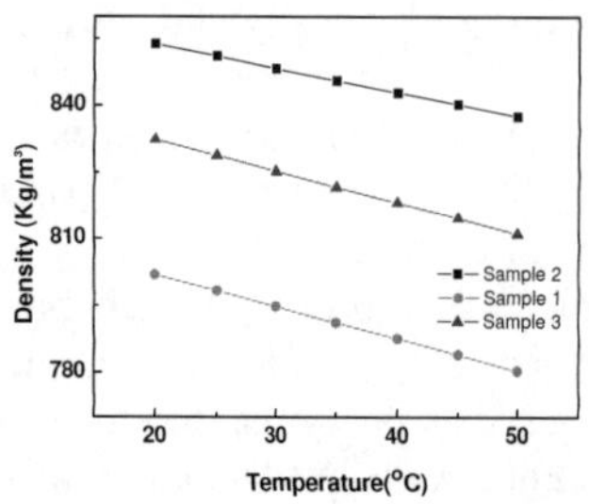

Fig. 1: Density as a function of temperature.

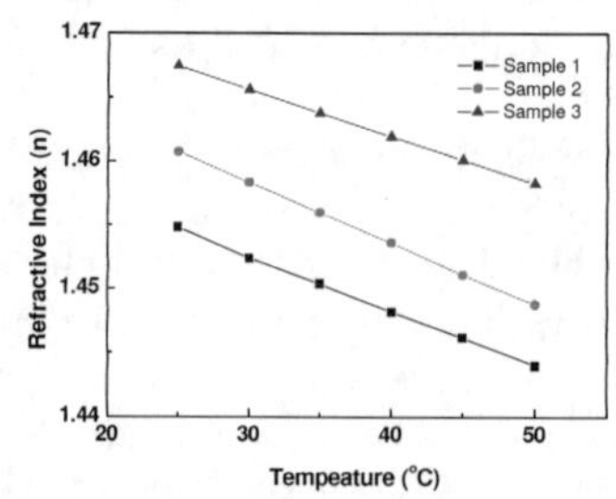

Fig. 2: Refractive index as a function of temperature.

The polarization (P) is defined as the dipole moment per unit volume, averaged over the volume of a cell. The dielectric constant (ε) is related to electronic polarizability (α) through the famous Clausius - Mossotti relation [8],

$$\frac{\varepsilon - 1}{\varepsilon + 2} = \frac{4\pi}{3} \Sigma_j N_j \alpha_j \tag{1}$$

where N_j is the concentration and α_j is the polarizability of atom j. For $\epsilon = n^2$ one gets,

$$\frac{n^2 - 1}{n^2 + 2} = \frac{4\pi}{3} \Sigma_j N_j \alpha_j \tag{2}$$

IOP Conf. Series: Journal of Physics: Conf. Series **869** (2017) 012016 doi:10.1088/1742-6596/869/1/012016

In the particular case of hydrocarbon and crude oil systems, where polar interactions are weak, intermolecular attractions are determined by the polarizability that can be related to the refractive index through Lorentz-Lorentz equation [16,17],

$$\alpha = \left(\frac{3}{4\pi N_A}\right) R_m \text{ with } R_m = \left(\frac{n^2-1}{n^2+2}\right) V_m \tag{3}$$

Here, R_m is the molar refractivity, N_A is the Avogadro's number, n is the refractive index and V_m is the molar volume. The values of the electronic polarizability (α) are presented in Table 1.

Table 1: Electronic polarizability as a function of temperature and the ratio F_{RI}/ρ

Temperature (°C)	Electronic Polarizability, $\alpha \times 10^{-24}$ (cm^3/mol)			F_{RI}/ρ		
	Sample 1	Sample 2	Sample 3	Sample 1	Sample 2	Sample 3
25	2.4958	2.4826	2.4493	0.3375	0.3372	0.3368
30	2.4909	2.4907	2.4526	0.3376	0.3372	0.3369
35	2.4859	2.4988	2.4558	0.3376	0.3374	0.3369
40	2.4808	2.5071	2.4590	0.3378	0.3375	0.3372
45	2.4756	2.5153	2.4622	0.3379	0.3375	0.3370
50	2.4703	2.5237	2.4654	0.3380	0.3377	0.3371

3.4: Testing of the One-third rule

Equation (3) of molar refractivity can further be expanded in terms of the molecular weight (M) and the density (ρ), i.e.,

$$R_m = \left(\frac{n^2-1}{n^2+2}\right)\frac{M}{\rho}, \text{ or } R_m = \left(\frac{F_{RI}}{\rho}\right) M \text{ where } F_{RI} = \left(\frac{n^2-1}{n^2+2}\right) \tag{4}$$

a fraction of refractive index. The computed values of (F_{RI}/ρ) for liquid hydrocarbons are tabulated in table 1.

The results suggest that (F_{RI}/ρ) is constant over a wide range of temperature of investigations. This often called in the literature [18] as *one-third rule* and is widely used for correlating the density and refractive index of crude oil. The average value of (F_{RI}/ρ) for light crude oil is around 0.337 which is very close to 1/3.

In addition to the above, the density data was used to calculate the refractive index as a function of temperature using the Lorentz-Lorentz expansion [3]:

$$\frac{F_{RI}(T)}{\rho(T)} = \left(\frac{n^2(T)-1}{n^2(T)+2}\right)\frac{1}{\rho(T)} = \frac{1}{\rho_o(T)} = 0.5054 - 0.3951\rho(T) - 0.2314\rho^2(T) \tag{5}$$

where $F_{RI}(T)$ is the fraction of the refractive index, $\rho_o(T)$ is the mass density of a molecule and $\rho(T)$ is the density. Refractive index (n) computed from measured density via equation (5) is close to the experimentally determined values of n. Near the room temperature it is found to differ by about 1.6%, 0.26% and 0.20% for samples 1,2 and 3 respectively. Similar variations are found at other temperatures.

3.5: Dielectric studies

The variation of the dielectric constant (ε') estimated from the observed capacitance with temperature is plotted in Fig 3. It has been observed that all three samples, the dielectric constant decreases with increasing temperature. The variation of the dielectric loss ε'' studied as a function of temperature is plotted in Fig 4. It has been observed that ε'' increases linearly with temperature. The gradient

$d\varepsilon''/dT \approx 4.85\times10^{-3}$ is the same for all three samples of light oils where the API varies in the range of 37.9° to 44.3°

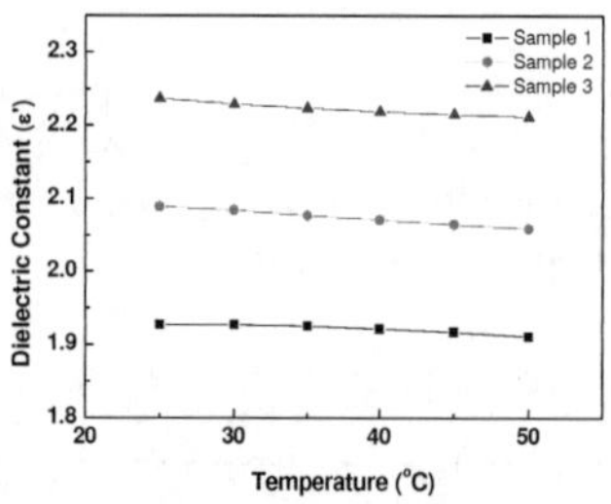

Fig. 3: Experimental values of the dielectric constant

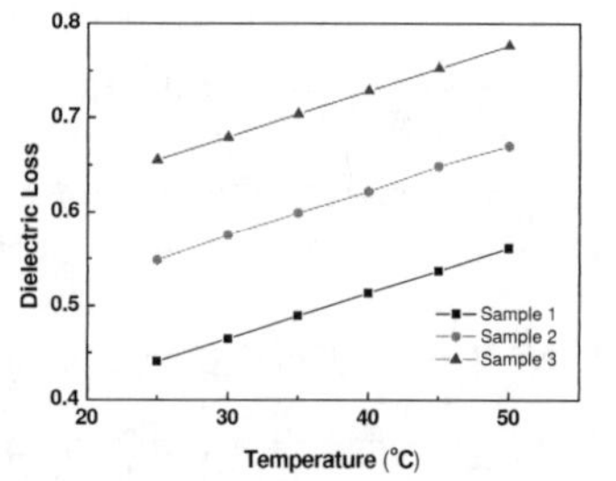

Fig. 4: Variation of the dielectric loss with temperature

For the sake of comparison, we also used the Lorentz-Lorentz equation:

$$\frac{\varepsilon'(T)-1}{\varepsilon'(T)+2}=\frac{n^2(T)-1}{n^2(T)+2}, \qquad (6)$$

to compute the dielectric constant from the refractive index. We found that ε' determined from equation (6) agree reasonably well with the directly measured value; the largest variation is in the case of Sample 1 (around 8%) at higher temperature. In the case of Sample 2 and 3, it differs by about 2.8% and 4.2% respectively.

4. Conclusions

The present study suggests that the density, refractive index and dielectric constant can readily be used to characterize crude oil samples. Since the three measured parameters complement each other, our results are promising as it offers a simple and reliable method for estimating crude oil properties, even when sufficient data is not available.

References

[1] *Petroleum Manufacturing Orientation Course* 1997 Mobil Research and Engineering (Paulsboro, New Jersey, USA)
[2] Riazi M R and Roomi Y A 2001 *Ind. Eng. Chem.* **40** 1975
[3] Achtermann H J, Hong J, Wagner W and Pruß, A. 1993 *J. Chem. Eng. Data* **37** 414
[4] Hadrich J 1975 *Appl. Phys.* **7**, 209
[5] Iglesias-Otero M A, Troncoso Carballo J E and Romaní L 2008 *J. Chem. Thermodyn.* **40**, 949
[6] Krishnaswamy R K and Janzen J 2005 *Polym. Test.* **24** 762
[7] Yarranton H W, Okafor J C, Ortiz D P and van den Berg F G A 2015 *Energy & Fuels* **29** 5723
[8] C. Kittel 1996 *Introduction to Solid State Physics*" 7th *Edition* (John Wiley, New York)
[9] George A K, Singh R N, Arafin S, Ryan T Abraham and Majda Al-Balushi 2014 *Petroleum Science and Technology* **32** 889
[10] Saraev D V, Lunev I V, Yusupova T N, Tagirzyanov M I, Yakubov M R, Gusev Y A and Romanov G V 2005 *Oil and Gas Business*, Available at: http://www.ogbus.ru/eng/.
[11] Sheu E Y and Acevedo S 2006 *Fuel* **85**, 1953
[12] Sheu E Y, Storm D A and Shields M B 1994 *Energy & Fuels* **8** 552
[13] Carlson J E, Tomren A L and Folgeø, Barth T 2014 *Chemometrics and Intelligent Laboratory Systems* **139** 1
[14] Chow R S, Tse D L and Takamura K 2008 *Can. J. Chem. Eng.* **82** 840
[15] Evdokimov I N and Losev A P 2010 *Energy & Fuels*. **24,** 3959
[16] Singh R N, George A K, Arafin S 2006 *Journal of Physics D: Applied Physics* **39** 1220
[17] Eugene Hecht 2002 *Optics*, 4th *Edition* (Addison Wesley, New York)
[18] Vargas F M and Chapman W G 2010 *Fluid Phase Equilibria* **290** 1

IOP Conf. Series: Journal of Physics: Conf. Series **869** (2017) 012017 doi:10.1088/1742-6596/869/1/012017

Thermoelectric power of ($Cu_{0.5}Tl_{0.5}$)-1223 superconducting phase added with $BaSnO_3$ nanoparticles

A Srour[1], W Malaeb[1], S Marhaba[1] and R Awad [1]

[1]Physics Department, Faculty of Science, Beirut Arab University (BAU), Lebanon
E-mail address: ali_sr@outlook.com

Abstract. In this study, we report the thermoelectric power (TEP) measurements of $Cu_{0.5}Tl_{0.5}Ba_2Ca_2Cu_3O_{10-\delta}$ added with $BaSnO_3$ nanoparticles. $BaSnO_3$ nanoparticles were prepared by chemical co-precipitation method, while $(BaSnO_3)_x$/(CuT1)-1223 superconducting samples with $0.00 \leq x \leq 1.50$ wt% were prepared using the solid-state reaction method. The standard four-probe technique was applied to measure DC electrical resistivity in the temperature range from 300 to 77 K. Superconducting transition temperature (T_c) increases up to 117.5 for x = 0.25 wt.% and then it decreases with further x addition. The TEP coefficient was measured as a function of temperature in a wide temperature range from ~77 K (liquid nitrogen temperature) up to 280 K using a standard differential technique. The behavior of the obtained TEP coefficient is suit with high-T_c copper-oxide superconductors. The results were investigated according to two-band model with an extra linear term. Several parameters such as the pseudo-gap temperature (T*), Fermi energy (E_F) and Fermi temperature (T_F) values were calculated and discussed in terms of nanoparticles $BaSnO_3$ addition.

1. Introduction

Doping of nanostructures was considered as an effective approach for improving the current capacity of high-T_c superconductors (HTSCs). Among these HTSCs, CuTl-1223 superconductor phase is one of the most candidates in the field of power transmission electric energy and superconducting magnetic field sensors [1]. The study of transport properties on this phase such as electrical resistivity and thermoelectric power is important to explore its conduction mechanisms. Moreover, it provides information about the band width, scattering mechanism, nature of charge carries and electronic transport properties [2-4]. TEP sign and coefficient are strongly depended on the oxygen content present in the compound [5, 6] and on the hole concentration within the CuO_2 planes [7]. This work studies the effect of adding different concentration of barium tin oxide nanoparticles to CuTl-1223 superconductor phase on its thermoelectric power.

2. Experimental and Investigation Techniques

The prepared superconducting samples added by x $BaSnO_3$ nanoparticles (prepared by chemical Co-precipitation method) with different concentrations (x = 0.00, 0.25, 0.50, 0.75 and 1.50 wt.%) using solid-state reaction technique at ambient pressure, reported in our previous work [8]. DC electrical resistivity was used to characterize the superconducting properties of the samples using standard four-probe DC technique, in the temperature range (77 to 300) K. TEP of the prepared superconductor samples was measured using the standard differential technique in the temperature range from 80 to 280 K. The sample, of dimensions $1.5 \times 0.3 \times 0.2$ cm^3, was clamped between two copper plates and each one supplied with a metal resistor heater. Two T-type thermocouples were used to monitor the temperatures of both blocks with accuracy ±0.1 K and record the thermoelectric motive force between the samples terminals using an accurate digital micro-voltmeter and the data are carefully taken after removing the offset. Finally, the data was collected using the "DATA Studio" program and an interface (Pasco CI-7500).

3. Results and discussion

The prepared samples were characterized using X-ray Powder diffraction (XRD) and scanning electron microscope (SEM) in our previous work [8]. As a results from these characterizations, the $BaSnO_3$ nanoparticles set at the grain boundaries (no variation of lattice parameters was observed), and the addition of low content of nanoparticles enhanced the formation of $(Cu_{0.5}Tl_{0.5})$-1223 phase [9]. Thus, increase in the inter-grain connectivity and grain size. The DC electrical resistivity as function of temperature for $(BaSnO_3)_x$/(CuTl)-1223 phase, with x = 0.50 wt.%, is shown in figure 1. The values of T_c for all samples are listed in table 1. The enhancement in T_c can be discussed according to the improvement in the volume fraction and inter grain connectivity for x = 0.25 wt.% [8]. The hole-carriers concentration per Cu ion (P) is listed in table 1, calculated through the following relation [10]

$$p = 0.16 - [(1 - \frac{T_c}{T_c^{max}})/82.6]^{0.5}\ , \qquad (1)$$

where T_c^{max} is taken as 120 K for $(Cu_{0.5}Tl_{0.5})$-1223 phase [11]. One can notice that both T_c and P enhance by increasing x up to x = 0.25% followed by a suppression in their values with further increase in x. The Pseudogap temperatures were also determined from the variation of dρ/dT with temperature for x = 0.50 wt.% sample as shown in the inset of figure 1, and the values for all samples are listed in table 1. T* represents the deviation point of resistivity from linear behavior at certain temperature, corresponding to the opining of pseudogap. It is also noticed from table 1, that the highest T_c corresponds to lower T*. This indicates that T* is probably affected by the oxygen content or the scattering of carriers at the grain boundaries and the change of carrier density in the CuO_2-planes [12, 13].

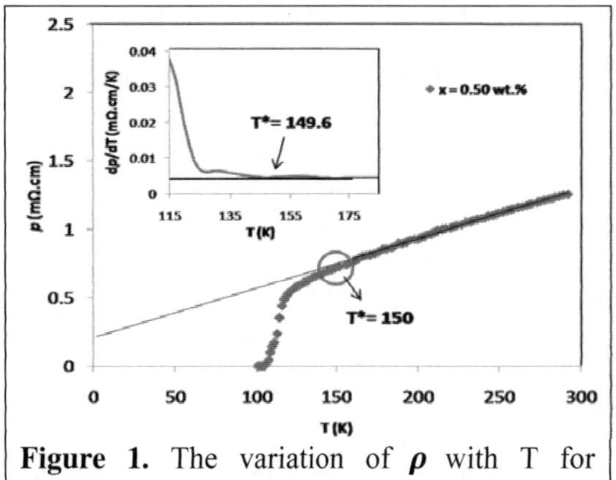

Figure 1. The variation of $\boldsymbol{\rho}$ with T for $(BaSnO_3)_{0.50wt.\%}$/CuTl-1223 phase. The inset figure correspond to the $\boldsymbol{d\rho/dT}$ as function T.

The temperature dependence thermoelectric power for the investigated set of $(BaSnO_3)_x$/(CuTl)-1223 samples with x = 0.00, 0.50, 0.75 and 1.50 wt.% superconductor samples is shown in figure 2. The inset of figure 2 represents the variation of TEP_{max} as function of x addition of $BaSnO_3$ nanoparticles. It is noticed that all the samples have positive TEP, suggesting that they are p-type conductive materials (holes in CuO_2-planes) [14]. It is well known that the $BaSnO_3$ nanoparticles possess n-type conduction character [15]. This means that $BaSnO_3$ does not affect the conduction mechanism of HTCSs [16]. TEP increases linearly as temperature decreases up to 150 K with a negative slope, resulting in the flow of the thermal excitation of extrinsic charge carriers at high temperatures. After that it has an optimum value at T*, and their values are listed in table 1. Finally it decreases as the temperature decreases, consisting with the Mott formula [17].

A similar behavior of TEP was reported for several series of Bi-, Hg- and Tl cuprate superconductors [18-20]. TEP_{max} decreases as x increases up to 0.75 wt.%, followed by an increase in TEP at x = 1.5 wt.%. At low addition concentration (0.25 wt.%), the decrease in TEP is due to change in hole concentration to higher value (0.142), after that the followed decrease may be due to the negative conduction character (n-type) of the added nanoparticles. Then, TEP increases for x = 1.5 wt.%, which could be due to the growing of impurities and inhomogeneous distribution with higher concentrations of nanoparticles. The measured TEP values are very small (5-10 μV/K), confirming the metallic behavior of $(BaSnO_3)_x/(CuTl)$-1223 phase [21].

Table 1. The variation of T_c, P and T*, fitting parameters (A, B and α), E_F, T_F and T* estimated from TEP measurements.

x (wt.%)	Electrical Resistivity			Thermoelectric power					
	T_c (K)	P	T* (K)	A (μV)	B (K)	α (μV/K²)	E_F (eV)	T_F (K) x10⁴	T* (K)
0.00	115.5	0.138	156.3	4000	50	-0.06	0.4	0.4	154.1
0.25	117.0	0.142	147.1	900	70	-0.0075	3.2	3.7	150.9
0.50	114.0	0.135	149.6	800	110	-0.0085	2.8	3.3	153.3
0.75	112.5	0.132	152.1	720	128	-0.009	2.7	3.1	154.7
1.50	111.0	0.129	155.4	1500	135	-0.014	1.7	1.9	155.6

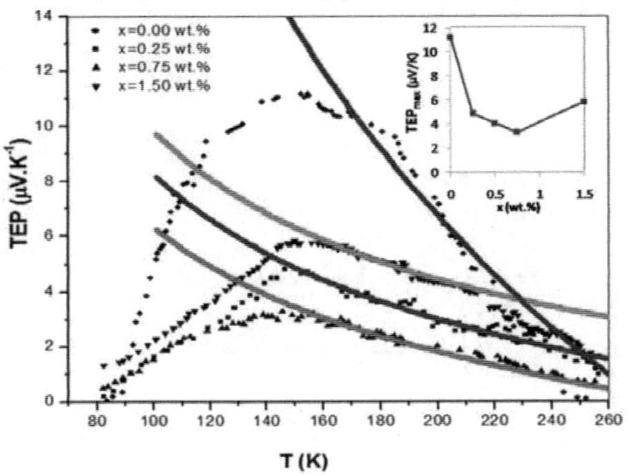

Figure 2. Variation of the TEP as function of T for $(BaSnO_3)_x$/CuTl-1223 phase for x = 0, 0.25 and 0.75 wt.%, fitted according equation (2).

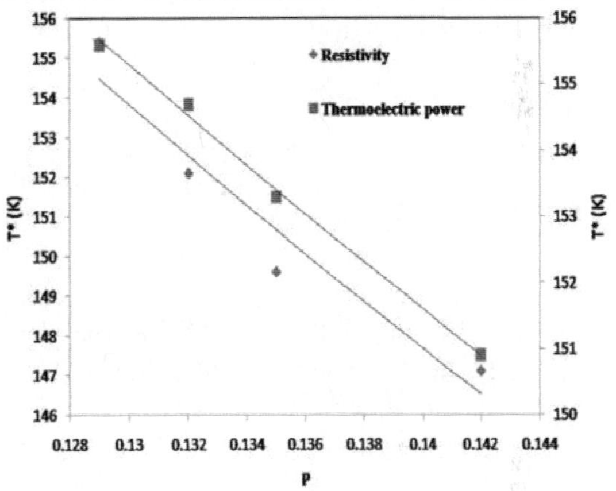

Figure 3. Variation of T* versus P for $(BaSnO_3)_x$/CuTl-1223 phase estimated from Resistivity and TEP measurements.

The measured TEP data were analyzed through the basis of the two-band model with an extra linear term on the decreasing part of the curves as shown in figure 2, which is developed by Forro *et al.* [22, 23] using

$$S = \frac{AT}{B^2+T^2} + \alpha T, \qquad (2)$$

on the decreasing part of the curves, where A is related to the contribution from the mobile holes in the Cu–O planes. The fitting parameters A, B and α are listed in table 1. It is observed that the variation of A and TEP_{max} have the same trend. Assuming the term αT, is the contribution of diffusion-like, the Fermi energy E_F and temperature [24] are given as

$$E_F = \frac{\pi^2 k_B^2}{3e\alpha} \qquad (3) \qquad\qquad T_F = \frac{E_F}{k_B}, \qquad (4)$$

and their values are calculated and listed in table 1. The values of E_F is consistent with other publish results for cuprates [25]. Also, the calculated values of T_F are in the range of 10^4 K, similar to those obtained for metals. According to the free electron theory, it was found that E_F as well as T_F are related to carrier concentration through equation 5:

$$E_F = \frac{\hbar^2}{2m}\left(\frac{3N\pi^2}{V}\right)^{\frac{2}{3}}, \tag{5}$$

where N/V is the carrier concentration at the Fermi surface, which confirmed the variation of P with both E_F and T_F. The variation of Pseudugap temperature estimated from resistivity and TEP data as function of P is shown in figure 3. T* is well fitted according to the relation $T^* = aP^{-b}$, by which a and b were determined to be 49.86 and 0.55 for resistivity, 79.96 and 0.42 for TEP, respectively.

4. Conclusion

The TEP of $(BaSnO_3)_x/(CuTl)$-1223 ($0.25 \le x \le 1.5$ wt.%) was measured from 80-280 K using the standard differential technique. For all samples, TEP had positive value, which indicated the hole-like trend. Moreover, TEP decreased as the concentration of $BaSnO_3$ increased up to 0.75 wt.%, and then it increased with further increase in x. T* determined from TEP was consistent with that determined from resistivity. TEP results were well fitted according to two-band model with an extra linear term. Fermi Energy and Fermi temperature were estimated as function of $BaSnO_3$ concentrations. It was found that the variation of both E_F and T_F were confirmed the variation of P with x.

References

[1] Jabbar A, Qasim I, Khan K, Ali Z, Nadeem K, and Mumtaz M 2015 *J. Alloys Compd.* **618** 110
[2] Tokura Y, Torrance P, Huang T, and Nazzal A 1988 *Phys. Rev.* B **38** 7156
[3] Kes P, Van den Berg J, and Narlikar A 1990 *Nova Science Publ, New York.* 228
[4] Okram G, Muralidhar M, Jirsa M, and Murakami M 2004 *Physica* C **402** 94
[5] Lee S, Lee J, Suh B, Moon S, Lim C, and Khim Z 1988 *Phys. Rev.* B **37** 2285
[6] Rao V, Rangarajan G, and Srinivasan R 1984 *J. Phys.* F **14** 973
[7] Batlogg B, Hwang H, Takagi H, Cava R, Kao H, and Kwo J 1994 *Physica* C **235** 130
[8] Srour A, Malaeb W, Barakat M, and Awad R 2016 (submitted to *Journal of temperature physics*) "Physical properties of $(BaSnO_3)_x/Cu_{0.5}Tl_{0.5}Ba_2Ca_2Cu_3O_{10-\delta}$ superconductor"
[9] Awad R, Abou-Aly A, Isber S, Malaeb W 2006 *J Phys.: Conf. Ser.* **43** 474
[10] Presland M, Tallon J, Buckley R, Liu R, and Flower N 1991 *Physica* C **176** 95
[11] Ihara H, Tanaka K, Tanaka Y, Iyo A, Terada N, Tokumoto M, Ariyama M, Hase I, Sundaresan A, Hamada N, and Miyashita S 2000 *Physica* C **341** 487
[12] Francois I, Jaekel C, Kyas G, Dierickx D, Van der Biest O, Heeres R, Moshchalkov V, Bruynseraede Y, Roskos H, Borghs G, and Kurz H 1996 *Phys. Rev.* B **53** 12502
[13] Mohammadizadeh M and Akhavan M 2003 *Physica* B*: Condensed Matter* **336** 410
[14] Lin C, Chen M, Huang R, Cheng Y, and Lee P 2015 *Energies* **8** 12573
[15] Duarte T, Buzolin P, Santos I, Longo E, and Sambrano J 2016 *Theor. Chem. Acc.* **135** 1
[16] Al-Rasoul K 2012 *Asian Trans, Basic. Appl. Sci.* **2** 1
[17] Fan X, Yang J, Zhu W, Bao S, Duan X, Xiao C, and Li K 2008 *J. Alloys Compd.* **461** 9
[18] Chanda B, Ghatak S, and Dey T 1994 *Physica* C **232** 136
[19] Orlando M, De Mello E, Passos C, Caputo M, Martinez L, Zeini B, Yugue E, Vanoni W, and Baggi-Saitovitch E 2001 *Physica* C **364** 350
[20] Mitra N, Trefny J, Yarar B, Pine G, Sheng Z, and Hermann A 1988 *Phys. Rev. B:Condens. Matter* **38** 7064
[21] Gul I and Maqsood A 2008 *J. Supercond. Nov. Magn* **21** 399
[22] Gottwick U, Gloss K, Horn S, Steglich F, and Grewe N 1985 *J. Magn. Magn. Mater.* **47** 536
[23] Forro L, Lukatela J, and Keszei B 1990 Solid State Commun. **73** 501
[24] Casquilho J and Teixeira P 2014 *Introduction to Statistical Physics. Cambridge University Press*
[25] Ghorbani S, Lundqvist P, Andersson M, Valldor M, and Rapp Ö 2001 *Physica C* **353** 77

IOP Conf. Series: Journal of Physics: Conf. Series **869** (2017) 012018 doi:10.1088/1742-6596/869/1/012018

Scrutiny of structural disorder using Raman spectra and Tauc parameter in $GeTe_2$ thin films

R T Ananth Kumar[1], Hussein A Mousa[2], P Chithra Lekha[3], Saleh T Mahmoud[1] and N Qamhieh[1, *]

[1] Department of Physics, UAE University, Al Ain, P.O. Box 15551, United Arab Emirates
[2] Department of Petroleum & Chemical Engineering, UAE University, Al Ain, United Arab Emirates
[3] Department of Physics, Indian Institute of Technology Madras, Chennai, Tamil Nadu, India
E-mail: nqamhieh@uaeu.ac.ae

Abstract. The optimization of post deposition annealing in $GeTe_2$ thin films towards structural rearrangements is reported. From X-ray diffraction, the prominent Bragg's peak is identified, and the full width at half maximum (FWHM), grain size, crystal structure, and microstrain are correlated with the annealing temperature. The process of chemical bonding energy from the Raman spectra confers the homopopolar and heteropolar bond formation. However, the assessed structural disorders are discussed in terms of the Tauc parameter.

1. Introduction

Binary materials based on tellurium are playing a remarkable role in the field of switching devices [1]. A clear thoughtful of the physical structure for the change from an amorphous to crystalline phase is essential to have an elaborating idea of the switching devices. To tackle this problem, a comprehensive study of samples based on Ge-VI (chalcogenides) materials has been carried out. To achieve this goal, a detailed work has been done on Ge-Te to cram the short and long-range order. Research on optical studies will divulge more on the electronic and band structure characteristics of these materials. In fact, the Tauc parameters will provide knowledge on the bands' tails that are a result of the disorder in the materials [2]. Annealing the film as a post deposition treatment will bring a clear idea about the structural defects resulted in the amorphous to crystalline transition process. The persistence of the present work is to investigate the disorder in the structure of $GeTe_2$ thin films and its dependence on the annealing temperature. The study is accomplished by post deposition annealing and performs the X-ray diffraction (XRD), Raman spectra and transmittance analysis.

2. Experiment

Germanium telluride thin films were deposited onto glass substrates by vacuum evaporation technique using Torr International system at a vacuum of 10^{-6} Torr. Films of 100 nm thickness were deposited, and the thickness were controlled and measured by a quartz crystal monitor. High purity bulk $GeTe_2$ (99.999 %, Testbourne Ltd, England) was used as precursor material for the evaporation. The stoichiometric ratio was checked using the JEOL JSM-6010 scanning electron microscope, and are highly stoichiometric. The films prepared were annealed for two hours each at different temperatures ranged from 60 – 200 °C. X-ray diffraction analysis were done using Shimadzu 6100 operated at 30 KV using Ni filter Cu K_α radiation of wavelength 1.5405 Å. For the optical investigations, the transmission spectra were done using (Jasco V-670) UV-Visible spectrophotometer. The Raman analysis was performed using Horiba Jobin Yuov (HR800UV) with a power of 2 mW.

3. Results

Figure 1 shows the XRD spectra for $GeTe_2$ thin films which were annealed at different temperatures as 60, 80, 100, 120, 160 and 200 °C. The as-prepared film is found to be amorphous in nature. The films partially start to crystallize by increasing the annealing temperature. At the annealing temperature of 200 °C the sample shows higher crystallinity, and the diffraction peaks are observed at 23.01°, 26.13°, 27.5°, 29.92°, 40.45°, 43.56°, 49.72° and 53.78° corresponds to (2 2 0), (3 1 1), (2 2 2), (3 2 1), (5 0 0), (5 1 1) and (5 4 1) planes, respectively. The prominent peak is found at 23° belongs to hkl plane (2 2 0).

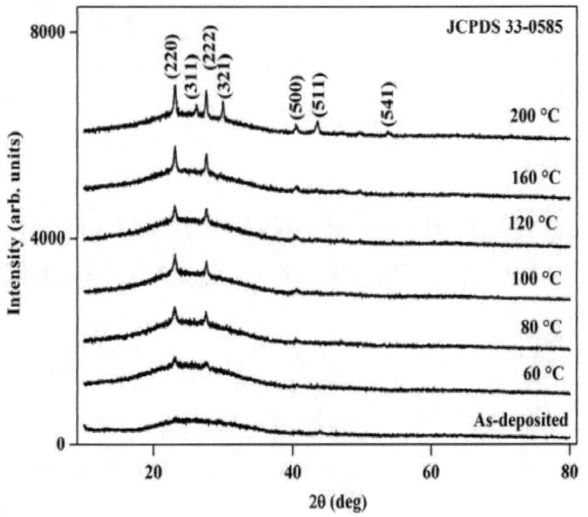

Figure 1. XRD spectra of $GeTe_2$ thin films at different annealing temperatures.

The grain size was found using Scherer's formula D= kλ/βcosθ, where k denotes the shape factor constant (0.94), λ the wavelength and β the full width at half maximum. Table 1 shows the values of FWHM of (2 2 0) diffraction peak and the grain size as calculated from the Scherer's formula. The decrement in the FWHM is a signature of an increment in the grain size which signposts the increase in crystallinity with raising the annealing temperature. The microstrain, ∈, is calculated using the formula ∈= β cosθ/4

Figure 2(a) shows the optical transmittance spectra. The as-prepared film of thickness 100 nm shows a maximum of about 50 % of transmission. Annealing at higher temperature leads to an increase in the transmittance to reach 75% at 80 °C, and then decreased to reach 42% at annealing temperature of 200 °C. Moreover, with the increase in the annealing temperature a red shift is noticed for the absorption edge. In amorphous semiconductors a strong absorption corresponds to transitions between extended states in the conduction and valance bands, which are characterized by the Tauc optical band gap relation given by

$$(\alpha h\nu)^{1/2} = B^{1/2}(h\nu - E_g), \quad (1)$$

where $B^{1/2}$ corresponds to Tauc parameter, *hv* indicates the incident photon energy, and E_g denotes the optical band gap. Figure 2(b) shows the plot of $(\alpha h\nu)^{1/2}$ vs photon energy. The extraction of the linear fit and energy intercept gives the value of E_g, and the slope crops the value of the Tauc parameter ($B^{1/2}$). Table 1 summarizes the values of E_g and $B^{1/2}$. The as-deposited film shows a band gap of 0.85 eV and it decreases to 0.70 eV when the temperature increases to 200 °C. The measured value of the Tauc parameter is 892 $cm^{-1/2}eV^{-1/2}$ for the as-prepared film, and it is found to increase by increasing the annealing temperature. On the other hand, the optical band gap decreases when the annealing temperature increases.

The inset of figure 3 shows the Raman spectra for all the annealed films with two broad Raman bands from 110-135 cm^{-1} and 135-145 cm^{-1} with a maximum at 121 cm^{-1} and 141 cm^{-1}. Moreover, figure 3 displays the Raman spectra for the film annealed at 200 °C by using the Gaussian fitting and an additional Raman band is noticed at 156 cm^{-1}.

4. Discussion

Figure 1 shows that when annealing the prepared $GeTe_2$ thin film at 60 °C, two small crystalline peaks are obtained at 2θ = 23° and 27° indicating the initiation of crystallization phase. This result is in agreement with the critical temperature (75 °C) that has been estimated for $GeTe_2$ thin films by the capacitance-voltage analysis reported earlier [3]. The X-ray diffraction pattern of $GeTe_2$ thin film that have been annealed at 200 °C (Figure 1) confirms the cubic crystal structure with unit cell constant of a = 11.10 Å which agrees well with the standard JCPDS data of a = 11.09 Å [4].

The decrement in band gap with the increase in annealing temperature (Table 1) is due to the increase in grain size. Moreover, with raise in temperature the adatom mobility increases and consequently the crystallization increased as shown in the red shift observed in transmittance spectra that results in decrease in band gap [7]. The Tauc parameter determined for the as-deposited $GeTe_2$ thin film (892 $cm^{-1}eV^{-1}$) is considered as a measure of disorder, which is found to increase with the disorder as evident from Table 1, and agrees with data in literature [8]. With increase in annealing temperature the order gets increase and structural rearrangements taking place, which leads to transition from amorphous to crystalline phases. Tellurium which is twofold coordinated makes $GeTe_2$ a more stoichiometric compound upon annealing which ease structural rearrangement. Germanium and tellurium with 4 and 2 valences, respectively, easily saturate the bond requirement upon rearrangement. This structural rearrangement can be explained by chemical bond energy process [9] that says: (i) The formation of heteropolar bonds favor more likely than the homopolar bonds (ii) depending on the valance requirement the atoms prefer to bond among each other and (iii) the higher bonding energy is preferable. In $GeTe_2$ system, Ge-Ge, Te-Te and Ge-Te bonds are formed depending on the structural changes taking place at the host network. The bond energies of homopolar and heteropolar bonds are as follows: Ge-Ge (157.3 KJ/mol), Te-Te (138.1 KJ/mol) and Ge-Te (157.3 KJ/mol) [9]. Stronger bonds of Ge-Ge and Ge-Te are formed initially and later the weaker bonds of Te-Te are formed in excess as of chalcogenides. These results are also in agreement with Raman spectroscopy.

From the inset of figure 3 the two broad Raman bands around 121 and 141 cm^{-1} for the as-deposited $GeTe_2$ thin film may be due to the availability of amorphous Ge-Te phases that existed in a large-scale atomic distribution. Increasing the annealing temperature from 60 °C to 200 °C, the obtained Raman bands turn out to be slightly stronger and their width turn into narrower. Figure 3 shows the Gaussian fitting for the film

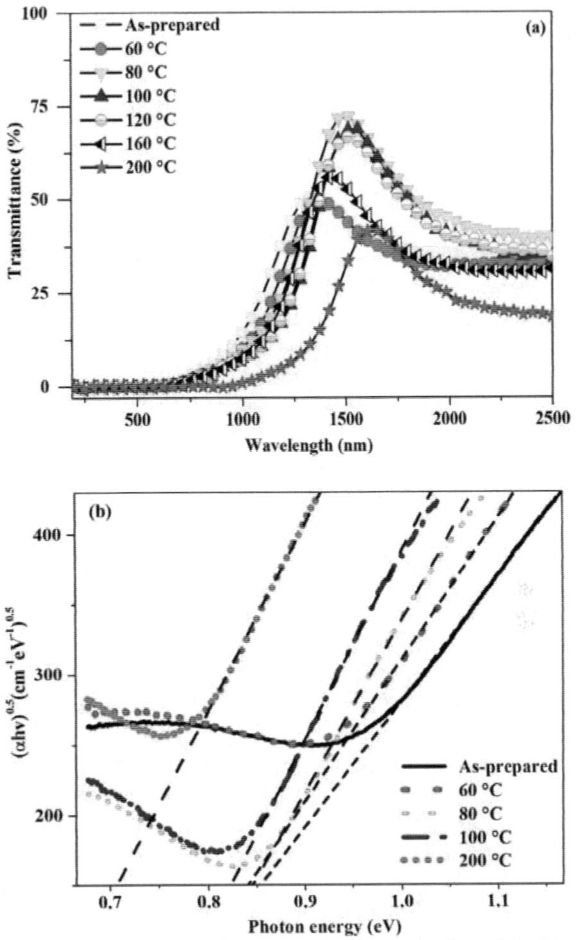

Figure 2 (a) Transmittance and (b) Tauc plot of $GeTe_2$ thin films.

Table 1. Values of D, FWHM, ε, E_g and $B^{1/2}$ of $GeTe_2$ thin films.

$GeTe_2$ Thin film	Grain size D (nm)	FWHM (deg.)	ε x 10^{-3}	E_g eV	$B^{1/2}$ $cm^{-1/2}$ $eV^{-1/2}$
As-prep	--	--	--	0.85	892
60 °C	18	0.4723	1.89	0.84	963
80 °C	29	0.2931	1.17	0.84	1009
100 °C	31	0.2755	1.11	0.82	1113
120 °C	34	0.2513	1.0	0.81	1221
160 °C	36	0.2362	0.947	0.81	1330
200 °C	44	0.196	0.786	0.70	1340

annealed at 200 °C which has the crystalline nature. The Raman band obtained at 121 cm^{-1} is assigned to A_1 mode of $GeTe_4$ tetrahedral unit [10]. Moreover, the Raman band obtained at 141 cm^{-1} is assigned to E mode in crystalline Te [11]. From figure 3, the shoulder obtained around 151 cm^{-1} is due to the symmetric stretching vibration of homopolar bonds of Te-Te [12]. The appearance of the shoulder enables the least presence of homopolar bonds of Te-Te. Normally for amorphous Te the Te-Te bonds appear around 151 cm^{-1}, while in crystalline nature the bonds are formed around 121 cm^{-1}.

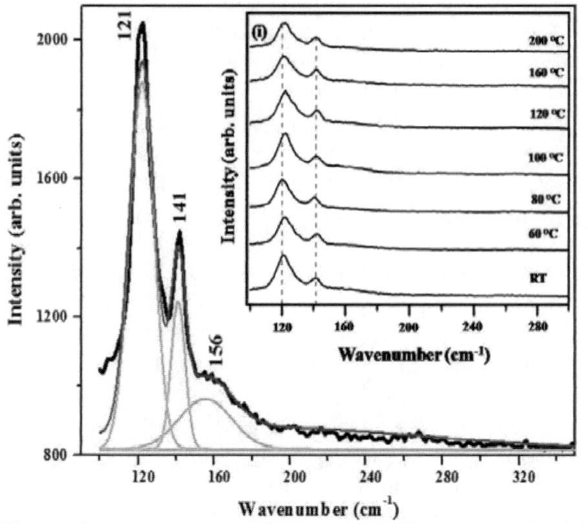

Figure 3. Raman spectra for annealed film at 200 °C with Gaussian fitting. Inset (i) shows Raman spectra of annealed films at different temperatures.

However, in respect to the presence of disorder, this homopolar Te-Te bonds appear at higher wavenumber. However, the formation of Ge-Ge is not noticed in the present study due to the chalcogen rich compound. The preferable formation of heteropolar over homopolar bonds was enhanced by the increase in the annealing temperature [12]. By chemical bond approach (CBA) process the Te becomes as 3-fold coordinated. Thus by increasing the annealing temperature from 60 °C to 200 °C, the formation of crystalline nature is enhanced and as a result Te rich films remain unsatisfied with homopoloar Te-Te bonds as evidenced by the shoulder around 150 cm^{-1}.

5. Conclusion

Structural disorder has been investigated by temperature enhancement in $GeTe_2$ thin films prepared by vacuum evaporation technique. The observed red shift by optical parameter coincides well with the increase in crystallization by XRD analysis. Moreover the order of the films get enhanced which is elucidated by Tauc parameter and Raman spectra. Twofold coordinated tellurium ease structural rearrangement upon annealing and is being correlated with the chemical bond energy process. The presence of homopolar and heteropolar bonds are being discussed by the structural rearrangement upon annealing and its presence is confirmed from the Raman analysis.

Acknowledgements

This work was supported by the UAEU Program for Advanced Research under grant No. 31S112-UPAR (5) 2013.

References

[1] Pumlianmunga and Ramesh K 2017 *Mater. Res. Bull.* **86** 88
[2] Cody G D, Tiedje T, Abeles B, Brooks B and Goldstein Y 1981 *Phys. Rev. Lett.* **47** 1480
[3] Ananth Kumar R T, Saleh T. Mahmoud, Khadija Said, Pathinettam Padiyan D and Qamhieh N 2017 *J. Alloys Compd.* **694** 163, JCPDS Card No. 33-0585
[4] Hassanien A S, Aly K A and Akl A A 2016 *J. Alloys Compd.* **685** 733
[5] Kale R B and Lokhande C D 2004 *Mater. Res. Bull.* **39 (12)** 1829
[6] Dheepa J, Sathyamoorthy R and Subbarayan A 2005 *J. Cryst. Growth* **274** 100
[7] Ananth Kumar R T, Chithra Lekha P, Sundarakannan B and Pathinettam Padiyan D 2012 *Phil. Mag.* **92** 1422
[8] Bicerano J and Ovshinsky S R 1985 *J. Non Cryst. Solids* **74** 75
[9] Tsunetomo K, Sugishima T, Imura T and Osaka Y 1987 *J. Non Cryst. Solids* **95** 509
[10] Torrie B H 1970 *Solid State Commun.* **8** 1899
[11] Sharma P and Katyal S C 2008 *J. Non Cryst. Solids* **354** 3836

IOP Conf. Series: Journal of Physics: Conf. Series **869** (2017) 012019 doi:10.1088/1742-6596/869/1/012019

Charge Effect on Current-Voltage Characteristics of MoS_2 Nanoribbon

B Dianat, N Taghizade and M Afshar

Materials Simulation Laboratory, Department of Physics, Iran University of Science and Technology, Tehran, Iran

E-mail: behnood.dianat@gmail.com

Abstract. Charge effect on I-V characteristics of single layer zigzag nanoribbon MoS_2 (Z-NR-MoS2) have been studied using first-principles quantum transport calculations. We show that the internal electric polarization is strongly dependent on dominating edge atoms; hence, two structures with different edge atoms are investigated. Our calculations show that due to this internal polarization a large negative differential resistance (NDR) has been observed.

1. Introduction

In recent years, low dimensional nanomaterials have drawn many attentions due to their physical, mechanical and chemical properties. Among them graphene is one of the most investigated one with its diverse applications in electronics[1], but the existence of an indirect band-gap in pristine graphene turned focuses on alternative materials such as boron nitride, tungsten disulfide, molybdenum disulfide (MoS_2), and others.

Molybdenum disulfide, a transition metal dichalcogenide, consists of three layers which are bounded by weak Van der Waals forces, though covalently interacting internally. In bulk form, MoS_2 is a semiconductor with an indirect band gap of 1.2 eV, however, there is a transition to 1.9 eV direct band gap in monolayer form. There have been reports about some unusual electronic and magnetic properties: the zigzag nanoribbon was ferromagnetic and metallic, while the armchair nanoribbon was nonmagnetic and semiconducting with a width-dependent band gap. Both experiment and theory show that the zigzag MoS_2 nanoribbons are generally more stable than the armchair MoS_2. Absorption and defects may also affect their physical properties [2].

An important requirement in material applications is the control of electronic and magnet properties by means of an external force in a given material. Therefore, effects of an external field on the electronic structures of graphene nanoribbon has been studied [3]. This result also can be applied to the modulation of MoS_2 nanoribbon. It has been shown that the electronic properties of zigzag MoS_2 nanoribbon are also sensitive to electric field [4], which stem from the energy level shifts induced by an internal electric polarization and the competing covalent/ionic interactions.

In present work, we report a first principles study of transmission properties of Zigzag Molybdenum disulfide nanoribbon passivated by Hydrogen (Z-MoS2-NR-H). Our results on electronic properties of the compound show promising functionalities for future FETs.

2. Computational method

First-principles calculations based on the Density functional theory (DFT) have been performed using SIESTA/TRANSIESTA package [5] with pseudopotentials implemented in the SIESTA code.

IOP Conf. Series: Journal of Physics: Conf. Series **869** (2017) 012019 doi:10.1088/1742-6596/869/1/012019

The generalized gradient approximation (GGA) in the form of the Perdew-Burke-Ernzerhof (PBE) exchange-correlation functional is implemented. 1 × 1 × 12 Monkhorst-Pack k-point grids are used in the structural optimizations and energy calculation. A mesh-cutoff of 300 Ry is chosen and all atoms are relaxed to a force of <0.02 eV/A°. To isolate the nanoribbons a vacuum of 4 nm and 5 nm is chosen in Z and Y directions respectively. Results are consistent with previous works on MoS_2 nanoribbon, hence ensuring their validity [6].

Unit cell is extracted from a sheet of MoS_2 nanoribbon then the structure is made by repeating it in Z direction, both sides are passivated by hydrogen to eliminate dangling bonds. As shown in figure 1 dashed lines show electrodes.

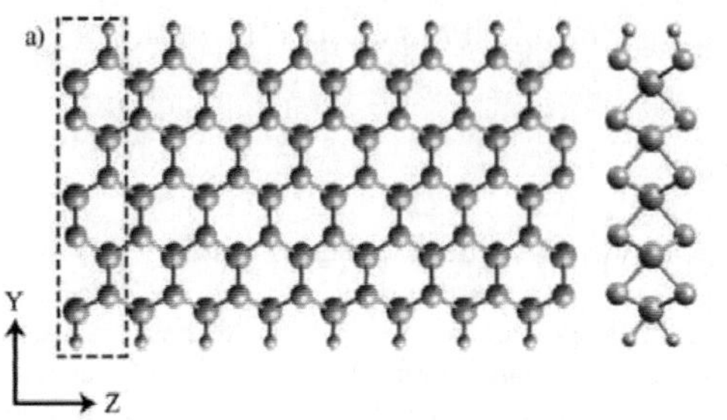

Figure 1. Two structures with different edges, (a) Zigzag MoS_2 nanoribbon with one edge dominated by Mo and the other dominated by S (Z-MoS2-SM), (b) a Zigzag MoS_2 nanoribbon with both edges dominated by S (Z-MoS2-SS)

3. Results

Previous studies on Zigzag MoS_2 nanoribbon showed that it has an internal electric dipole due to the asymmetrical edges [4]. In order to investigate the effects of internal electric dipole we used two types of structures, first a Zigzag MoS_2 nanoribbon with one edge dominated by Mo and the other dominated by S (Z-MoS2-SM), second structure is a Zigzag MoS_2 nanoribbon with both edges dominated by S (Z-MoS2-SS).

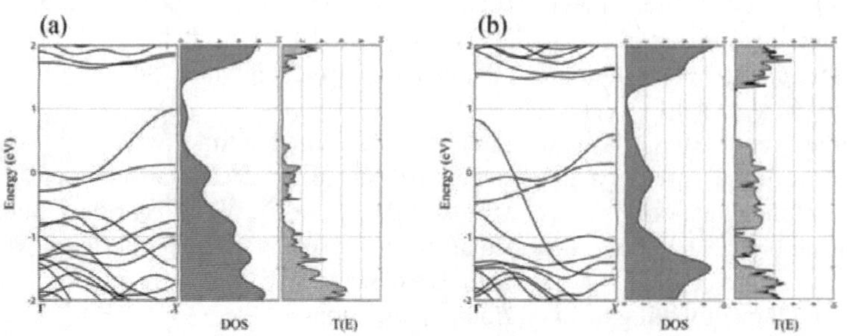

Figure 2. Band structure, Density of states and Transmission spectrum for (a) (Z-MoS2-SM) and (b) (Z-MoS2-SS).

8 rows of atoms are used for scattering region, any more than that had little effect on transmission spectrum. Band structure calculations are done using a unit cell with boundary conditions in Z direction.

Figure 2 shows the band structure of a metal with two bands crossing the Fermi level (E_f) which is consistent with density of states (DOS) for both structures. We also calculated transmission spectrum of the system at 0 Voltage, near E_f there are two transmission channels (two bands crossing E_f), it indicates that these channels contribute most to the current.

IOP Conf. Series: Journal of Physics: Conf. Series **869** (2017) 012019 doi:10.1088/1742-6596/869/1/012019

As can be seen from figure 3, by increasing Voltage, most contributing channels stay near Fermi level while the higher level channels move away from it.

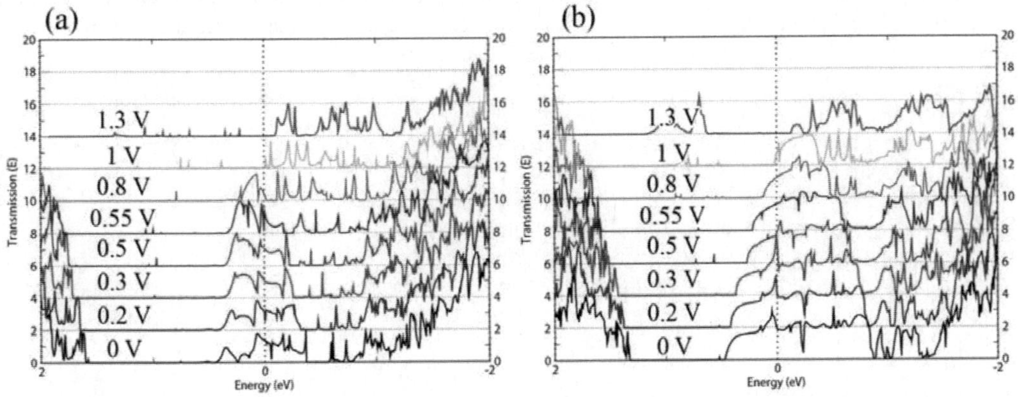

Figure 3. Transmission spectrum for (a) (Z-MoS2-SM) and (b) (Z-MoS2-SS), colors indicate different Bias Voltages ranging from 0 to 1.3 V.

In this figure 3, it is observed that the transmission channel near E_f vary dramatically after 0.5V, as for the other structure, It's transmission near E_f decreases gradually with increasing voltage (figure 3b). As a result of these variations one observes negative differential resistance (NDR) effect around 0.5 V in the I-V characteristics (figure 4). , We believe, observed NDR is caused by an internal electric dipole which strongly depends on edge atoms. This can be evidence from comparison of internal electric dipoles: internal electric dipole for Z-MoS2-SM is almost 7 times stronger than that of Z-MoS2-SS. Notice that for Z-MoS2-SM an increase in the current is observable after 1V which suggests the saturation of transmission channels, hence showing ohmic behavior in the diagram. Furthermore, I-V characteristics revealed larger number of transmission channels for Z-MoS2-SS which in return means larger conductivity in this configuration.

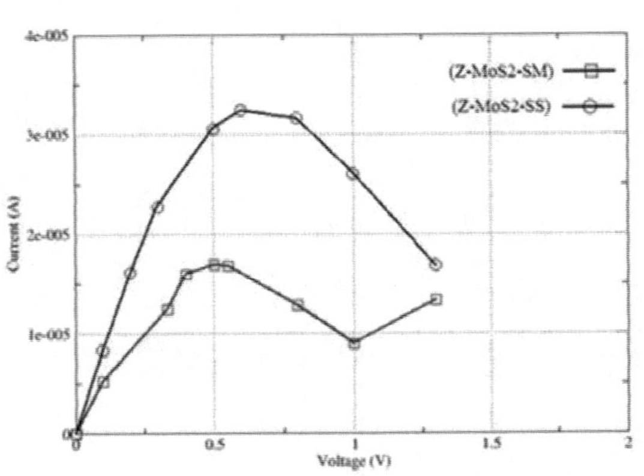

Figure 4. Current-Voltage diagram for (Z-MoS2-SM) and (Z-MoS2-SS).

4. Conclusion

Charge effect on current-voltage characteristics of MoS_2 were investigated using TRANSIESTA package, findings suggest that Internal electric dipole created by edge atoms, plays an important role in electron transmission and hence its overall conductivity. A negative differential resistance effect has been observed. The origin of this effect has been discussed.

References

[1] Chowdhury R *et al* 2011 *Phys. Rev.* B **83**(4) 045401.
[2] Erdogan E *et al* 2012 *Eur. J. Phys.*B **85**(1): 1-4.
[3] Kan E J *et al* 2007 *Appl. Phys. Lett* **91**(24): 243116.
[4] Kou L *et al* 2012 *J. Phys. Chem. Lett.* **3**(20): 2934-2941.
[5] Soler J M *et al* 2002 *J. Phys. Condens. Matter* **14**(11): 2745.
[6] Peng L *et al* 2014 *J. Appl. Phys.* **115**(22): 223705.

IOP Conf. Series: Journal of Physics: Conf. Series **869** (2017) 012020 doi:10.1088/1742-6596/869/1/012020

Generation of spin waves by fs-laser pulses in transparent magnetic films: role of the laser beam diameter

A I Chernov[1,2], M A Kozhaev[1,2], A K Zvezdin[1,2] and V I Belotelov[1,3]

[1]Russian Quantum Center, 100 Novaya str., 143025, Skolkovo, Moscow Region, Russia
[2] Prokhorov General Physics Institute, RAS, 38 Vavilov str., 119991, Moscow, Russia
[3] Faculty of Physics, Lomonosov Moscow State University, Leninskie Gory, 119234, Moscow, Russia
E-mail: v.belotelov@rqc.ru

Abstract. Magnetostatic spin waves (MSWs) in magnetic dielectrics are perspective candidates for various applications including data processing, spin wave logic and quantum information. They can propagate hundreds of microns, some have nonreciprocal character and can be localized near the film interfaces. Excitation of MSWs by fs-laser pulses is beneficial since it provides local excitation and tunability. We demonstrate a method for the optical pumping of MSWs due to the inverse Faraday effect in magnetic dielectrics. Its essence is in the excitation with a nonuniform distribution of the optical power through the film depth and in optimization of the pump beam size. Propagation of the spin waves is demonstrated by detecting the magnetization dynamics tens microns away from the excitation spot. Variation of the pump beam diameter allows tuning between different types of the excited spin waves.

1. Introduction

Magnonics deals with manipulation and control of spin waves that are very promising for information and image processing since spin wave amplitude and phase can be manipulated with relatively low powers. Magnetostatic surface spin waves (MSSWs) demonstrate nonreciprocal propagation [1, 2] and therefore are of prime importance for various applications [1, 3, 4]. Nonreciprocity of the MSSWs appears in relatively thick films so that spin waves have pronounced localization at the interface. A kind of asymmetry between both surfaces of a magnetic film should be present. In the case of magnetic dielectrics the nonreciprocity effect is manifested if one of the film surfaces is metallized. The metallic cover suppresses magnetic field of the MSSW localized at the metal/dielectric interface and thus influences its properties. In some cases, the nonreciprocal behavior is due to the excitation method [2]. Initially optical excitation and control over the magnetostatic spin waves has been reported for backward volume magnetostatic spin wave (BVMSW) patterns [5]. Possibility to optically excite different types of spin waves [6] is crucial for specific applications.

Here we demonstrate a method for the optical generation of the MSWs in magnetic dielectric films and tuning between different MSW types. A method is based on creation of a nonuniform distribution of the circularly polarized optical field through the film depth and on optimization of the pump beam size. In contrast to the spin waves conventionally excited by a microwave field, laser pulses allow local generation of MSWs.

2. Materials and methods

Epitaxial films of various iron garnet compounds have been studied. The average thickness of the films was 4 μm. However magnetostatic spin waves were also registered in the samples with the thickness below 1 μm. The magnetostatic spin waves were excited and detected in the two-color pump-probe experiment using optical pumping by femtosecond laser pulses with duration of 200 fs (Newport Mai Tai HP Ti:Sapphire laser, 80.68 MHz repetition rate combined with Spectra-Physics Inspire Auto 100 optical parametric oscillator) at wavelength of λ = 545 nm (2.28 eV). Variations in the out-of-plane magnetization component caused by the excited magnetization precession in an external magnetic field were detected by the Faraday rotation of the linearly polarized probe beam at λ = 821 nm (1.51eV) that illuminated the sample at some time delay with respect to the pump beam. Delay line is based on the retroreflector mounted on the linear stage (Newport IMS-500). The time delay between pump and probe pulses was varied from -0.5 to 2.6 ns, where zero time delay corresponds to the simultaneous propagation of the pump and probe pulses through the sample. The bias magnetic field of 100 mT was applied along x-axis (The sample surface is in x-y plane). The probe beam can be spatially separated from the pump beam in order to study the spin wave propagation. The Faraday rotation, θ, was measured with a balanced diode detector using a lock-in technique at a reference frequency of 46 kHz.

3. Results

Figure 1 demonstrates the artistic representation of the experimental technique that was used to study the propagating MSWs. Optical excitation can result in generation of different types of MSWs. Depending on the bias magnetic field direction several types and combinations of

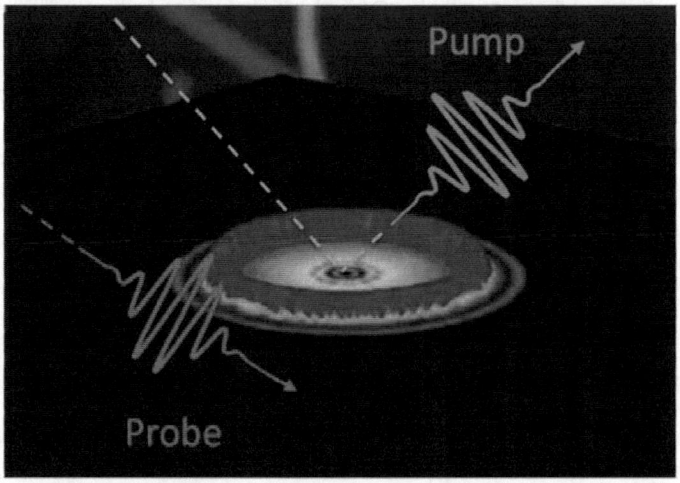

Figure 1. Artistic representation of the spin wave propagation in the iron garnet film induced by the excitation with the pump laser beam (circular polarization). The probing is performed in spatially different location with another laser beam.

MSWs can be excited, i.e. BVMSW, MSSW, and forward volume magnetostatic spin waves (FVMSW). In our experiment the bias magnetic field was applied within the sample surface, which rendered only two possible MSW types: BVMSWs and MSSWs. Time-resolved Faraday rotation signals (Figure 2 a, b) demonstrate the evolution of the magnetization precession depending on the distance between pump and probe spots. The shift of the probe beam results in decrease of the signal intensity, however, the magnetization precession can be detected on the distances larger than 25 μm. One can notice the beating behavior from the curves shape, which is due to the existence of several spectral components. Frequencies of the spectral components depend on the pump beam diameter (Figure 3). Increase of the pump diameter diminishes frequency of the high-frequency mode and increases frequency of the low-frequency mode. Along with that, adjustment of the optical power distribution across the sample volume allows modifying the ratio between the excited BVMSWs and MSSWs. Variation of the pump beam diameter allows gaining additional control and results in tuning possibility between two types of MSWs in our experimental configuration. The propagation pattern of spin waves depends on the direction of the bias field. Moving the bias magnetic field out of the sample surface can result in the excitation of other wave types combination: FVMSWs and BVMSWs.

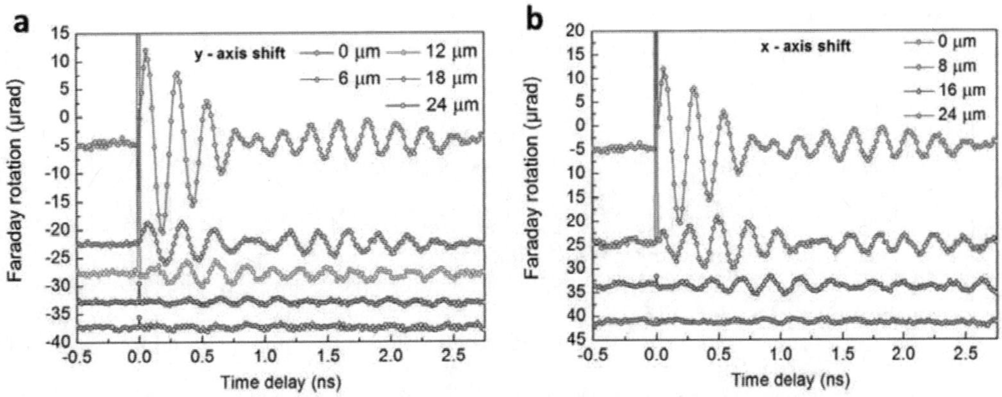

Figure 2. Time-resolved Faraday rotation signals produced by optically excited spin waves in the sample of rare-earth iron garnet with Bi ions and observed at different distances between pump and probe spots: 0, 6, 8, 12, 16, 18 and 24 µm. The pump beam is shifted with respect to the probe along y-axis, i.e. perpendicular to the bias magnetic field **(a)**, or along x-axis, i.e. parallel to the bias magnetic field **(b)**. The composition of the epitaxial film is $Bi_{0.8}Lu_{2.2}Fe_5O_{12}$.

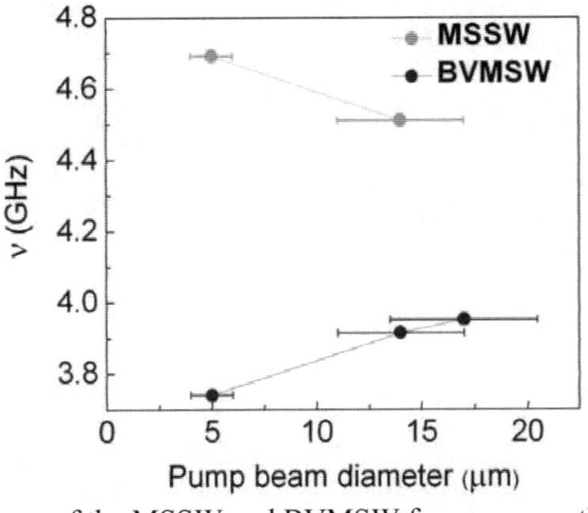

Figure 3. Dependence of the MSSW and BVMSW frequency on the pump beam diameter.

4. Conclusions

In this work we demonstrate local optical generation of the MSWs in magnetic dielectrics, i.e. in thin epitaxial films of rare earth iron garnet, by illuminating the sample with circularly polarized laser pulses at the wavelengths corresponding to moderate optical absorption. Tuning between several types of MSWs can be performed via changing the pump beam diameter. The optical losses provide asymmetric in depth distribution of the optical energy that is a key factor in the MSWs generation. Optical impact on the magnetization state in our experiments is due to the inverse Faraday effect, which may nonlinearly depend on the pump fluence [7]. Optical pumping provides rather broad tunability of the spin waves in frequencies and relative amplitude. Thus, variation of the pump beam diameter by several times allows to excite MSSWs together with BVMSWs or to switch off one of the contributions. As a result, it provides different space distribution of the magnetization dynamics inside the magnetic film.

Selective excitation of the exact MSW type can be performed, including MSSWs. The nonreciprocal character of the MSSWs makes such spin waves valuable in respect to their applications for logic devices, data processing and storage, quantum calculations. Optical excitation of MSWs provides the way to locally probe the magnetic properties of thin and layered materials [8].

Acknowledgements
This work is supported by Russian President Grant (Project No. MD-1615.2017.2).

References

[1] Jamali M, Kwon J H, Seo S-M, Lee K-J and Yang H 2013 *Scientific Reports* **3** 3160
[2] Schneider T, Serga A A, Neumann T, Hillebrands B and Kostylev M P 2008 *Phys. Rev.* B **77** 214411
[3] Lisenkov I, Kalyabin D, Osokin S, Klos J W , Krawczyk M and Nikitov S 2015 *Journal of Magnetism and Magnetic Materials* **378** 313-319
[4] Mruczkiewicz M, Pavlov E S, Vysotsky S L, Krawczyk M, Filimonov Yu A and Nikitov S A 2014 *Phys. Rev.* B **90** 174416
[5] Satoh T, Terui Y, Moriya R, Ivanov B A, Ando K, Saitoh E, Shimura T and Kuroda K 2012 *Nat Photon* **6** 662-666
[6] Chernov A I, Kozhaev M A, Savochkin I V, Dodonov D V, Vetoshko P M, Zvezdin A K and Belotelov V I 2017 *Optics Letters* **42** 279-282
[7] Kozhaev M A, Chernov A I, Savochkin I V, Kuz'michev A N, Zvezdin A K and Belotelov V I 2016 *JETP Letters* **104** 851–855
[8] Chernov A I, Kozhaev M A, Vetoshko P M, Dodonov D V, Prokopov A R, Shumilov A G, Shaposhnikov A N, Berzhanskii V N, Zvezdin A K and Belotelov V I 2016 *Physics of the Solid State* **58** 1128-1134

Physical properties of ZnO nanoparticles doped with Mn and Fe

D Moussa[1], D El-Said Bakeer[2], R Awad[1], A M Abdel- Gaber[3]

[1]Physics Department, Faculty of Science, Beirut Arab University, Lebanon
[2]Physics Department, Faculty of Science, Damanhour University, Damanhour, Egypt
[3]Chemistry Department, Faculty of Science, Beirut Arab University, Lebanon
E-mail address: dam684@student.bau.com

Abstract. This study reports the effect of magnetic ions doping on the structural, optical and magnetic properties of ZnO nanoparticles. Samples of $Zn_{0.97-x}Mn_{0.03}Fe_xO$, $0.0 \leq x \leq 0.1$ wt. %, nanoparticles are synthesized by Co-Precipitation route. The synthesized samples are characterized by X-ray diffraction analysis (XRD), Transmission electron microscopy (TEM), Fourier transform infrared (FTIR) spectrometer and UV spectroscopy. XRD results reveal the hexagonal wurtzite structure of ZnO with the formation of secondary phase in the Fe-doped samples corresponding to Zinc Iron Oxide. The crystallite size is calculated by X-ray peak broadening using Debye- Scherer's formula, and the results are in good agreement with TEM. The FTIR reveals two major peaks that are shifted toward a lower value when the concentration of Fe ions becomes higher than that of Mn ions. The UV results indicate a shift in the band-gap energy toward lower value upon increasing Fe-content. The study of magnetization hysteresis loop measurements infers that the samples of $Zn_{0.97-x}Mn_{0.03}Fe_xO$ show a well-defined hysteresis loop, reflecting the paramagnetic behavior.
Keywords: Diluted magnetic semiconductor, FTIR, UV, M-H loop.

1. Introduction

zinc oxide (ZnO) is considered to be one of the most important multifunctional semiconductor material for technological applications such as light emitting diodes, gas sensors, drug delivery and transparent field effect transistors due to the wide band gap (3.37 eV), large excitation binding energy (60 meV) at room temperature and high transmission coefficient in the visible and near infrared spectral range [1, 2]. Fe doped ZnO has been synthesized for optical properties [3] since Fe is an optical emitter in doped ZnO. (Cr)- (Fe) co-doped ZnO was an efficient doping element to improve the structural and optical properties of ZnO nanomaterials [4]. In the present work, we study the effect of magnetic ions (Mn and Fe) doping on the structural, optical and magnetic properties of ZnO nanoparticles.

2. Experimental Techniques

ZnO and $Zn_{0.97-x}Mn_{0.03}Fe_xO$ nanoparticles are prepared using the Co-Precipitation method using $FeCl_3 \cdot 6H_2O$, $MnCl_2 \cdot 4H_2O$ and $ZnCl_2$ as the raw materials. The required metal salts were mixed and dissolved in distilled water depending on the percentage of Fe and Mn doping to form aqueous solutions. The pH value of the solution was controlled by adding 4M sodium hydroxide solution under constant magnetic stirring to obtain a pH= 13. Samples were then heated at 60°C for 2 hours, washed and filtered several times with distilled water to ensure that excess reactants and impurity ions have been removed from the product. Then, they were dried in a box furnace at 100^0C for 24 hours. After complete dehydration, the obtained samples were grinded and calcinated at 500^0C for 4 hrs. XRD analysis for the prepared samples are performed, using Bruker D8 advance powder diffractometer with CuK_α radiation ($\lambda = 1.54056$ Å) in the range $20^o \leq 2\ \theta \leq 70^o$. In order to examine the morphology of the samples, we used Jeol transmission electron microscope (TEM) JEM-100CX, operated at 80kV. Functional groups are studied using the FTIR-8400S. The UV-visible optical spectroscopy, between 300nm and 800nm, are measured by V-670 NIR Spectrophotometer. The room temperature magnetic hysteresis loops are collected using vibrating sample magnetometers (VSM) Lakeshore 7410.

3. Results and discussions

Figure1 shows the XRD patterns of ZnO and $Zn_{0.97-x}Mn_{0.03}Fe_xO$, with $x = 0.0$, 0.02, 0.04, 0.08 and 0.1. All samples show hexagonal wurtzite crystal structure. The sharp and narrow peaks illustrate that $Zn_{0.97-x}Mn_{0.03}Fe_xO$ nanoparticles are with high crystalline quality and purity. Similar results were obtained by *Mote et al* [5] for Mn doped ZnO samples prepared by sol- gel route and by Xu [6] for Fe doped ZnO. Upon increasing x, doped samples reveal the formation of secondary phase Zinc Iron Oxide [7] of cubic spinal structure due to coexistence of Fe^{3+} ions with Fe^{2+} ions in our samples. The lattice parameters, calculated, using the d-spacing equation [8], are listed in table1 and found to be a=0.325 nm and c=0.520 nm, which are similar to those of bulk ZnO [9]. The lattice parameters go on increasing upon doping Mn because of higher ionic radius of Mn^{2+} ions in comparison to the Zn^{2+} ions, and decrease with increasing Fe content due to existence of Fe^{3+}, with ionic radius greater than Zn^{2+}, instead of Fe^{2+}. The average crystallite sizes of the samples are calculated from XRD using Debye- Scherer's formula [10] and are also listed in table1. Crystallite sizes increase from 40 nm to 62 nm upon doping ZnO with Mn, and decrease with increasing Fe content to 34 nm.

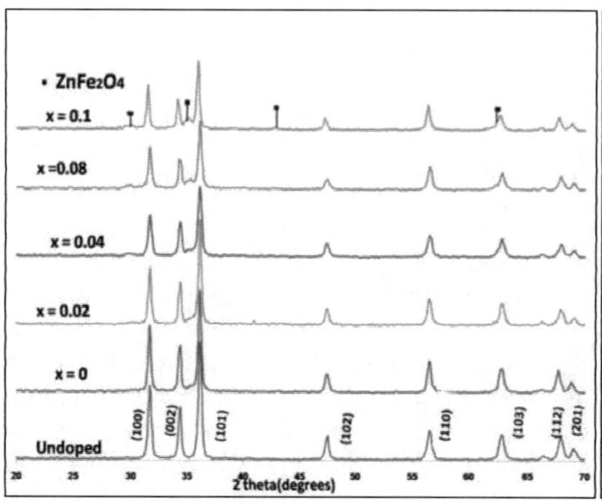

Figure 1.The XRD spectra of undoped ZnO and $Zn_{0.97-x}Mn_{0.03}Fe_xO$ samples.

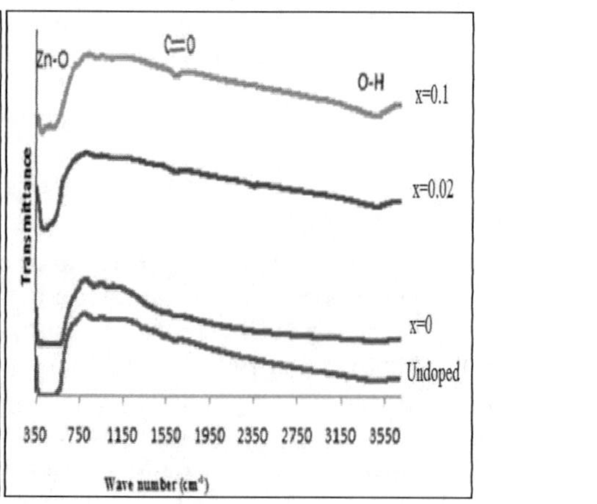

Figure 2.FTIR of undoped ZnO and $Zn_{0.97x}Mn_{0.03}Fe_xO$, with x=0.0, 0.02 and 0.1.

The TEM images of undoped ZnO and $Zn_{0.97-x}Mn_{0.03}Fe_xO$, $x = 0.0$, 0.02 and 0.1 are shown in figures 3a-3d.The figures show that the shape of the particles is spherical for all samples. The average grain size of the particles is measured and listed in Table1. It is observed that the average grain size decreases with increasing Fe content. The result is consistent with those obtained by XRD measurements. The FTIR spectra of the prepared samples with different Fe content are shown in figure 2. The broad absorption peak at 3467.56 cm^{-1} represents the stretching vibration of the O-H group of water molecule whereas the second at 400.3 cm^{-1} corresponds to Zn-O stretching vibration [11] which is Shifted to lower value as a result of Fe doping suggesting that micro structural changes taking place in the ZnO matrix. This can prove that the Zn–O–Zn network was perturbed by the presence of Fe. The slight shift in the Zn-O stretching peak might be due to the change in the parameters and the bond properties of Zn affected by doping.

Table1. Lattice parameters (*a* and *c*), Crystallite size using Sherrer equation (max peak), TEM and optical band gap energy of $Zn_{0.97-x}Mn_{0.03}Fe_xO$ nanoparticles.

Sample	Lattice parameters(Å)		Crystallite size (nm)		UV
	a	c	Sherrer equation	TEM	Optical band energy (eV)
Undoped ZnO	3.248	5.204	40.7	41.2	3.1
x=0.00	3.253	5.211	62.0	63.1	3.08
x=0.01	3.253	5.208	40.9	42.3	3.08
x=0.02	3.251	5.204	40.3	41.5	3.04
x=0.04	3.250	5.207	36.1	38.0	3.04
x=0.08	3.250	5.205	33.9	35.1	2.97
x=0.10	3.247	5.206	34.5	34.7	2.97

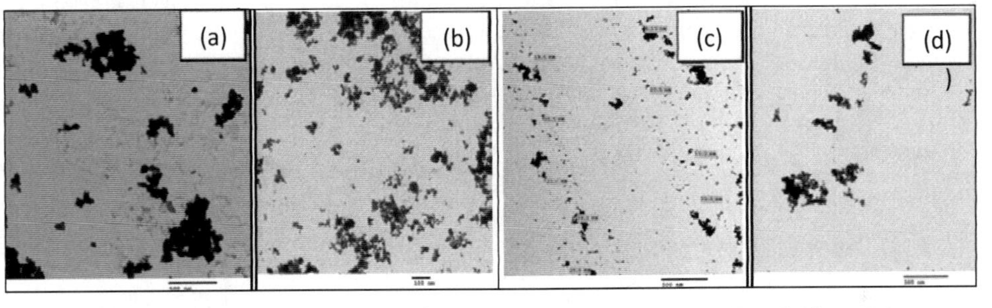

Figure 3.TEM images of (a) undoped ZnO, (b) x=0, (c) x=0.02 and (d) x= 0.1.

Optical study of ZnO and $Zn_{0.97-x}Mn_{0.03}Fe_xO$, x =0.0, 0.02 and 0.1 are shown in figure 4. The absorption peak of the doped ZnO samples showed a blue shift towards a smaller wavelength. This shift might be due to surface effects, change in the crystallite size, morphology or an indication that the electronic structure of undoped ZnO is altered by Mn and Fe doping[18]. The optical band gap energy of prepared samples, for direct band gap, is calculated by Tauc relation [12]. The obtained optical band gap energies are listed in table 1. As the Fe content increases optical band gap energies decrease. These exchange interactions arise between band electrons of ZnO and the localized d electron of Fe atoms when Fe ions substituted some of the Zn ions in ZnO. The decrease in optical band gap energy could be due to the increase of density of localized state in the conduction band. Magnetic hysteresis loops for $Zn_{0.97-x}Mn_{0.03}Fe_xO$, x =0.0, 0.02 and 0.1, at room temperature with the maximum applied field of 20 kOe, are shown in figure 5. Room temperature paramagnetic is observed for all the samples, we did not detect any evidence of a ferromagnetic response. Similar paramagnetic behavior was observed for ZnMnO nanoparticles [13]. *Wójcik* et al [14] demonstrated that highly uniform ZnMnO films are paramagnetic. The absence of ferromagnetism in Mn-doped ZnO nanoparticles can be explained with theoretical ab initio calculations [15]. Mn 3d states in the ZnO lattice, with a cubic crystal-field, split into an e doublet and a t_2 triplet. e Levels have a lower energy than t_2 levels. t_2 Levels are fully occupied leading to the absence of a hole for the coupling. It can be ferromagnetic if a double exchange via the localized holes is predominant. The paramagnetic contribution in Fe doped samples may be due to disordered Fe^{3+} spins on the nanoparticle surface. From figure 5, it can be observed that the value of the magnetization increases with increasing Fe content due to the fact that individual Fe ions act as centers of magnetization.

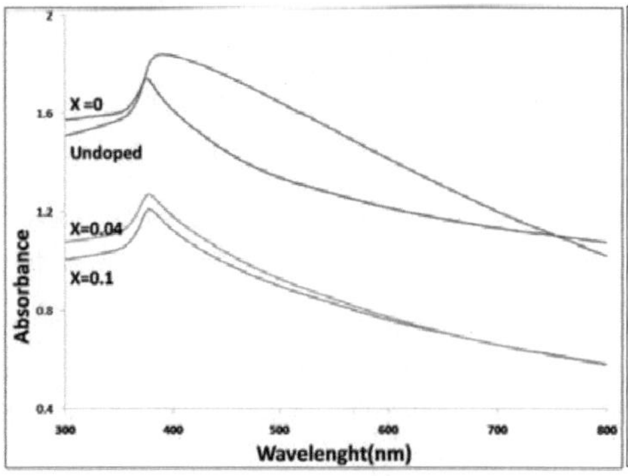

Figure 4.The UV spectra of undoped ZnO and $Zn_{0.97-x}Mn_{0.03}Fe_xO$, x=0.0, 0.02 and 0.1.

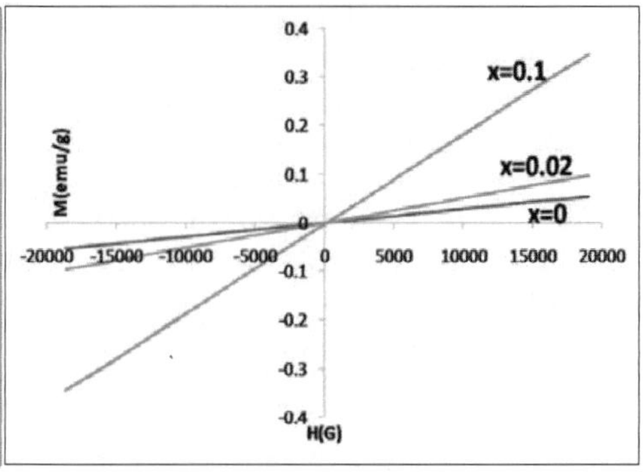

Figure 5.(M–H) curves of $Zn_{0.97-x}Mn_{0.03}Fe_xO$ nanoparticles.

4. Conclusion

In this work, ZnO and $Zn_{0.97-x}Mn_{0.03}Fe_xO$ nanoparticles, of nominal composition x= 0.00, 0.02, 0.04, 0.08 and 0.10 have been successfully synthesized using Co-Precipitation technique. Samples characterization showed that co-doping did not influenced the wurtzite hexagonal structure of ZnO, while it varied the lattice parameters a and c. Moreover, crystalline size, determined from both XRD and TEM, decreased with increasing Fe content. UV analysis indicated that the absorption peaks are shifted to smaller wavelength and Eg decreased as Fe contents increase. The M-H hysteresis loop revealed the absence of ferromagnetic behavior.

References

[1] Qian L, Liu Z, Mo Y, Yuan H and Xiao D 2013 *Mater. Lett.* **100**, 124–126.
[2] Wang Z L *ACS Nano* 2008 **2** 1987–92.
[3] Rattana T, Suwanboon S, Amornpitoksuk P, Haidoux A, and Limsuwan P *Journal of Alloys and Compounds* 2009 **480** 603–607.
[4] Prakash C, Anurag G and Ashavani K 2014 *International Journal of Chemical, Molecular, Nuclear, Materials and Metallurgical Engineering* **8** 1321–1324.
[5] MoteV D, Dargad J S and Dole B N 2013 *Nanoscience and Nanoengineering* **1** 116–122.
[6] Xu L and Li X 2010 *J Cryst Growth* **312** 851–855.
[7] Mandal S K, Das A K, Nath T K and Debjani K 2006 *Appl. Phys. Letts.* **89** 144105; Vaseem M, Umar A and HahnY B 2010 *Metal Oxide Nanostructures and Their Applications* **5** 1.
[8] Reeber R R 1970 *Journal of Applied Physics* **41** 5063.
[9] Metin H and Esen R 2003 *Semicond.Sci.Technol.* **18** 647.
[10] Bhatti K P, Chaudhary S, Pandya D K and Kashyap S C 2005 *Solid state commun.* **136** 384–388.
[11] Azam A, Ahmed A S, Habib S S and Naqvi A H 2012 *Journal of Alloys and Compounds* **83** 523.
[12] Tomaszewska-Grzęda A, Opalińska A, Grzanka E, Łojkowski W, Gedanken A, Godlewski M, Yatsunenko S, Osinniy V and Story 2006 *Appl. Phys. Lett.* **89** 242102.
[13] Pankove J N 1971 Optical Processes in Semiconductors, Dover, New York.
[14] Wójcik A, Kopalko K, Godlewski M, Guziewicz E, Jakieła R, Minikayev R and Paszkowicz W 2006 *Appl. Phys. Lett.* **89** 51907.
[15] Amit D L, Rajesh C, Khan M, Ali N, Gopalakrishnan I K, Kulshreshtha S K and Mahamuni S 2007 J. *Appl. Phys.* **101** 103906–103910.

IOP Conf. Series: Journal of Physics: Conf. Series **869** (2017) 012022 doi:10.1088/1742-6596/869/1/012022

Impurity effect on entanglement asymptotic state in one-dimensional Ising system coupled to a dissipative environment

G Sadiek[1,2]

[1]Department of Applied Physics and Astronomy, University of Sharjah, Sharjah 27272, UAE
[2]Physics Department, Ain Shams University, Cairo 11566, Egypt

E-mail: gsadiek@sharjah.ac.ae

Abstract. We consider a finite one-dimensional Ising spin chain under the influence of a dissipative Lindblad environment obeying the Born-Markovian constrain in presence of an external magnetic field with open boundary conditions. We study the effect of a single impurity, located at the terminal or center of the chain, on the time evolution and asymptotic steady state of the bipartite entanglement in the chain starting from a maximally entangled initial state. We found that the impurity has a significant effect on the bipartite entanglement of its nearest spins and can be used to tune their steady state value but has almost no noticeable impact on the far ones. At finite temperature, the thermal excitations suppress the dynamics of the system and reduce the value of the steady state and may completely wipe it out as the temperature is increased, which eliminates the effect of the impurity in that case.

1. Introduction

Quantum entanglement represents the main barrier between the quantum and classical worlds with no classical analog [1]. It is considered as a crucial resource in quantum teleportation, cryptography, and quantum computation and many other fields of interest [2]. Deep analysis and good quantification of entanglement are needed to explain the different behaviors of many body quantum systems. Particularly, there has been great interest in studying the different sources of errors in quantum computing and their effect on quantum gate operations [3]. Different approaches have been proposed for protecting quantum systems during the computational implementation of algorithms, such as quantum error correction [4] and decoherence-free subspace [5]. Therefore, studying the effect of naturally existing sources of errors, such as impurities and dissipative environments in systems of interest is a must. In a previous work we have shown how the entanglement dynamics and ergodicity in Heisenberg spin systems can be controlled using impurities and anisotropy [6]. Very recently, we have investigated the impact of a thermal dissipative environment on the dynamic behavior of one-dimensional Heisenberg spin systems [7]. We demonstrated that the system entanglement reaches asymptotically a steady state that varies significantly depending on the degree of anisotropy and the environment temperature. In this paper, we show how we can control the asymptotic behavior and the entanglement steady state value in a one-dimensional open boundary Ising spin system coupled to a dissipative environment at finite temperature using magnetic impurities.

IOP Conf. Series: Journal of Physics: Conf. Series **869** (2017) 012022 doi:10.1088/1742-6596/869/1/012022

2. The Model

We consider a set of 7 localized spin-$\frac{1}{2}$ particles in a one-dimensional open boundary Ising chain coupled through exchange interaction $J_{i,j}$ in the x-direction and subject to an external magnetic field of strength h applied in the z-direction. All the particles are identical except one, which is considered as an impurity. The Hamiltonian for such a system is given by

$$H = -\sum_{<i,j>} J_{i,j}\sigma_i^x\sigma_j^x - h\sum_i \sigma_i^z, \tag{1}$$

where σ_i^α ($\alpha = x$, y or z) are the Pauli operators (for convenience we set $\hbar = k = 1$) and $< i, j >$ is a pair of nearest-neighbors sites on the lattice, with $J_{i,j} = J$ for all sites except the sites nearest to an impurity. The coupling between the impurity and its neighbors is $J_{i,j} = J' = (\alpha+1)J$, where α is a parameter that tunes the strength of the impurity. The dynamics of an isolated quantum system is described by the time evolution of its density matrix $\rho(t)$ according to the quantum Liouville equation $\dot{\rho}(t) = -i[H,\rho]$. For an open quantum system interacting weakly with an environment that has a short relaxation time for its excitation modes, the Born-Markovian approximation can be applied and the time evolution of the system is best described by the Lindblad Master equation [8], which is given by

$$\dot{\rho}(t) = -i[H,\rho] + \mathcal{D}_\rho\,, \tag{2}$$

where $\mathcal{D}_\rho$ is the extra term that describes the dissipative dynamics and is represented in the Lindblad form as

$$\mathcal{D}_\rho = -\frac{1}{2}\sum_{k=1}\left\{[L_k\rho, L_k^\dagger] + [L_k, \rho L_k^\dagger]\right\}\,, \tag{3}$$

where the Lindblad operator L_k represents all the effects of the considered environment on the system site k. It is more convenient to work in the Liouville space, where Eq. (2) can be reformulated into the matrix equation form $\vec{\dot{\rho}}(t) = (\hat{\mathcal{L}}^H + \hat{\mathcal{L}}^D)\vec{\rho} = \hat{\mathcal{L}}\vec{\rho}$, where $\hat{\mathcal{L}}^H$ and $\hat{\mathcal{L}}^D$ are superoperators acting on the vector ρ in the Liouville space. The exact numerical solution of the last equation yields the density vector $\vec{\rho(t)}$, which can be rearranged into the customary density matrix form $\rho(t)$. The effect of the dissipative thermal environment on the considered spin system is given by the local Lindblad operator [8]

$$L_k = \Gamma\left\{\frac{(\bar{n}+1)}{2}\,S_k^- + \bar{n}\,S_k^\dagger\right\}\,, \tag{4}$$

where S^+ and S^- are the spin raising and lowering operators, $S^\pm = S^x \pm iS^y$. Γ is a phenomenological parameter that determines the strength of the coupling between the environment and the system and is assumed to be the same for all spins. The thermal parameter $\bar{n}$ is proportional to the temperature of the environment. We consider $0 \le \bar{n} \le 0.1 (\approx 41$ mK). Due to the Born-Markovian approximation, we consider values of Γ and J such that Γ and $J << \omega$. We adopt the concurrence $C_{i,j}$ as a measure of the bipartite entanglement between any two spins i and j in the system [9]. We study the time evolution of the system starting from the maximally entangled state, $|\psi_m\rangle = \frac{1}{\sqrt{2}}(|\uparrow\downarrow\rangle + |\downarrow\uparrow\rangle)\,|\downarrow\downarrow\cdots\downarrow\rangle$.

3. Results and discussion

We study the bipartite entanglements along the 7 spins Ising chain for two different impurity location scenarios. In the first one, we consider an impurity at the terminal (spin 1), which is discussed and illustrated in figure 1. In the second case the impurity is assumed to be in the center (spin 4), shown in figure 2. In both cases, the system is initiated in a global

IOP Conf. Series: Journal of Physics: Conf. Series **869** (2017) 012022 doi:10.1088/1742-6596/869/1/012022

maximally entangled state where the two spins 1 and 2 are maximally entangled with each other but are disentangled from the rest of the spins in the chain, where they are in a separable state themselves. To briefly discuss the overall dynamics we focus on and illustrate only the bipartite entanglement of the terminal pair (1 and 2), the central pair (3 and 4) and the other terminal pair (6 and 7) at zero and finite temperature. For convenience we consider the time evolution of the system in terms of the dimensionless time $T = \omega\, t$. As can be noticed in figure 1(a), C_{12}

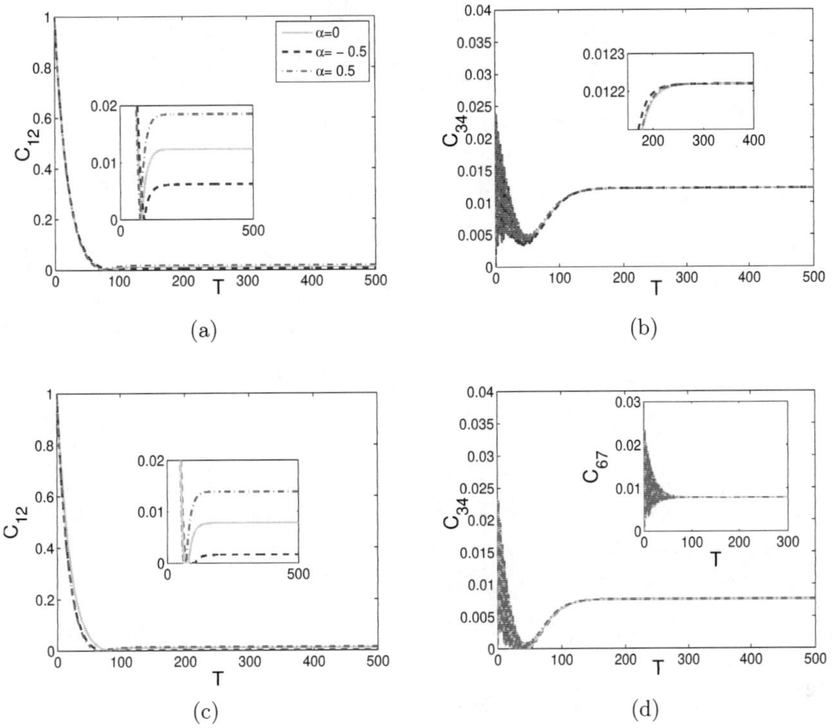

Figure 1. Dynamics of the entanglements C_{12} and C_{34} with a terminal impurity in the Ising chain at $\bar{n} = 0$ in (a) and (b) and $\bar{n} = 0.05$ in (c) and (d). In the inner panel of (d) C_{67} is depicted versus T. The legend is as shown in panel (a).

at $\bar{n} = 0$ starts with a maximum value but vanishes within a short time before reviving again and reaching asymptotically a steady state value. The steady state value depends significantly on the impurity strength, represented by α and is enhanced as it increases. The central pair entanglement C_{34}, illustrated in figure 1(b) behaves differently where it starts from zero value then strongly oscillates before reaching a steady state value that is very slightly affected by the terminal impurity. C_{67} was found to show a very similar behavior to C_{34}. The effect of the temperature on C_{12} and C_{34} is explored in figure 1(c) and (d) respectively. As expected the steady state values of C_{12} are reduced due the thermal excitations but the impurity impact is still effective. The thermal effect on C_{34} is very similar to what was observed for C_{12}, where the steady state values is reduced but the impurity effect is still slim. A similar behavior to C_{34} is shown by the far terminal pair C_{67}. The effect of a central impurity is explored in figure 2. As one can see, it has no noticeable impact on C_{12} neither at zero temperature (panel (a)) nor at finite temperature (panel(c)). On the other hand, the central spin pair are the most affected one by the central impurity as shown in panel (b) for zero temperature, where higher impurity strength enhances the steady state value. At finite temperature, the thermal excitations again reduces the steady state values but impurity is still effective as can be concluded from panel (d).

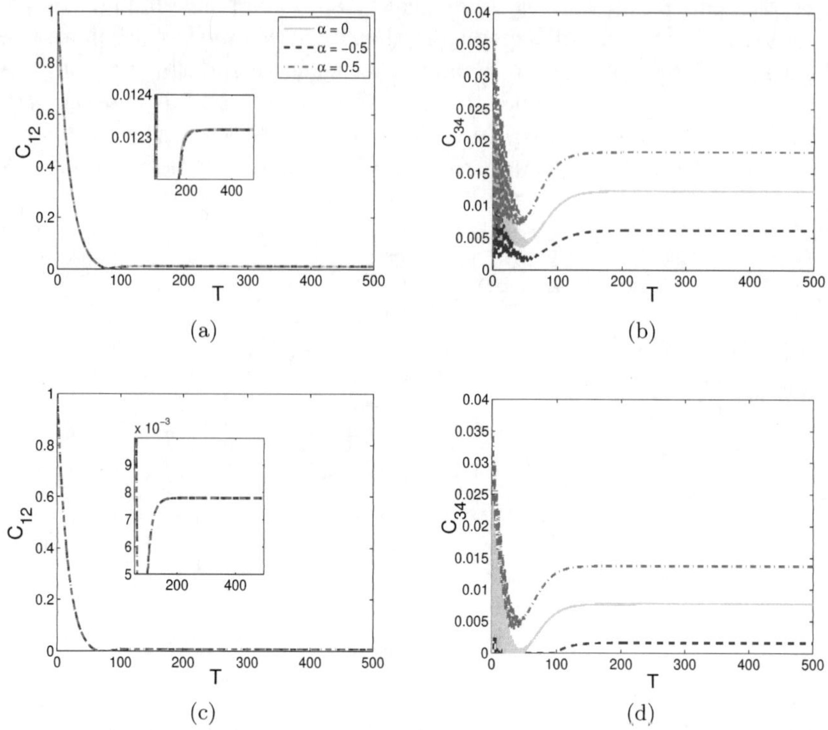

Figure 2. Dynamics of the entanglements C_{12} and C_{34} with a central impurity in the Ising chain at $\bar{n} = 0$ in (a) and (b) and $\bar{n} = 0.05$ in (c) and (d). The legend is as shown in panel (a).

4. Conclusion

We studied the effect of an impurity in a one-dimensional Ising spin chain in presence of a Lindblad environment on the asymptotic behavior of the system. The asymptotic steady state value of the bipartite entanglement of the nearest neighbor spins of the impurity was found to be strongly dependent on the impurity strength and is enhanced as it increases regardless of the impurity location. As the temperature is raised the steady state value is suppressed and the impurity effect is reduced significantly.

Acknowledgment

This work was supported by the University of Sharjah seed grant number G.R.C./S.R. 83/2015.

References

[1] Peres A 1993 *Quantum Theory: Concepts and Methods* (Kluwer, Dordrecht)
[2] Nielsen M and Chuang I 2000 *Quantum Computation and Quantum Communication* (Cambridge University Press, Cambridge, UK)
[3] Jones J A 2003 *Phys. Rev.* A **67** 012317; Cummins H K *et al* 2003 *Phys. Rev.* A **67** 042308
[4] Shor P W 1995 *Phys. Rev.* A **52** R2493
[5] Bacon D *et al* 2000 *Phys. Rev. Lett.* **85** 1758
[6] Sadiek G *et al* 2012 *Phys. Rev.* A **85** 042313
[7] Sadiek G and Almalki S 2016 *Phys. Rev.* A **94** 012341
[8] Breuer H P and Petruccione F 2002 *The Theory of Open Quantum Systems* (Oxford University Press, Oxford)
[9] Wootters W K 1998 *Phys. Rev. Lett.* **80** 2245

IOP Conf. Series: Journal of Physics: Conf. Series **869** (2017) 012023 doi:10.1088/1742-6596/869/1/012023

Multiscale processing of loss of metal: a machine learning approach

G De Masi[1], M Gentile[1], R Vichi[1], R Bruschi[1] and G Gabetta[2]

[1]Advanced Engineering (ADVEN) Dept. Saipem, Fano, Italy
[2]SIAV Dept., eni, San Donato Milanese, Italy
Email: giulia.demasi@saipem.com

Abstract. Corrosion is one of the principal causes of degradation to failure of marine structures. In practice, localized corrosion is the most dangerous mode of attack and can result in serious failures, in particular in marine flowlines and inter-field lines, arousing serious concerns relatively to environmental impact. The progress in time of internal corrosion, the location along the route and across the pipe section, the development pattern and the depth of the loss of metal are a very complex issue: the most important factors are products characteristics, transport conditions over the operating lifespan, process fluid-dynamics, and pipeline geometrical configuration. Understanding which factors among them play the most important role is a key step to develop a model able to predict with enough accuracy the sections more exposed to risk of failure. Some factors play a crucial role at certain spatial scales while other factors at other scales. The Mutual Information Theory, intimately related to the concept of Shannon Entropy in Information theory, has been applied to detect the most important variables at each scale. Finally, the variables emerged from this analysis at each scale have been integrated in a predicting data driven model sensibly improving its performance.

1. Introduction

CO_2 corrosion, or the so called "sweet corrosion", is a major concern in the application of carbon and low alloy steels, which are still the principal construction materials, offering economy, availability and strength. External sea water corrosion is prevented by the use of an external coating (passive protection) and by cathodic protection (active protection). While for long transportation pipelines the gas is normally dry and the internal corrosion is not an issue, for flowlines and inter-field lines transporting untreated fluids, internal corrosion plays a crucial role for structure integrity. This is a growing and challenging problem for Oil and Gas industry, since the age of plants and components is worldwide increasing. It is necessary to prevent failures of all these components.

Internal corrosion shows a very complex phenomenology due to interaction of different mechanisms playing a role at different spatial scales. Water and electrochemistry, protective scales, flow velocity, steel composition and localized bacteria attacks are the most relevant.

While it is reasonably easy to understand a corrosion event "retrospectively" with failure analysis methods, a large degree of uncertainty is associated with the attempt of quantifying a prediction for the future evolution of damage.

Relevant physics analysis is multi-disciplinary and includes different tasks: analysis of products, transport conditions over the operating lifespan, process fluid-dynamics and pipeline configuration on the real seabed [1].

An effective corrosion modelling methodology would be based on fundamental laws and first principles since analyses could be performed over a wide range of conditions with some degree of confidence. But corrosion mechanisms are not yet understood well enough to develop purely first-principles mechanistic models, although this is the final goal of research on corrosion. For this reason semi-empirical and empirical models have been studied. Unfortunately, due to its complexity, the localized corrosion process is hardly reproduced even by semi-empirical models proposed in literature [2, 3]. Furthermore, differently from other publications focusing on prediction of generalized corrosion, the scope of the present paper is to identify along the same pipeline where the sections with high risk of corrosion are located.

Field data, nowadays increasingly available, e.g. from internal line inspections (ILI) by intelligent pigs, allows the development of data-driven fully empirical models that can be useful to process corrosion data [4]. Given that the uncertainties related to measurements of remaining wall thickness in an intelligent pig are very low, this allows to accurate estimate volume loss and corrosion rate along a specific pipeline. Data driven model can also shed light on which factors are more relevant for the localized corrosion process, with the final aim to develop an analytic model.

This paper offers a methodology to identify the crucial factors playing the most important role in the corrosion process under study. The methodology is based on Mutual Information Theory strictly related to the Shannon Entropy theory[5]. The mutual information (MI) of two random variables is a measure of the mutual dependence between the information content of two variables, estimated by their Entropies H(X), H(Y) like in figure 1.

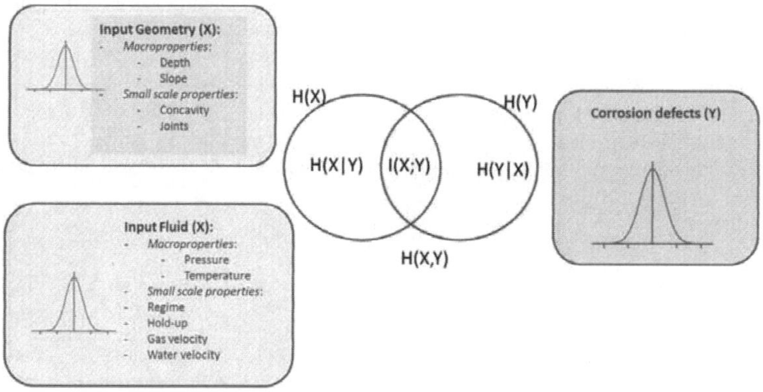

Figure 1. Scheme of the Mutual Information Analysis.

More specifically, it quantifies the "amount of information" obtained about one random variable, through the other random variable. In our particular case it is used to understand how the observed corrosion pattern depends on each possible external condition (from material property, to fluid transport conditions, to pipeline geometry, to microbiology). Analyses have been performed at different spatial scales, being the phenomenon partially global on the whole length of the pipe (generalized corrosion), partially medium scale (corrosion due to increase/decrease of fluid-velocity or passage of slugs), partially very local scale (for instance, stagnant water regions).

Finally the most important variables are fed into a predictive model. Among several expert systems and data driven modelling, in this paper artificial neural networks (ANN) have been chosen [6]. While application of ANNs has already been demonstrated to be very useful for internal corrosion prediction[7, 8], the authors apply the MI method to improve a predictive model based on a set of cascading ANNs which has been demonstrated to be particularly effective to identify the sections with higher corrosion up to a detailed scale of 10m [9].

The MI analysis strongly improves the prediction performance trough the identification of the most important variables at each scale, compared with past papers where all the contribution were considered equivalent[9].

2. Methodology

2.1. Product characterization and historical data collection
First step is collection and organization of all existing, relevant, essential, historic and current operating data about the pipeline segments and/or regions relevant to corrosion distribution.
From design and construction records, operating and maintenance histories, corrosion survey records, gas and liquid analysis reports, and inspection reports from prior integrity evaluations or maintenance actions.

2.2. Multiphase flow regimes
Once identified a representative scenario of fluid-dynamic boundary conditions, multiphase flow modelling is done using OLGA Multiphase Flow Simulator software. As the operating conditions vary along the route, this program determines temperature profile along the pipeline, pressure profile, velocity profiles of each phase, hold-ups and flow regimes, given boundary pressure, temperature values and flow composition.

2.3. Geometric profile
The choice of a significant profile and consequently inclination and concavity to set as input for ANN is a very crucial issue. Generally, a profile, i.e. KP vs. Elevation, provided by an ILI inspection is quite accurate. In order to obtain significant variations of the geometrical variables, a filtering of the profile has been performed. The filter is chosen such that the final profile is representative of main spatial variations, without having too much detail.

2.4. Mutual Information Theory
As evident, one crucial point in the corrosion dataset processing is to identify which are the most relevant variables for the predictive model. In order to evaluate the dependence between the observed corrosion and the other variables, the Mutual Information (MI) analysis has been implemented. MI or (formerly) trans-information of two random variables X and Y is a measure of the variables mutual dependence (Fig.1). Not limited to real-valued random variables like the correlation coefficient, MI is more general and determines how similar the joint distribution p(X,Y) is to the products of factored marginal distribution p(X)p(Y) and it is given by

$$MI(X,Y) = \sum_i p(x_i, y_i) log \frac{p(x_i,y_i)}{p(x_i)p(y_i)} \quad (1)$$

It can be demonstrated that the relation of MI with Information Entropy of single variables H(X), H(Y) and their cross Entropy H(X,Y) is given by:

$$MI(X,Y) = H(X) + H(Y) - H(X,Y) \quad (2)$$

In contrast to the linear correlation coefficient, it is sensitive also to dependencies which do not manifest themselves in the covariance. In the present study, the variable Y is the corrosion volume loss while the variable X comes from four different categories:

- Hydrocarbon characteristics (CO_2 partial pressure)
- Geometrical pipeline characteristics (elevation, inclination, concavity)
- Fluid dynamic multiphase variables (flow regime, hold-up, pressure, gas flow, total flow, liquid velocity, gas velocity)
- Deterministic models (de Waard and NORSOK)

2.5. Predictive model
In order to represent multi-scale phenomenology and to obtain prediction of regions more exposed to corrosion also at fine scale (for our purposes 10m), a set of cascading ANNs, each one representing the phenomenon at a different spatial scale (1000m, 100m, 10m) has been designed [9].

3. Results and Conclusions

The MI analysis has been performed at different spatial scales (1000m, 100m and 10m). The results are reported in table 1.

Table1. Results of Mutual Information Analysis. Only variables with Mutual Information over 0.5 are reported.

1000m		100m		10m	
Elevation	0.75	Pressure	0.63	Hold-up	0.84
KP	0.59	Elevation	0.63	Concavity	0.64
Liquid Velocity	0.57	Gas Velocity	0.63	Liquid Velocity	0.60
Pressure	0.57	Hold-up	0.59	GasVelocity	0.50
Gas Flow	0.57	Concavity	0.53		
Hold-Up	0.57	Gas Flow	0.53		
Regime	0.53				

The variables well represent the dynamics at different spatial scale. Feeding the variables selected by MI analysis on the predictive hierarchical model, the final prediction at fine scale (10m) strongly improves the prediction performance like evident in the example reported in Fig.2.

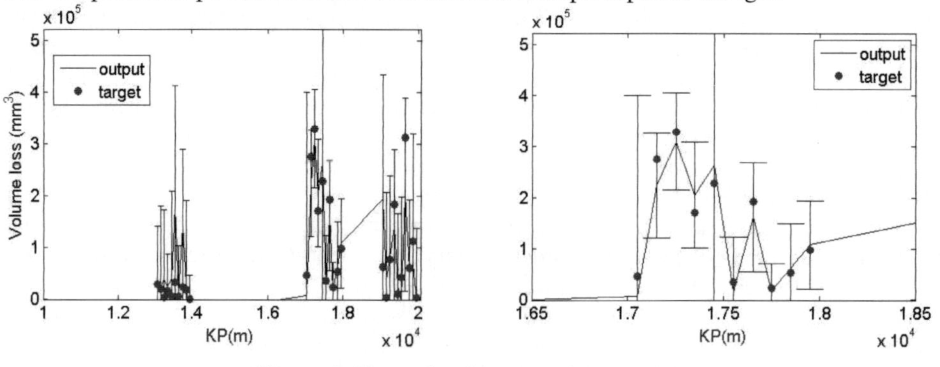

Figure 2. Example of output of the model.

Despite the fact that the observed corrosion pattern is very complex, the integration of multiscale MI analysis and the predictive model seems to capture main evidences of local corrosion, providing a promising methodology for crucial assessment. Analysis and predictions can be further improved in the future considering larger datasets (with several pipeline cases and different flow conditions).

References

[1] Bruschi R, Cherubini P, Vichi R, Gentile M 2013 *Integrity Management On Pipeline Affected by Corrosion* IBP1471_13 Rio Pipeline Conference & Exposition 2013 Rio de Janeiro (Brazil)
[2] De Waard C and Lotz U 1993 *NACE Corrosion 1993 Proc.* (Houston) Paper N°69
[3] *NORSOK Standard N°M-506 Rev1 Norwegian Technology Standards Institution* (http://wwwntsno/norsok)
[4] Li X, Zhang D, Liu Z, Li Z, Du C and Dong C 2015 *Nature* **527** 441-442
[5] Gray R M 2008 *Entropy and Information Theory* (Springer)
[6] Haykin S 1999 *Neural Networks A comprehensive foundation* (Pearson Prentice Hall)
[7] Gabetta G, De Masi G, Gentile M, Vichi R and Scapin M 2014 *ADIPEC Proc.* SPE **17**1919
[8] De Masi G, Vichi R, Gentile M, Bruschi R and Gabetta G 2015 *Oceans 2015 Proc.* (Genova)
[9] De Masi G, Vichi R, Gentile M, Bruschi R and Gabetta G 2016 *Compit 2016 Proc.* (Lecce)

EPR studies for $GdBa_2Cu_3O_{7-\delta}$ added with nanosized ferrite $CoFe_2O_4$

H Basma[1], S Isber[2], S Noureldeen[3], M Anas[4] and R Awad[1]

[1]Physics Department, Faculty of Science, Beirut Arab University (BAU), Beirut.
[2]American University of Beirut, Department of Physics, Beirut, Lebanon.
[3]Physics Department, Faculty of Science, Lebanese University, Beirut.
[4]Physics Department, Faculty of Science, Alexandria University, Egypt.

Abstract. EPR measurements were conducted at different temperatures $100K \leq T \leq 295K$ for $GdBa_2Cu_3O_{7-\delta}$ superconducting samples prepared by the conventional solid state reaction technique, added with x wt.% nanosized ferrite of $CoFe_2O_4$,prepared by chemical co precipitation method, with $0 \leq x \leq 0.40$ wt.% . A broad isotropic symmetric EPR line with g factor $g \approx 2$ is detected for the $(CoFe_2O_4)_xGdBa_2Cu_3O_{7-\delta}$ samples, corresponding to Gd^{3+} ions. The broadness is attributed to the superposition of signals resulting from the clustering of Gd^{3+} ions and the dipole interaction between these ions. The analysis of the absorption curve and its first derivative allowed to obtain several magnetic parameters including the number of spins N participating in EPR resonance which showed a decrease with increasing temperature and the activation energy E_a, which experienced a decrease with x.

1. Introduction

Electron paramagnetic resonance (EPR) serves as an effective tool to investigate the magnetic properties of high temperature superconductors. Awad *et al.* [1] conducted EPR studies of $GdBa_2Cu_{3-x}Ru_xO_{7-\delta}$. The results showed that the line symmetry is independent on both Ru-content and temperature while the magnetic parameters showed an increase with the Ru-content. Guskos *et al.* [2] measured the EPR spectra of $GdBaCu_3O_{7-\delta}$ (Gd-123) superconducting phase at different temperatures (4.2–120 K) and detected a signal from Gd^{3+} ions with g = 1.99 and a signal from Cu^{2+} ions with $g_x = 2.074$, $g_y = 2.015$ and $g_z, = 2.202$. Alekseevskii *et.al* [3-5] conducted EPR studies for $GdBa_2Cu_30_{7-\delta}$ with $\delta = 0.15$ and 0.5. A single line spectrum of Gd^{3+} ions was observed for all samples. EPR measurements showed a metallic character for samples with δ =0.5, while resistivity measurements showed a semiconducting behavior. The EPR signal consisted of two signals with nearly equal g values (g -~ 1.99) but with two different line widths. For the first line, the line width linearly depends on the temperature. For the second one, the line width does not depend on the temperature and is equal to 1000 Oe. In this work, EPR measurements for $(CoFe_2O_4)_x$,Gd-123 with $0 \leq x \leq 0.4$ wt. % prepared by the conventional solid state reaction technique were conducted. The analysis of the absorption curve and its first derivative of the samples allowed the calculation of several magnetic parameters. The variations of these parameters are discussed with respect to the amount of nano ferrite addition and the variation of temperature.

2. Experimental techniques:

Nanosized $CoFe_2O_4$ powder and $(CoFe_2O_4)xGdBa_2Cu_3O_{7-\delta}$, $0.0 \leq x \leq 0.4$wt % samples were prepared by chemical Co-precipitation method and conventional solid-state reaction technique, respectively as reported previously [6]. EPR measurements were performed using a Bruker microwave controller ER 048 spectrometer, operating at the X-band frequency (9.24 GHz). The magnetic field was scanned in the range 0–10000 G and a microwave power of 15mW. Powder sample of 100 mg was taken in a quartz tube for EPR measurements. The EPR spectra of the samples were recorded at temperature range (100–295K).

3. Results and Discussions:

The first derivative EPR absorption spectra at different temperatures (T=295, 263,233, 193, 163, 133 and 100K) for $GdBa_2Cu_3O_{7-\delta}$ are shown in figure 1. The EPR spectra contain a broad isotropic line corresponding mainly to Gd^{3+}ions [7] which is symmetric about its central position for all samples.

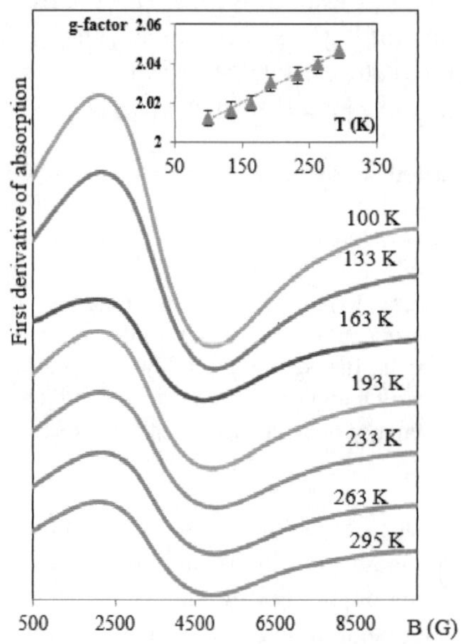

Figure.1: First-derivative absorption spectra for $GdBa_2Cu_3O_{7-\delta}$ versus the magnetic field at different temperatures. The inset represents the variation of g-factor versus temperature.

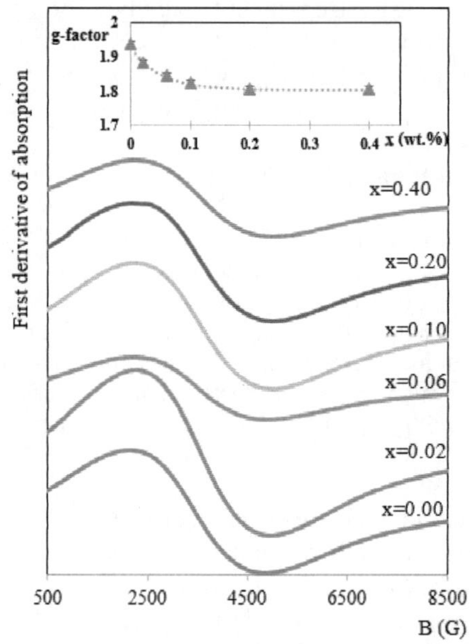

Figure.2: First-derivative absorption spectra versus magnetic field for $(CoFe_2O_4)_x GdBa_2Cu_3O_{7-\delta}$phase at T = 100 K. The inset represents the variation of g-factor versus x.

The symmetric shape indicates the absence of skin effect [8]. The intensity of the EPR lines at liquid-nitrogen temperature is stronger than that at room temperature. The broadness of the EPR spectra for all samples can be attributed to one of three possible mechanisms. It may result from the dipole–dipole interactions between the Gd^{3+}ions or the modification of the degree of short-range ordering in the environment of the Gd^{3+} sites or the departure from a random spatial distribution of these sites in the glass network (clustering etc) [9].The values of the g-factor at different temperatures for $GdBa_2Cu_3O_{7-\delta}$ are shown in the inset of figure 1. A linear increase of the g- factor with temperature is observed for all samples. The slight temperature dependence of the g- factor, and its shift to lower values at lower temperatures can be related to the dynamic Jahn Teller (JT) effect [10,11]. Figure 2 shows the EPR Spectra for $(CoFe2O4)_xGdBa_2Cu_3O_{7-\delta}$ (x=0.01, 0.04, 0.06, 0.20 and 0.40 wt. %) samples at T =100 K. The inset of figure 2, shows the variation of the g- factor of $(CoFe2O4)_xGdBa_2Cu_3O_{7-\delta}$ for different values of x at T=100K. The values of the g-factor showed a decrease with x. This decrease can be attributed to the

IOP Conf. Series: Journal of Physics: Conf. Series **869** (2017) 012024 doi:10.1088/1742-6596/869/1/012024

environment changes around Gd^{3+} which contains Cu^{2+} magnetic ions and $BaCuO_2$ impurities detected from XRD measurements [12]. It is important to mention that the ferromagnetic resonance line FMR corresponding to Co^{2+} ions at $T \approx 100$ K, originated from Co moments in $CoFe_2O_4$ nanoferrites for $(CoFe_2O_4)_xGdBa_2Cu_3O_{7-\delta}$ doesn't not appear in the spectrum. The ferromagnetic order is caused by an antisymmetric exchange coupling between neighboring Co moments. The disappearance of FMR line confirms the electrical resistivity data which did not show a ferromagnetic order at $T = 100$ K [12].

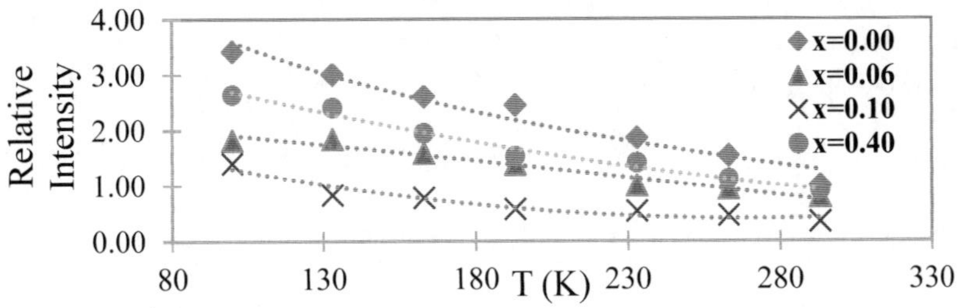

Figure 3: Temperature dependence of the EPR relative intensity for $(CoFe_2O_4)_x$.

The variation of The EPR relative intensity as a function of temperature is shown in Figure 3 for x=0.0, 0.06, 0.10 and 0.40.The EPR line intensity increases as the temperature decreases from room temperature to 100 K. The EPR line intensity is proportional to the magnetic susceptibility of the localized moments spin system. The magnetic susceptibility obeys Curie–Weiss law for a temperature higher than T_c for Gd-123 phase [13]. The EPR relative intensity decreases with the increase in x indicating that the addition of Co Fe_2O_4 nano ferrite suppressed the unpaired electrons in Gd-123 phase and consistent with the XRD results that showed a reduction in the Gd-123 phase and an enhancement in the $BaCuO_2$ impurity phase with the increase of x [12].

Figure 4 shows the temperature dependence of the peak-to-peak line width (ΔH_{pp}) for $(CoFe_2O_4)_xGdBa_2Cu_3O_{7-\delta}$. The line width shows a linear increase when the temperature increases from 100 to 295 K. The temperature dependence of the peak-to-peak line width can be fitted according to the relation:

$$\Delta H_{PP} = A_{PP} + B_{PP}T \quad (1)$$

Where A_{pp} is the residual line width and B_{pp} is the thermal broadening, corresponding to the Korringa-type interaction between the paramagnetic ions with the host conduction carriers through the exchange interactions J_s [14]. The values of A_{pp} and B_{pp} are listed in table 1. The values of B_{pp} show a decrease with the increase in x. The fitting parameter B_{pp} is proportional to the density of state which is proportional to the number of carriers [15]. A similar trend was observed for the variation of the hole carriers' concentration at CuO_2 planes with $CoFe_2O_4$ addition from previous RBS measurements [16].The number of spins (N) participating in the resonance was calculated by comparing the area under the absorption curve of the virgin samples at different temperatures with that of the standard $CuSO_4.5H_2O$ [15]. Figure 5 shows the logarithmic plot of N as a function of the reciprocal of absolute temperature. A linear relation between log(N) and $1/T$ is found as expected from Boltzmann law. The activation energy E_a was calculated from the slope of the lines and the values of E_a are also listed in table 1. The activation energy shows a similar trend to that of B_{pp} resulting from the suppression of N with the addition of $CoFe_2O_4$.

IOP Conf. Series: Journal of Physics: Conf. Series **869** (2017) 012024 doi:10.1088/1742-6596/869/1/012024

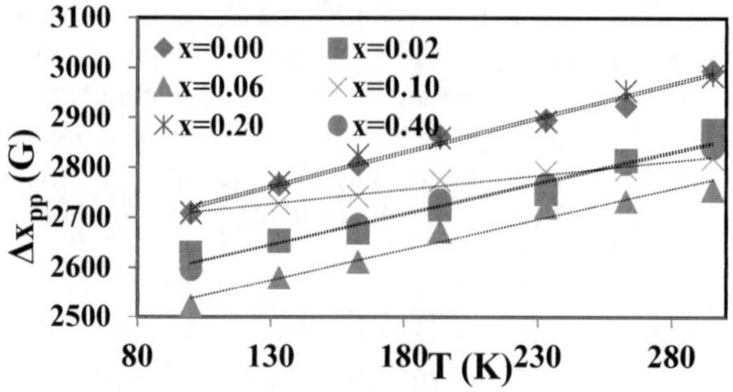

Figure 4: Variation of ΔH_{pp} with temperature.

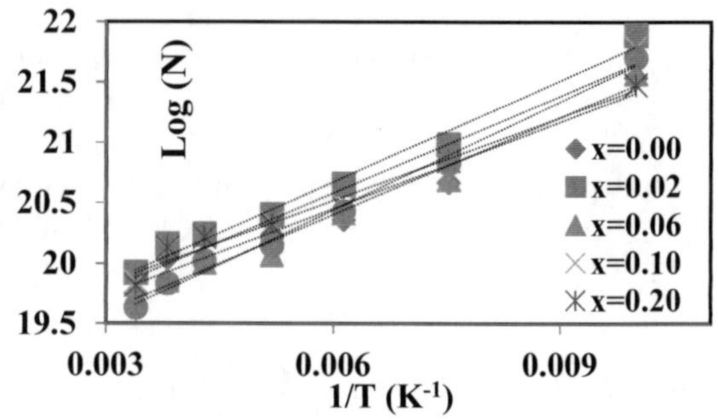

Figure 5: Log (N) as a function of (1/T).

Table1: The variation of A_{pp}, B_{pp} and E_a as a function of x.

x	0.00	0.02	0.06	0.10	0.20	0.40
Bapp (G/K)	1.378	1.242	1.225	1.568	1.178	1.128
App (G)	2579.2	2448.8	2415.2	2653.2	2585.2	2484.4
Ea (eV)	0.0258	0.0248	0.0231	0.023	0.0197	0.0208

4. Conclusion

The EPR studies of Gd-123 superconducting samples added with x wt. % of $CoFe2O_4$ nano ferrites were conducted. The line width showed an increase with temperature, the logarithmic plot of the number of spins with (1/T) showed a linear dependence. The magnetic parameters including the thermal broadening (B_{pp}), the number of spins (N) and the activation energy (E_a) were suppressed with the increase in x.

References

[1] Awad R *et al* 2012 *Physica C: Superconductivity*, **477** 74-83
[2] Guskos N *et al* *Phys.* 1990 *Stat. Sol. (b)* **162** 243
[3] Alekseevskii N E *et al* 1988 *JETP Lett.* **48** 37
[4] Alekseevskii N E *et al* 1989 *J. Low Temp. Phys.* **77** 87
[5] Garifullin I A 2015 *J. Low Temp. Phys.* **178** 243
[6] Awad R *et al* 2015 *Journal of Superconductivity and Novel Magnetism* **28(2)** 535-539
[7] Nakamura F *et al* 1989 *Physica C: Superconductivity*, **162** 1287-1288
[8] Bejjit L and Haddad M 2002 *Physica C: Superconductivity* **371(4)**, 339-343
[9] Kliava J *et al* 2003 *Journal of Physics: Condensed Matter* **15(40)** 6671
[10] O'Brien M C, In Proceedings of the Royal Society of London (Vol. **281**, No. 1386, pp. 323-339).
[11] Riley M *et al* 1986 *Chemical physics*, **102(1)** 11-28
[12] Awad R *et al* 2014 *Journal of Superconductivity and Novel Magnetism* **27** no 7 1757-1767.
[13] Kikuchi H *et al* 1988 *Journal of the Physical Society of Japan* **57(6)** 1887-1890
[14] Barnes S E 1981 *Advances in Physics*, **30(6)** 801-938
[15] Weil J A and Bolton J R John 2007 Wiley & Sons.
[16] Basma H *et al* 2015 *Materials Sciences and Applications*, **6 (09)** 828

IOP Conf. Series: Journal of Physics: Conf. Series **869** (2017) 012025 doi:10.1088/1742-6596/869/1/012025

Transport properties of asymmetric zigzag graphene nano-ribbon built by carbon vacancy in edge

H Jafari, B Dianat and M Afshar

Materials Simulation Laboratory, Physics Department, Iran University of Science and Technology, Tehran, Iran

E-mail: j.august91@ymail.com

Abstract. Using nonequilibrium Green's functions in combination with density functional theory, we investigated the electronic transport behaviors of single layer zigzag graphene Nano-ribbon (ZGNR) with a Carbon vacancy in edge. The results show that electronic transport properties of the asymmetry ZGNR can be modulated by vacancy in edges, and prominent negative differential resistance (NDR) can be observed. These results may be useful for designing practical devices based on graphene Nano-ribbons.

1. Introduction

Since the experimental synthesis of stable single graphene layer at room temperature was achieved, graphene has been the subject of numerous experimental and theoretical investigations due to the high electron mobility and the potential for sizing the electronic devices to nanoscales [1]. In particular, much effort is focused on obtaining graphene nano-ribbon (GNR) with tailored properties that could be key elements in future electronic applications. Two basic nano-ribbons are mostly studied, which are the edge of GNR pointing along the C-C bond referred to as the armchair GNR, and the edge perpendicular to the C-C bond referred to as the zigzag GNR.

It is well known that the C atoms of GNR are sp^2 bonded. Hence an edge atom would have dangling bonds, which would be saturated by atoms or molecular species. Recently, much research have been done concerning the modification of edges because it can be used to control the electronic properties of GNRs for different purposes. The sp^3 bonded edge atoms with 2H atoms at the edge can be stabilized over 1H atom terminated edge at high temperatures and pressures.

In this paper, we present a theoretical description of transport properties of a 6-ZGNR, including vacancy in edge in order to show its impact on the whole structure's conductivity. The geometrical structures of two primitive (one Carbon and its adjacent hydrogen at edge picked up) and its relaxed and optimized models have been shown in figure 1. The electronic transport properties have been studied by the first-principles calculation, differences compared.

As shown in figure 1. (a)-(b), the model configuration, called 6-ZGNR in some previous references, are outstandingly differ from each other specifically in edges.

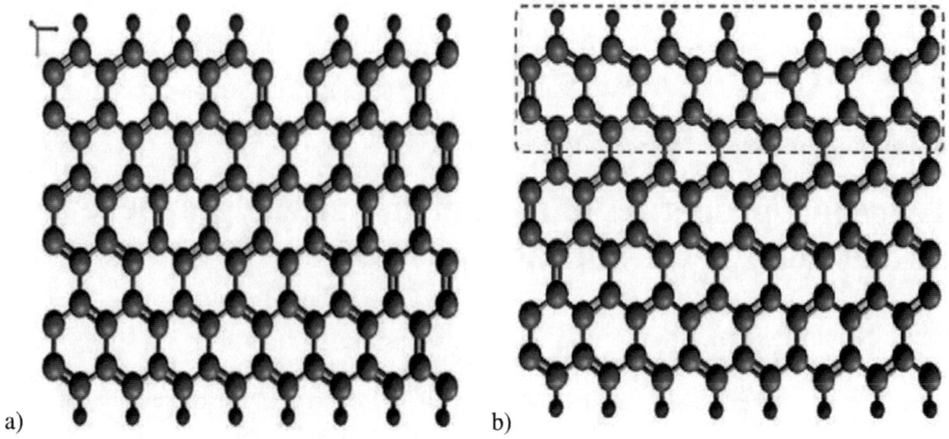

Figure 1. The geometry structure of a single layer zigzag graphene nano-ribbon passivated by Hydrogens when it (a) lacks of Carbon at edge (b) is relaxed and optimized.

2. Computational method and details

Transport calculations are based on density functional theory (DFT) combined with, nonequilibrium Green's function (NEGF) with SIESTA package. In our calculations, the vacuum layers between ZGNRs in neighboring along x and y directions (defined in figure 1) are 20 Å. The Perdew-Zunger exchange and correlation function within the generalized gradient approximation (GGA) is used. A cutoff energy of 350 Ry and a Monkhorst-Pack k-mesh of $1 \times 1 \times 12$ are chosen and double-ζ plus polarized basis set is adopted for electron wave function which has been proved to be capable of well describing the π-conjugated bond system and the transport calculation including transmission and current-voltage characteristics [2]. Geometry optimization is performed until the force is less than 0.001 eV/Å. The NEGF-DFT self-consistency is controlled by a numerical tolerance of 10^{-4}eV. The electron temperature is set to 300 K in the transport calculations.

3. Results and discussion

Calculating band structure, we find the small negative direct band gap, which refers to conductivity of the structure, as shown in figure 2. Hence, we focused on central atoms in scattering region as shown in figure 3. First and foremost, we analyzed the total density of states of scattering region which as shown in figure 4 indicates outstanding number of states to be occupied in some intervals of energy, compared to their neighbors.

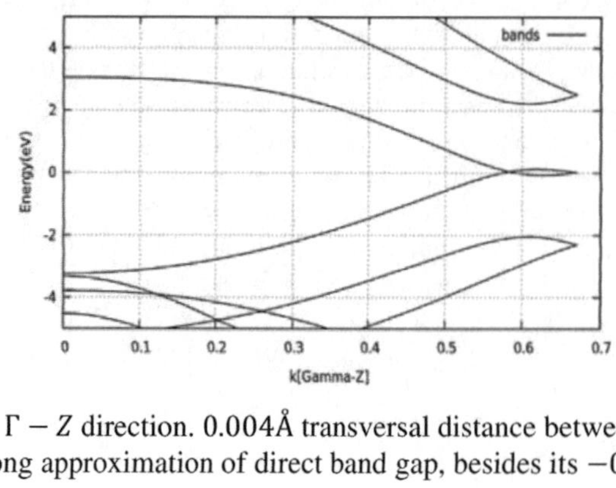

Figure 2. Band structure in $\Gamma - Z$ direction. 0.004Å transversal distance between conduction and valence bands indicates strong approximation of direct band gap, besides its -0.193 eV energy gap.

IOP Conf. Series: Journal of Physics: Conf. Series **869** (2017) 012025 doi:10.1088/1742-6596/869/1/012025

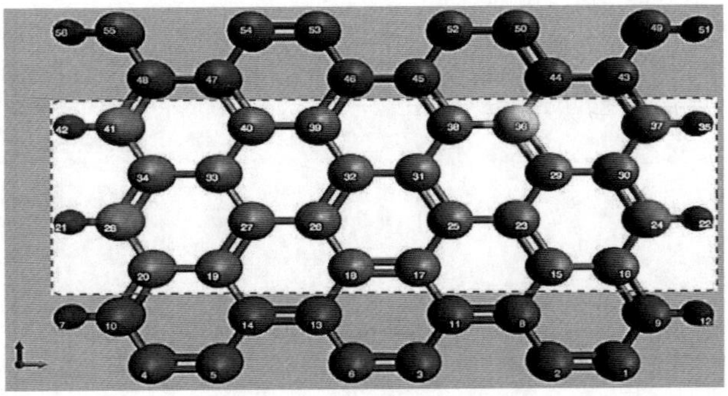

Figure 3. Scattering region focused on central atoms. PDOS of red (Carbons at edges) and green (Carbons in the middle) Carbon has been calculated.

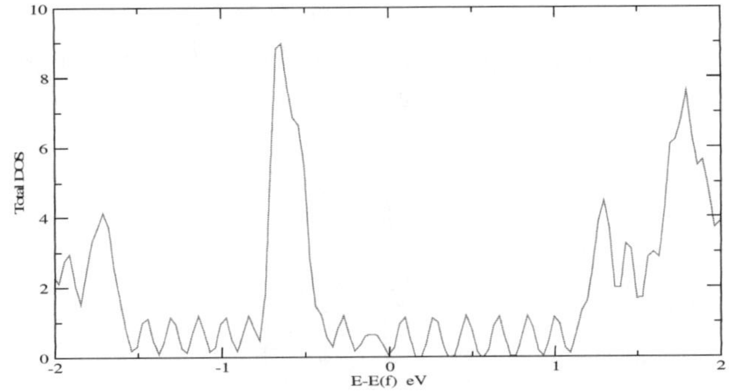

Figure 4. Total density of states (DOS) of the entire scattering region.

We studied the central atoms in more details, i.e. looking up the behavior of spin DOS for specific atoms depending on their positions. Except Carbons at edges, all the other Carbons in the middle of the scattering region have same treating in their spin DOS for different spatial directions and orbitals, as shown in figure 5. (Atom #36 in figure 3 is the green one which all the other middle Carbons have the same behavior).

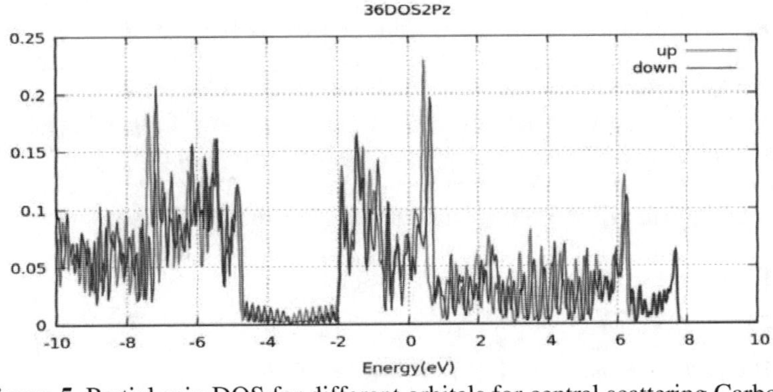

Figure 5. Partial spin DOS for different orbitals for central scattering Carbons.

When we calculated PDOS for edge Carbons, we figured out interesting spin separation in their DOS indicating which one does conducting more, shown in figure 6. Finally we computed the spin kind of I-

IOP Conf. Series: Journal of Physics: Conf. Series **869** (2017) 012025 doi:10.1088/1742-6596/869/1/012025

V characteristics for perfect and defect structure in figure 7, showing reduction in conduction in defect one and of course negative differential resistance (NDR), although totally the current is much low because of the structure's geometry [3].

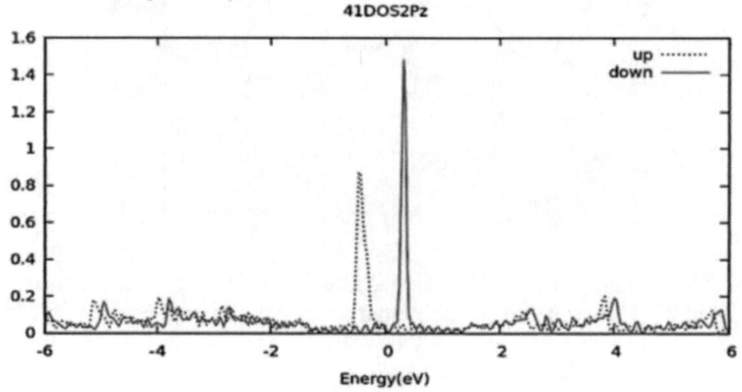

Figure 6. Atom #41 PDOS for orbital P in Z direction. All other edge Carbons (red ones) behave the same.

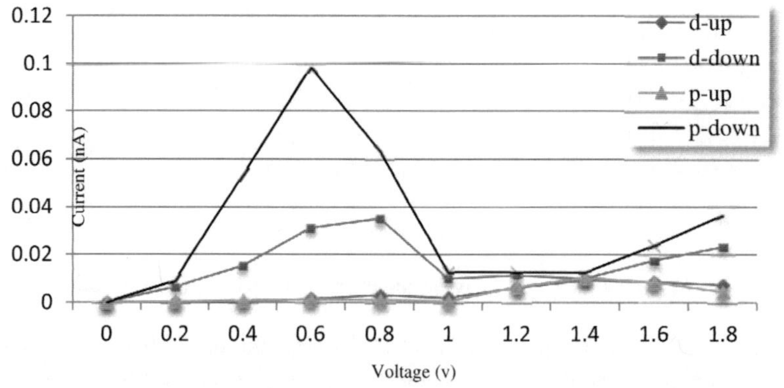

Figure 7. I-V spin graph of defect and perfect structures. Up spin current whether for defect one or perfect one has the same trending of current as voltage rising.

Since we observed spin filtering effect resulting from pdos polarization as shown in Figure 6, it leads us forward to using in spintronic devices. Besides, Negative Differential Resistance (NDR) which is showed in I-V curve is of great importance that occurred due to the electron tunneling into one of the neighboring islands (valleys) in conduction band i.e. into a system of states of reduced dimensionality. It means high energy electrons transmitting between two terminals don't have enough space (dimensional confinement) in collector (conduction band) so that they are tunneling to second barrier in collector and finally we observe decreasing current. Our results on transport properties of defected graphene and defect free graphene indicate a great potential in applications such as spin sensitive electronic devices.

References

[1] Prins F *et al* 2011 *Nano Lett.* **11**(11) 4607-4611.
[2] Kim W Y and K S Kim 2008 *Nat. Nanotechnol.* **3**(7) 408-412.
[3] Li Z *et al* 2008 *Phys. Rev. Lett.* **100**(20) 206802.

IOP Conf. Series: Journal of Physics: Conf. Series **869** (2017) 012026 doi:10.1088/1742-6596/869/1/012026

THERMOPOWER of NiO/(Bi, Pb)-2223 composite

H T Rahal[1], R Awad [2] and A M Abdel-Gaber[1, 3]
[1]Department of Chemistry, Faculty of Science, Beirut Arab University, Lebanon
[2]Department of Physics, Faculty of Science, Beirut Arab University, Lebanon
[3]Department of Chemistry, Faculty of Science, Alexandria University, Egypt
E-mail: hrahal@student.bau.edu.lb

Abstract: Solid-state reaction method was used to prepare superconductor samples with nominal composition of $(NiO)_x(Bi_{1.6}Pb_{0.4})Sr_2Ca_2Cu_3O_{10-\delta}$,where $0.0 \leq x \leq 0.2$ wt%. The prepared samples were characterized by X-ray diffraction (XRD) as well as scanning electron microscopy (SEM). XRD patterns indicated that the addition of nano-NiO particles to (Bi, Pb)-2223 did not affect the tetragonal structure and their lattice parameters. The thermopower coefficient was measured as a function of temperature from 250 K down to 77 K, using a standard differential technique. The results were analyzed according to the two-band (Fermi-liquid) and the two-band with an extra linear term models. Pseudo gap temperature (T^*), Fermi energy (E_F) and Fermi temperature (T_F) values were calculated and discussed in terms of nano-NiO additions.

1. Introduction

The study of thermoelectric power is important for exploring the conduction mechanisms of superconductors. This is because nowadays, superconductors are regarded as thermoelectric materials that can replace semiconductors and can be used in refrigeration and electrical power generation [1, 2]. Intense research efforts [3-5] have been devoted to understand the thermoelectric power of superconductors because it gives information about scattering mechanisms, as well as the concentration and sign of carriers. One of the strategies to overcome this issue is to introduce nanomaterials into the bulk matrix, which is known to be nano inclusion [3]. The idea of nanostructures aimed to reduce thermal conductivity through scattering phonons, and to sustain or improve electrical conductivity and Seebeck coefficient at the same time. Joo *et al* [6] investigated the thermoelectric properties of $Bi_2Te_{2.7}Se_{0.3}$ alloys embedded with MgO nanoparticles. It was found that the addition of MgO nanoparticle resulted in an increase in the maximum Seebeck coefficient and the electrical resistivity and an overall decrease in the thermal conductivity in the temperature range of 298 K - 573 K.The present work aims to explore effect of nano-NiO addition on the thermopower of (Bi,Pb)-2223 superconductor.

2. Experimental techniques

EDTA-capped nano-NiO particles and superconducting samples of type $(NiO)_x(Bi_{1.6}Pb_{0.4})Sr_2Ca_2Cu_3O_{10-\delta}$, where $0.0 \leq x \leq 0.2$ wt. %, were prepared and characterized in previous work [7,8].The prepared sampled were characterized using X-ray diffraction analysis (XRD) using the Bruker D8 advance powder diffractometer. However, their microstructure morphology was identified using Jeol scanning electron microscope JSM-5300-7500.The thermoelectric power (TEP) of the prepared superconductor samples was measured using the standard differential technique in a temperature range from 250 to 77 K in liquid nitrogen. Sample of dimensions1.5 × 0.2 × 0.2 cm^3 was clamped between two copper blocks serving as temperature reservoirs; one block is heated with a metal resistor, and two T-type thermocouples were used to monitor the temperatures of both blocks with accuracy ± 0.1 K. Finally, the variation of thermoelectric voltage with the temperature difference was recorded using the "DATA Studio" program and an interface (Pasco CI-7500).The electrical resistivity of the prepared samples was measured from room temperature down to zero-resistivity temperature (*T*) with a closed cryogenic refrigeration system.

3. Results and Discussion

Figure 1 reveals the XRD patterns for $(NiO)_x(Bi_{1.6}Pb_{0.4})Sr_2Ca_2Cu_3O_{10\delta}$,superconductor samples. The diffraction patterns indicate that all the samples are well indexed with the tetragonal indices of (Bi, Pb)-2223 dominant phase with a space group P4/mmm [8], with a small amount of impurities such as (Bi, Pb)-2221 and Ca_2PbO_4 phases. The values of the lattice parameters, a= 36.96± 0.03Åand c = 5.42 ± 0.02Å do not show systematic variation with respect to the pure sample. Such results signifies that NiO particles did not participate in the crystal structure of (Bi,Pb)-2223 phase [8].

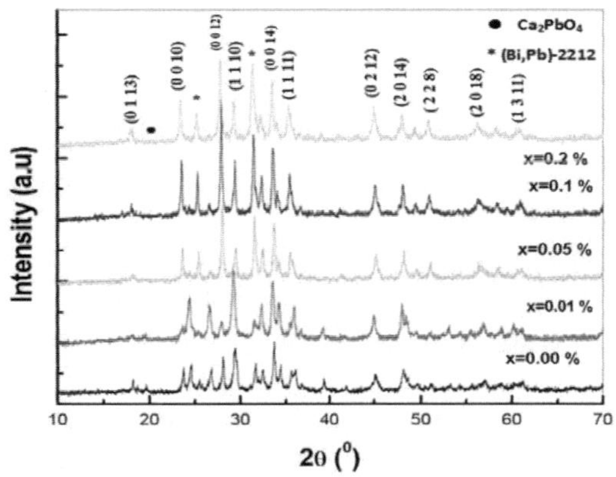

Figure 1: Typical XRD-patterns for $(NiO)_x$(Bi,Pb)-2223 phase superconductors added with different percentages of NiO nanoparticles.

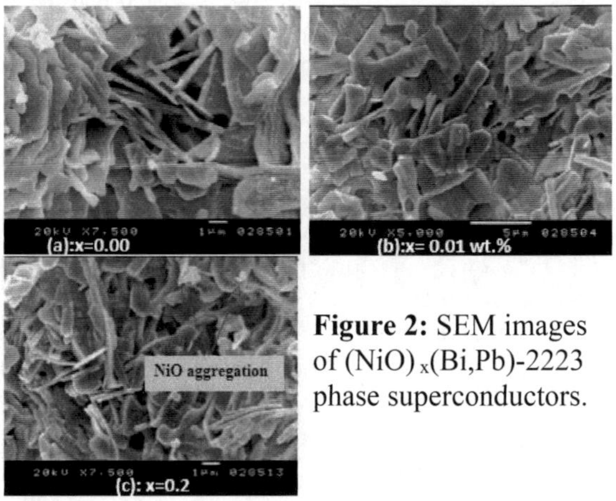

Figure 2: SEM images of $(NiO)_x$(Bi,Pb)-2223 phase superconductors.

Figure 2 shows typical SEM images of $(NiO)_x$ (Bi,Pb)- 2223 phase superconductors with (a) x= 0.00 (b) x = 0.01, and (c) x = 0.20 wt%. The pure sample consists of randomly oriented plate-like grains, indicating the phase formation of (Bi,Pb)-2223 [8] . However, figures 2(b) and 2(c) show the existence of some white patches for samples added with NiO nanoparticles. These are due to the sticking of NiO nanoparticles on grain boundaries, leading to an increase in grain connectivity and filling up of cracks and voids. Figure 2(b) shows that the inter-grain voids decreases upon adding 0.01 wt%. NiO. However, the addition of 0.2 wt%. NiO nanoparticles (Figure 2(c)) causes aggregation and agglomeration of NiO nanoparticles.

The temperature dependence of the thermopower coefficient, TEP, for $(NiO)_x(Bi_{1.6}\ Pb_{0.4})\ Sr_2Ca_2Cu_3O_{10-\delta}$, superconductor are shown in Figure 3. All samples have a positive seebeck coefficient (S), suggesting a predominant hole-type conduction of superconductors [9].It is obvious that S initially increases rapidly till 145-184K, consisting with Mott formula [10], then it decreases with rising temperatures due to thermal excitation of extrinsic charge carriers at high temperatures. As seen, S_{max} increases with NiO nanoparticles addition up to 0.05 wt. % and then it decreases with further addition of NiO nanoparticles. This may be attributed to the fact that low NiO additions reduces the number of pores and voids present in the superconductor matrix, which speeds up the heat transfer within the superconducting grains. Whereas, high

NiO concentrations result in aggregation of nanoparticles causing an increase in grain boundary resistance which decreases the heat transport in the matrix. Moreover, such behavior may be related to carrier concentration. In our case, it is noted that the NiO additions (up to $x = 0.05$) induces a slight decrease in carrier concentration *n,* leading to an increases in S values [11]. However, the decrease of S at high concentrations of NiO nanoparticles may be attributed to their agglomeration, thus providing large carrier concentration for x>0.05 wt. % which is consistent with the lower S in such samples [12]. The obtained thermoelectric power values in our samples are very small, indicating that the behavior of $(NiO)_x$(Bi,Pb)-2223 samples is typically metallic [1].

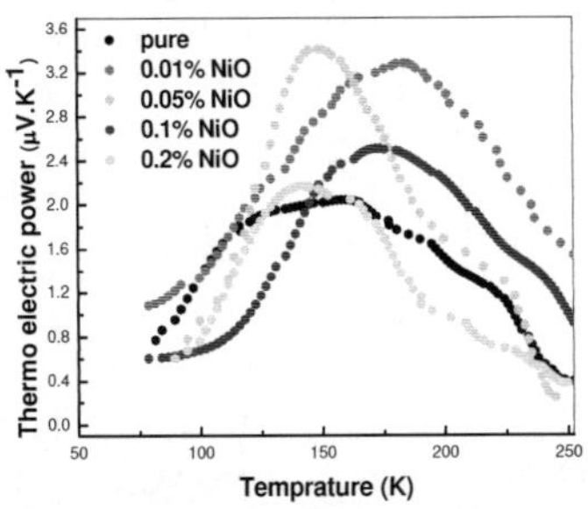

Figure 3: Seebeck coefficient of the $(NiO)_x$(Bi,Pb)-2223 superconductors in the temperature range from 77 to 250 K for various nano-NiO addition.

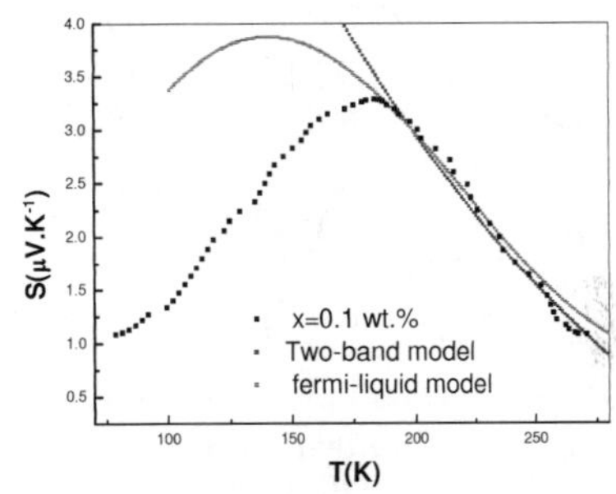

Figure 4: Fermi liquid model and two band model fitting for $(NiO)_x$(Bi,Pb)-2223.

The fitting of TEP data was carried on the decreasing according to the two-band (Fermi-liquid) and the two-band with an extra linear term models shown in equations 1 and 2 respectively [13]

$$S = AT + (B\lambda + C\,T)e^{(-\lambda/T)} \quad (1)$$

$$S = aT\,(b^2 + T^2)^{-1} + \alpha T \quad (2)$$

The fitting data according to both equations for x=0.1wt.% sample are shown in figure 4. The fitting parameters, as well as the calculated T_c obtained from and the room temperature electrical resistivity (the maximum point of the $d\rho/dT$ -T plot) and J_c for the prepared ssamples are listed in Table 1. The tabulated data showed that the parameter *a*, which is related to the contribution from the mobile holes in the Cu–O planes, increases then decreases with increasing nano-NiO addition. Large positive values of *a* imply a decrease in the hole concentration with temperature [14]. Values of the energy gap λ increases and then decreases with the increase of nano-NiO additions. T_c and J_c are improved up to $x = 0.01$ wt. % followed by a decrease with further increase in *x*. Such improvement can be discussed according to the enhancement in inter-grain connectivity, confirming with the results observed from SEM micrographs. The Fermi energy E_F and temperature T_F are given by equations 3 and 4 [15]:

$$E_F = \frac{\pi^2 k\,B^2}{3\,e\,\alpha}, \quad (3)$$

$$T_F = \frac{T_F}{k_B}. \quad (4)$$

The values of E_F, T_F, and $T^*$are given in table 1.It is well known that both E_F and T_F are related to carrier concentration [16]. As noticed from table 1, both E_F and T_F decrease with low nano-NiO additions till 0.05 wt. % and then increase for further additions. Such behavior may be attributed to the fact that N, T_F and E_F

are proportional, thus the decrease in T_F is accompanied with the decrease in carrier concentration. This confirms the increase in S for low NiO additions and vice versa for higher nano-NiO additions.

Table 1: The fitting parameters according to different models for $(NiO)_x (Bi_{1.6} Pb_{0.4}) Sr_2Ca_2Cu_3O_{10-\delta}$.

	Fermi-liquid model		**two-band model with an extra linear term**						
NiO wt.%	**A (μV.K^{-1})**	**λ (μV.K^{-1})**	**a (μV)**	**α (μV.K^{-2})**	**E_F (eV)**	**T* (K)**	**T_Fx10^4 (K)**	**T_c (K)**	**J_c (A/cm^2)**
0	0.029	635	560	-0.007	3.49	151	4.02	107	183
0.01	0.035	710	950	-0.009	2.72	184	3.15	111	190
0.05	0.033	700	745	-0.01	2.45	150	2.83	104	178
0.1	0.037	610	790	-0.009	2.72	174	3.15	102	174
0.2	0.0315	583	525	-0.007	3.26	145	4.02	99	169

4. Conclusion

Superconducting samples with nominal composition $(NiO)_x(Bi_{1.6}Pb_{0.4})Sr_2Ca_2Cu_3O_{10-\delta}$,($0.0 \leq x \leq 0.2$ wt%) were successfully prepared using solid state method. It was found that NiO nanoparticles did not alter the crystal lattice of (Bi, Pb)-2223 phase superconductors. However, nano- NiO additions proved their efficiency to enhance the seebeck coefficient of (Bi,Pb)-2223 phase at low concentrations due to the fact that the carrier concentration decreased, whereas high nanoparticles concentration may result in a decrease in *S* due to aggregation of NiO nanoparticles. Both E_F and T_F decreases with low nano-NiO additions till 0.05 wt. % and then increases for further additions. Such behavior may be attributed to the fact that *N*, T_F and E_F are proportional, thus the decrease in T_F is accompanied with the decrease in carrier concentration.

References

[1] Gul I and Maqsood A 2008 *J. Supercond. Nov. Magn* **21** 399
[2] Martins B P 2007 *Recent developments in superconductivity research.* (Hauppauge, New York Nova Publishers) p.128
[3] Tokura Y Torrance P Huang T and Nazzal A 1988 *Phys. Rev.* B **38** 7156
[4] Awana V Moorthy V and Narlikar A 1994 *Phys. Rev.* B **49** 6385
[5] Cooper J Obertelli S Carrington A Loran J and Liang W 1991 *Physica* C **185** 1265
[6] Joo S Son J Min B Lee J Kim B Ryu B and Lee H 2016 *J Korean* Phys Soc. **69** 1314
[7] Rahal H, Awad R, Abdel Gaber A and EL-Said Bakeer D 2017 *J Nanomater* **2017** https://doi.org/10.1155/2017/7460323.
[8] Rahal H, Awad R, Abdel Gaber A and Roumie M 2016 *J. Supercond. Nov. Magn* 1
[9] Lin C, Chen M, Huang R, Cheng Y and Lee P 2015 *Energies* **8** 12573
[10] Fan X, Yang J, Zhu W, Bao S, Duan X, Xiao C and Li K 2008 *J. Alloys Compd.* **461** 9
[11] Xiao Y, Yang J, Jiang Q, Fu L, Luo Y, Zhang D and Zhou Z 2016 *J. Electron. Mater.* **45** 1266
[12] Nandhakumar I, White N and Beeby S 2016 *Thermoelectric Materials and Devices Royal Society of Chemistry* **17**
[13] Xin Y, Wong K, Fan C, Sheng Z and Chan F 1993 *Phys. Rev.* B **48** 557
[14] Ibrahim E, Saleh M M, and Ahmed S A 2008 *Supercond. Sci. Technol* **21** 075001
[15] Casquilho J and Teixeira P 2014 *Introduction to Statistical Physics. Cambridge University Press*
[16] Onsager L 1931 *I. Phys. Rev.* **37** 405

Synthesis and characterization of Cu-doped TiO_2 thin films produced by the inert gas condensation technique

H A Ahmed[1], S I Abu-Eishah[2,*], A I Ayesh[3] and S T Mahmoud[4]

[1] Materials Science & Eng. Program, UAE University, Al-Ain, UAE
[2] Department of Chemical & Petroleum Eng., UAE University, Al-Ain, UAE
[3] Department of Mathematics, Statistics and Physics, Qatar University, Doha, Qatar
[4] Department of Physics, UAE University, Al-Ain, UAE
Email: s.abueishah@uaeu.ac.ae

Abstract. The bandgap of thin films Cu-doped TiO_2 nanoclusters prepared using the inert gas condensation (IGC) technique have been investigated at various Cu contents. The samples were characterized using XRD, SEM/EDS and UV-visible spectrophotometer. It was found that doping of TiO_2 thin film nanoclusters with Cu enhance its optical activity and shift it to the visible region; which makes it useful in photocatalytic applications.

1. Introduction
Nanostructured materials can be synthesized by the most widely inert gas condensation (IGC) technique, which can control the size, distribution and purity of the nanoparticles by using a thermal evaporation source, an electron beam evaporation device or a sputtering source in an inert gas atmosphere of argon or helium[1]. IGC can produce particles of well-defined grain size and with a narrow size distribution and can also have the capability to modify the nano-sized morphology, crystallinity, and particle size [2-4].

2. Materials and research methods

2.1 Materials Used
Argon gas (99.999% purity) was used along with a mass flow controller (MKS Instruments, Andover, MA1810, USA) in the range of 0–100 sccm. Ti (99.9% pure) and Cu-Ti alloys sputtering targets (with 1.6, 3.2, 4.8, 6.4, and 8 wt.% Cu) supplied by Testbourne Ltd. (USA) with 50.8 mm diameter x 6.35 mm thickness.

2.2 Preparation and Characterization of the Cu-TiO_2 thin films
An ultra-high vacuum (UHV) compatible system with a nanocluster source of magnetron plasma sputtering and inert gas aggregation condensation (Nanogen-50, Mantis Deposition Ltd., Oxfordshire, UK) was used to produce the Cu-Ti nanoclusters. The procedure followed for sputtering is the same as that published earlier [5] but with some variation in the parameters. The nanocluster-based thin film samples were characterized by using X-ray diffraction (Shimadzu 6100 X-ray Diffractometer with Ni and Cu-Kα radiation (λ = 0.15406 nm)), Scanning Electron Microscope (variable pressure Tescan VEGA XM SEM) Energy Dispersive Spectrometer (EDS) (Oxford Instruments X-Max 50 EDS detector (LN2 free system) with 125 eV resolution) and UV–visible absorption spectra (UV/Vis NIR spectrophotometer (JASCO 670) in the wavelength range of 200 to 3200 nm (photon energy of 0.39 eV to 6.2 eV)).

3. Results and Discussion

3.1 Structure Analysis

The deposited films were annealed at 400°C for 1h to form crystalline phase for the XRD analysis. Figure 1 illustrates the XRD patterns of the annealed undoped and Cu-doped TiO_2 nanoparticles based thin films. The patterns clearly show peaks of rutile TiO_2 for the prepared TiO_2 and all Cu-doped TiO_2 thin films, the lattice planes of rutile phase (110), (101), (111), (211) at 2θ values 27.4°, 36°, 41.1°, 54.3°. For different contents of Cu-doped samples, no Cu, Cu_2O or CuO were observed. The formation of rutile phase could be due to higher concentration of the excited clusters producing from the high electron temperature in the plasma, or due to higher energy of the particles impinging on the growing film surface [4].

3.2 Surface morphology, chemical composition, elemental analysis and size distribution

SEM analysis has been conducted to observe the morphology of sample with different contents of Cu-TiO_2. As seen in Figure 2, the shape of nanoparticles is almost spherical and the thin films are porous with some agglomeration.

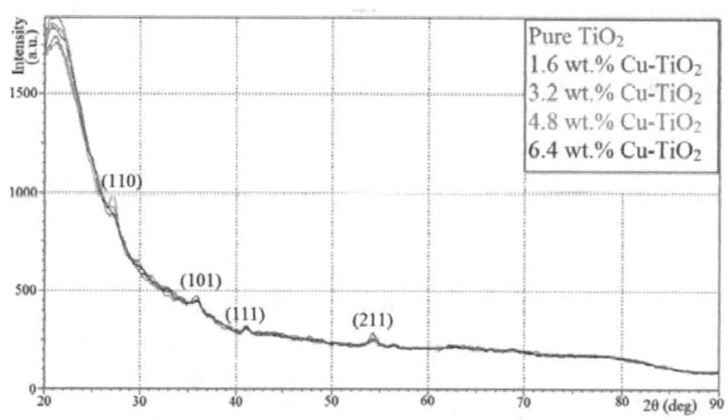

Figure 1. XRD patterns of pure TiO_2, 1.6 wt.% Cu-TiO_2, 3.2 wt.% Cu-TiO_2, 4.8 wt.% Cu-TiO_2, and 6.4 wt.% Cu-TiO_2.

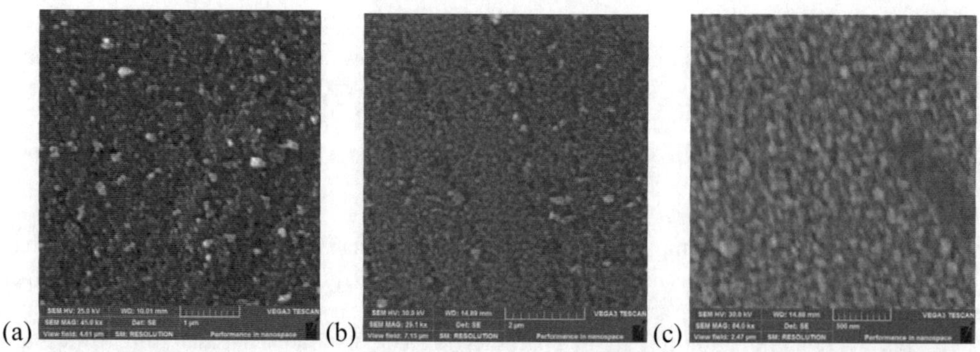

Figure 2. SEM images (a) pure TiO_2, (b) 3.2 wt.% Cu-TiO_2, and (c) 6.4 wt.% Cu-TiO_2.

On the other hand, Figure 3 illustrates the EDS analysis for 8% Cu-TiO_2 samples. For thin film Cu, Ti, O, C and Si were detected. Si shows the highest peak which is referred to the substrate (quartz glass). The appearance of C peak is due to the stub used. Both Si and C peaks were ignored in analysis. However, the presence of O reveals that the samples are completely oxidized as soon as they are exposed to air (at room temperature). Thin films have very strong cohesive and adhesive strength, so they can't be wiped out. This makes the prepared films very useful in photocatalytic applications which they can be used repeatedly for several times compared to the coated samples in the sol-gel method, which can be used only once. The small amount of Cu detected could be due to the low concentration in the sample.

Size distribution of Ti-Cu nanoclusters with different Cu contents (1.6, 3.2, 4.8, 6.4, and 8) wt.% measured using the QMF is shown in figure 4. The nanoclusters have a size distribution within the range of 5.4 to 7.1 nm. The intensity of the size distributions at various Cu contents is very low due to the small production yield of nanoclusters. The parameters that have great influence on nanocluster size are the Ar flow, f_{Ar}, and the aggregation length, L. In a previous work on pure Ti [5], we found that the changes in the Ar flow and aggregation length affect the nanocluster size and yield. Furthermore, the collision probability plays an important role in controlling the clusters nucleation and growth. The collision probability is detected by the pressure inside the aggregation chamber and by the clusters' residence time inside the cluster source, as reported elsewhere [6]. Therefore, Ar flow rate increases, the clusters will be removed faster, so their residence time in the aggregation chamber will be reduced. Thus when the interaction between the Ar and the clusters and metal atoms is reduced, it enhances a decrease in the clusters size [5]. Moreover, the amount of clusters and aggregates increase with increasing of the deposition time of the Cu-Ti target alloy.

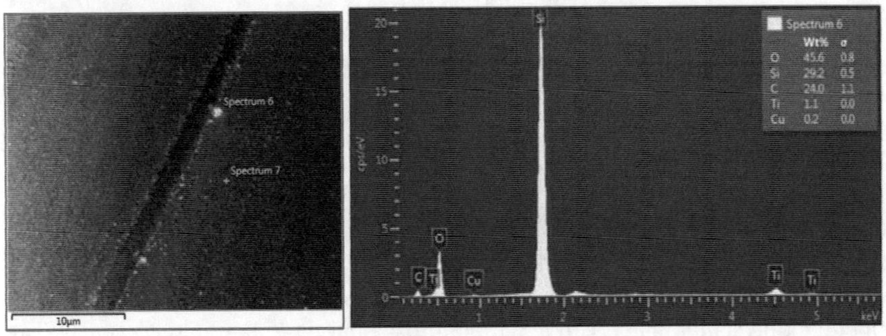

Figure 3. EDS analyses for 8 wt.% Cu-TiO_2.

3.3 Optical analysis

Figure 5 shows the UV-Vis absorption spectra of Cu-TiO_2 nanoparticles based thin films of various Cu contents (wt.%). Pure TiO_2 reveals absorption at 389 nm which lies in the range of UV region. Introducing Cu dopant with (1.6, 3.2, 4.8, 6.0, and 8.0 wt.%) resulted in shifting the absorption to the visible region. Absorption spectrum shifts up to 485 nm for the 8wt.% Cu content, which indicates the improvement of the optical activity of Cu-TiO_2 nanoparticles based thin films. The calculated band gap energy for pure TiO_2 and various Cu contents in Cu-TiO_2 thin films are presented in table 1. The band gap energy varies between 3.19 eV (0.0 wt.% Cu or pureTiO_2) and 2.56 eV (8.0 wt.% Cu). The presence of acceptor states in the band gap could be reason of this change [7]. In addition, transition metals narrow the electronic properties and alter the optical responses of TiO_2, causing a decrease in the bandgap energy and resulting in wide range absorber of visible light, which can be used in photocatalytic applications.

Table 1. Wavelength and band-gap of Cu-doped TiO_2 with various Cu contents.

Cu-TiO_2 (wt.%)	Wavelength (nm)	Band gap (eV)
0.0	389	3.19
1.6	400	3.10
3.2	415	2.99
4.8	431	2.88
6.4	460	2.70
8.0	485	2.56

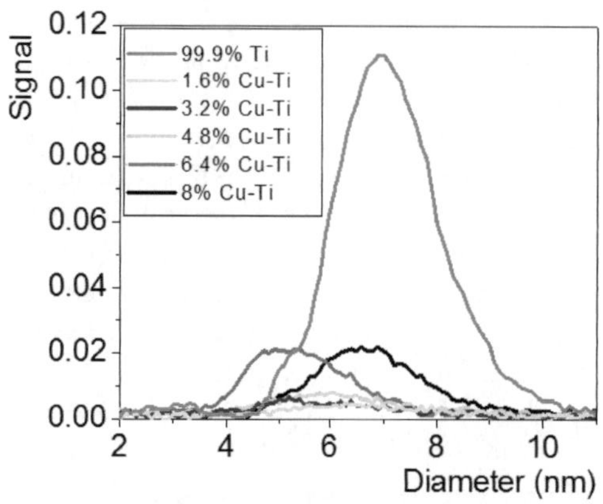

Figure 4. Size distribution of Ti-Cu nanoclusters with different Cu contents produced using I = 200 mA, f_{Ar} = 35 sccm, L = 90 mm and U/V = 0.12.

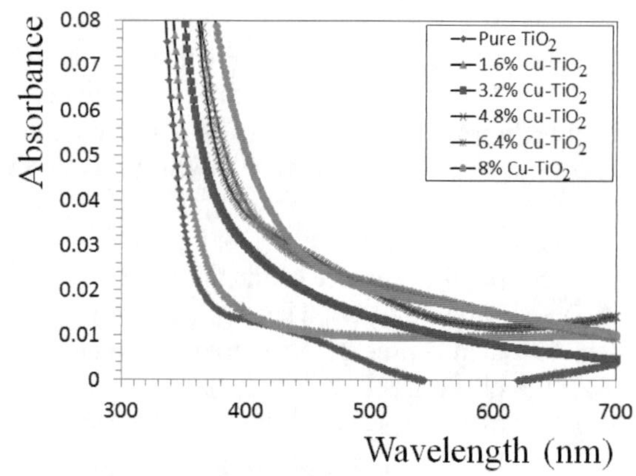

Figure 5. UV-Vis absorption spectra of Cu-TiO_2 thin films at various Cu contents (wt.%).

4. Conclusion

The IGC technique was used in this work to prepare thin films of Cu-TiO_2 nanoparticles with various Cu contents (0, 1.6, 3.2, 4.8, 6.0, and 8.0 wt.%). XRD analysis showed a rutile phase of TiO_2. SEM analysis showed homogeneous nanocluster films with almost spherical nanoclusters. EDS analysis insured the presence of Ti, O and Cu in the Cu-TiO_2 thin film samples. UV-Visible absorption spectra indicated that the presence of Cu in the Cu-TiO_2 thin film samples had shifted the light absorbance to the visible region. Increasing Cu content decreased the band gap energy form 3.19 eV to 2.56 eV. The Cu-doped TiO_2 thin films produced by sputtering are expected to have high activities when used in photocatalytic applications.

Acknowledgments

Funded by UAE University (Project #: 31N093) and Emirates Foundation– Grant #: 2011/177 (UAE).

References

[1] Raffi M, Mehrwan S, Bhatti T M, Akhter J I, Hameed A, Yawar W and Ul Hasan M M 2010 *Ann Microbiol* **60** 75–80.
[2] Guillou N, Nistor L C, Fuess H and Hahn H 1997 *Nanostruct. Mater.* **8** 545-57.
[3] Mattox D M 2010 Atomistic film growth and some growth-related film properties *Handbook of Physical Vapor Deposition (PVD) Processing* (Elsevier Inc.) 333-98.
[4] Yamagishi M, Kuriki S, Song P K and Yuzo Shigesato Y 2003 *Thin Solid Films* **442** 227–31.
[5] Ayesh A I, Ahmed H A, Awwad F, Abu-Eishah S I and Mahmood S T 2013 *J. Mater. Res.* **28** 2622-28.
[6] Manninen N K, Figueiredo N M, Carvalho S and Cavaleiro A 2014 *Plasma Process. Polym* **11** 629–38.
[7] Liu B, Zhao X, Zhao Q, He X and Feng J 2005 *J. Electron Spectroscopy & Related Phenomena* **148** 158–63.

IOP Conf. Series: Journal of Physics: Conf. Series **869** (2017) 012028 doi:10.1088/1742-6596/869/1/012028

The need for control of magnetic parameters for energy efficient performance of magnetic tunnel junctions

I A H Farhat[1], E Gale[2], C Alpha[3] and A F Isakovic[1]

[1]Khalifa University of Science and Technology, Dept. of Physics, Abu Dhabi, UAE
[2]University of Bristol, 12a Priory Road, Bristol, UK, BS8 1TU
[3]Cornell University, Center for Nanofabrication, Ithaca, NY, USA 14850
E-mail:abdel.isakovic@kustar.ac.ae

Abstract. Optimizing energy performance of Magnetic Tunnel Junctions (MTJs) is the key for embedding Spin Transfer Torque-Random Access Memory (STT-RAM) in low power circuits. Due to the complex interdependencies of the parameters and variables of the device operating energy, it is important to analyse parameters with most effective control of MTJ power. The impact of threshold current density, J_{co}, on the energy and the impact of H_K on J_{co} are studied analytically, following the expressions that stem from Landau-Lifshitz-Gilbert-Slonczewski (LLGS-STT) model. In addition, the impact of other magnetic material parameters, such as M_s, and geometric parameters such as t_{free} and λ is discussed. Device modelling study was conducted to analyse the impact at the circuit level. Nano-magnetism simulation based on NMAG™ package was conducted to analyse the impact of controlling H_K on the switching dynamics of the film.

1. Introduction and motivation

STT-RAM memory technology shows potential in improving the performance and scaling challenges of current memory technologies [1, 2]. Optimizing the energy performance of MTJ is crucial to further incorporation of MTJs into STT-RAM and related low power spintronics nanoelectronics devices platforms. Figure 1 shows an illustration of an MTJ, where geometrical parameters are defined on the device shown. The electrical energy required for switching the MTJ state, E_W, is defined as [3-5]:

$$E_W = J_c^2 \delta A t = J_{co}[1 - (k_B T/\Delta E)\,(\ln[\tau_p/\tau_o])]\,\delta A t, \tag{1}$$

where J_c is the critical current density, J_{co} is the threshold current density, k_B is Boltzmann constant, T is temperature, ΔE is energy barrier, τ_p/τ_o is the temporal signature of the switching process, δ is the resistance-area product (RA), A is the cross-sectional area of the junction and t is the duration of the applied voltage. For in-plane MTJs (IMTJ), J_{co} is defined as [6]:

$$J_{co} = (2e\alpha/\hbar\eta) t_{free} M_s (H_{applied} + H_K + 2\pi M_s), \tag{2}$$

where e is the charge of an electron, α is the Gilbert damping coefficient, $\hbar$ is the reduced Planck constant, η is spin polarization, t_{free} is the free layer thickness, M_s is the magnetization saturation, $H_{applied}$ is the external applied field and H_K is the effective anisotropy field. For perpendicular-to-plane MTJ (PMTJ), J_{co} is defined as [7, 8]:

$$J_{co} = (2e\alpha/\hbar\eta) t_{free} M_s (H_{applied} + H_K). \tag{3}$$

Magnetic material parameters and geometrical parameters are considered. In addition, we focus on an additional control on the energy performance of the device by controlling effective anisotropy field H_K. Due to the quadratic dependence of E_W, equation (1), on J_{co}, equations (2-3), this report focuses on optimizing E_W through optimizing J_{co}.

In section 2, the expressions of J_{co}, equations (2-3), are analysed as functions of geometric and magnetic parameters. Model predictions were compared with points from literature. Section 3 discusses circuit level device modelling of MTJ, where the impact of H_K on the device performance is studied at the circuit level. In section 4, nano-magnetic dynamics simulations are performed for simple MTJ devices, illustrating the impact of controlling anisotropy energy on the device nanomagnetic dynamics. Section 5 concludes with summary and future work.

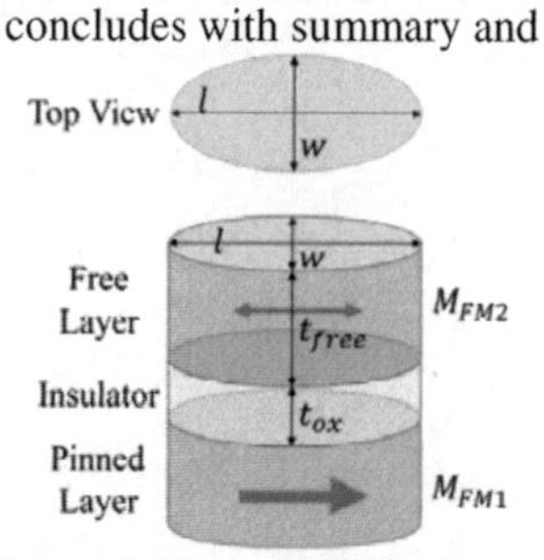

Figure 1. Basic MTJ device and parameters/notations used in this paper.

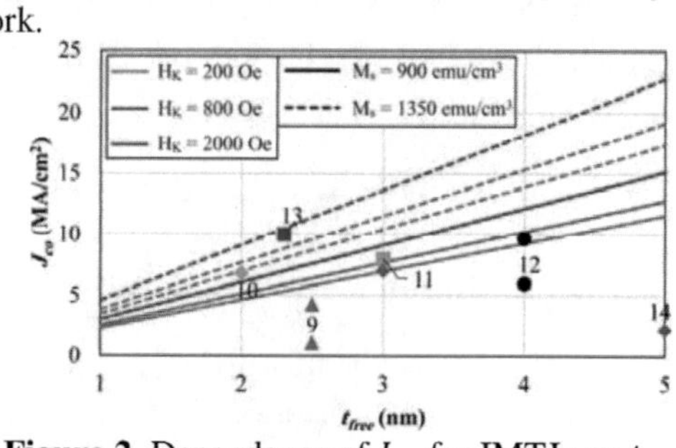

Figure 2. Dependence of J_{co} for IMTJs on t_{free} with varied H_K and M_s.

2. J_{co} analytical expression

Given the multivariate and multiparametric nature of this problem, it is understandable that some features of the overall findings from the nano-magnetism community on optimizing such nanodevices are not readily presentable in simple 2D graphs. We therefore used either parametric 2D plots or equivalent 3D plots. In figure 2, we present the dependence of $J_{co}$for IMTJ devices as a function of t_{free}, while controlling H_K and M_s. It is found that as t_{free} increases, J_{co} increases, where controlling H_K results in lowering J_{co} by ~20%. M_s impact on J_{co} is stronger, where ~30% reduction in J_{co} can be obtained by controlling M_s. The data points presented in figure 2 are from [9-14].

In figure 3 the behaviour of J_{co} (t_{free}, H_K) for both types of MTJs is shown. Although MTJ devices such as those in references here are bounded by model curves in figure 2, one can see in figure 3 that the model predictions can only be considered semi-quantitatively valid. Figure 3 (a) predicts that as t_{free} increases, the dependence of J_{co} on H_K increases, reaching ~20% decrease in J_{co} for high t_{free} values. Similarly, higher M_s values result in stronger dependence of J_{co} on t_{free} and H_K (faded plane in figure 3 (a)).

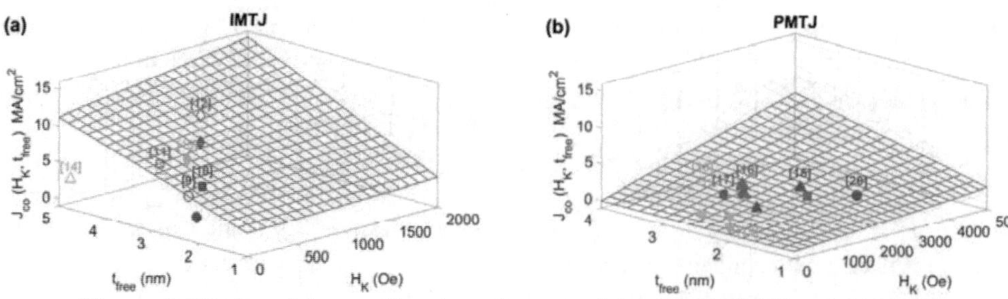

Figure 3. 3D plot of J_{co} as a function of t_{free} and H_K for (a) IMTJ and (b) PMTJ.

Figure 3 (b) depicts the behaviour of J_{co} for PMTJs, where similar behaviour as in figure 3 (a) is observed. However, a stronger impact of H_K on J_{co} is observed for PMTJs. The magnetic properties, such as M_s, have a strong impact on J_{co}. Controlling M_s is achieved by changing the materials used. However, in figure 4, we considered M_s as a more broadly controllable variable. A stronger dependence of J_{co} on M_s is found compared to its dependence on H_K for IMTJ devices. However, PMTJ devices show linear dependence on M_S, with the H_K impact being stronger. Even for very broad range of values of t_{free}, H_K and M_S, only a limited validity of the model predictions is achieved.

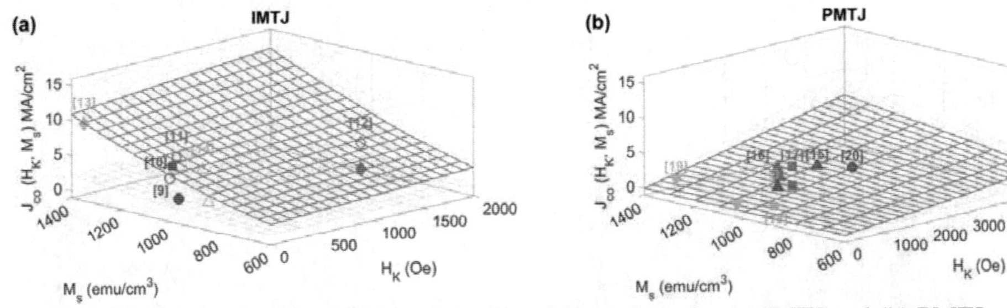

Figure 4. 3D plot of J_{co} as a function of M_S and H_K for (a) IMTJ and (b) PMTJ.

Due to the limited validity predictions of J_{co} equations (2-3) stemming from LLGS-STT approach, the utilization of J_{co} equations (2-3) is limited to predicting the range of the current density, and the trend of J_{co} behavior when controlling different parameters, as shown in figures 2-4.

3. MTJ circuit device modeling

We have modelled the I-V curves of the MTJ to explore the effect of controlling H_K on the overall power (energy) needed to operate the device. Zhao *et al.* MTJ circuit model was modified to account for the impact of H_K [21]. We modelled the device for several different parameter families. Some salient features are shown in figure 5, where the change in resistance-area product (δ) and H_K induces a change in switching current (I$_{SW}$) and voltage (V$_{SW}$) of the device. Figure 5 shows typical I-V hysteresis curves of the device, for two values of δ, 25 and 100 $\Omega\mu m^2$ and two values of H_K, 100 and 1000 Oe. Manipulating δ has greater impact on V$_{SW}$ compared to I$_{SW}$.

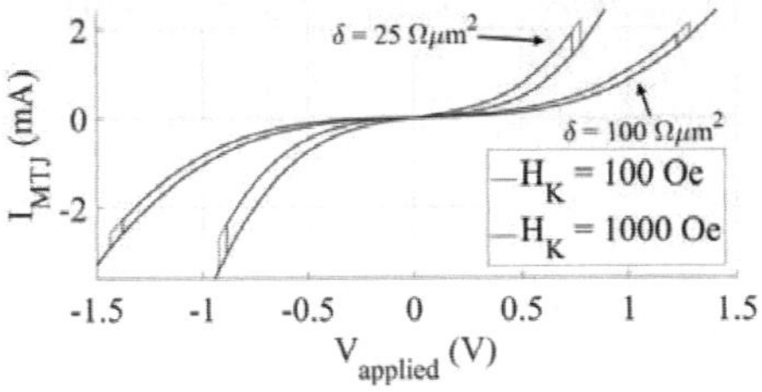

Figure 5. I-V hystersis loop of MTJ device model.

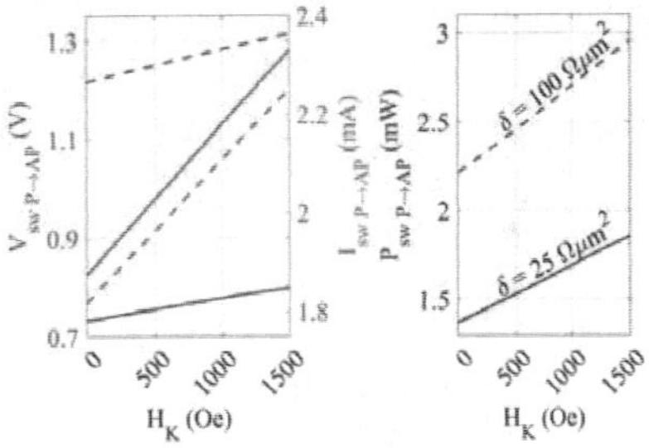

Figure 6. Effect of H_K and δ on V$_{SW}$, I$_{SW}$ and P$_{SW}$.

A set of simulations have been performed, where V$_{SW}$ and I$_{SW}$ are extracted and plotted as a function of H_K, (figure 6). We find that V$_{SW}$ shows stronger dependence on H_K. The switching power (P$_{SW}$ (H_K)) is calculated by multiplying V$_{SW}$ and I$_{SW}$ curves, figure 6. Since the impact of δ is stronger, we accounted for the impact of δ (solid and dashed lines are for δ = 25 and 100 $\Omega\mu m^2$, respectively.).

Controlling δ also affects the H_K, where larger δ values results in stronger impact on the control of H_K, and both are partially interface-controlled. For δ = 100 $\Omega\mu m^2$, changing H_K by ~40% results in decreasing V$_{SW}$ by < 5 %, while I$_{SW}$ decreases by ~10%. Therefore, controlling H_K, in addition to the control of δ, should result in further controlling the power consumption and energy performance.

4. Nano-magnetic simulations of magnetization reversal

Given that the need for a better control of H_K is one of the main suggestions of this report, we examined the influence the changes in anisotropy have on the elementary process of magnetization reversal in an elliptically shaped disk that is a model of one ferromagnetic layer of an MTJ. We used NMAG package

[21], and, we modelled devices $20 \times 10\ nm^2$ in size ($\lambda = 2$). The initial results of the magnetization dynamics are shown in figure 7. Two panels of the figure differ from each other only in the change of the effective second order anisotropy (by 30%). We note that both devices have completely “flipped”.

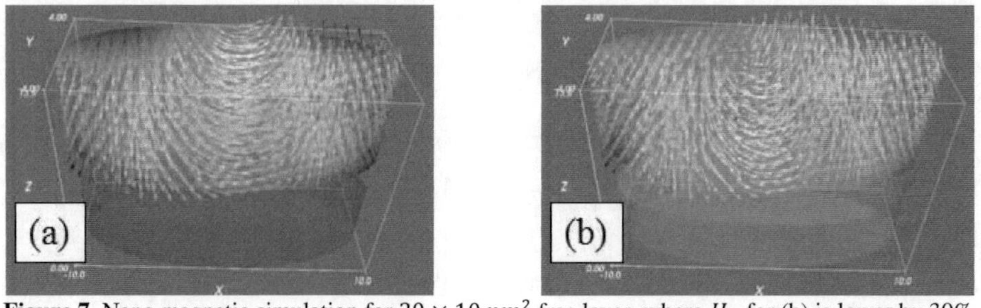

Figure 7. Nano-magnetic simulation for $20 \times 10\ nm^2$ free layer, where H_K for (b) is lower by 30%.

5. Conclusions and future work

It is shown that geometric and magnetic parameters can be tuned to minimization operating energy of MTJs. The role of H_K, t_{free} and M_s on the threshold current density, J_{co}, are demonstrated in the light of the need to control J_{co} for a more efficient energy performance of MTJs. We further emphasize the role of H_K as the key operating parameter that is tuneable, potentially via surface nanoengineering of MTJ layers.

Acknowledgement

The work in this paper has been supported by the Mubadala-SRC grant 2012-VJ-2335. Part of the work has been supported by Brookhaven National Laboratory, BNL-CFN, supported by the US DoE, Office of Basic Energy Sciences. We acknowledge support from A. Stein (BNL), K. Ng (SRC) and A. Jacob (GlobalFoundries). We are thankful to I. Zutic (SUNY-Buffalo) for useful advice in the early stages of the report.

References

[1] Chen E *et al* 2010 IEEE Trans Mag **46 (6)** 1873-1878.
[2] Slaughter J M *et al* in 2012 IEEE-IEDM.
[3] Lee S *et al* 2010 *Solid State Electron* **54(4)** 497-503.
[4] Koch R H *et al* 2004 *Phys. Rev. Lett.* **92(8)** 088302-088302.
[5] Lacour D *et al* 2004 *Appl. Phys. Lett.* **85(20)** 4681-4683.
[6] Sun J Z 2000 *Phys. Rev.* B **62(1)** 570-570.
[7] Zhang Y *et al* 2012 *IEEE Trans Electron Devices* **59(3)** 819-826.
[8] Ikeda S *et al* 2010 *Nat. Mater* **9(9)** 721-724.
[9] De Castro M M *et al* 2012 *J. Appl. Phys.* **111(7)** 07C912-07C912.
[10] Hayakawa J *et al* 2006 *Jpn J Appl. Phys.* **45(10L)** L1057-L1057.
[11] Huai Y *et al* 2004 *Appl Phys. Lett.* **84(16)** p 3118-3120.
[12] Finocchio G *et al* 2006 *Appl. Phys. Lett.* **89** 262509-262509.
[13] Huai Y *et al* 2005 *IEEE Trans Mag* **41(10)** 2621-2626.
[14] Sievers S *et al* 2015 *IEEE Trans Mag* **51(1)** 1400804-1400804.
[15] Amiri P K *et al* 2011 *Appl. Phys. Lett.* **98** 112507-112507.
[16] Chien C-W *et al* 2014 *IEEE Electron Device Lett.* **35** 738-740.
[17] Ikeda, S *et al* in 2014 *IEEE IEDM.*
[18] Huai Y *et al* 2006 *Jpn J Appl. Phys.* **45(5R)** 3835-3835.
[19] Hayakawa J *et al* 2005 *Jpn J. Appl. Phys.* **44(9L)** L1267-L1267.
[20] Bandiera S *et al* 2011 *Appl. Phys. Lett.* **99** 202507-202507.
[21] Zhao W *et al* in 2006 *IEEE Inter Behavioral Modeling and Simulation Workshop.*
[22] Hans F T F and Matteo F NMAG (2012) (URL: http://nmag.soton.ac.uk/nmag/)

IOP Conf. Series: Journal of Physics: Conf. Series **869** (2017) 012029 doi:10.1088/1742-6596/869/1/012029

The investigation of the mechanical properties of Mn doped $BaSnO_3$ nanoparticles

Iman Kazah[1] and R.Awad[1]

[1] Physics Department, Faculty of Science, Beirut Arab University, Beirut, Lebanon
E-mail: ekazah93@gmail.com

Abstract. This study investigates the effect of Mn substitution on the structural and mechanical properties of nano $BaSnO_3$ using X-ray diffraction analysis (XRD), transmission electron microscopy (TEM), Scanning electron microscope (SEM) and Vickers microhardness. For this investigation, samples of type $BaSn_{1-x}Mn_xO_3$ with $0.04 \leq x \leq 0.4$ were prepared by co-precipitation method. XRD patterns indicate the formation of single phase cubic structure $BaSnO_3$ with space group (Pm3m) for $0 \leq x \leq 0.1$. $BaMnO_2$ hexagonal phase appears as impurity for $x > 0.1$. The crystallite size, increases as x increases. Vickers microhardness H_v was carried out at different applied loads varying between 0.98 N and 9.8 N at dwell time 40 seconds. The results show that H_v increases as the Mn-content increases, whereas it decreases as the applied load increase. The H_v results were analysed using Elastic Plastic Deformation and the Modified proportional specimen resistance models. The Vickers microhardness analysis indicates that Modified Proportional Specimen Resistance model is the most suitable one for describing the load independent microhardness region of the investigated samples.

1. Introduction

One of the interesting compounds, in materials science, for practical application as a sensor material for host gases [1,2], photocatalytic applications [3,4] and optical devices [5] is $BaSnO_3$. In the last few years, many preparation methods were reported to obtain nanoparticles of $BaSnO_3$ with cubic perovskite structure such as solid state reaction technique [6], sol gel method [7] and hydrothermal method [1,8]. One of the most suitable methods to prepare nanoparticles is the chemical co-precipitation method because of its simplicity and control of the crystalline size.

Balamurugan. et al reported the effect of transition elements such as Fe [9] and Mn [10] on the magnetic properties of $BaSnO_3$. They found that these samples exhibited a ferromagnetic behaviour at room temperature. Moreover, the mechanical properties of $BaSnO_3$ were studied by Maekawa et al. [11], where Vickers microhardness was found to be 5 Gpa. To our knowledge, there are no reports on the mechanical properties of $BaSnO_3$ doped by transition elements. Hence, the aim of this work is to study the effect of Mn doping on the mechanical properties of $BaSnO_3$ through Vickers microhardness investigation.

2. Experimental details

The nano-samples of nominal composition $BaSn_{1-x}Mn_xO_3$, with $0.04 \leq x \leq 0.4$, were prepared using co-precipitation method [12] where the starting materials were $BaCl_2.2H_2O$, $SnCl_4.5H_2O$ and

$MnCl_2.4H_2O$. The nano-samples $BaSn_{1-x}Mn_xO_3$ were characterized by (XRD) analysis using Brucker D8 advanced power diffractometer with $20° < \theta < 80°$ range and Cu-Kα radiation (λ = 1.54056 Å). The surface morphology was examined using Jeol transmission electron microscope JEM 100CX and Scanning electron microscope (SEM) ASC-200, respectively. Vickers microhardness measurements were performed using a digital microhardness tester H_v MHVD-1000IS by using equation (1) [13]

$$H_v = 1854.4 \times \frac{F}{d^2} \quad , \tag{1}$$

where F is the applied load in Newton and d is the diagonal length of the indentation mark in μm.

3. Results and discussion

Figure 1 displays XRD patterns for $BaSn_{1-x}Mn_xO_3$ with $0 \leq x \leq 0.4$. The major peaks indicate the formation of $BaSnO_3$ cubic perovskite with space group #221 (pm3m) up to x=0.1. For x > 0.1, $BaMnO_2$ peak was observed at $2\theta = 31.2°$. The lattice parameter *a* and the crystalline size *D* were estimated from Bragg's equation and Debye Scherrer's relation respectively [12], a and D are listed in table1. TEM and SEM for $BaSnO_3$ and $BaSn_{0.9}Mn_{0.1}O_3$ are presented in figure 2.

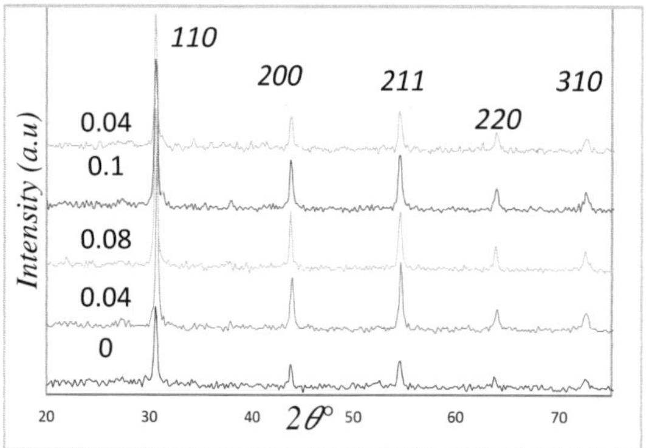

Figure 1. X-Ray diffraction patterns for $BaSn_{1-x}Mn_xO_3$ for x=0,0.04,0.08,0.1,0.4.

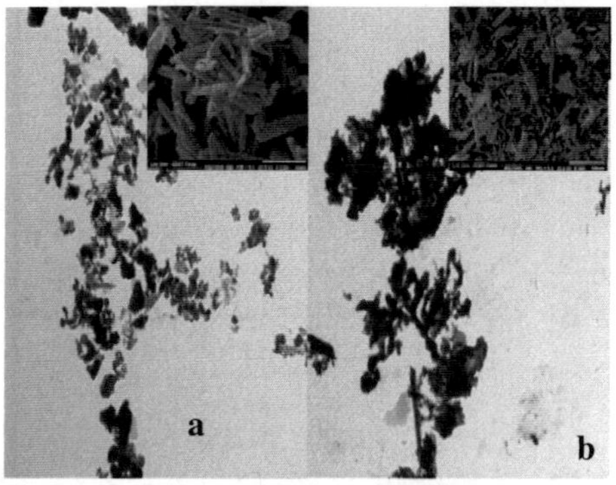

Figure 2. TEM and inserted SEM images for $BaSn_{1-x}Mn_xO_3$ samples for(a) x=0 and (b) 0.08.

The nano crystallites are of cuboidal shape with size ranging from 41-60 nm. The lattice parameter *a* fluctuates as x increases, where similar results were obtained by Ekaphan et al. [7] for $BaSnO_3$ doped by Fe. Figure 3 shows the variation of Vickers microhardness number H_v with the applied load F for $BaSn_{1-x}Mn_xO_3$ with x=0,0.04,0.08,0.1and 0.4. H_v decreases non-linearly as the applied loads increase till about 3.00 N, and it tends to attain saturation at F < 3.00 N. This behaviour can be explained on the basis of penetration depth of the indenter by Foerster et al. [14]. It is also noticed that H_v increases with the increase of x. This improvement can be attributed to the enhancement of resistance to crack propagation and due to the increase of cluster formation formed by Mn substitution.

The H_v results were analysed according to Elastic Plastic Deformation (EPD) and Modified Proportional Specimen Resistance (MPSR) models which are described by equations (2) and (3) respectively [15,16], the plots of the two models are represented in figures 4 a and b respectively. where A_1 is a constant, d_0 is the elastic component, $\alpha_1$corresponds to the minimum applied load to produce and indentation, $\alpha_2$and α_3 are related to the elastic and plastic properties of the material.

$$F = A_1(d + d_0)^2 \tag{2}$$

$$F = \alpha_1 + \alpha_2 d + \alpha_3 d^2, \tag{3}$$

Table 1. Lattice parameter (a), Crystallite size from XRD (DXRD), average particle size from TEM (DTEM) of $BaSn_{1-x}Mn_xO_3$ (x = 0,0.04,0.08,0.1 and 0.4) nanoparticles.

x	*a* (Å)	D_{XRD} (nm)	D_{TEM} (nm)
0	4.1118	50.0	60.0
0.04	4.1106	46.4	51.0
0.08	4.1148	44.2	48.0
0.1	4.1153	43.7	45.0
0.4	4.1176	42.4	41.0

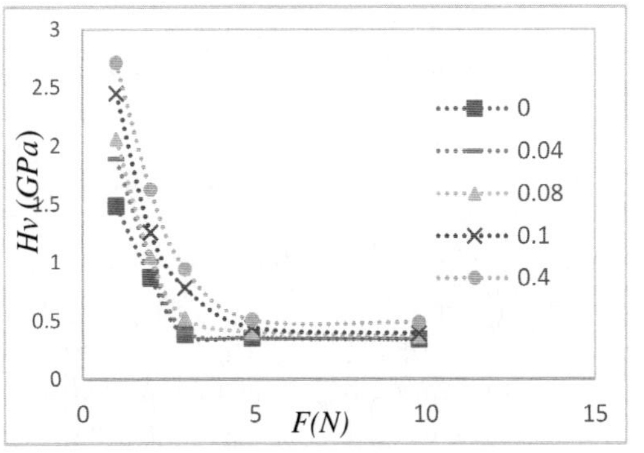

Figure 3. Variation of apparent H_v *with x=* *0.00*, 0.04, 0.08, 0.1 and 0.4 at different applied loads.

We notice from the figures that the experimental results are well fitted according to the 2 models. The fitting parameters for all models are listed in table 2. It is noticed from table 2 that the values of α_1 are positive for all samples, which indicates that the applied load is enough to create both the elastic and plastic deformations [15]. Moreover the values of d_0 are positive for all values of x which indicates that elastic deformation is present for all values of x. The True microhardness values obtained from EPD and MPSR approaches are calculated according to equations (4) and (5) respectively, and listed in table 2.

It is clear from table 2 that the values of H_{MPSR} are comparable with those obtained from experimental results. Whereas, it is noticed that EPD model displays much lower values than the experimental values H_v. Thus, the MPSR Model is the best fitted model for $BaSn_{1-x}Mn_xO_3$ samples.

$$H_{EPD} = 1854.4 \times \frac{F}{(d_0+d^2)} \quad (4)$$

$$H_{MPSR} = 1854.4 \times \frac{\alpha_1+\alpha_2 d+\alpha_3 d^2}{d^2} \quad (5)$$

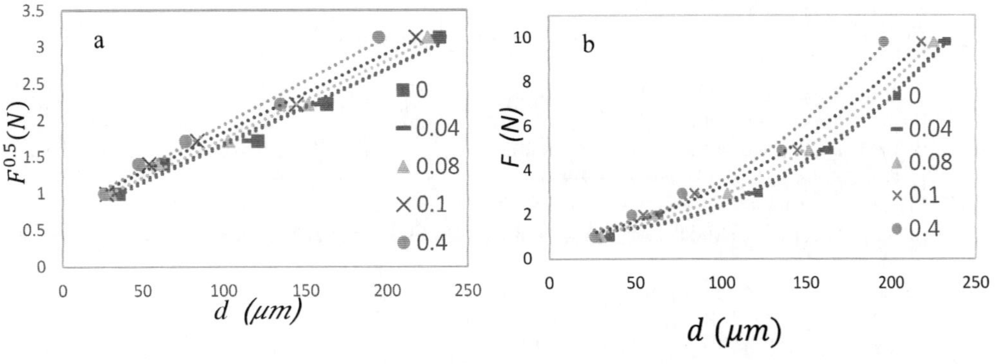

Figure 4. a. √F against d according to EPD model, b. F against d according to MPSR model.

Table2. Microhardness analysis results for different models and parameters for $BaSn_{1-x}Mn_xO_3$ samples with x= 0.00, 0.04, 0.08, 0.1, and 0.4; variation of Vickers microhardness values with respect to different applied models from experimental values calculated at t = 40 s.

	EPD model		MPSR model			Vickers microhardness values		
x	A_1 $(\frac{N}{\mu m^2})$	d_0 (μm)	α_1 (N)	$\alpha_2(\frac{N}{\mu m})$	$\alpha_3(\frac{N}{\mu m^2})$	Hv experimental	Hv (EPD)	Hv (MPSR)
0.0	0.00011	68.77670	0.00004	0.0291	0.0002	0.3555	0.1864	0.3632
0.4	0.00010	73.00980	0.00002	0.0330	0.0002	0.3668	0.1833	0.3953
0.8	0.00011	70.41509	0.00002	0.0343	0.0002	0.4286	0.2019	0.4273
0.1	0.00012	72.63889	0.00003	0.0361	0.0001	0.5373	0.2143	0.5106
0.4	0.00013	70.73684	0.00002	0.0398	0.0002	0.6442	0.2617	0.6037

4. Conclusion

$BaSn_{1-x}Mn_xO_3$ where $0.00 < x < 0.4$ were successfully prepared using co-precipitation method. XRD results indicated that the substitution of Mn to $BaSnO_3$ did not affect the cubic crystal structure but slightly increased the crystallite size. TEM and SEM images assure the increase of grain size of the particles and showed the Mn clusters that appear at high doping. The Vickers microhardness values increased as x increased. By analyzing Vickers microhardness according to different models, it was noticed that H_v results obeyed the Modified Proportional Specimen Resistance model.

References

[1] Maekawa T, Kurosaki K and Yamanaka S 2006 *J. Alloys Compd.* **416** 214–17
[2] Deepa A S, Vidya S, Manu P C, Solomon S, John A and Thomas J K 2010 *J. Alloys Compd.* **509** 1830–35
[3] Köferstein R, Jäger L, Zenkner M and Ebbinghaus S G 2009 *J. Euro. Ceram. Soc.* **29** 2317–24.
[4] Borse P H, Joshi UA, Ji SM, Jang JS, Lee J S, Jeong E D and Kim H G 2007 *Appl. Phys. Lett.* **90** 034103–11.
[5] Mizoguchi H, Woodward P M, Park C H and Keszler D A 2004 *J. Am Chem Soc* **126** 9796–9800.
[6] Fu-an Guo, Guoqiang Li and Weifeng Zhang 2010 *International Journal of Photoenergy* doi:210.1155/2010/105878.
[7] Ekaphan Swatsitang, Attaphol Karaphun, Sumalin Phokha and Thanin Putjuso 2016 *J Sol-Gel Sci Technol* **77** 78–84.
[8] Lu W and Schmidt H 2005 *J. Euro. Ceram. Soc.* **25** 919–925.
[9] Balamurugan K, Kumar H N, Chelvane J A and Santhosh P N 2008 *J. Alloys Compd* **472** 9–12.
[10] Balamurugan K, Kumar H N, Ramachandran B, Ramachandra Rao M S, Chelvane A J and Santhosh P N 2009 *Sol. Stat. Commun.* **149**: 884–87.
[11] Maekawa T, Kurosaki K and Yamanaka S 2006 *J. Alloys Compd.* **416** 214–17.
[12] Rahal H T, Awad R, Abdel-Gaber A M and EL-Said Bakeer D 2016 *Journal of Nanomaterials* ID 7460323.
[13] Leenders A, Ullrich M and Freyhardt H C 1997 *Phys.* C **279** 173.
[14] Foerster C E, Lima E, Rodrigues Jr P, Serbena F C, Lepienski C M, Cantao M P, Jurelo A R and Obradors X 2008 *Braz. J. Phys.* **38** 341.
[15] Awad R, Abou Aly AI , Kamal M and Anas M 2011 *J. Supercond Nov Magn* **24** 1947–1956
[16] Rahal H T, Awad R, Abdel Gaber A M and Roumie M 2016 *J. Supercond Nov Magn* DOI 10.1007/s10948-016-3654-4

IOP Conf. Series: Journal of Physics: Conf. Series **869** (2017) 012030 doi:10.1088/1742-6596/869/1/012030

Investigation of physical and mechanical properties of $(BaSnO_3)_x(Bi,Pb)$-2223 composite

K Habanjar[1], M M E Barakat[2] and R Awad[1]

[1] Physics Department, Faculty of Science, Beirut Arab University, Beirut, Lebanon
[2] Physics Department, Faculty of Science, Alexandria University, Alexandria, Egypt
E-mail address: khabanja@student.bau.edu.lb

Abstract: The effect of $BaSnO_3$ nanoparticles addition on the structural and mechanical properties of (Bi,Pb)-2223 superconducting phase by means of X-rays diffraction analysis (XRD), scanning electron microscope (SEM), electrical resistance and Vickers microhardness measurement was studied. $BaSnO_3$ nanomaterial and $(BaSnO_3)_x$(Bi,Pb)-2223 superconducting samples were prepared using co-precipitation method and standard solid-state reaction techniques, respectively. From XRD data, the addition of $BaSnO_3$ into (Bi,Pb)-2223 phase does not affect the tetragonal structure and the lattice parameters. SEM images indicate that the microstructure of (Bi,Pb)-2223 was enhanced by adding $BaSnO_3$ nanoparticles by filling its pores and voids. The superconducting transition temperature T_c as well as the critical transport current density J_c, estimated from electrical resistivity measurements, are increased up to $x = 0.5$ *wt%*, then decreased with further increase in x. Vickers microhardness measurements H_v were carried out at room temperature as a function of applied. The experimental H_v results were analysed in view of Meyer's law, Hays and Kendall (HK) approach, elastic/plastic deformation (EPD) and proportional specimen resistance (PSR). All samples exhibit normal indentation size effect (ISE), in addition to that, the analysis shows that the Hays and Kendall model is the most suitable one to describe the load independent microhardness for $(BaSnO_3)_x$(Bi,Pb)-2223 superconducting samples.

1. Introduction

(Bi,Pb)-2223 based compound appears to be the most effective candidate for power and technological applications because of its high superconducting transition temperature ($T_c \approx 110$ K), atmospheric stability and its high current and magnetic carrying capacity [1,3]. On the other hand, the major limitations of (Bi,Pb)-2223 phase are related to its ceramic nature, quite brittle, anisotropic and difficult to produce as a single phase [1]. The addition of various nanoparticles into (Bi,Pb)-2223 phase improved its physical properties such as microhardness, elastic modulus, yield strength and fracture toughness [4-6].

In this work, the effect of $(BaSnO_3)$ nanoparticles addition with different weight percentages on the structural and mechanical properties of (Bi,Pb)-2223 is investigated.

2. Experimental Techniques

Co-precipitation method was used to synthesis $BaSnO_3$ nanoparticles. Stoichiometric amounts of $BaCl_2.2H_2O$ and $SnCl_4.5H_2O$ were dissolved in 50 mL of distilled water. Simultaneously, the solution of $BaCl_2.2H_2O$ was heated and stirred till the temperature reached 75°C. A solution of NaOH is added to get pH=13, followed by the addition of $SnCl_4.5H_2O$. The mixture was then carried by a simultaneous heating and stirring until the temperature reached 95°C for 1h. The solution was washed with distilled water, after aged overnight. The resultant filtrate of $BaSnO_3$ was dried at 100°C for 24 h. The substrate was finally calcinated at 900°C for 2 h. Superconducting samples of $(BaSnO_3)_x(Bi_{1.6}Pb_{0.4})Sr_2Ca_2Cu_3O_{10-\delta}$, $0.0\% \leq x \leq 1.5\%$, were prepared by solid-state reaction technique as reported by *Rahal et al.* [5]. Resistivity-temperature and transport current density data were obtained by using four probe DC method. Vickers microhardness H_v is defined as the ratio of an applied load F in (N) to the pyramidal contact area d^2 in (μm^2) of indentation defined by:

$$H_v = 1854.4 \frac{F}{d^2} \quad (1)$$

Vickers microhardness of the samples were measured in atmospheric air at room temperature using a digital microhardness tester (MHVD-1000 IS) with a pyramidal indenter and loads ranging between 0.49N and 9.8 N.

3. Results and Discussion:

XRD patterns for $(BaSnO_3)_x$(Bi,Pb)-2223 superconducting samples indicate that all the major peaks belong to the tetragonal indices of (Bi,Pb)-2223 with a space group $P_{4/mmm}$ where the minor peaks correspond to (Bi,Pb)-2212 and Ca_2PbO_4 phase.

The calculated lattice parameters, $a = 5.394 \pm 0.002$ Å and $c = 37.14 \pm 0.04$ Å, do not show a significant change by adding $BaSnO_3$ nanoparticles. Moreover, the volume fraction of (Bi,Pb)-2223 is increased with increasing $BaSnO_3$ addition from 93.42% to 95.01% for *x*=0.5 *wt%*. SEM micrographs showed an enhancement in connection between the superconducting grains in the form of platelets which are also found by adding $BaSnO_3$ nanoparticles, indicating that nanoparticles do not participate in the crystal structure, they are sited between the superconducting grains [6-8].

The temperature dependence of the electrical resistivity for $(BaSnO_3)_x$(Bi,Pb)-2223 samples is shown in Figure 1a. The dc electrical resistivity measurements show a well-defined metallic behavior and superconducting transitions for all samples. The superconducting transition temperature is enhanced by increasing the addition of $BaSnO_3$ nanoparticles up to 0.5 *wt%* as well as the critical current density in Figure 1(b). This result is consistent with the behavior of the volume fraction and grain connectivity versus *x*.

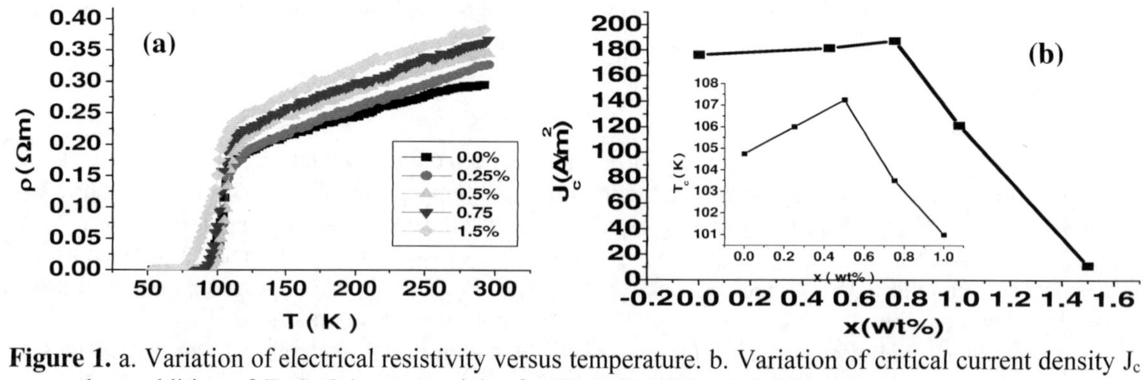

Figure 1. a. Variation of electrical resistivity versus temperature. b. Variation of critical current density J_c versus the x addition of $(BaSnO_3)$ nanoparticles for $(BaSnO_3)_x$(Bi,Pb)-2223 samples.

The room temperature Vickers microhardness H_v are calculated according to Equation (1) and plotted as function of the applied load in Figure 2. It is obvious that H_v is decreased non-linearly as the applied load is increased up to 3 N, beyond which the curves tend to saturate. This non-linear behavior, knowing as indentation size effect (ISE) can be explained qualitatively on the basis of penetration depth of the sample [5,9,10]. It is also observed that H_v increases with increasing $BaSnO_3$ nanoparticles addition from 0.0 to 0.05 *wt%*. This improvement can be attributed to the reduction of porosity, enhancement of resistance to crack propagation among the grains and the increase in grain connectivity.

The normal *ISE* behaviour could be explained according to Meyer's law which represents the relation between the load *(F)* and the indentation diagonal length *(d)*:

$$F = Ad^n \quad (2)$$

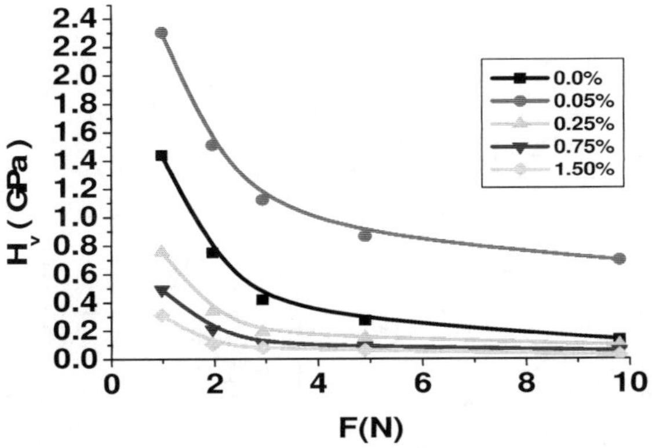

Figure 2. Variation of H_v with applied loads for $(BaSnO_3)_x$(Bi,Pb)-2223 samples at t=30 s.

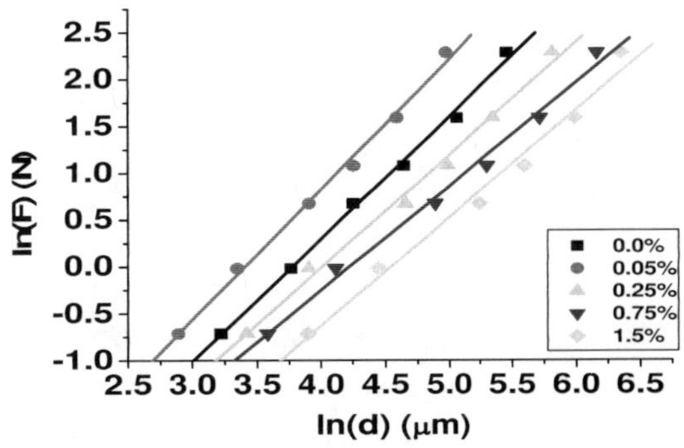

Figure 3. Variation of lnF with $\ln(d)$ according to Meyer's Model for $(BaSnO_3)_x$(Bi,Pb)-2223 samples.

The exponent n is the Meyer's index which is less than 2 for normal *ISE* and greater than 2 for reverse *ISE* [9,10]. It is noted that the values of n $(1.1 \leq n \leq 1.4)$ confirm the normal *ISE* behavior for all prepared samples. Several models were applied in order to estimate the true microhardness value. Hays and Kendall model revealed the existence of a minimum applied test load W_{HK} to initiate plastic deformation from which Hays and Kendall load-independent microhardness H_{HK} can be estimated [11]:

$$F = W_{HK} + A_1 d^2 \qquad ; \; H_{HK} = 1854.4 \times \frac{F - W_{HK}}{d^2} \qquad (3)$$

The Elastic/Plastic model assumed that the indentation contains an elastic portion, so to estimate the independent elastic/plastic microhardness H_{EPD}, an elastic component (d_0) should be added to the measured semi-diagonal plastic indentation [10-12]:

$$F = A_2 (d + d_0)^2 \qquad ; \; H_{EPD} = 1854.4 \; \frac{F}{(d + d_0)^2} \qquad (4)$$

Moreover, the proportional specimen resistance model assumed that during an indentation process, the external work applied by the indenter is changed into surface energy component (α_s) and strain energy component (β). Thus, PSR load independent microhardness H_{PSR} can be estimated [13]:

$$\frac{F}{d} = \alpha_s + \beta d \qquad ; \; H_{PSR} = 1854.4 \times \frac{\alpha_s d + \beta d^2}{d^2} \qquad (5)$$

Figures 4(a-c) display the variation of load variation with the indentation according to various models. The fitting parameters A_1, W_{HK}, A_2, d_0, α_s and β are listed in Table 1 versus x. It is clear that W_{HK} is positive indicating that the applied load is sufficient to create both elastic and plastic deformation. Also, the positive values of d_0 also indicate that the elastic deformation is observed along with the plastic deformation so elastic relaxation is present for all the samples. However, the increase in the value of surface energy α_s with increasing x up to 0.05% is due to dissipation of the energy from cracks at the

Table 1. Fitting parameters of HK, EPR and PSR for $(BaSnO_3)_x$(Bi,Pb)-2223 samples.

		HK Model			EPD Model			PSR Model		
x $(wt\%)$	Hv (GPa)	$A_1.10^{-4}$ (GPa)	W_{HK} (GPa)	H_{HK} (GPa)	$A_2.10^{-4}$ (GPa)	d_0 (μm)	H_{EPD} (GPa)	$\alpha_s.10^{-2}$ $(^N/_{\mu m})$	$\beta.10^{-4}$ $(^N/_{\mu m^2})$	H_{PSR} (GPa)
0.00	0.277	1.61	0.987	0.257	1.19	52.3	0.219	1.94	0.91	0.168
0.05	0.894	4.38	0.745	0.891	3.39	24.78	0.629	2.43	3.1	0.574
0.25	0.146	0.797	1.09	0.146	0.578	80.67	0.106	1.59	0.42	0.077
0.75	0.085	0.391	1.17	0.081	0.25	141.1	0.046	1.35	0.19	0.035
1.50	0.054	0.271	0.872	0.052	0.185	137.9	0.034	0.96	0.12	0.022

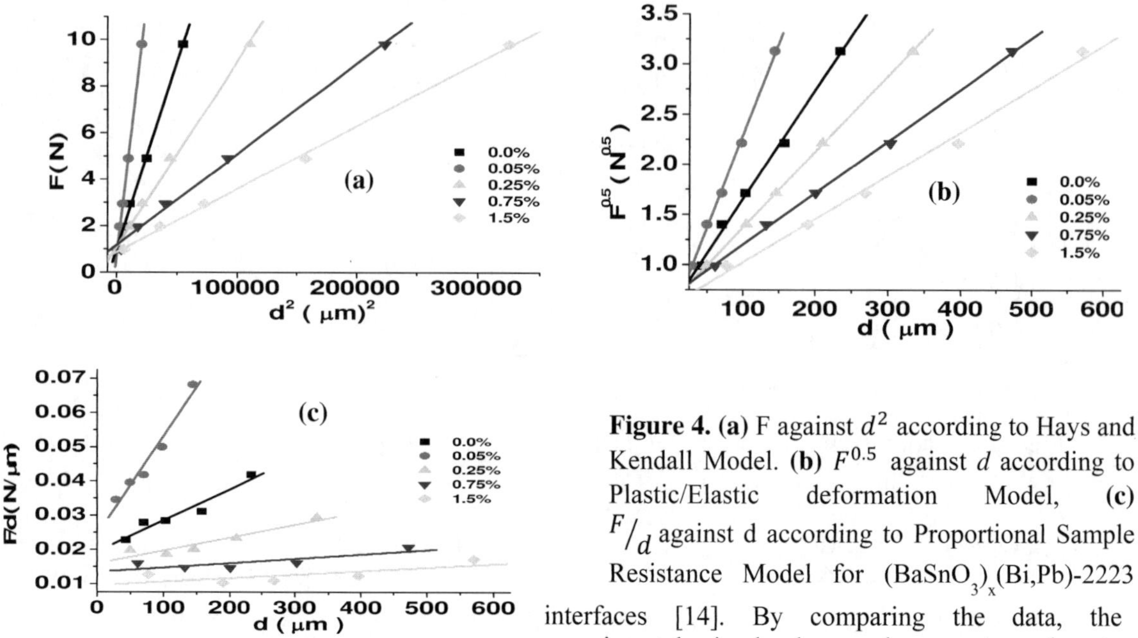

Figure 4. **(a)** F against d^2 according to Hays and Kendall Model. **(b)** $F^{0.5}$ against d according to Plastic/Elastic deformation Model, **(c)** F/d against d according to Proportional Sample Resistance Model for $(BaSnO_3)_x$(Bi,Pb)-2223 interfaces [14]. By comparing the data, the experimental microhardness values are very close to those obtained from HK model with deviation of about 7% which is the best fitting model. On the other hand, the values of H_v obtained from EPD and PSR models are much lower than the experimental values with deviations 21% and 40%, respectively.

4. Conclusion:

Series of $(BaSnO_3)_x$(Bi,Pb)-2223 samples were successfully prepared via conventional solid-state reaction technique. XRD shows that the tetragonal structure and the lattice parameters of (Bi,Pb)-2223 were not affected by the addition of nanosized $(BaSnO_3)$. The grain connectivity was enhanced and the voids were filled up with $(BaSnO_3)$ as the micrographs of SEM indicate. Vickers microhardness was measured and analysed using Meyer's law, HK approach, EPD and PSR models. The calculated values of H_v by the HK approach were more suitable than the other models, where H_v increased up to $x = 0.05\%$, and then it decreased for further increase in *x*.

References:

[1] Zelati A, Amirabadizadeh A, Kompany A and Salamati A 2014 *Indian J. Sci. Technol.* **7(2)** 123-134.
[2] Abdeen W, Marhaba S, Awad R and Abou A I 2016 *Advaned Ceramics.* **5(1)** 54-69
[3] Musa Abbas M 2015 *Int. J. Curr. Eng. Technol.* **5** No.3
[4] Ozkurt B, Madre M A, Sotelo A and Diez J C 2013 *J. Mat. Sci. Mat. Electron.* **24** 1158-67
[5] Rahal H T, Awad R, Abdel Gaber A M and Roumie M 2016 *Supercond. Nov. Mag.*
[6] Mohammad N H, Awad R, Abou A I, Ibrahim I H and Hassan M H 2012 *Mater. Sci. Appl.* **3** 224
[7] Ghazanfaria N, Kilic A, Gencerb A and Ozkana H 2007 *Solid. State.Commun.* **144** 210-14
[8] Roumie M, Abdeen W, Awad R, Korek M, Hassan I and Mawassi R 2014 *J. Low. Temp. Phys.* **174** 45
[9] Mott B W 1956 Microindentation Hardness Testing **206**
[10] Bull S J, Page T F and Yoffe E H 1989 *Philos. Mag. Lett.* **59** 281
[11] Hays C and Kendall E G 1973 *Metallurgy* **6** 275
[12] Upit G P and Varchenya S A 1966 *Phys. Status. Solidi (b)* **17** 831-35
[13] Li H and Bradt R C 1993 *J. Mater. Sci.* **28** 917
[14] Leenders A and Mich M 1997 *Phys.* C **279** 173-80

IOP Conf. Series: Journal of Physics: Conf. Series **869** (2017) 012031 doi:10.1088/1742-6596/869/1/012031

Electrical and structural properties of the Ta/Ag thin films prepared by DC magnetron sputtering

M A Moghri Moazzen[1], P Taiebyzadeh[2] and S M Borghei[3]

[1]Young Researchers and Elite Club, Karaj Branch, Islamic Azad University, Karaj, Iran
[2]Department of Physics, Ibn Sina College, Tehran, Iran
[3]Department of Physics, Karaj Branch, Islamic Azad University, Karaj, Iran
E-mail: m_moghri_moazzen@yahoo.com

Abstract. Tantalum on silver (Ta/Ag) thin films have quickly increased into high research for applied science with the promise of suitable for high temperatures environments and micro-systems for electronics applications. Ag and Ta/Ag thin films were deposited on silicon substrates by dc magnetron sputtering method. We choose the dc magnetron sputtering method because it has many advantages, such as high growth rate, the possibility of large area deposition, and low cost. X-ray diffraction (XRD) analysis and four point probe (FPP) were used for determining the prepared samples. For Ag thin film deposited in room temperature, there are no peaks corresponding to Ag in the XRD pattern which demonstrates amorphous structure. Also, the XRD pattern of Ta/Ag thin film illustrates that the peak of Ta has grown to the crystal direction (002), which shows that the structure of deposited Ta layer on Ag thin film becomes a crystalline state from amorphous state. The relationship between thin film resistivity and Ta/Ag film thicknesses are investigated in this paper.

1. Introduction

The electrical and structural properties of thin metal films can be effected by a large number of parameters such as time, surface, thickness and different kind of substrates. A composition of two or more thin metal film phases can improve the structural, electrical and optical properties. The wide use of tantalum thin films is in the knowledge of microelectronics, capacitors and transistors, because of their excellent physical and electrical properties [1-3]. Silver has many interesting attributes, like excellent high frequently virtues, highest electrical conductivity, low resistivity, and a rather high melting point [4]. The inset of Ag particles into thin films cans enhancement film conductivity [5].

With numerous studying the articles and the books, is reached this conclusion that Ta/Ag has been almost reviewed seldom. With these descriptions, the most papers existing in such a research field, are reviewing the electrical properties of Ag/Ta layers. Mardani et al. prepared Ag/Ta and Ag/TaN thin films using a DC magnetron sputtering method. They studied the thermal stability of Ag on SiO_2 with layer of Ta or TaN as a diffusion barrier. They found that Ag on Ta can be used permanent high-temperature metallization [5]. Thus, we predict that the professional applications of Ta/Ag layer are more practical in electrical devices.

In this work, we applied our research investigated the electrical and structural properties of Ta and Ta/Ag thin films that prepared by DC magnetron sputtering method. Eventually, we prepare a summary of our results and present some inferences.

IOP Conf. Series: Journal of Physics: Conf. Series **869** (2017) 012031 doi:10.1088/1742-6596/869/1/012031

2. Materials and methods

The substrates used in the experiments were (100) p-type silicon wafers. Ag thin films with 20, 30, 40 and 50 nm thicknesses were deposited on silicon (Si) substrate by DC magnetron sputtering of pure Ag in a room temperature. Substrates were cleaned, ultrasonically, in ethanol and acetone for 15 min each and blown dry using N_2 gas before being were introduced into the chamber. Thicknesses of prepared samples were controlled by a quartz crystal mass monitor in the chamber. The deposition parameters of Ag and Ta/Ag layers are shown in table 1.

Table 1. Deposition parameters of Ag and Ta/Ag layers.

Samples condition	Base pressure (mbar)	Work pressure (mbar)	Voltage (V)	Current (mA)	Deposition time (s)	Deposition rate (nm/s)	Thickness (nm)
Ag	4×10^{-5}	5×10^{-2}	70	15	100, 150, 200, 250	0.2	20, 30, 40, 50
Ta/Ag	5×10^{-5}	5×10^{-2}	105	65	420	0.48	20

Then the phase and crystalline structure of the prepared samples were characterized by using X-ray diffraction (XRD; Philips pw 1800) analyses using Cu Kα X-rays ($\lambda = 0.154$). Conventional four-point probe (FPP5000) resistivity measurements investigated the electrical properties of Ag and Ta/Ag films. The structural and resistivity of thin films will be briefly discussed.

3. Results and discussion

There are no peaks corresponding to Ag in the XRD pattern of the Ag thin film deposited on the Si substrate by 20 nm thickness in room temperature. The XRD pattern demonstrates amorphous structure for all Ag thin films. It is probably due to two reasons that no peaks are seen in Ag samples. Since the thickness of Ag is very low, so the X-ray easily penetrates into the substrate and effects on it. Just the peak of the silicon substrate can be observed in this spectrum. The second reason could be due to an adsorption between the substrate surface and Ag atoms. An adsorption phenomenon occurs because of the properties on the surface of the silicon substrate [6]. These properties have two origins. First, the discontinuity phenomenon which means that the properties of substrate surface are different from the properties of the rest of the substrate points and in fact, the surface atoms are different from the bulk atoms. Second, the unsaturated solid surface atoms which have created a surface bond with neighboring atoms below the surface of solid atoms, but at the level of atoms of the above are free and have the ability to create bonds with other atoms [7-8]. So there is a desire to reach a state of saturation in the molecules of the surface.

The XRD pattern of the Ta/Ag thin film deposited on the Si substrate is shown in 'figure 1'. As can be seen, The peak of the substrate observed at $2\theta = 69.41°$, in addition at $2\theta = 33.11°$, the peak of Ta has grown to the crystal direction (002), which shows that the structure of deposited Ta layer on Ag thin film becomes a crystalline state from amorphous state. So as it was mentioned above, the free Ag atoms have bonded to Ta atoms.

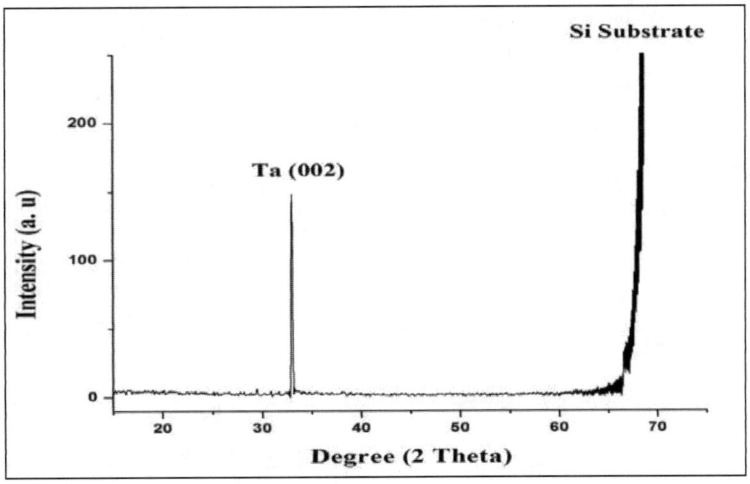

Figure 1. XRD pattern of Ta/Ag thin film.

The electrical properties of Ag and Ta/Ag thin metal films with different thicknesses of Ag from 20 to 50 nm were investigated by FPP method. The result shows the dependence of the Ag films resistivity on thickness. As can be seen, all the Ag thin films had amorphous structure, thus certain changes were not observed in their resistivity. The resistivity of thin films (ρ) can be expressed as [9]

$$\rho = \rho_p + \rho_m + \rho_f + \rho_i + \rho_s \quad (1)$$

Where ρ_p, ρ_m, ρ_f, ρ_i and ρ_s represents the resistivity caused by photon, impurity, defects, grain boundary and the surface scattering, respectively. The photon can be assumed to have the same influence on resistivity. The resistivity caused by impurity and the defects should be the same under the same deposition conditions except varying deposition time [2, 10]. Also, the resistivity due to surface scattering should not play its role in the decreasing resistivity trend in our experiment due to film thicknesses of our Ag films ranging from 20 to 50 nm are much higher than the electron mean free path of Ag [2, 10]. 'Figure 2' shows the dependence of the Ta/Ag films resistivity on Ag films thickness. So, the Ta/Ag thin films resistivity decreases with increasing Ag film thickness to 50 nm, in which the decreasing trend can be attributed to the inverse linear dependence of film resistivity increase on grain size.

Orientation of atoms in crystalline structures is very important. The materials have different crystalline orientation that shows they have different properties. In this project, crystalline orientation of Ta/Ag thin film is very significant for studied electrical and structural properties.

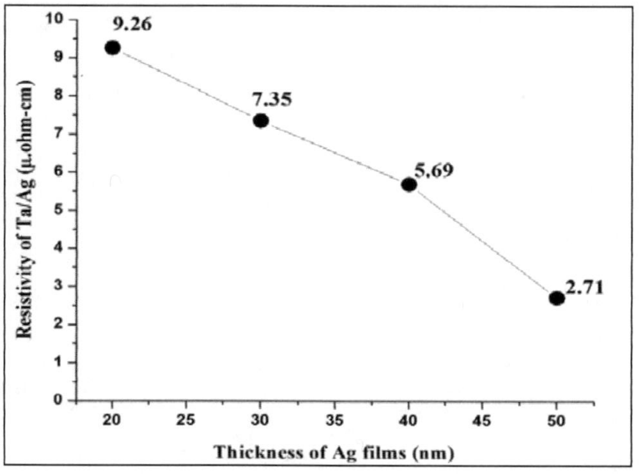

Figure 2. Dependence of the Ta/Ag resistivity on the Ag films thickness.

4. Conclusion

Ag and Ta/Ag thin metal films were deposited on Si substrates by DC magnetron sputtering method and studied in order to understand the influence of Ag thickness and Ta layer on electrical and structural properties. The results show the Ta layer effectively decreased the Ag agglomeration that's improvement in crystallinity. As the Ag thin films thicknesses increased, Ta/Ag thin films resistivity decreased, which illustrated that the Ta/Ag thin films are good choices in microelectronics devices, compared to Ag thin films.

References

[1] Park S H, Lee K S and Redy A S 2011 *J. APP. Phys.* **110** 063508
[2] Angha N 2001 *Expansion and contraction within Being* (Washington DC: M.T.O Publications) pp 359-408
[3] Ashcroft N W and Mermin D N 1976 *Solid state physics* (Brooks Cole: Harcourt, Inc.) pp 726-755
[4] Bock F X, Christensen T M and Rivers S B et al 2004 *Thin Solid Films* **468** 57-64
[5] Mardani S, Vallin O and Timo J 2014 *App. Phys. Lett.* **105** 071604
[6] Rouquerol J, Rouquerol F, Llewellyn P, Maurin G and Kenneth S.W 2013 *Adsorption by Powders and Porous Solids, Principles: Methodology and Applications* (London: Academic Press) pp 25-60
[7] Bruch L W, Cole M W and Zaremba E 2007 *Physical Adsorption: Forces and Phenomena* (New York: Dover Publications) pp 27-79
[8] Masel R I 1996 *Principles of Adsorption and Reaction on Solid Surfaces* (New York: Wiley-Interscience; 1 edition) pp 108-147
[9] Moazzen M A M, Hojabri A and Hajakbari F 2012 *JNS* **2** 107-112
[10] Foadi F, Darabi E and Hantezadeh M R 2014 *Eur. Phys. J. Plus.* **129**80

IOP Conf. Series: Journal of Physics: Conf. Series **869** (2017) 012032 doi:10.1088/1742-6596/869/1/012032

Effect of cobalt doping in $Nd_{1-x}Sr_xMn_{1-y}Co_yO_3$

M Kh Hamad [1], I A Abdel-Latif [2,3] and Kh A Ziq[1]

[1]Physics Department, King Fahd University of Petroleum and Minerals, Saudi Arabia
[2]Physics Department, College of Sciences and Arts and Advanced Materials and Nano-Research Centre, Najran University, P.O. Box 1988, Najran 11001, Saudi Arabia
[3]Reactor Physics Department, NRC, Atomic Energy Authority, Abou Zabaal P. O. 13759 Cairo, Egypt
Email: g201310510@kfupm.edu.sa

Abstract. We investigate the effects of Co substitutions the associated magnetic phase changes and the orthorhombic lattice distortions in $Nd_{0.6}Sr_{0.4}Mn_{1-x}Co_xO_3$.The effect of Co substitutions on the magnetic state and the orthorhombic lattice distortions has been studied. The orthorhombic distortion is found to be maximum (7.377×10^{-3}) for the pure manganite phase and decreases almost linearly with Co doping, reaching its minimum value 1.044×10^{-3} 1.044×10^{-3} for the pure cobaltites. Similar behavior has been found for unite cell volume. Since Co may have several oxidation states, it may also affect the (Mn^{4+}/Mn^{3+}) ratio, the double magnetic exchange interaction, the magnetic state and metal insulator transition.

1. Introduction

Manganite are perovskite materials that lend itself to several interesting practical applications [1-3]. The rich magnetic phase diagram arising from the interplay between the spin, orbital, charge, and lattice degrees of freedom results in considerable changes in their physical properties, such as magnetoresistance (MR) and magnetocaloric effect (MCE) [4].

Perovskite manganite compounds are represented as $R_{1-x}M_xMnO_3$, where R is trivalent rare-earth elements (Nd, La, …etc), and M is divalent alkaline earth metal like Sr, Ca, and Ba [5]. Upon cooling below its transition temperature, manganites commonly undergoes two kinds of transitions: electric; metal-insulating, and magnetic transition (FM or AFM) to paramagnetic states depending on the substituted elements and level. Variations of these transitions with substitution is a key idea in exploring their rich phase diagram. Our interest in $Nd_{1-x}Sr_xMnO_3$ system is mainly due to its possible use in near room temperature magnetic refrigeration [6]. Cobaltites on the other hand, have additional spin and lattice degrees of freedom; namely Co ion may exist in more than one oxidation state with multiple spin state in which Hund's coupling energy is comparable to the crystal field splitting energy. Hence, it is interesting to study a material that contains manganite and cobaltites, and investigate the transition from one system to the other [7-9]. In this paper, we investigate changes in magnetic states of $Nd_{0.6}Sr_{0.4}MnO_3$ upon Co substituting at the Mn sites.

2. Experimental Details

One set of $Nd_{0.6}Sr_{0.4}Mn_{1-x}Co_xO_3$ (x= 0, 0.3, 0.7, and 1) was prepared using co-precipitation method (see ref. [10] for details) and another set was prepared by standard solid-state reaction. High purity (4*N* to 5 *N*) powder of Nd_2O_3, $SrCO_3$, MnO_2, and CoO were used. Stoichiometric ratios of the elements were mixed, grind and pressed into pellets and calcinated in air for 10 hours at 1000 °C. Samples were grinded, pressed and re-annealed at 1000 °C and 1300 °C for another 10 hours at each temperature. The samples were furnace cooled to room temperature.

A homemade ac-susceptometer was fitted in closed cycle refrigerator (Model SHI-4T). Real and imaginary parts of the susceptibility were measured using SR850 Lock-in-amplifier. The temperature was monitored using Lake Shore 336 temperature controller along with LABVIEW software to control all measuring systems. The X-ray powder diffraction patterns were obtain using Rigaku XRD (MiniFlex-600).

3. Results and discussions

3.1. X-ray analyses

XRD-diffraction patterns for $Nd_{0.6}Sr_{0.4}Mn_{1-x}Co_xO_3$ are presented in figure 1. XRD-patterns were analyzed using Fullprof and Maud software, the goodness of fit x was in the range 5–6. The crystalline structure of $Nd_{0.6}Sr_{0.4}Mn_{1-x}Co_xO_3$ is orthorhombic with Pnma-space group (No 62). Traces of non-reacted Nd_2O_3 are found near 30 degrees for samples prepared by solid state reaction. Samples prepared using chemical rout and annealed at lower temperatures (850 °C) did not show traces of non-reacted Nd_2O_3 and were in all subsequent XRD- analyses. All calculated lattice parameters and the unit cell volume are listed in table 1. Data belongs to samples prepared by solid state reaction are marked as (a) in the table 1. The observed variations in the lattice parameters reflects the sensitivity of the lattice parameters to the annealing temperatures and cooling process. The orthorhombic lattice distortion (d) is an important parameter that is correlated with electrical and magnetic properties. The unit cell volume and the D-values are given in table 1 and plotted in figure 1(b). The lattice distortion and the volume of the unite cell are maximum for the pure manganite sample $Nd_{0.6}Sr_{0.4}MnO_3$ and decrease with increasing Co-substitution, reaching their minimum values in the pure cobaltites sample $Nd_{0.6}Sr_{0.4}CoO_3$.

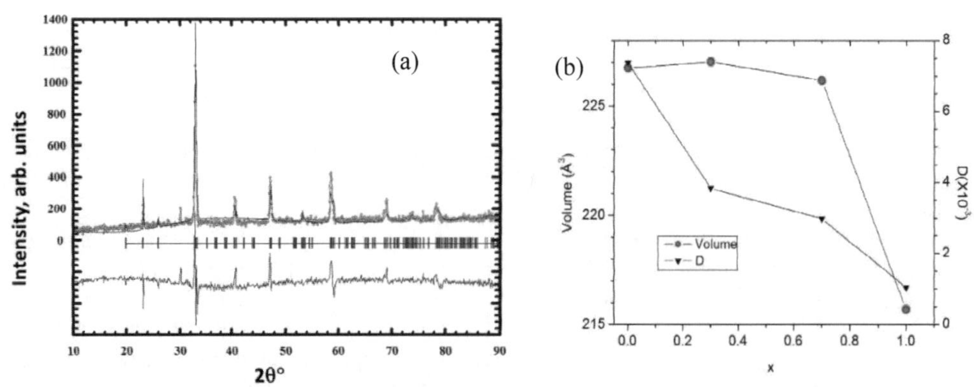

Figure 1. (a) Room temperature XRD-pattern for $Nd_{0.6}Sr_{0.4}MnO_3$. (b) Variations of unit cell volume and the orthorhombic distortion with Co-concentrations.

3.2. AC-susceptibility

The variations of real part of the ac-susceptibility with temperature is presented in figure 2. The figure shows that he pure manganite sample $Nd_{0.6}Sr_{0.4}MnO_3$ is ferromagnetic with a wide maximum centered around 180 K, followed by a sharp transition to the paramagnetic state near 260 K. The figure also reveals the presence of another maximum near 45 K, in agreement with previously published data and was attributed to the charge ordered state. The cobaltite $Nd_{0.6}Sr_{0.4}CoO_3$ susceptibility reveals a sharp peak at 190 K. This is reminiscent of the sharp peak often observed in spin glass material and has been confirmed by ZFC and FC susceptibility shown in figure 2(b) [10-12]. The transition temperatures T_C are presented in table 1.

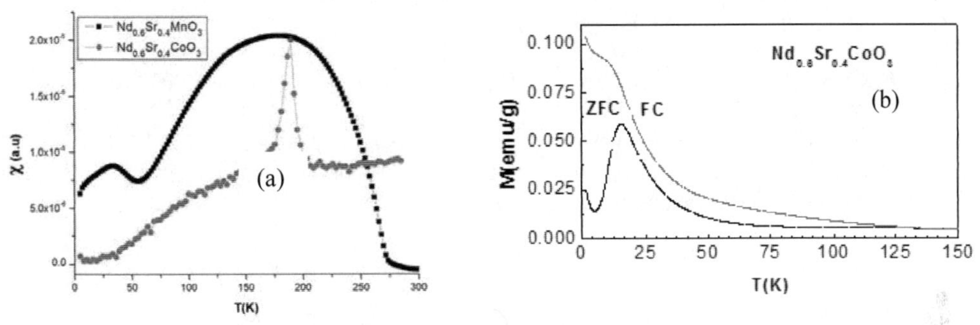

Figure 2. (a) The ac-susceptibility for $Nd_{0.6}Sr_{0.4}Mn_{1-x}Co_xO_3$polycrystalline samples. (b) Changes of magnetization under ZFC and FC protocols with temperature.

Table 1. Lattice parameters for samples with different Co concentration.

x	a (Å)	b (Å)	c (Å)	V (Å^3)	D ($\times10^{-3}$)	T_c(K)
1.0	5.3632(5)	7.5499(10)	5.3266(7)	215.68(5)	1.044	190
0.7	5.4524(5)	7.6682(6)	5.4094(6)	226.17(4)	2.98	147
0.3	5.4605(4)	7.6944(6)	5.4035(4)	227.03(3)	3.839	119
0.0	5.4646(7)	7.6519(13)	5.4107(10)	226.74(6)	7.377	
0.0 (a)	5.4787(5)	7.6859(8)	5.4439(7)	229.24(2)		180
Goodness of fit	$\chi = 5-6$					

4. Conclusion

Cobalt substitutions in $Nd_{0.6}Sr_{0.4}Mn_{1-x}Co_xO_3$ dramatically affects its magnetic state and structural properties. The unsubstituted $Nd_{0.6}Sr_{0.4}MnO_3$ material showed wide superparamagnetic peck near 175 K and sharp drop to paramagnetic phase near 250 K, while the cobaltites $Nd_{0.6}Sr_{0.4}CoO_3$ showed sharp antiferromagnetic peck near 180 K. Moreover, the orthorhombic distortion is found to decreases almost linearly with Co substitutions, reaching its minimum value the pure cobaltites.

5. Acknowledgments

We would like to acknowledge the help and support provided by King Fahd University of Petroleum and Minerals (FUPM) through funding the project number IN121002.

References

[1] Tokura Y and Nagaosa N 2000 *Science* **462** 288
[2] Goodenough J B 1972 *Prog. Solid State Chem.* **5** 145

[3] Wang H, Zhang X, Hundley M F, Thompson J D, Gibbons B J, Lin Y, Arendt P N, Foltyn S R, Jia Q X, MacManus Driscoll J L 2004 *Appl. Phys. Lett.* **84** 1147
[4] Ben Jemaa F, Mahmood S H, Ellouze M, Halouani F 2015 *Ceramics International* **41** 8191
[5] Manh-Huong Pand Seong-Cho Y 2007 *J. Magnetism and Magnetic Materials* **308** 325
[6] Mansour A I, Ziq K H, Salem A F, Haman R 2014 *J Supercond Nov Magn* **27** 183
[7] Kundu S and Nath T K 2013 *Journal of Magnetism and Magnetic Materials* **325** 1–6
[8] Zobel C, Kriener M, Bruns D, Baier J, Grüninger M, Lorenz T, Reutler P and Revcolevschi A 2002 *Phys. Rev.* B **66** 020402
[9] Imada M 1998 *Reviews of Modern Physics* **70** 1039
[10] Hassan M M, Ahmed A S, Chaman M, Khan W, Naqvi A H, Azam A 2012 *Mater. Res. Bull.* **47** 3952–3958.
[11] Thaljaoui R, Pękała M, Fagnard J F, Vanderbemden Ph 2016 *Physica* B **482** 8–13
[12] Hamad M Kh., PhD Thesis, Department of Physics, King Fahd University of Petroleum and Minerals, under preparation.

IOP Conf. Series: Journal of Physics: Conf. Series **869** (2017) 012033 doi:10.1088/1742-6596/869/1/012033

EPR studies of $SmBa_2Cu_3O_{7-\delta}$ /$MnFe_2O_4$ superconducting composites

R Awad[1], S Isber[2], A I Abou Aly[3], WAbdeen[3], A El-Tahan[4], E M El-Maghraby[5], A Khalaf [5] and S Noureddine[6]

[1]Department of Physics, Faculty of Science, Beirut Arab University, Beirut, Lebanon
[2]Department of Physics, Faculty of Science American University of Beirut, Beirut, Lebanon
[3]Department of Physics, Faculty of Science, Alexandria University, Ibrahimia, P.O. Box 426, Alexandria 21321, Egypt
[4]Department of Physics, Faculty of Science, Tanta University, Tanta, Egypt
[5]Department of Physics, Faculty of Science, Damanhour University, Damanhour, Egypt
[6]Department of Physics, Faculty of science, Lebanese University, Beirut, Lebanon

Abstract. Electron paramagnetic resonance (EPR) studies are carried out on composite samples of $SmBa_2Cu_3O_{7-\delta}/(MnFe_2O_4)_x$ superconductors with $0.00 \leq x \leq 0.20$ wt% at different temperatures $100K \leq T \leq 295K$. An EPR signal of Cu^{2+} ions in the orthorhombic local symmetry is observed. Moreover, the EPR signal intensity increases as the temperature decreases from room temperature down to 100 K. The temperature dependence of g-factors indicates the pseudogap onset temperature (T*), corresponding to the peak of the observed curves between 127 K and 162 K. The number of spins (N) participating in EPR resonance are calculated as a function of both nanosized $MnFe_2O_4$ addition and temperature. In particular, it is shown that any EPR signal for $SmBa_2Cu_3O_{7-\delta}/(MnFe_2O_4)_x$ is probably due to small amount of spurious phases typically present in the samples such as $BaCuO_2$.The presence of $BaCuO_2$ is confirmed by magnetization-field (M-H) hysteresis measurements

1. Introduction

Electron paramagnetic resonance (EPR) is a powerful tool in high temperature superconductors (HTSCs) for examining their magnetic origin, and clarifying the nature of the superconducting mechanism [1]. However, a large number of investigations on HTSCs of the type $REBa_2Cu_3O_{7-\delta}$ (RE-123, RE rare-earth ions) using the EPR method have been reported [2]. The majority of these investigations exhibit a characteristic EPR signal of Cu^{2+}ions with orthorhombic local symmetry. Up to now, a number of authors have suggested that this spectrum of Cu^{2+} ions arises from other phases, e.g. green phase or $BaCuO_2$ [3]. However, this only enhances its importance as a tool for the detection of Cu^{2+} containing impurities. In the present work, we report on the EPR studies of Sm-123 added with nanosized $MnFe_2O_4$, with $0.00 \leq x \leq 0.20$ wt%.

2. Experimental Techniques

Nanosized superparamagnetic ($MnFe_2O_4$) and $SmBa_2Cu_3O_{7-\delta}/(MnFe_2O_4)_x$, x= 0.0, 0.02, 0.08, 0.15 and 0.20 wt%, superconducting samples were prepared by the co-precipitation method and the conventional solid state reaction respectively, as previously reported [4]. EPR measurements were performed with a Bruker ER 041 X-band spectrometer operating at 9.24 GHz using 100 kHz field modulation in the temperature range 100–295 K. A vibrating sample magnetometer (VSM), Lakeshore 7410, was used to investigate the magnetic properties at 77K.

3. Results and Discussion

In our previous work [4] effects of nanosized $MnFe_2O_4$ addition on superconducting properties of $SmBa_2Cu_3O_{7-\delta}$ (Sm-123) has been investigated in details. The value of T_c decreases form 91.50 K to 79.01 K as x increases from 0.00 to 0.20 wt% which consistent with that the magnetic impurities generally suppress superconductivity. Figure 1 shows the first derivative EPR absorption spectra of Sm-123 pure sample at temperatures $100K \leq T \leq 295K$. In the all temperature range, the EPR spectra consist of a single strongly asymmetric line corresponding to an anisotropic g-factor with three principal g-values g_x, g_y and g_z characteristic of an orthorhombic symmetry around the Cu^{2+}ion [3].

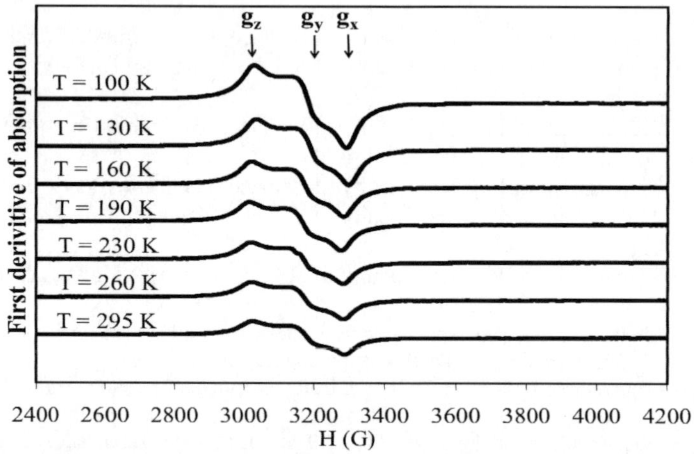

Figure 1. First derivative absorption spectra for $SmBa_2Cu_3O_{7-\delta}$ at different temperatures.

The variation of the anisotropic g-values with T for Sm-123 pure sample is shown in figure 2. It is observed that g_x, g_y and g_z decrease with increasing temperature up to 130K then increase slightly again. The observed peak of the anisotropic g-factors variation with T is corresponding to the opening of pseudo-gap (PG) [5]. PG is defined as the onset of negative deviation for linear temperature dependence in-plane resistivity at high T. Then, PG onset temperature T* is established by the derivative $D_i(T) = (g_{i,j} - g_{i,k})/(T_j - T_k)$ and plotted as shown in figure 3, where $T_j > T_k$ for the anisotropic g_i (i = x, y and z). One can find that T* (≈130 K) for Sm-123 ($T_c \approx 91.5$ K) coincides with the phase boundary established by resistivity (T*= 127±2 K) [4]. The variation of T* determined from resistivity and EPR with nanosized $MnFe_2O_4$ is shown in figure 4. It is observed that T* obtained from the both methods are nearly close to each other and exhibiting the same trend, T* increases with nanosized $MnFe_2O_4$ addition up to x= 0.20 wt%. This result suggests the existence of orthorhombically elongated octahedral Cu^{2+} (1) sites suffering the Jan-Teller effect via the vibration interactions [8].

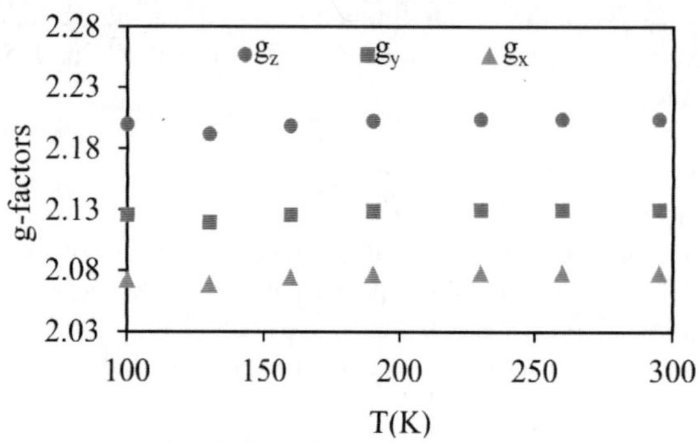

Figure 2. Variation of g-factors with T for Sm-123.

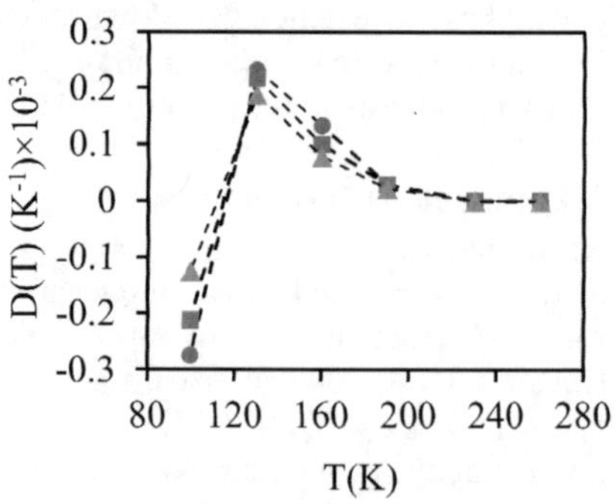

Figure 3. Derivatives of g_i (i = x, y and z).

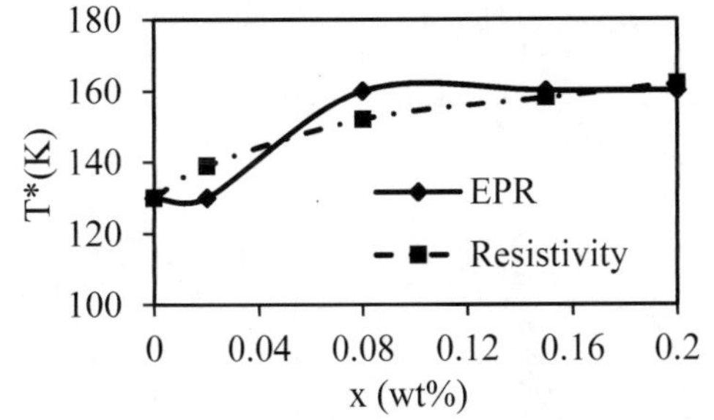

Figure 4. Variation of T* as a function of nanosized $MnFe_2O_4$ content.

However, the EPR relative intensity decreases as the nanosized $MnFe_2O_4$ addition increases up to 0.02 wt% and then increases with further addition x> 0.02 wt% as shown in figure 5. This is proportional to the amount of $BaCuO_2$ typically present in the samples as supported by our X-ray diffraction (XRD) results [4]. Based on this observation, we assume that this EPR spectrum originates mainly from delocalized d electrons of Cu^{2+} ions of the $BaCuO_2$ compound [7]. However, the relative EPR line intensity for Sm-123/$(MnFe_2O_4)_x$ composites increases as the temperature decreases from 295K down to 100 K as approximately Curie-Weiss law at T > T_c [8].

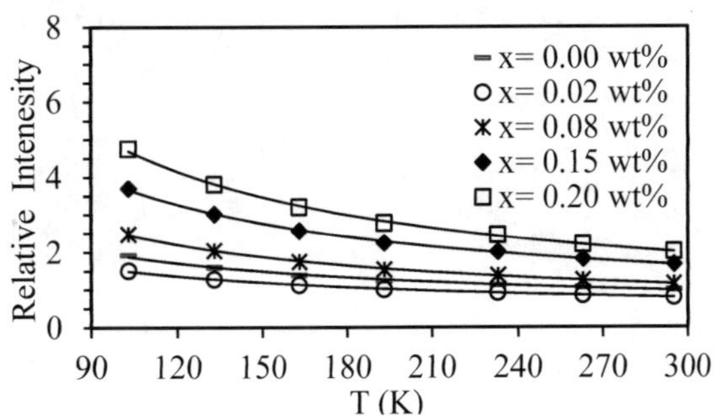

Figure 5. Variation of EPR relative intensity with T for $SmBa_2Cu_3O_{7-\delta}/(MnFe_2O_4)_x$.

The number of spins N participating in the resonance can be calculated at different temperatures by comparing the area under the absorption curve with that of a standard $CuSO_4.5H_2O$ [1]. As expected from Boltzmann's law, a linear relationship between log (N) and 1/T for Sm-123/$(MnFe_2O_4)_x$ composites is shown in figure 6. It is found that N decreases as the nanosized $MnFe_2O_4$ addition increases up to 0.02 wt% then increases for x>0.02 wt%. Furthermore, it increases with temperature decreasing from 295K down to 100K. The activation energy E_a as a function of nanosized $MnFe_2O_4$ addition are calculated from the slope of lines as shown in figure 7. The enhancement in E_a for x > 0.02 wt%, is consequently the result of increasing N with nanosized$MnFe_2O_4$ addition for *x*> 0.02 wt%. Figure 8 shows the M–H curves of Sm-123/$(MnFe_2O_4)_x$ at 77 K. As it is seen all the samples show a paramagnetic character. Magnetization increases linearly with the magnetic field and does not saturate even at 20kG. However, in this case, superconductivity seems to be dominant, especially at low fields. The M–H curves show small hysteresis at 77 K. Such a hysteresis, which is proportional with the critical current density J_c, is commonly encountered at the M–H curves of typical ceramic superconductors. The possible reason of paramagnetism like behavior may be accounted for $BaCuO_2$ impurity phase present in the samples. Similar paramagnetism like behavior observed by Tobal [9] for Y-123 superconductor. These results confirmed that our EPR measurements are mainly due to small amounts of $BaCuO_2$ present in the samples. The sample with 0.02 wt% $MnFe_2O_4$ addition has the largest ΔM and the weaker Cu^{2+} EPR signal. Therefore, It is suggested that an amount (~0.02wt %) of $MnFe_2O_4$ addition in the Sm-123 superconductor can increase J_c due to enhancement of flux pinning properties and phase purity. Beyond x>0.02 wt% the hysteresis width and consequently, the flux pinning and J_c decreased.

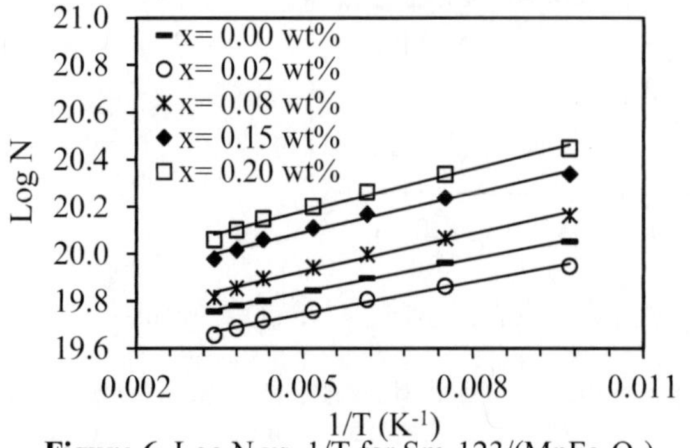

Figure 6. Log N vs. 1/T for Sm-123/$(MnFe_2O_4)_x$.

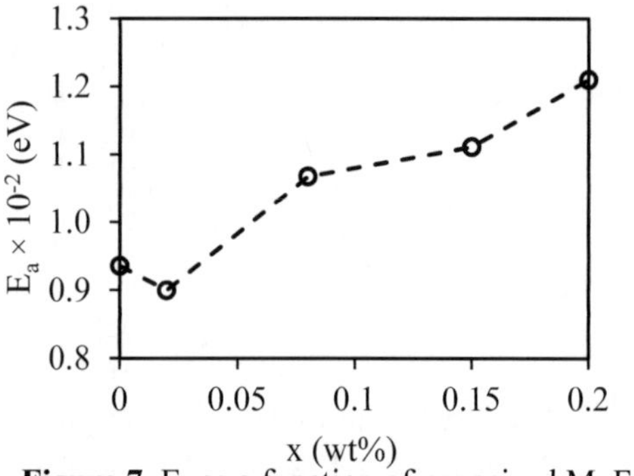

Figure 7. E_a as a function of nanosized $MnFe_2O_4$.

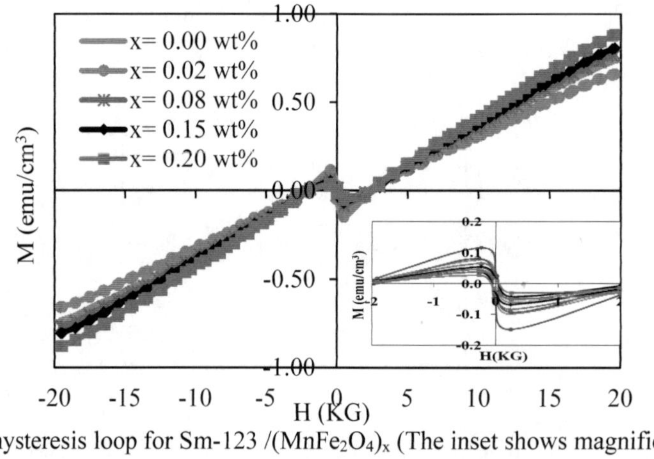

Figure 8. M-H hysteresis loop for Sm-123 /$(MnFe_2O_4)_x$ (The inset shows magnified M-H curves).

4. Conclusion

The EPR signal of Sm-123/ $(MnFe_2O_4)_x$ composites comprises a dominant Cu^{2+} anisotropic pattern, which is mainly due to the impurity phase $BaCuO_2$. However, the addition of nanosized $MnFe_2O_4$ up to x=0.02 wt% improved the flux pinning and J_c.

References

[1] Awad R, Abou Aly A I, Mahmoud S A and Barakat M M E 2011 *Appl. Magn. Reson.* **41** 79.
[2] Hwang S and So H 1989 *Bull. Korean Chem. Soc* **10** 23.
[3] Morton J R 1989 *J. Chem. Soc., Faraday Trans. I.* **85** 4091.
[4] Abdeen W, El-Tahan A, Roumié M, Awad R, Abou-Aly A.I, El-Maghraby E.M and Khalaf A 2016 *J. Magn. Magn. Mater.* **419** 354.
[5] Kuang M Q, Wu S I, Han H K, Wang L D, Duan S K, Kuang A L and Li G Q 2017 *J. Alloy. Compd.* **690** 169.
[6] Xu Y Q, Wu S Y and Zhang Z H 2015 *Int. J. Mod. Phys* B **29** 1542016.
[7] Guskos N *et al* 1995 *J. Solid State Chem.* **119** 50.
[8] Genossar J, Shaltiel D, Zevin V, Grayevski A and Fisher B 1989 *J. Phys. Condens. Matter* **19** 471.
[9] Topal U 2010 *Phys. Status Solidi A* **207** 1196.

Thermally activated flux flow in $FeSe_{0.5}Te_{0.5}$ superconducting single crystal

R M Hamad[1], T S Kayed[2], S Kunwar[1] and Kh A Ziq[1]

[1] Physics Department, King Fahd University of Petroleum and Minerals, Saudi Arabia.
[2] Physics Department, Dammam University, Dammam, Saudi Arabia.
Email: g201305850@kfupm.edu.sa

Abstract. The current–voltage *(J-E)* isotherms of single crystal $FeSe_{0.5}Te_{0.5}$ sample have been measured at several temperatures near the transition temperature (T_c) and under applied magnetic fields (H). A power law ($E \sim J^{\beta}$) has been used to fit the data and evaluate the activation energy U_o (T) using $\beta = U_o/k_BT$. At low current density ($J \ll J_c$), the initial behaviour is associated with thermally activated flux Flow *(TAFF)* while at $J \gg J_c$ vortex flux flow (*FF*) behavior is expected. The effects of applied magnetic field on *FF* and *TAFF* also been investigated. We found that U_o(*FF*) was reduced with by about an order of magnitude in magnetic fields as low as ~1.5 Tesla -the reduction in U_o(*TAFF*) is even faster than in U_o(*FF*)- hence reflecting the low pinning nature (defects, vacancies etc.) of $FeSe_{0.5}Te_{0.5}$ superconductor.

1. Introduction

The vanishing electrical resistance and diamagnetism below T_c is one of the main characteristics of superconductor. In superconductors, current transports occur with no energy loss for the currents lower than the critical current (J_c). However, it also depends on the temperature and the applied magnetic field. But in type-II superconductors, as the applied magnetic field increases above certain critical field H_{c1}, the vortex state complicates the picture. The vortex motion activated thermal energy may lead to "flux flow resistance "which could cause the energy dissipation.

Transport studies and dissipative properties in type II superconductors are usually limited to the region near T_c. Flux creep and vortex motion close to T_c can be explained using the Kim-Anderson model [1, 2], if the pinning potential was taken into account. Vortex dynamics, the irreversibility line and a vortex glass transition have been studied in high temperature superconductors [1,3-7]. From *(I-V)* measurements of thin films and single crystals high T_c superconductors several researchers suggested the continuous vortex glasses transition at temperatures below T_c. Fisher *et al.* suggested a scaling behaviour for the *(I-V)* curves obtained at various temperatures close to T_c [8-10] and Koch *et al.* [9] showed experimentally the first evidence for the continuous vortex glass transition at temperatures below T_c by scaling the nonlinear *(I-V)* curves in YBCO thin films. Similar results have been found in YBCO single crystal by Gammel *et al.* [11]. Sun *et al.* [4] have suggested a power-law relation as $V = \alpha I^{\beta}$ to describe the *(I-V)* curves. Where the exponent β is related to the pinning barrier U_o. Based on the status of vertices the mixed state in type II superconductor contains different regimes; thermal activated flux flow regime (TAFF), flux creep regime (FC) and flux flow regime (FF) according to status of vortices. Although there is no clear-cut distinction between each regime, the variations of the current density with temperature is used as a guide in this respect [12]. The newly discovered Fe-based superconductors have an intermediate transition temperature where thermal fluctuations are in between conventional and high T_c superconductors. Hence, it will be interesting to investigate the effects of thermally activated flux flow on various transport properties of these new materials. These properties are strongly influenced by the dynamic behaviour of vortices in the superconductor and can be studied by measuring the current–voltage curves [11,13].

In this paper, we report our *(I-V)* isotherms of single crystal $FeSe_{0.5}Te_{0.5}$ near T_c (~14.5K) in the presence of applied magnetic field. We discuss our results in terms of the power law form, then present the variation of the activation energies U_o (pinning potential barrier) with temperature.

2. Experimental Details

High purity (≥99.99 %) elements Fe, Se and Te have been used in 1:0.5:0.5 atomic ratios to prepare $FeSe_{0.5}Te_{0.5}$. The elements were grinded, pressed and sealed in a quartz tube partially filled with Argon gas. The tube was annealed at 500 ºC for 2 hours and then at 1070 ºC for 24 hours. The temperature was reduced at a rate of 4ºC/h to 700ºC and kept for 5 hours, then reduced to 400ºC, at kept and annealed 2 hours, then furnace cooled to room temperature. Four-probe (*dc*) technique was used to measure the current-voltage *(I-V)* characteristics using Lake Shore 7507 measuring system. Keithley 220 constant current source has been used to supply the current, and the voltage was measured using Keithley 2182 nano-voltmeter. The sample was cooled in a closed cycle cryostat (Advanced Research Systems) and the temperature was controlled using Lake Shore 340 temperature controller.

3. Results and discussions

From x-ray diffraction pattern for $FeSe_{0.5}Te_{0.5}$ single crystal, the structure of the sample is a tetragonal structure with cell parameters a = 3.797 Å, b = 3.797 Å and c = 5.984 Å. The lattice parameters are agreeing with the published data.

(I-V) isotherms of single crystal $FeSe_{0.5}Te_{0.5}$ have been measured at two different values of magnetic field B=0T and B=1.5T. Figure 1 (a) and (b) shows a representative (*I-V*) isotherms of the sample at magnetic field B=0T and B=1.5T respectively. The data is presented on a log-log graph that simplify evaluating the α in the power-law behaviour $V = \alpha I^{\beta}$ where α is a constant, $\beta = U_o/k_BT$ and U_o is the pinning barrier [4-6]. The exponent β is calculated from the slope of linear part of the (*I-V*) isotherms obtained at H=0T and 1.5T and was used to evaluate the activation energies U_o(*TAFF*) and U_o(*FF*). The activation energy is essentially the pinning barrier provided by defects and pinning centres in the sample. The results are presented in Figure 2 (a) and (b) as a variation of 1/T for both TAFF and FF regimes at H=0T and H=1.5T. Thermal activated flux flow (*TAFF*) occurs near 1μV, while flux flow (*FF*) regime is at much higher voltage. The figures reveal that as the temperature decreases below T_c, U_o continuously increases. The increase in TAFF regime is much faster than in FF regime. In other words, the gradual increase in exponent follow gradual increase in the pinning barrier with decreasing temperature. The exponent β approaches unity (Ohmic behaviour) near T_c. The deviation from Ohmic behaviour occurs at a non-linear temperature T_{nl} ~ 14.2K, which is about 0.3K below the T_c (~14.5K). The pinning potential at H=1.5T for both cases drops to about one tenths of its initial value obtained at H=0T. Above T_c the exponents β~1.4, for both regimes; which illustrates that the I-V curves are nonlinear material is still not in the normal state. Fluctuations and thermal effects near critical temperature may cause such deviations and pose limitations for Sun et al. model

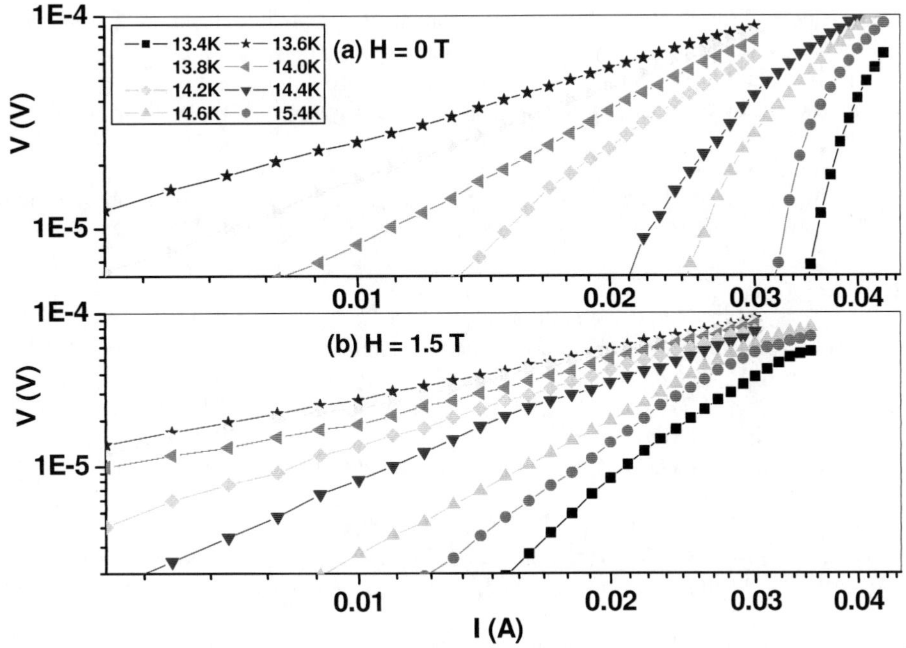

Figure 1. Selected (I-V) isotherms for single crystal $FeSe_{0.5}Te_{0.5}$ (a) at B = 0T and (b) at B = 1.5T around its corresponding superconducting transition temperature.

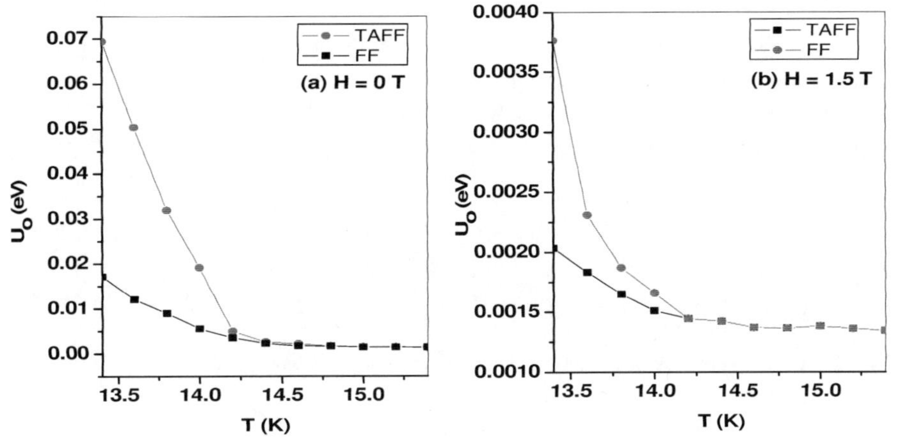

Figure 2. The pinning potential U_o obtained for H=0T (a) and for H=1.5T (b) are presented as variations with temperature in both thermally activated flux flow (TAFF) and flux flow (FF) regions.

IOP Conf. Series: Journal of Physics: Conf. Series **869** (2017) 012034 doi:10.1088/1742-6596/869/1/012034

4. Conclusion

Current-voltage isotherms of the single crystal $FeSe_{0.5}Te_{0.5}$ exhibit a power law behaviour in both regimes namely, TAFF and flux flow. The exponents β obtained of (from- TAFF and flux flow (FF) regime are found to decreases linearly as T increases. The pinning barrier Uo for TAFF is about 0.85eV and for FF is about 0.11eV. Both energy barriers decrease by about an order of magnitude in 1.5T. These values indicate that the pinning forces are relatively week in $FeSe_{0.5}Te_{0.5}$ single crystals.

Acknowledgments
We would like to acknowledge the support of King Fahd University of Petroleum & Minerals for this work. This project has been funded by NSTIP under project number 11-ADV1631-04.

References

[1] Kim Y B, Hempstead C F and Strand A R 1965 *Phys. Rev.* A **139** 1163.
[2] P.W. Anderson and Y. B. Kim, Rev. Mod. Phys. 36, 39, (1964).
[3] Yamamoto K, Mazaki H, and Yasuoka H 1993 *Phys. Rev.* B **47** 915.
[4] Sun J A, Eom C B, Lairson B, Bravman J C and Geballe T H 1991 *Phys. Rev.* B **43** 3002.
[5] Tinkham M 1988 *Phys. Rev. Lett.* **61** 1658.
[6] Ziq Kh A, Abdelhadi M M, Hamdan N M and Al Harthi A S 1996 *Supercond. Sci. Technol.* **9** 192.
[7] Yeshurn Y, Malozemoff A P 1988 *Phys. Rev. Lett.* **60** 2202.
[8] Fisher M P A 1989 *Phys. Rev. Lett.* **62** 1415.
[9] Koch R H, Foglietti V, Gallagher W J, Koren G, Gupta A and Fisher M P A 1989 *Phys. Rev. Lett.* **63** 1511.
[10] Fisher D S, Fisher M P A and Huse D A 1991 *Phys. Rev.* B **43** 130.
[11] Gammel P L, Schneerneyer I F and Bishop D J 1991 *Phys. Rev. Lett.* **66** 953.
[12] Waldram J R 1996 Superconductivity of Metals and Cuprates. Institute of Physics Publishing.
[13] Blanca G R, Ronulo J, Dumlao G, and Sarmago R 2004 *Science Diliman* **16** No.2 26.

Conduction mechanism and dielectric properties of pure and composite resorcinol formaldehyde aerogels doped with silver

S M Attia[1,2], M S Abdelfatah[1] and M M Mossad[1]

[1]Physics Department, Faculty of Science, Kafrelsheikh University, Kafr El-Sheikh, Egypt
[2]Physics Department, Faculty of Applied Sciences, Umm Al-Qura University, PB 7296, Makkah 21955, Saudi Arabia
Email: saidattia2004@yahoo.com

Abstract. Pure and composite Resorcinol–Formaldehyde (RF) aerogel samples were prepared by sol-gel process using KOH as a catalyst and doped with silver nanoparticles at different concentrations (1.2×10^{-4}, 2.4×10^{-4}, 3.6×10^{-4}, and 4.8×10^{-4} wt.% at catalyst ratio 0.024 wt.%). DC electrical conductivity σ_{dc}, AC electrical conductivity σ', and the dielectric properties of the prepared samples have been measured at different frequencies and temperatures. The results show that σ' increases with increasing frequency. The values of σ' range from $\sim10^{-4}$ $\Omega^{-1}m^{-1}$ to around unity at room temperature. The analysis of the results of $\sigma'(\omega,T)$ reveals that the large overlapping polaron (OLP) is the most favorable mechanism to describe the conduction mechanism in these samples. The behavior of the dielectric constant with the frequency of these samples is normal, where it decreases with increasing frequency, while the behavior of dielectric loss tangent tanδ exhibits a peaking behavior at relatively higher temperature.

Keywords: RF aerogels, porous materials, composite materials, conduction mechanism, dielectric constant.

1. Introduction

Composites are two or more materials combined on a macroscopic scale to form a useful material, and exhibit the best qualities of their constituents and often some qualities that neither constituent possesses. The synthesis of nano scale composite materials has attracted a great deal of attention due to their physical and chemical properties which are suitable for a wide range of industrial applications [1]. The properties of these materials depend strongly on the size, distribution and specific concentration of metal clusters within the host porous matrix as well as the properties of the support [2, 3]. Because of their unique properties such as low density, high surface area continuous porosity and high cross linking structure, aerogels are widely used as thermal insulation, nuclear particle detection, light-guides, electronic device [4, 5], super-capacitors [6], battery electrodes [7], adsorbents for gas separation [8, 9], catalysts supports [10, 11], electrode material for double layer capacitors [12, 13], and energy storage device [14]. The objective of this work is to investigate the dielectric properties and the conduction mechanism of a novel composite material, made of RF aerogel doped with silver nanoparticles at different concentrations.

2. Material and methods

Pure aerogel was prepared by adding resorcinol ($C_6H_4(OH)_2$) with formaldehyde (HCOH) with molar ration 1:2. Potassium hydroxide (KOH) was used as a catalyst (0.024 wt. %). After stirring the solution for one hour by magnetic stirrer, the solution was poured into glass tube which sealed by high temperature flame. Then the samples were stored in an oven at 90°C for a week. Then the samples were placed in acetone bath. Finally, the samples were supercritically dried using supercritical dryer, type–Quorum Technologies E3100 Series. A silver nano particles were prepared by dissolving $AgNO_3$ in the

IOP Conf. Series: Journal of Physics: Conf. Series **869** (2017) 012035 doi:10.1088/1742-6596/869/1/012035

water under the action of NaOH as a catalyst. Then a dark brown precipitate of Ag_2O was formed immediately. Ag_2O precipitate was dispersed in a sodium dodecyle sulfate solution to form a colloidal. Finally the colloidal of Ag_2O was reacted with a solution of formaldehyde to produce a silver nano particle colloidal. After preparing nano-silver colloidal, it has been added to pure aerogel during the preparation process of RF aerogel at different Ag concentrations (1.2×10^{-4}, 2.4×10^{-4}, 3.6×10^{-4}, and 4.8×10^{-4} wt.%). The samples were poured in tube glass and left in an oven for one week at 90°C then they were cut in a disk shape using a diamond saw, and place in the acetone path before the final step, the supercritical drying. After preparing the sample, they have been characterized using FTIR spectroscopy, UV-Vis spectroscopy, and scanning electron microscopy. The electrical conductivity, σ_{dc}, was measured at different temperature using an electrometer (Keithley model 617). The frequency and temperature dependence of dielectric constant, ε', AC electrical conductivity, σ', and dielectric loss tangent, tanδ, was determined using an RLC meter (type HIOKI 3532-50).

3. Results and Discussions

DC and AC electrical conductivities were measured for pure and composite RF aerogels doped with different concentration of silver (1.2×10^{-4}, 2.4×10^{-4}, 3.6×10^{-4} and 4.8×10^{-4} wt. %). The measurements were carried out in the temperature range from room temperature ~25°C up to 200°C and cover the frequency range from 100 Hz to 5MHz for AC electrical conductivity. The frequency dependence of the measured AC conductivity, σ', at different selected temperatures is shown in figure 1. It is shown that σ' increases with increasing frequency. Also the values of AC electrical conductivity at room temperature increase with increase of Ag content. The values of these samples vary from 3×10^{-5} to 10 $\Omega^{-1}m^{-1}$. The behavior of σ' for the investigated samples can be expressed by the relation [15]

$$\sigma_{ac}(\omega) = \sigma'(\omega) - \sigma_{dc} = A\,\omega^{s}, \quad (1)$$

where, A is a constant depending on temperature, ω is the angular frequency ($=2\pi f$), $\sigma_{ac}(\omega)=A\omega^s$ is the pure AC conductivity arises due to the bound charges [16], and the frequency exponent, s, can be determined by the relation; $s = d(\ln\sigma_{ac}(\omega))/d(\ln(\omega))$.

Figure 2 shows the variation of DC conductivity (σ_{dc}) with the temperature as $10^3/T$, for pure and composite RF aerogel doped with silver at different concentration. The results show that the DC electrical conductivity of aerogel samples decreases with increase of temperature till a certain temperature after which, the conductivity increases with further increase of temperature. This means that the DC electrical conductivity behaves as semiconductor at relatively low temperature region and exhibit a metallic behavior at relatively high temperature region. The values of DC electrical conductivity vary from 2×10^{-8} to 10^{-5} $\Omega^{-1}m^{-1}$ which are less than AC conductivity in the order 10^{-3}. In order to explain the behavior of $\sigma_{ac}(\omega)$, with both temperature and frequency, different theoretical models have been proposed to correlate the conduction mechanism of AC conductivity with s(T) behavior [16-17]. The classical barrier model for single electron hopping between two localized states over a potential barrier separating the sites as well as the correlated barrier hopping (CBH) model suggested that s should decrease with increasing the temperature. The quantum mechanical tunneling (QMT) model implies that s is temperature independent. In the case of small polaron (SP) model, s should increase with increasing the temperature. Finally, the large overlapping polaron (OLP) model suggested that s should decrease with increasing the temperature up to a certain temperature, after that s increases with further increase of the temperature [15]. Figure 3 shows the behavior of the exponent, s, with the temperature for all samples. It should note that the behavior of s(T) exhibits a minimum value for all samples with the temperature. Therefore, the results shown in figure 3 suggest that the large overlapping polaron (OLP) conduction mechanism is the most favorable mechanism to describe the conduction mechanism for the samples under investigation.

The dielectric constant ε' and loss tangent tanδ were measured in the temperature range from room temperature ~25°C up to 200°C and cover the frequency range from 100 Hz to 5MHz. The Variation of dielectric constant ε' with the frequency at various temperatures for some RF aerogel doped with silver at different concentrations are shown in figure 4. It is shown the behavior of the dielectric constant is normal, *i.e.*, ε' decreases with increasing frequency. The same behavior was observed for other polymers [18]. The dispersion in both σ' and ε' reveals a strong correlation between the AC electrical conductivity and the dielectric constant in these materials.

IOP Conf. Series: Journal of Physics: Conf. Series **869** (2017) 012035 doi:10.1088/1742-6596/869/1/012035

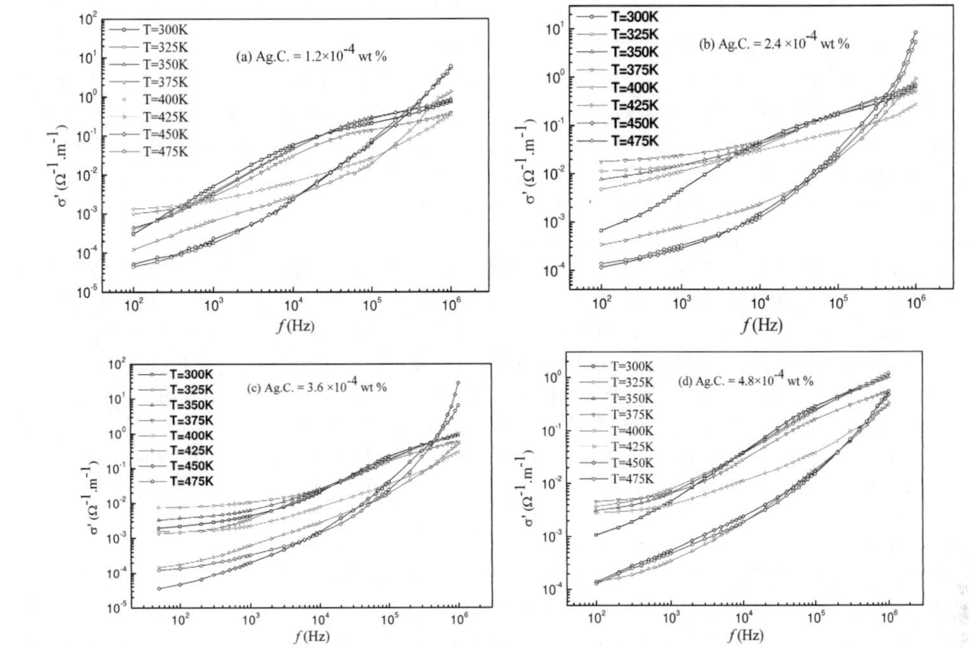

Figure 1. The frequency dependence of AC conductivity σ′ at different temperatures for RF aerogel doped with silver at different concentrations.

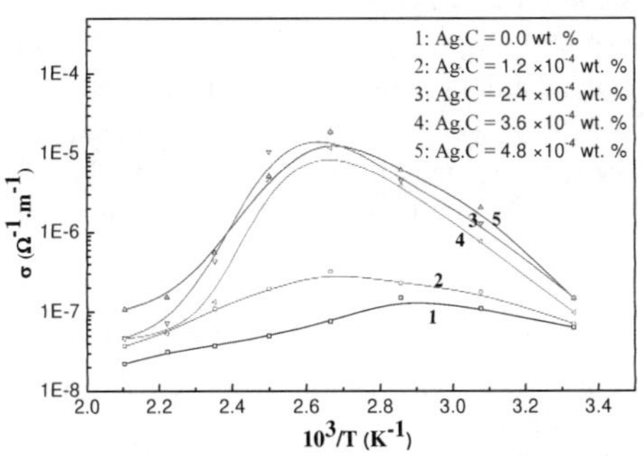

Figure 2. The temperature dependence of DC conductivity for pure and composite RF aerogels doped with silver.

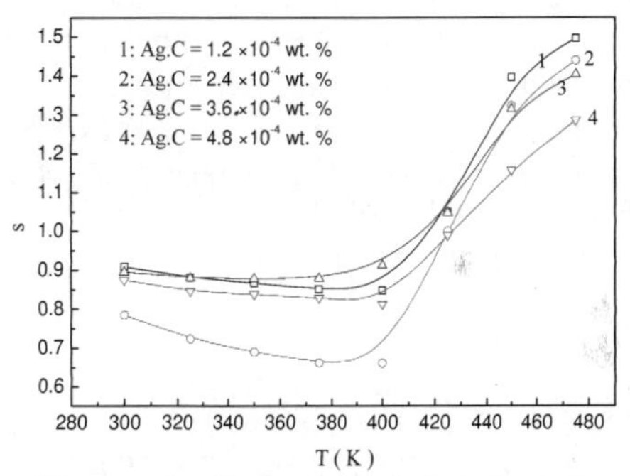

Figure 3. Variation of the frequency exponent, s, with the temperature T(K) for RF aerogel doped with silver.

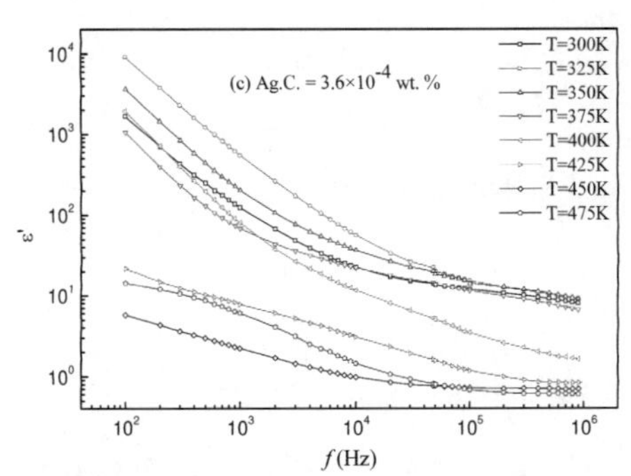

Figure 4. The frequency dependence of ε' at different temperatures for some RF aerogel doped with silver.

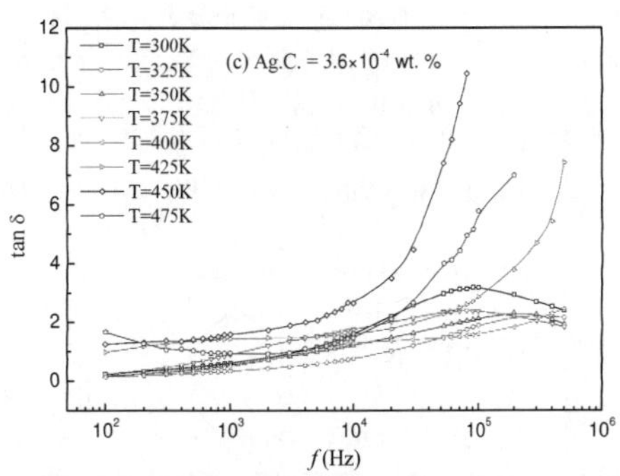

Figure 5. The frequency dependence of tanδ at different temperatures for RF aerogel doped with silver.

It is well known that the complex dielectric constant ($\varepsilon^* = \varepsilon' + j\varepsilon''$) of the medium vary with the angular frequency ω according to the following equation [19]; $\varepsilon^* = \varepsilon_\infty + [(\varepsilon_0 - \varepsilon_\infty)/(1 + j\omega\tau)]$, where τ is the relaxation time, ε_o is the static field dielectric constant, and ε_∞ is the high-frequency (optical) dielectric constant of the material. The real part of the complex dielectric constant ε' can be expressed as $\varepsilon' = \varepsilon_\infty + [(\varepsilon_0 - \varepsilon_\infty)/(1 + \omega^2\tau^2)]$. This equation gives an explanation for the decrease of the dielectric constant with the frequency. From the other hand, the behavior of dielectric constant with the frequency can be explained on the fact that the hopping of the charge carriers can not follow the frequency of the external applied electric field beyond a certain frequency value [20].

The imaginary part of the dielectric constant ε" represents the dielectric loss through the material. A very important quantity is the dielectric loss tangent; tanδ, which measures directly the phase difference due to loss of energy within the sample at a particular frequency, tanδ is given by $\tan\delta = \varepsilon''/\varepsilon'$. The Variation of dielectric loss tangent (tanδ) with the frequency at different temperatures for RF aerogels doped with silver is shown in figure 5. It observed that dielectric loss tangent exhibits a peaking behavior at relatively low temperature. The peak is shifted toward a higher frequency with increasing the temperature. By increasing temperature the peaking behavior disappears and the dielectric loss tangent increase rapidly with the frequency.

4. Conclusion

Different samples of the RF-aerogels doped with silver nanoparticles were prepared by the sol-gel process using KOH as catalyst. DC electrical conductivity, σ_{dc}, AC electrical conductivity, σ′, dielectric constant, ε', and dielectric loss tangent, tanδ were measured for all prepared samples. The measurements were carried out in the frequency range from 10^2 to 10^6 Hz and at different temperature from 300 to 475 K. The results can be concluded as follows; the values of σ′ are higher than those of σ_{dc} for all samples; σ′ increases with increasing frequency; σ_{dc} exhibits a semiconducting behavior at relatively lower temperature region, while it exhibits a metallic behavior at relatively higher temperature region; Large overlapping conduction mechanism is the predominant conduction mechanism in these materials; ε′ exhibits a normal behavior with the frequency, where it decreases with increasing frequency; tanδ(ω) increases with increasing frequency at relatively lower temperature range, while it exhibits a peaking behavior at relatively higher temperature region. The peaks are shifted towards lower frequency with increasing temperature.

References

[1] Saquing C D, Kang D, Aindow M and Erkey C 2005 *Micropor. Mesopor. Mater.* **80** 11
[2] Stakheev A Y and Kustov L M 1999 *Appl. Catal. A* **188** 3
[3] Beecroft L L and Ober C K 1997 *Chem. Mater.* **9(6)** 1302
[4] Hrubesh L W 1998 *J. Non-Cryst. Solids* **225** 335
[5] Zhou J, Ji Y, He J, Zhang C and Zhao G, 2008 *Microporous and Mesoporous Materials* **114** 424
[6] Mayer S T, Pekala R W and Kaschmitter J L 1993 *J. Electrochem. So.* **140** 446
[7] Schaefer D W, Pekala R and Beaucage G 1995 *J. Non-Cry. Solids* **186** 159
[8] Sircar S, Golden T C and Rao M B 1996 *Carbon* **34** 1
[9] Yamamoto T, Endo A, Ohmori T and Nakaiwa M 2004 *Carbon* **42** 1671
[10] Polo M S, Utrilla J R and Gunten U V 2006 *Wat. Res.* **40** 3375
[11] Castilla C M and Hodar F J 2005 *Carbon* **43** 455
[12] Li W, Reichenauer G and Fricke J 2002 *Carbon* **40** 2955
[13] Pekala R W, Farmer J C, Alviso C T, Tran T D, Mayer S T, Miller J M and Dunn B 1998 *J. Non-Cryst. Solids* **225** 74
[14] Frackowiak E and Beguin F 2001 *Carbon* **39** 937
[15] Elliott S R 1987 *Adv. Phys.* **36** 135
[16] Attia S M, Abo El Ata A M and El Kony D 2004 *J. of Magn. Magn. Mater.* **270** 142
[17] Hazara S and Chosh A 1996 *Phil. Mag. B* **74** 235
[18] Attia S M, Sharshar T, Abd-Elwahed A R and Tawfik A 2013 *Mater. Sci. Eng. B* **178** 897
[19] Cole K S and Cole R H 1941 *J. Chem. Phys.* **9** 341
[20] Abo El Ata A M and Attia S M 2003 *J. Magn. Magn. Mater.* **257** 165

IOP Conf. Series: Journal of Physics: Conf. Series **869** (2017) 012036 doi:10.1088/1742-6596/869/1/012036

Characterization of pure and composite resorcinol formaldehyde aerogels doped with silver

S M Attia[1,2], M S Abdelfatah[1] and M M Mossad[1]

[1]Physics Department, Faculty of Science, Kafrelsheikh University, Kafr El-Sheikh, Egypt
[2]Physics Department, Faculty of Applied Sciences, Umm Al-Qura University, PB 7296, Makkah 21955, Saudi Arabia
Email: saidattia2004@yahoo.com

Abstract. A series of Resorcinol Formaldehyde (RF) aerogels composites with nanoparticles of sliver were prepared by the sol-gel method at different concentrations doped silver. FTIR spectra of pure and composite RF aerogels show six absorption bands attributed to –OH groups bonded to the benzene ring, stretching of $-CH_2-$ bonds and aromatic ring stretching. FTIR results ensured that sliver particles do not interact with aerogel network. UV-visible spectrum of pure silver show an absorbance peak at 420 nm attributed to the surface plasmon excitation of sliver Nano spheres. UV-visible spectral of pure and composite RF aerogels shows a steep decrease of absorption with wavelength after 500 nm, making sample's color reddish brown. TEM and SEM images of pure and composite RF aerogels revealed that the textural arrangement of RF aerogels can be described as densely packed small nodules.

Keywords: RF Aerogels, Porous Materials, Composite Materials, Sol-Gel, Infrared Spectroscopy.

1. Introduction

Aerogels are classified as nano-porous materials. The most interesting characteristics of aerogels are low bulk density, high surface area, continuous porosity and high crosslinking structure [1, 2]. Organic aerogels have most frequently been obtained by sol-gel polymerization of Formaldehyde and another reactive monomer Resorcinol, which is called RF aerogels [3]. Applications as thermal insulators, electrodes for super capacitors, nuclear particle detection, light guides and electronic devices have been suggested for RF aerogels [4-6]. The properties of RF aerogel as well as their application are related to synthesis and processing conditions [7]. Composites are two or more materials combined on a macroscopic scale to form a useful material. One of the utilized methods for synthesis the metal-aerogel composite involves mixing the desired metal (in the form of its salt) with the sol-gel mixture. An alternative method for preparation metal-aerogel composite involves replacement of resorcinol by its derivative containing an ion-exchange moiety. This technique generates an organic polymer with each repeat unit containing a binding site which can be used for incorporation of the desired metal ions through an ion-exchange process [8]. In this work we introduce an investigation on the synthesis and characterization of pure and composite RF aerogel doped with silver nanoparticles as a novel material.

2. Material and methods

2.1. Preparation of pure RF aerogel

The first step of preparing RF aerogel is the aqueous polycondensation of resorcinol (1, 3-dihydroxybenzene $C_6H_4(OH)_2$) with formaldehyde (HCOH) at molar ratio 1:2 and KOH as a catalyst (0.024 wt. %). After stirring the solution for one hour, It was poured into glass tube which sealed by high temperature flame. Then the samples were stored in an oven at 90°C for seven days. During gelation, the solution progressively changed color from clear to yellow to orange to deep red as a function of the reaction time. In general the solution gels within 12-16 hours, but the extended cure is necessary to complete crosslinking as much as possible. RF aerogels are dark red in color as a result of oxidizing products formed during the polymerization. Then the samples are placed in acetone bath and make several exchanges with fresh acetone to remove water before supercritical drying. In supercritical drying, samples are placed in a pressure vessel which is filled with liquefied CO_2 at a temperature of 15°C and a pressure of 900 psi. After a suitable time (an hour for each 1mm sample thickness), the system is closed and raised to 40°C, where it is left for about 10 minutes. Excess CO_2 is vented during heating. The chamber is bled slowly to atmospheric pressure.

2.2. Preparation of silver nanoparticles
A colloidal dispersion of silver nanoparticles was prepared by adding 75 ml of 1 M solution of NaOH to 125 ml of a 0.1 M solution of $AgNO_3$. $AgNO_3$ was transferred to the intermediate Ag_2O colloidal dispersion as follows:

$$2Ag^+ + 2NO^{-3} + 2Na^+ + 2OH^- \rightarrow Ag_2O\downarrow + 2NaNO_3 + H_2O \quad (1)$$

A dark brown precipitate of Ag_2O was formed immediately. Then, Ag_2O precipitate was dispersed in 125 ml deionized water containing 1.25 ml of 1 M solution of sodium dodecyle sulfate (SDS), after 10–15 min, a stable colloidal dispersion was formed. A 37% solution of formaldehyde was used as reduction agent. The reaction of Ag_2O with formaldehyde can be written as follows:

$$Ag_2O + HCOH + H_2O \rightarrow 2Ag\downarrow + CO_2 + 2H_2O \quad (2)$$

The formation process of silver colloidal dispersion of reaction (2) proceeds slowly. In fact, it takes several days at room temperature. However, the formation time can be reduced to a several hours by increasing the temperature of the solution and then conform its formation.

2.3. Preparation of RF aerogel composite with Ag
After preparing nano-silver colloidal as mentioned before, it has been added to aerogel during the preparation process of RF aerogel (at catalyst ratio 0.024 wt. %) at different Ag concentrations (1.2×10^{-4}, 2.4×10^{-4}, 3.6×10^{-4}, and 4.8×10^{-4} wt. %). After stirring, for more than 30 min, the mixture was kept in the ultrasonic bath for 30 min. Finally, the mixture was poured into a test tube, which sealed by high temperature flame. The sealed tubes were kept for 7 days at a temperature 90°C. Then, the prepared composite aerogel was cut using diamond saw in a disk shape. The supercritical dryer, type–Quorum Technologies E3100 Series was used in order to get rid off any solvent for all samples.

2.4. Characterization of the materials
FTIR spectra of the samples were recorded using Jasco FT/IR 4100 spectrometer in the range 400 - 4400 cm^{-1}. UV-Vis spectrum was recorded in the wavelength range 200-900 nm using Shimadzu UV-2450 Spectrophotometer. Morphology of the samples was performed using scanning electron microscopy (SEM) (Model JEOL JSM 6360 LA Analytical Scanning Electron Microscope).

3. Results and Discussions
Figure 1 shows FTIR spectra of pure and composite RF aerogel samples in wave number ranges from 400 to 4000 cm^{-1}. The results show the existence of six absorption bands, namely ν_1 at 3400 cm^{-1}, ν_2 at 2940 cm^{-1}, ν_3 at 1606 cm^{-1}, ν_4 at 1473 cm^{-1}, ν_5 at 1222 cm^{-1}, and ν_6 at 1090 cm^{-1} respectively, which are in good agreement with that reported by others [9-11]. The broadband at 3400 cm^{-1} can be attributed mainly to –OH groups bonded to the benzene ring, but may also be due to $-CH_2OH$ groups connected to the resorcinol molecule, which did not take part in network formation [10]. The bands at 2940 and 1473 cm^{-1} are attributed to stretching of $-CH_2-$ bonds. The band at 1606 cm^{-1} is attributed to aromatic ring stretching. The bands at 1090 and 1222 cm^{-1} confirm the methylene ether C–O–C linkage stretching between two resorcinol molecules (phenyl rings), which is expected in the polycondensation reaction between resorcinol and formaldehyde [10, 11]. The change in transmission intensity can be attributed to aerogel concentration with respect to KBr. The differences of transmission intensity can be attributed to the concentration of the materials with KBr. This ensures that the silver particles do not affect the position of the absorption band of IR spectra. Figure 2 shows UV-visible spectra for pure Resorcinol with concentration 0.074 g per 250 ml of water. It is shown that the maximum absorption peak occurs at 273 nm representing an aromatic ring, which is in good agreement with literature [12], where the

maximum absorption for resorcinol occurs within wavelength range 273 - 294 nm depending on pH values.

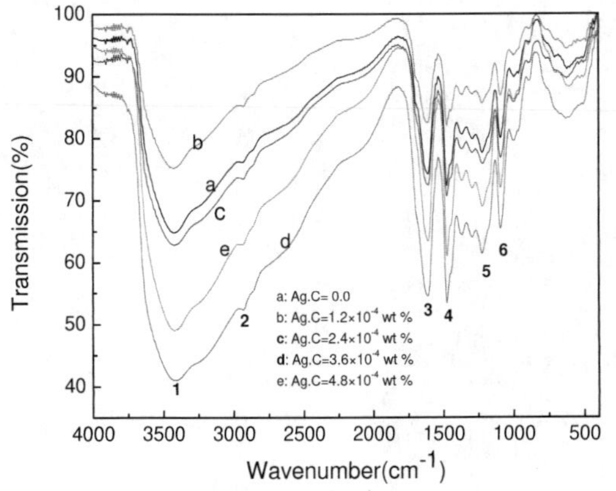

Figure 1. FTIR spectra of pure and composite RF aerogels doped with Ag.

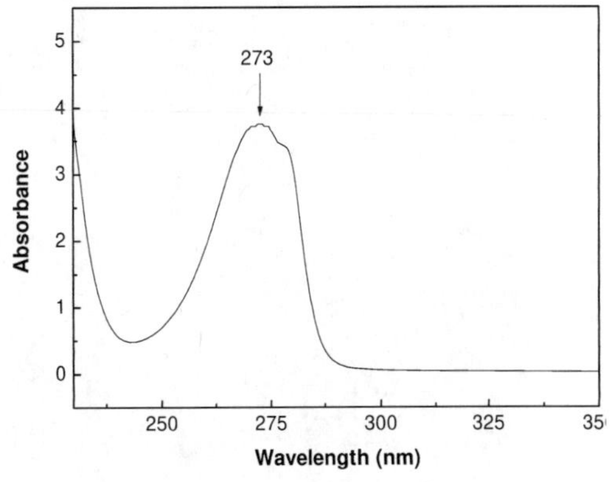

Figure 2. UV-visible spectra of pure Resorcinol.

UV-visible spectra of pure and composite RF aerogel doped with different concentration of Ag are represented in figure 3. A shoulder occurs at ~500 nm after which the absorption decreases with increasing wavelength. Therefore, RF aerogel appears to be in reddish brown [13, 14]. The absorbance peak, which appears around 300 nm can be attributed to a lone pair of electron in oxygen atoms, while the peak at 250 nm is due to aromatic ring [13]. This means that the shift of the absorbance peak of aromatic ring in pure resorcinol (273 nm) to 250 nm (blue shift) in RF aerogel is an indication of formation of RF aerogel. The change in the intensity of absorbance can be attributed to catalyst concentration. After preparation of Ag nanoparticles under reduction of $AgNO_3$ by formaldehyde, a UV-visible spectrum was recorded as shown in figure 4. The spectrum shows an absorption peak at 410 nm which can be attributed to the surface Plasmon excitation of silver Nano spheres, and could be taken as an indication of formation of silver nanoparticles [15,16].

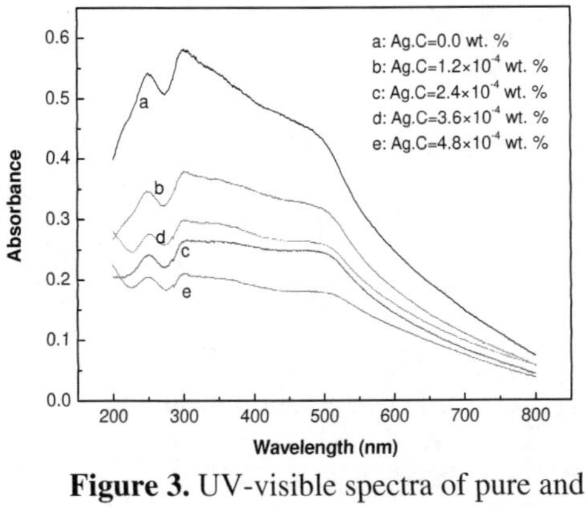

Figure 3. UV-visible spectra of pure and composite RF aerogels doped with Ag.

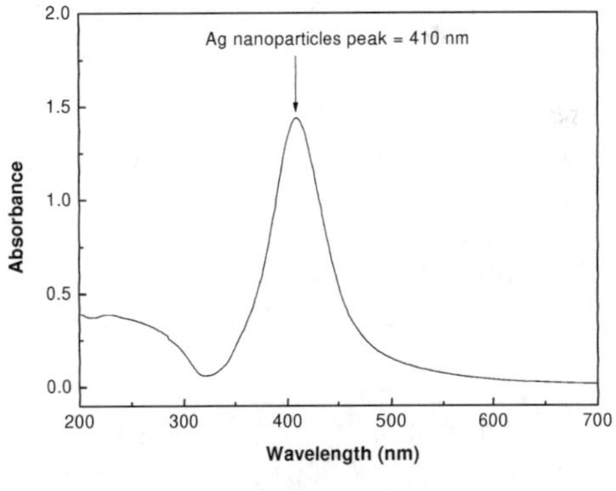

Figure 4. UV-visible spectra of silver nanoparticles.

Figure 5(a) shows SEM image of pure RF aerogel. The morphology of pure RF aerogel has an open cell structure with continuous porosity with particle size ranging from 50 nm to about 150 nm. The result agrees with literature [17]. Figure 5(b-e) shows SEM micrographs of RF aerogel doped with silver. It is shown that the interconnection occurs in all directions of the network and there are many mezzo and macro pores among the interconnected channels which increase by increasing silver. The observed particle size ranges from 65 nm to 300 nm with large pore size which agrees with other [18]. These results reveal that the materials have a high surface area, which make them to be a good candidate as electrodes for supercapacitors.

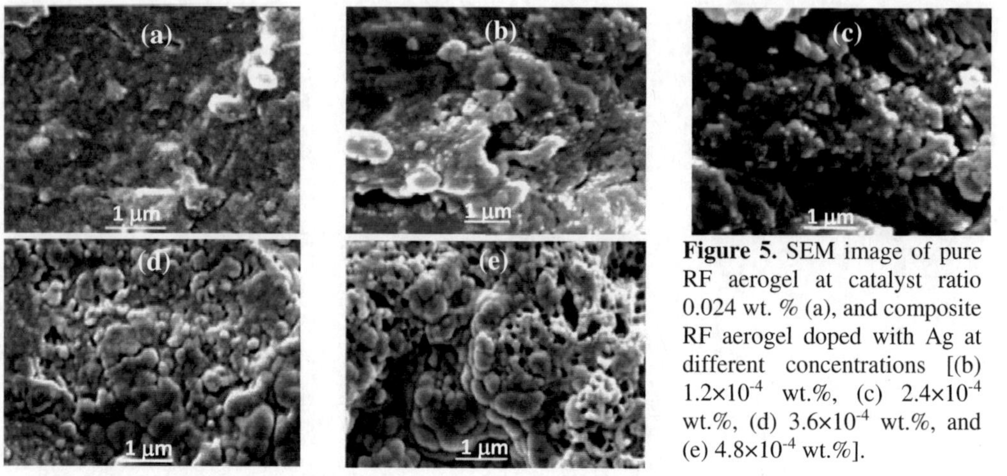

Figure 5. SEM image of pure RF aerogel at catalyst ratio 0.024 wt. % (a), and composite RF aerogel doped with Ag at different concentrations [(b) 1.2×10^{-4} wt.%, (c) 2.4×10^{-4} wt.%, (d) 3.6×10^{-4} wt.%, and (e) 4.8×10^{-4} wt.%].

4. Conclusion

Nano composite RF aerogels doped with Silver nanoparticles were prepared by the sol-gel process. The results of FTIR spectra show existence of six absorption bands for pure and composite RF aerogels and ensured that the silver particles do not affect IR spectra. UV-visible spectra for pure resorcinol show a maximum absorption peak at 273 nm which represent aromatic ring. UV-visible spectra for pure and composite RF aerogels show a steep decrease of absorption with wavelength after 500 nm and the peak of aromatic ring (273 nm) is shifted to 250 nm in RF aerogel. The shift of the absorbance peak of aromatic ring can be taken as an indication of formation of RF aerogel. UV-visible spectrum of Ag Nano particles shows an absorbance peak at 410 nm attributed to the surface plasmon excitation of silver Nano spheres. TEM images of pure and composite RF aerogels show that the textural arrangement of RF aerogels can be described as densely packed small nodules.

References

[1] Yang J, Li S, Yan L, Liu J and Wang F 2010 *Micropor. Mesopor. Mater.* **133** 134
[2] Scherer G W 1998 *Adv. Colloid Interface Sci.* **76–77** 321
[3] Carrott P J M, Marques L M and Ribeiro Carrott M M L, 2012 *Micropo. Mesopor. Mater.* **158** 170
[4] Hrubesh L W 1998 *J. Non-Cryst. Solids* **225** 335
[5] Zhou J, Ji Y, He J, Zhang C and Zhao G 2008 *Micropor. Mesopor. Mater.* **114** 424
[6] Amaral-Labat G, Szczurek A, Fierro V, Masson E, Pizzi A and Celzard A 2012 *Micropor. Mesopor. Mater.* **152** 240
[7] Attia S M, Sharshar T, Abd-Elwahed A R and Tawfik A 2013 *Mater. Sci. Eng.* B **178** 897
[8] Baumann T F, Fox G A, Satcher J H, Yoshizawa N, Fu R and Dresselhaus M S, 2002 *Langmuir* **18** 7073
[9] Lee J Y, Lee K, Lee H and Kim J H 2002 *J. Ind. Eng. Chem.* **8(6)** 546
[10] Hebalkar N, Arabale G, Sainkar S R, Pradhan S D, Mulla I S, Vijayamohanan K, Ayyub P and Kulkarni S K 2005 *J. Mater. Sci.* **40** 3777
[11] Mulik S, Leventis C S and Leventis N 2006 *Polymer Preprints* **47(2)** 364
[12] Blanco S E, Almandoz M C and Ferretti F H, 2005 *Spectrochimica Acta Part A* **61** 93
[13] Pretsch E, Buhlmann P and Badertscher M 2009 *Structure Determination of Organic Compounds* 4th edition, Springer
[14] Pavia D L, Lampman G M, Kriz G S and Vyvyan J R 2009 *Introduction to Spectroscopy* 4th edition, Brooks/Cole Cengage Learning
[15] Dubas L, Kulsiriwiwat T, Samneingjam K, Sirtaveesinsub V, Wangyao P and Dubas S T 2007 *Acta Metallurgica Slovaca* **13(2)** 147
[16] Song K C, Lee S M, Park T S and Lee B S 2009 *Korean J. Chem. Eng.* **26(1)** 153
[17] Fairén-Jiménez D, Carrasco-Marín F and Moreno-Castilla C 2006 *Carbon* **44(11)** 2301
[18] Luyjew K, Tonanon N and Pavarajarn V 2008 *J. American Ceramic Society* **91(4)** 1365

IOP Conf. Series: Journal of Physics: Conf. Series **869** (2017) 012037 doi:10.1088/1742-6596/869/1/012037

Magnetic phenomena in exchange coupled spin systems

S Sakhi and A Patel

Department of Physics, American University of Sharjah, United Arab Emirates, POBOX 26666
Email: ssakhi@aus.edu

Abstract. We study theoretically a cluster of several interacting spins modelling a hypothetical single-molecule magnet. Starting from a spin Hamiltonian, we obtain the individual energy levels and their evolution in an applied magnetic field. By substituting the energy levels into the partition function, we compute macroscopic observables like the magnetisation and the susceptibility of the system.

1. Introduction

Low-dimensional magnetic materials [1] attract interest because their fascinating physical properties show potential applications in areas such as high-density information storage, quantum computing, and biomedical applications such as magnetic resonance imaging [2]. Of particular interest are single-molecule magnets [3] which can be synthesized in a bottom-up approach using an increasing number of magnetic centres [4] such as transition metals or rare earth ions with spins equalling 5/2 or 7/2. These nano-magnets reveal a slow relaxation of the magnetization at low temperature and offer a controllable approach to nanoscale magnetism, where the energy barrier to magnetization reorientation is derived from the anisotropy of the molecular spin rather than the movement of domain walls, as in bulk magnets.

Theoretically these nano-magnets are remarkable because they offer enough complexity to yield new types of properties while simple enough to study in detail. They proved to be testing grounds for theories of the coexistence of quantum and classical phenomena [5]. The magnetic interactions in molecular systems are in principle the same as those in continuous lattices. However, the pair interactions are strongly localized in molecular systems with a predominantly electrostatic origin.

In this paper, we theoretically investigate a cluster of several interacting spins modelling a hypothetical single-molecule magnet. Starting from an effective spin Hamiltonian, we obtain the individual energy levels and their evolution in an applied magnetic field. In the effective spin Hamiltonian, all orbital coordinates are eliminated and replaced by spin coordinates while taking advantage of the symmetry properties of the system. We use a basis consisting of eigenfunctions of the total spin operator to ensure that the Hamiltonian is block-diagonal. This is straightforward in principle, but the problem rapidly grows in complexity as in general there are $(2S_i + 1)^N$ states for N spins. We also utilize an irreducible tensor method which exploits the symmetry of the full rotation group to derive the energy levels of the system.

2. The spin model

The interaction between localized spin magnetic moments in the strong exchange limit is described by the Heisenberg model.

$$H = -2\sum_{ij} J_{ij}\vec{S}_i \cdot \vec{S}_j + g\mu_B \vec{B}\cdot\sum_i \vec{S}_i \tag{1}$$

J_{ij} is a symmetric matrix containing the exchange parameters between spins at sites i and j. $J_{ij} < 0$ corresponds to antiferromagnetic and $J_{ij} > 0$ to ferromagnetic coupling. The first sum in (1) runs over all possible terms (i, j). The vector operators $\vec{S}_i$ are the single-particle spin operators. The last term (Zeeman) describes the interaction with an external magnetic field which is assumed to be along the z-axis. The dimensionless spin operators in (1) obey the commutation relations

$$\left[S_i^\alpha, S_j^\beta\right] = i\delta_{ij}\varepsilon^{\alpha\beta\gamma} S_i^\gamma \tag{2}$$

where α, β, γ refer to the Cartesian coordinates and $\varepsilon^{\alpha\beta\gamma}$ is the totally anti-symmetric Levi-Civita symbol. For a large cluster, the general problem of finding the eigenvalues of Hamiltonian (1) is challenging because of the lack of translation symmetry which prevents the reduction of the size of the matrix. Here we consider clusters comprising a reduced number of exchange-coupled spins and special topologies which allow exact solution of equation (1).

2.1. Two spin Cluster

The Hamiltonian for a pair of spins in the strong exchange limit *is*

$$H = -2J\vec{S}_i \cdot \vec{S}_j + g\mu_B B_z (S_i^z + S_j^z) \tag{3}$$

It can be exactly solved by using the total spin $\vec{S} = \vec{S}_i + \vec{S}_j$ whose square has eigenvalues $\vec{S}^2 = S(S+1)$ in the basis $|S_i, S_j; SM>$ with $0 \le S \le 2s$ and $S_i = S_j = s$. S_z has eigenvalues M with $-2s \le M \le 2s$. The eigenvalues of the pair Hamiltonian (3) are

$$E = -J\left[S(S+1) - 2s(s+1)\right] + g\mu_B B_z M \tag{4}$$

The magnetisation and the susceptibility are found from the partition function which is given by

$$Z = \mathrm{Tr} e^{-\beta\hat{H}} = \sum_{S=0}^{S=2s}\sum_{M=-S}^{M=+S} e^{\beta JS(S+1) - \beta g\mu_B B_z M} = \sum_{S=0}^{S=2s} \frac{\sinh(xS + x/2)}{\sinh(x/2)} e^{\beta JS(S+1)} \tag{5}$$

where the summation over M is performed and $x = \beta g\mu_B B_z$, $\beta = 1/(k_B T)$. The free energy is

$$F = -k_B T \ln Z \tag{6}$$

To find the magnetic susceptibility defined below in (8) as the second derivative with B_z as $B_z \to 0$, it is sufficient to expand the partition function in B_z and keep terms up to B_z^2. This gives

$$Z = \sum_{S=0}^{S=2s}\left[2S + 1 + S(S+1)(2S+1)B_z^2/6\right] e^{\beta JS(S+1)} \tag{7}$$

The magnetic susceptibility is

$$\chi = -\frac{\partial^2 F}{\partial B_z^2} = \frac{k_B T}{3} \frac{\sum_{S=0}^{S=2s}\left[S(S+1)(2S+1)\right] e^{\beta JS(S+1)}}{\sum_{S=0}^{S=2s}\left[2S+1\right] e^{\beta JS(S+1)}} \tag{8}$$

For a cluster of two spins s=1/2, the partition function reduces to

$$Z = 1 + e^{\beta J}(1 + 2\cosh x) \tag{9}$$

The average pair magnetization is

$$< \mathcal{M} >= \frac{4\sinh(x)}{e^{-\beta J} + 1 + 2\cosh(x)} \tag{10}$$

2.2. Three spin cluster

The Hamiltonian for a three spin cluster in the strong exchange limit is

$$H = -2J_1\vec{S}_1 \cdot \vec{S}_2 - 2J_2\vec{S}_1 \cdot \vec{S}_3 - 2J_3\vec{S}_2 \cdot \vec{S}_3 + g\mu_B B_z(S_1^z + S_2^z + S_3^z) \tag{11}$$

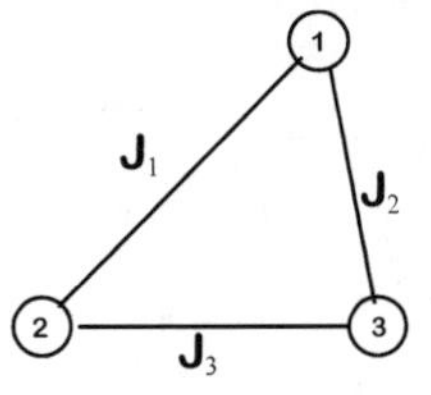

Figure1. Three spin cluster.

The eigenstates of (11) are found in the basis that diagonalizes $\vec{S}^2$ and S_z where $\vec{S}$ is the total spin operator. These states though are not uniquely determined by S and M_z, but require an additional quantum number which depends on the coupling scheme of the spins. Here, we first couple $\vec{S}_1$ with $\vec{S}_2$ to give $\vec{S}_{12} = \vec{S}_1 + \vec{S}_2$ and then $\vec{S}_{12}$ with $\vec{S}_3$ to give $\vec{S} = \vec{S}_{12} + \vec{S}_3$. The resulting states are labeled by the eigenvalues of the commuting set $\left\{\vec{S}_1^{\,2}, \vec{S}_2^{\,2}, \vec{S}_{12}^{\,2}, \vec{S}_3^{\,2}, \vec{S}^2, S_z\right\}$ as $|S_1S_2S_{12}S_3SM>$, and (11) reduces to

$$H = -J_1\left(\vec{S}_{12}^2 - \vec{S}_1^2 - \vec{S}_2^2\right) - J_2\left(\vec{S}^2 - \vec{S}_{12}^2 - \vec{S}_3^2\right) + g\mu_B B_z S^z + 2(J_2 - J_3)\vec{S}_2 \cdot \vec{S}_3 \tag{12}$$

For a symmetric spin arrangement for which $J_2 = J_3$, (12) is completely diagonal with eigenvalues

$$E = J_1\left(\tilde{S}_1 + \tilde{S}_2 - \tilde{S}_{12}\right) + J_2(\tilde{S}_3 + \tilde{S}_{12} - \tilde{S}) + g\mu_B B_z M_z \tag{13}$$

Here $\tilde{S}_\alpha = S_\alpha(S_\alpha + 1)$, $|S_1 - S_2| \le S_{12} \le S_1 + S_2$ and $|S_{12} - S_3| \le S \le S_{12} + S_3$. When $J_2 \neq J_3$, the last term in (12) is not diagonal. Its matrix elements can be obtained by Wigner-Eckart [6] theorem

$$\begin{aligned} < S_{12}'S'M' | \vec{S}_2 \cdot \vec{S}_3 | S_{12}SM >= \delta_{S'S}\delta_{M'M}(-1)^{2S-2M+1}\left[(2S+1)(2S'+1)(2S_{12}+1)(2S_{12}'+1)(2S_1+1)\right]^{1/2} \\ \times\left[(2S_2+1)S_2(S_2+1)(2S_3+1)S_3(S_3+1)\right]^{1/2} \begin{Bmatrix} S_1 & S_1 & 0 \\ S_2 & S_2 & 1 \\ S_{12} & S_{12}' & 1 \end{Bmatrix} \begin{Bmatrix} S_{12} & S_{12}' & 1 \\ S_3 & S_3 & 1 \\ S & S' & 0 \end{Bmatrix} \end{aligned} \tag{14}$$

The curly brackets are 9-j symbols whose numerical values are given in tables such as in Ref. [7], and some packages such as in Mathematica. For a cluster of three S =1/2 spins, S_{12} = 0, 1 and S = 3/2, 1/2. In this case we obtain compact expressions for the eigenvalues of (12) as follows

IOP Conf. Series: Journal of Physics: Conf. Series **869** (2017) 012037 doi:10.1088/1742-6596/869/1/012037

$$E(S=3/2,M)=-(J_1+J_2+J_3)/2+g\mu_B B_z M; \quad M=\pm 3/2,\pm 1/2 \tag{15}$$

$$E(S=1/2,M)=(J_1+J_2+J_3)/2\pm g\mu_B B_z M\pm\Delta; \quad M=\pm 1/2 \tag{16}$$

and $\Delta=[(J_1-J_2)^2+(J_2-J_3)^2+(J_1-J_2)(J_2-J_3)]^{1/2}$. The magnetic susceptibility for $B_z\to 0$ is

$$\chi=\frac{(g\mu_B)^2}{4k_BT}\frac{5+e^{-(J_1+J_2+J_3)/k_BT}\cosh(\Delta/k_BT)}{1+e^{-(J_1+J_2+J_3)/k_BT}\cosh(\Delta/k_BT)} \tag{17}$$

For a linear cluster, we put J_2=0, in equations (11-17).

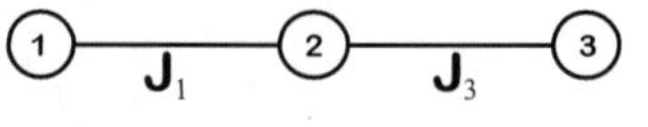

Figure 2. Three spin linear cluster.

3. Conclusion

Starting from an effective spin Hamiltonian, we have investigated the magnetic properties of a spin cluster. Using a spin coupling scheme and Wigner-Eckart theorem we computed exactly the energy levels, the free energy, and the magnetic susceptibility. These findings are useful for large systems made of weakly interacting small clusters (e.g. three-spin clusters), in which the system's magnetic susceptibility is that found in eq.17 times the total number of clusters. Moreover these findings form a good basis to treat approximately complex Heisenberg systems of many interacting degrees of freedom. In this case, the small cluster encompasses the short range effects and the spin operators outside the cluster are approximately replaced by their expectation values. The resulting reduced and more manageable problem is similar to the one studied here, and is solved in a self-consistent way.

References

[1] Morrish A H 2001 *The Physical Principles of Magnetism* (New York Wiley)
[2] Bartolome J, Fernando L and Fernandez J 2014 *Molecular Magnets* (Springer-Verlag)
[3] Lukaszew R A 2016 *Handbook of Nanomagnetism* (Taylor & Francis Group)
[4] Feltham H L C and Brooker S 2014 *Review of purely 4f and mixed-metal nd-4f single-molecule magnets containing only one lanthanide ion* Coordination Chemistry Reviews **276** p 1-33
[5] Barbara B, Imry Y, Sawatzky G and Stamp PCE 2008 *Quantum Magnetism* (Springer)
[6] Chaichian M and Hagedorn R 1997 *Symmetries in Quantum Mechanics* (Institute Of Physics Publishing)
[7] Schopper H 1968 *Numerical Tables for Angular Correlation Computations: 3j-, 6j-, 9j-Symbols, F- and Gamma-Coefficients* Landolt-Börnstein - Volume 3

Reflectivity modulator based on GaSb/GaAs heterostructure

S Rabbaa

Department of Physics, Arab American University-Jenin (AAUJ), Palestine

Email: sulaiman.rabbaa@aauj.edu

Abstract. A structure of gallium antimonide (GaSb) and gallium arsenide (GaAs) wafers is built to modulate light reflectivity at CO_2 laser wavelength. A quantum well composed of GaSb/GaAs heterojunction with highly doped GaAs up to 3×10^{18} cm^{-3} is inserted inside a layer structure. A grating of periodic structure of GaAs and gold layer is added just below the substrate. Gsolver software is used to determine the reflectivity of incident light with the existence of free carriers. A voltage is applied to the doped layer to deplete the free electrons and the reflectivity is determined again. The significant difference in reflectivity between the two cases can be used to build a light reflectivity modulator device.

1. Introduction

GaSb/GaAs heterostructure is used in photovoltaic applications, especially, in concentrator solar cells and thermophotovoltaics for space and terrestrial applications [1]. The heterostructure is also used in optoelectronic devices, such as quantum dot lasers [2, 3].

Reflectivity modulation process can be used to build Integrated Mirror Optical Switch (IMOS) and modulating retro-reflectors (MRR) for free space optical communications, which are characterized by security, high bandwidth and non-interference [4].

In this paper, Gsolver is used to optimize a structure of GaAs and GaSb layers to be used as IMOS for Q switch. A similar study was done in [5] and [6] but using only GaAs layers. The impact of using GaSb in the structure is given in the discussion.

2. Device structure

Figure 1 represents the adopted structure built as reflectivity modulator. GaAs substrate is followed by a thick layer of lowly doped n GaAs with doping density of 2×10^{17} cm^{-3}. The aim of this layer is to make the applied electric field perpendicular to the structure which makes equipotential surfaces parallel to layers. A thin n+ GaAs conductive layer with doping concentration of 1×10^{18} cm^{-3} is added to work as an ohmic contact. The following three layers compose a single quantum well of highly doped GaAs wafer sandwiched between two GaSb layers. A dielectric buffer layer of GaAs is followed by a diffraction grating composed of periodic structure of GaAs and gold (Au). The diffraction grating is covered by a thick layer of gold to prevent any transmission of light outside the device. Some applications of grating system are given in [7]. Description of grating structure and its elements is given in [8] and [9].

TM-polarized light is incident on the backside of the substrate. The light propagates through the structure and interacts with the diffraction grating. The incident light corresponds to a CO_2 laser radiation of wavelength λ=10.6 μm, which is in the mid-infrared spectrum. The structural parameters are given in table 1. Refractive index of GaSb is taken from [10] and refractive index of GaAs is taken from [6] and [11].

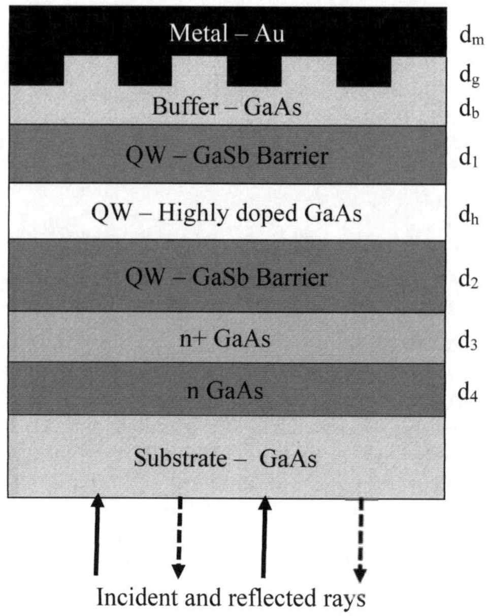

Figure 1. The device structure.

Table 1. The structural parameters.

Layer	Material	Doping density (cm^{-3})	Refractive index		Thickness	
			n	k	symbol	μm
Metal	Au		8.521	75.395	d_m	1.5
Grating	GaAs Au	undoped	3.3 8.521	0.0002 75.3946	dg	0.3
Buffer	GaAs	undoped	3.3	0.0002	d_b	0.1
Barrier	GaSb	undoped	3.791	0.000006	d_1	0.04
Highly doped	GaAs	3×10^{18}	2.35	0.59	d_h	0.0088
Barrier	GaSb	undoped	3.791	0.000006	d_2	0.04
Doped layer	n+ GaAs	1×10^{18}	3.072	0.0046	d_3	0.08
Lowly doped	n GaAs	2×10^{17}	3.276	0.00029	d_4	1.5
Substrate	GaAs	undoped	1.000	0		350

3. Results and discussions

The structure is optimized using the software by determination of dependence of light reflectivity on structure parameters in existence of free carriers in the quantum well. This is exhibited in graphs legend as "doped". The range of low reflectivity R_l, for example when $R_l < 0.3$, is determined. The reflectivity is determined again but with removing charge density from the quantum well. This is exhibited in graphs legend as "undoped". The range of higher reflectivity R_h, for example when $R_h > 0.6$, is extracted from graphs. Finally, the intersection between the two intervals range shows the optimized parameter of the structure. Quantum well free carriers can be depleted using suitable applied voltage by the means of the ohmic contact represented by the n GaAs layer [6].

IOP Conf. Series: Journal of Physics: Conf. Series **869** (2017) 012038 doi:10.1088/1742-6596/869/1/012038

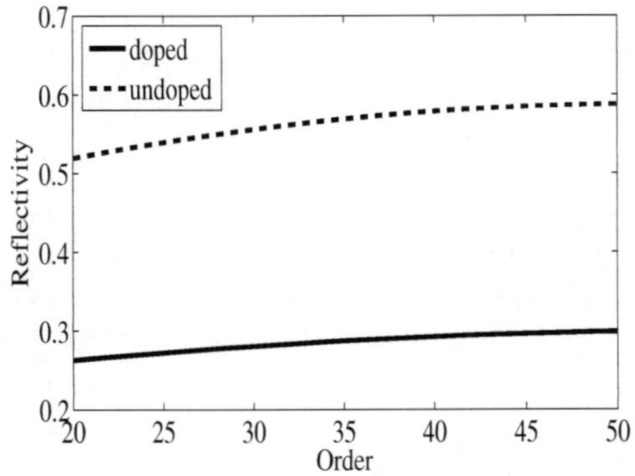

Figure 2. Reflectivity versus number of orders at P=3.15 μm, d_g=0.3 μm and DC=0.30.

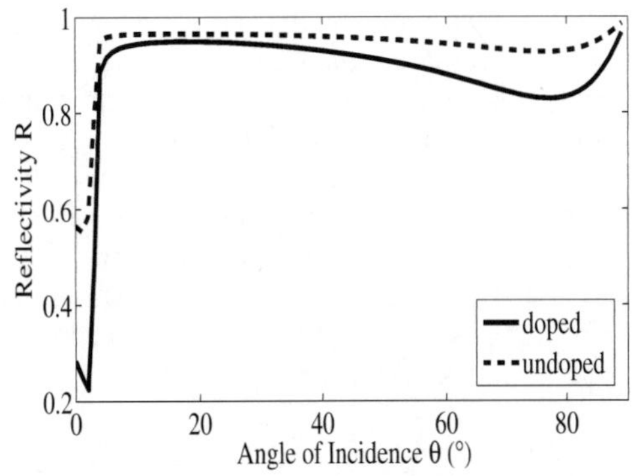

Figure 3. Reflectivity versus incident angle at P=3.15 μm, d_g=0.3 μm and DC=0.30.

Figure 2 represents the dependence of reflectivity R on number of orders adopted in the software. Generally, since we have a grating structure, number of orders should be relatively large. There is no high impact of number of orders on R for the range beginning from 20 orders. In our case, we choose orders of 30. The time of software running is larger if number of orders is increased.

The incident light has TM polarization, which is suitable to interact with grating. The relation between R and the angle of incidence is shown in "figure 3". There is no considerable difference between R of doped and undoped cases when $\theta > 3°$, and so we take normal incident case. There is no significant change in reflectivity when $\theta > 3°$.

The dependence of reflectivity on grating period is shown in "figure 4". The software is used many times to check the reflectivity with diffraction period at different values of duty cycle (DC) and grating height. The most suitable behaviour is taken to be dg=0.3 μm, and DC=0.30. At these parameters, we get $R_l = 0.64$ and $R_h = 0.23$. It is obvious that the figure exhibits a wide region of minimum reflectivity. For example at R = 0.5, the width of the U-shaped graph is $\Delta P = 0.18$ μm. The difference between reflectivity for the doped and undoped cases is maximum at P = 3.1 μm. Therefore, we consider 3.1 μm as the optimal value of period.

Figure 5 shows the relation between reflectivity and grating height d_g at P=3.15 μm and DC=0.30. At these parameters, we get $R_l = 0.61$ and $R_h = 0.28$. We notice that the figure exhibits a wide region of minimum reflectivity. For example at R = 0.5, the width of the U-shaped graph is $\Delta d_g = 0.10$ μm. The difference between reflectivity for the doped and undoped cases is maximum at $d_g = 0.30$ μm. Therefore, we consider 0.30 μm as the optimal value of grating height.

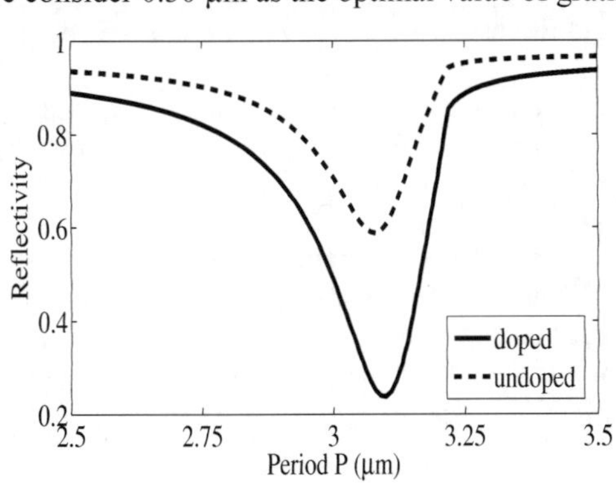

Figure 4. Reflectivity versus period at d_g=0.3 μm and DC=0.30.

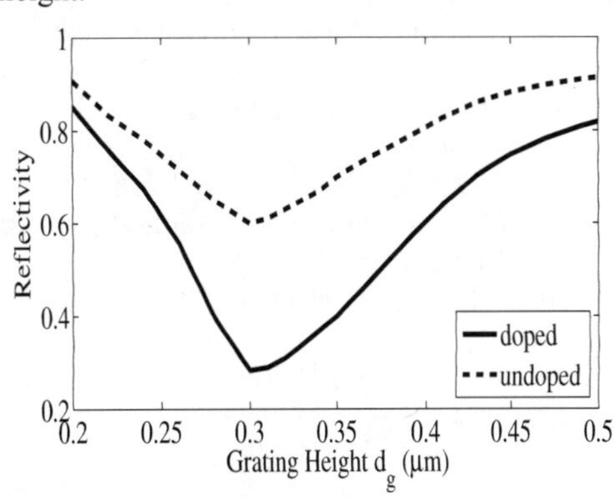

Figure 5. Reflectivity versus grating height at P=3.15 μm and DC=0.30.

Comparison between GaSb/GaAs structure in the present work and GaAs-based structure in [6], it is found that the results of reflectivity against period and grating height are more interesting in the current work. In the present work, minimum reflectivity of the doped and undoped case occurs at the same point of period and grating height (P = 3.1 μm and d_g = 0.3 μm).on the other hand, there is a difference in results of [6] between minimum reflectivity of doped and undoped cases in the graphs of period and grating height. It was found in [6] that reflectivity against period has minimum at P = 3.04 μm and 2.90 μm for doped and undoped case, respectively. In addition, reflectivity against grating height has minimum at d_g = 0.33 μm and 0.26 μm for doped and undoped case, respectively. It is more applicable for a device design if the minimum reflectivity in the case of doped and undoped layer occurs at the same value of structure parameter.

4. Conclusions

We have optimized a grating structure based on GaSb/GaAS quantum well for normal incidence of 10.6 μm. The parameters of the device can be considered as: grating period = 3.1 μm, grating height = 0.3 μm and duty cycle = 0.32. For the structure with theses parameters, we have R_l = 0.64 and R_h = 0.23, which gives a modulation depth of 0.41 and an extinction ratio 2.78. The existence of minimum reflectivity for doped and undoped curves at the same value of parameter (period or grating height) is a point of interest for GaSb / GaAs heterostructure.

Acknowledgment
This work has been funded by Bank of Palestine, the Arab Monetary Fund and Welfare Association through Zamalah fellowship program, and by the Arab American University- Jenin (AAUJ), Palestine.

References

[1] Andreev V M, Khvostikov V P, Paleeva E V, Sorokina S V, and Shvarts M Z 1996 GaAs and GaSb based solar cells for concentrator and thermophotovoltaic applications *Photovoltaic Specialists Conf., Conf. Record of the Twenty Fifth IEEE* (Washington DC) 5507448.
[2] Sukkabot W 2013 *Int. Journal of Applied Physics and Mathematics* **3** 178.
[3] Qiu F *et al* 2016 *Nanotechnology* **27** 065602.
[4] Rabinovich W 2012 Optical modulating refro-reflectors *Advanced Optical Wireless Communication System* ed Shlomi Arnon *et al* 1st edition (Cambridge University Press) chapter 13 p 328.
[5] Vandermeiren W 2011 Novel Semiconductor Based Devices for Detection and Generation of CO_2 Laser Radiation *PhD dissertation* (Uitgeverij VUBPRESS Brussels University Press: Brussels).
[6] Rabbaa S, Vandermeiren W, Shkerdin G and Stiens J 2016 *IEEE Int. Conf. on Semiconductor Electronics (ICSE)* (Kuala Lumpur) 16322030.
[7] Zhao B and Bouchard M 2008 Characterization and application of median- density gratings *New Trends in Lasers and Electro-optics Research* ed W Arkin 1st edition (Nova Science Publishers, Inc.: New York).
[8] Li L and Zeng L 2005 *Applied Optics* **44** 4494.
[9] Palmer C 2005 *Diffraction Grating Handbook* (Newport Corporation: New York) 6th edition.
[10] Adachi S 1999 *Optical Constants of Crystalline and Amorphous Semiconductors: Numerical Data and Graphical Information* (Springer Science + Business Media New York) pp 235.
[11] Vandermeiren W, Stiens J, Shkerdin G, De Tandt C and Vounckx R 2014 *J. Phys. D: Appl. Phys.* **47** 025104.

IOP Conf. Series: Journal of Physics: Conf. Series **869** (2017) 012039 doi:10.1088/1742-6596/869/1/012039

Generation of runaway electron beams in high-pressure nitrogen

V F Tarasenko, A G Burachenko and E Kh Baksht

Institute of High Current Electronics, Russian Academy of Sciences,
2/3 Akademichesky Ave., 634055 Tomsk, Russia
E-mail: VFT@loi.hcei.tsc.ru

Abstract. In this paper the results of experimental studies of the amplitude-temporal characteristics of a runaway electron beam, as well as breakdown voltage in nitrogen are presented. The voltage pulses with the amplitude in incident wave ≈120 kV and the rise time of ≈0.3 ns was used. The supershort avalanche electron beam (SAEB) was detected by a collector behind the flat anode. The amplitude-time characteristics of the voltage and SAEB current were studied with subnanosecond time resolution. The maximum pressure at which a SAEB is detectable by collector was ~1 MPa. This pressure increases with decreasing the voltage rise time. The waveforms of the discharge and runaway electron beam currents was synchronized with the voltage pulses. The mechanism of the runaway electron generation in atmospheric-pressure gases is analyzed on the basis of the obtained experimental data.

1. Introduction

The generation of runaway electron beams in high-pressure gases with sharp-ended electrodes or so-called electrodes with a small curvature radius is a fundamental physical phenomenon that determines the delay of breakdowns and the properties of discharges. By now, several scientific groups have managed to detect runaway electron beams with collectors downstream of anode foils in air and other gases at a pressure of ≅0.1 MPa [1, 2], and only one scientific group has reported on detection of such a beam with a collector at ≈0.2 MPa and higher and on measurements of the beam current amplitude and its pressure dependence in different gases [3–5]. The highest pressure at which a runaway electron beam was detected with a collector was 0.4 MPa for nitrogen [3], 0.2 MPa for SF_6 [4], and 1.2 MPa for helium [5]. There are three main factors that make difficult the detection of a runaway electron beam at a diode pressure higher than 0.2 MPa. First, increasing the gas pressure decreases the beam amplitude. Second, if we increase the gas pressure, we have to complicate the design of an experimental setup. Third, it is required to use generators with a voltage amplitude of hundred kilovolts and rise time shorter than a nanosecond. Note that in some studies, a runaway electron beam in nitrogen at up to 4 MPa was detected from luminescence of a scintillator [6, 7], but neither the beam current amplitude and duration nor their dependence on pressure and other parameters were measured.

Here we investigate the parameters of a supershort avalanche electron beam (SAEB), which is a runaway electron beam downstream of an anode foil [8], in nitrogen at a pressure of 0.2–1 MPa.

2. Experimental setup

The setup with new modification pulser SLEP-150M [9] was used in our experiments. To work at high pressures were reduced to 30 mm inner diameter of transmission line and gas filed diode. The output of pulser SLEP-150M (part of transmission line), a gas filed diode, and a collector are shown on the figure 1.

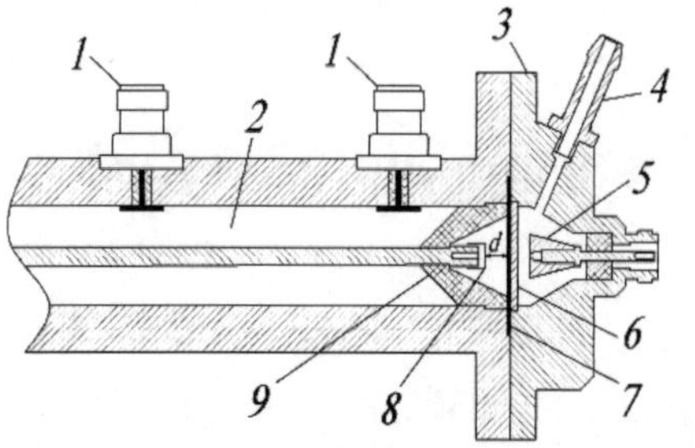

Figure 1. Schematic of the output of the SLEP-150M pulser, gas filed diode, and collector with the 13-mm-diameter receiving area: (*1*) capacitive voltage dividers, (*2*) transmission line of the pulser, (*3*) housing of the collector, (*4*) connection for gas pumping, (*5*) receiver part of the collector, (*6*) mesh, (*7*) foil, (*8*) tubular cathode in gas filed diode, (*9*) insulator.

SLEP-150M formed a voltage pulse with a FWHM of ~1 ns at a matched load. The voltage rise time was ~ 250 ps at a level of 0.1-0.9. The wave impedance of transmission line *2* was 100 Ohm. The voltage amplitude of incident wave in the transmission line was ~120 kV.

The parameters of a SAEB were measured with the cathode that was made as a stainless steel foil tube of diameter 6 mm and thickness 100 μm. The anode of the gas diode was AlBe foil *7* 60-μm-thickness reinforced with a grid *6* from the side of the collector *3*. The collector receiving part *5* was 13 mm in diameter. The time resolution of the collector reached 50 ps. The measuring equipment also included the capacitive voltage dividers *1*. The signals from the dividers and collector were transmitted to a DSO-X6004A real-time digital oscilloscope (bandwidth 6 GHz, sampling increment 50 ps) via 5D-FB PEEG high-frequency cables (Radiolab) of length 1 m. For signal of voltage attenuation, we used high-frequency attenuators (Barth Electronics, model 142-NMFP) with a bandwidth of up to 30 GHz. The voltage and the SAEB current were measured simultaneously in each pulse. The timing accuracy of SAEB and voltage pulses was no worse than 100 ps. The SAEB generation time with respect to the voltage pulse was determined from capacitive current of collector for revealing one the foil was removed. This procedure is described in detail in [10]; the timing accuracy of voltage and SAEB current pulses with a LeCroy WaveMaster 830Zi-A real-time digital oscilloscope (bandwidth 30 GHz, sampling increment 12.5 ps) was no worse than 10 ps.

3. Experimental results and discussion

In our experiments, we studied the generation of a SAEB at a nitrogen pressure of 0.2–1 MPa for three gap widths which were smaller than 12 mm optimal at a nitrogen pressure of 0.1 MPa [11]. The gap was decreased for increasing the parameter U_b/pd, where U_b is the maximum (breakdown) voltage across the gap, p is the nitrogen pressure, and d is the gap width. In the range of nitrogen pressures from 0.2 to 1 MPa, we recorded waveforms of the gap voltage, incident voltage wave in the transmission line, and SAEB current. The results are presented in figures 2, 3, and 4. Figure 2 shows typical waveforms of the voltage from the capacitive divider located near the gas diode and of the SAEB current. The SAEB current and the voltage were synchronized reasoning that the maximum SAEB amplitude under optimum conditions roughly corresponds to the breakdown (maximum) voltage across the gap [10]. It is seen from figure 2 that the SAEB has a FWHM of about 90 ps and that its decay is shorter than the decay of the gap voltage. This is because most of the runaway electrons are produced at the front of an ionization wave moving from the sharp-ended cathode to the anode. Once the ionization front reaches the anode, the electric field in the gap decreases to less than its critical value and the generation of runaway electrons ceases

IOP Conf. Series: Journal of Physics: Conf. Series **869** (2017) 012039 doi:10.1088/1742-6596/869/1/012039

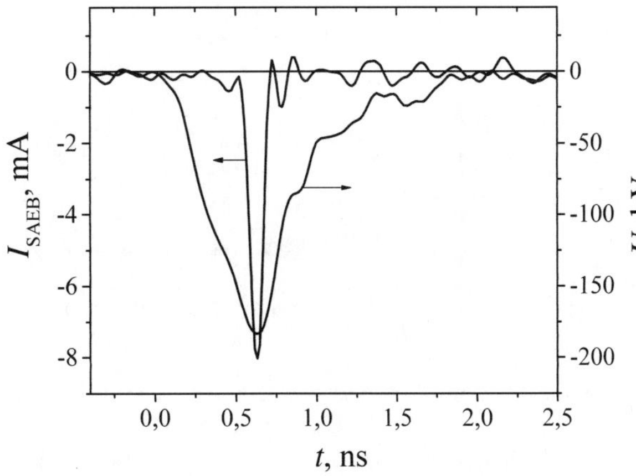

Figure 2. Voltage and SAEB current at $p = 0.8$ MPa and $d = 4$ mm.

The breakdown voltage as a function of nitrogen pressure at $d = 8$ mm, 4 mm, and 1 mm is shown in figure 3.

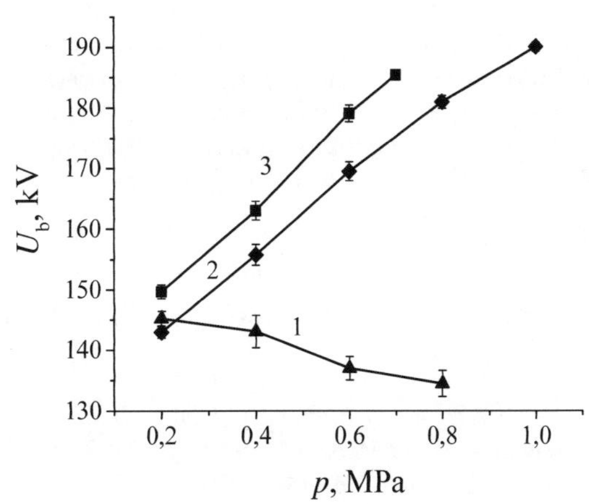

Figure 3. Breakdown voltage vs nitrogen pressure at $d = 1$ mm (*1*), 4 mm (*2*), and 8 mm (*3*).

It is seen from figure 3 that increasing the nitrogen pressure from 0.2 to 1 MPa increases the breakdown voltage at $d = 8$ mm and 4 mm. At $d = 1$ mm, the dependence is different: U_b decreases with increasing the pressure; however, at more than 0.5 MPa, the SAEB is also detected by the collector. As the pressure is increased from 0.2 to 1 MPa, the SAEB amplitude decreases for all three gaps (figure 4).

The presented dependences are rather typical for high pressures [3–6]. Although the breakdown voltage at $d = 4$ mm and 8 mm increases (figure 3), the SAEB amplitude decreases (figure 4). The highest SAEB amplitudes were obtained at $d = 8$ mm and 0.2 MPa, which is the minimum pressure in the experiments. Note that the collector detects only part of the runaway electrons, i.e., those arriving at its receiving area of diameter 13 mm. Moreover, the electrons with energies less than 40 keV are absorbed by the AlBe foil.

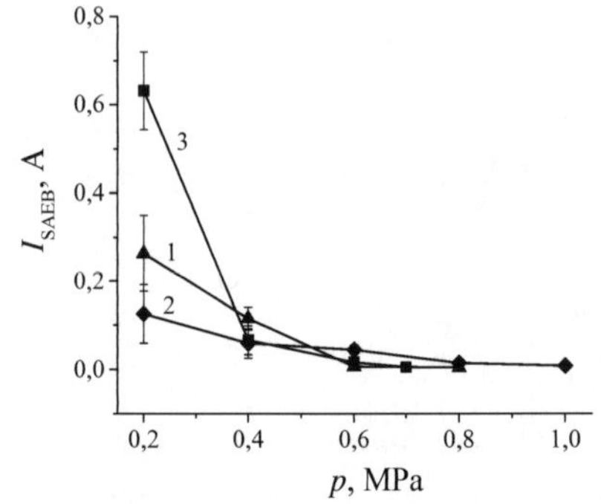

Figure 4. SAEB amplitude vs nitrogen pressure at $d = 1$ mm (*1*), 4 mm (*2*), and 8 mm (*3*).

Thus, our experiments demonstrate that at a nitrogen pressure higher than 0.2 MPa, the mechanism of SAEB generation remains the same. Most of the runaway electrons are produced between the ionization front, which bridges the gap, and the anode [8], and this is confirmed by experimental and theoretical data [1, 2].

4. Conclusion

Thus, we studied the generation of a super short avalanche electron beam (SAEB) in nitrogen at a pressure of 0.2–1 MPa for gap widths of 1, 4, and 8 mm. The beam parameters were measured for the first time at a nitrogen pressure higher than 0.5 MPa. It is shown that on the SLEP-150M generator with a transmission line diameter of 30 mm, the SAEB amplitude decreases with increasing the nitrogen pressure from 0.2 to 1 MPa, whereas the breakdown voltage under these conditions increases at $d = 4$ mm and 8 mm and decreases at $d = 1$ mm.

Acknowledgments

The work is performed in the framework of the State task for HCEI SB RAS, project #13.1.3.

References

[1] Tarasenko V F ed 2016 *Generation of Runaway Electron Beams and X-Rays in High Pressure Gases, Volume 1: Techniques and Measurements* (New York: Nova Science Publishers, Inc)

[2] Tarasenko V F ed 2016 *Generation of Runaway Electron Beams and X-Rays in High Pressure Gases, Volume 2: Processes and Applications* (New York: Nova Science Publishers, Inc)

[3] Alekseev S B, Orlovskii V M, Tarasenko V F, Tkachev A N and Yakovlenko S I 2005 *Tech. Phys.* **50**. 1623

[4] Tarasenko V F, Baksht E H, Burachenko A G, Kostyrya I D, Lomaev M I and Rybka D V 2008 *Laser Part. Beam.* **26** 605

[5] Baksht E H, Lomaev M I, Rybka D V, Sorokin D A and Tarasenko V F 2008 *Tech. Phys.* **53** 1560

[6] Ivanov S N 2013 *J. Phys.* D: *Appl. Phys.* **46** 285201

[7] Ivanov S N 2014 *Izv. Vyssh. Uchebn. Zaved. Fizika.* **57** 3/2 13

[8] Tarasenko V F, Orlovskii V M and Shunailov S A 2003 *Russ. Phys. J.* **46** 325

[9] Tarasenko V F, Kostyrya I D, Baksht E Kh and Rybka D V 2011 *IEEE Trans. Dielect. Electr. In.* **18** 1250

[10] Burachenko A G and Tarasenko V F 2010 *Tech. Phys. Lett.* **35** 1185

[11] Tarasenko V F, Baksht E H, Kostyrya I D, Lomaev M I and Rybka D V 2008 *J. Phys.: Conf. Ser.* **133** 012021

IOP Conf. Series: Journal of Physics: Conf. Series **869** (2017) 012040 doi:10.1088/1742-6596/869/1/012040

Cleaning of niobium surface by plasma of diffuse discharge at atmospheric pressure

V F Tarasenko[1,2], M V Erofeev[1,2], M A Shulepov[1] and V S Ripenko[1]

[1] Institute of High Current Electrons, 2/3 Akademichesky ave., Tomsk, 634055, Russia
[2] National Research Tomsk Polytechnic University, 30 Lenin ave., Tomsk, 634050, Russia
E-mail: VFT@loi.hcei.tsc.ru

Abstract. Elements composition of niobium surface before and after plasma treatment by runaway electron preionized diffuse discharge was investigated in atmospheric pressure nitrogen flow by means of an Auger electron spectroscopy. Surface characterizations obtained from Auger spectra show that plasma treatment by diffuse discharge after exposure of 120000 pulses provides ultrafine surface cleaning from carbon contamination. Moreover, the surface free energy of the treated specimens increased up to 3 times, that improve its adhesion property.

1. Introduction

Nowadays surface modification of metals and dielectrics by low-temperature plasma formed in different types of gas discharge are widely used in microelectronics, nano-technology and medicine. A number of methods have been developed for the surface modification using low-pressure gas discharge, such as plasma nitriding [1], plasma spraying [2], treatment by low energy high current electron-beam [3]. However, most of these plasma sources require special vacuum equipment, which limits integration in production lines and increasing its cost. In this work, runaway electron beam and X-rays formed in non-uniform electric field by applying high voltage pulses of nanosecond duration to the electrode with small radius of curvature was used to generate a diffuse discharge in a working gas of atmospheric pressure [4]. In this type of discharge called a runaway electrons preionized diffuse discharge (REP DD), the plane ground electrode experiences action of dense plasma, a shock wave, a super short avalanche electron beam, as well as optical radiation of different spectral regions, including UV, VUV, X-rays from the discharge plasma. Voltage pulse polarity, and consequently direction of ions and electrons motion, define type of the treatment on the ground electrode. Thus the use of REP DD, which can generate volume non-thermal plasma at atmospheric pressure, provides scientists another approach for surface modification with low cost, simple facilities and easy operation.

In our previous work [5], pulsed generator RADAN-220 was used for diffuse discharge formation. The main disadvantage of this generator is a low repetition pulse rate – less than 1 Hz, that demands of long exposure time to obtain any significant results of discharge plasma treatment. In this work, the experimental setup based on GIN-100 generator with pulse repetition frequency up to 1000 Hz was developed, that allows to shorten the exposure time for plasma treatment. The aim of the work was to study the effect of diffuse discharge plasma formed in nitrogen flow of atmospheric pressure on the surface of niobium. Due to the use of niobium in modern high-technological equipment, e.g., SQUID magnetometer in nuclear magnetic tomographs, high-frequency cell in elementary particle accelerators and superconducting solenoids, cleanness of Nb surface and thin Nb-films from oxides or other contaminants is very important.

IOP Conf. Series: Journal of Physics: Conf. Series **869** (2017) 012040 doi:10.1088/1742-6596/869/1/012040

2. Experimental setup and methods of measurements

Design of the discharge chamber has been reported in [6]. In the experiments, a REP DD was formed by applying negative voltage pulses from GIN-100-1 generator (Antares Ltd., Russia) with a FWHM of 4 ns with amplitude up to 80 kV to electrode of small curvature radius at a pulse repetition frequency of up to 100 Hz. Typical waveforms of voltage pulses at the discharge gap consists of a sequence of voltage pulses of both polarities, so plane electrode is affected by action of electron beam and positive ions, which were generated in the discharge. The distance between sharp cathode and plane anode, which was made from niobium of cylindrical shape (diameter 14 mm and height 0.4 mm), was 9 mm. At such experimental conditions, the mean input power in nitrogen plasma was about 20 W.

Niobium specimens were mechanically polished with emery paper, before final polishing with diamond suspensions until their surface was like a mirror. Then they were cleaned twice in an ultrasonic bath (Elmasonic S 10H, Elma Schmidbauer GmbH, Germany) with alcohol at 50 °C. Since cleaned metal specimens are highly prone to oxidation, inert gas such as nitrogen was used for plasma treatment. After surface modification by REP DD, which was initiated in nitrogen flow rate of 5 sl/min, the changes in concentration of the main chemical elements in the surface layers were measured with Auger-electron spectrometer Shkhuna-2 (NPO "Electron", Russia). The surface free energy was evaluated by contact angle measurements performed by optical microscope (LOMO MIKMED-1 Version 2, Russia) before and immediately after plasma treatment. The surface energies, the sum of the polar energy and the dispersion energy, were calculated using the Owen-Wendt method [7] by solving the system of the OWRK equation for two different liquids with the known dispersive and polar components. During contact angle measurement, 3 μl of deionized (DI) water and glycerine were, respectively, dropped onto the metal surface at room temperature. The surface roughness of specimens before and after treatment was studied with using of a 3D contactless profilometer Micro Measure 3D Station (STIL).

3. Results and discussion

Figure 1 shows the concentration profiles of carbon (fig. 1a), oxygen (fig. 1b) and nitrogen (fig. 1c) in surface layer of niobium before and after plasma treatment with 80000 and 120000 REP DD pulses. It is seen from the figure, that carbon concentration of contaminants decreased with increasing of discharge number of pulses. The carbon concentration on the top surface after exposure by 80000 pulses decreased from 24 % at. to 20 % at. and in the underlying layer, in the depth deeper than 50 nm, it decreased 1.5 times. After plasma treatment with 120000 REP DD pulses, the carbon concentration in the whole depth decreased lower than 10 % at. Since cleaned metal specimens are highly prone to oxidation, immediately after treatment the surface has covered with thin (less than 10 nm) oxidized film, in which oxygen concentration exceeds the concentration in the untreated specimen (fig. 1b). However, after plasma treatment, in the underlying layer in the depth deeper than 10 nm, it decreased almost 2 times. Moreover, slight increasing of nitrogen concentration was found in the top surface layer after treatment by 120000 pulses (fig. 1c). It should be noted, that nitrogen concentration is below 5 % at., that is close to limit of sensibility of Auger electron spectroscopy method.

As a consequence of the cleaning, the surface free energy of the treated specimens drastically increased in the initial stage of plasma treatment and slowly increased with increasing power and number of discharge pulses. Figure 2 shows the dependence of the surface free energy versus number of discharge pulses. Before plasma treatment, surface with high content of carbon contaminants has low surface free energy value due to the weak polarity of C-C/H bonds. After removing the carbon impurity through plasma treatment, the surface free energy increases significantly.

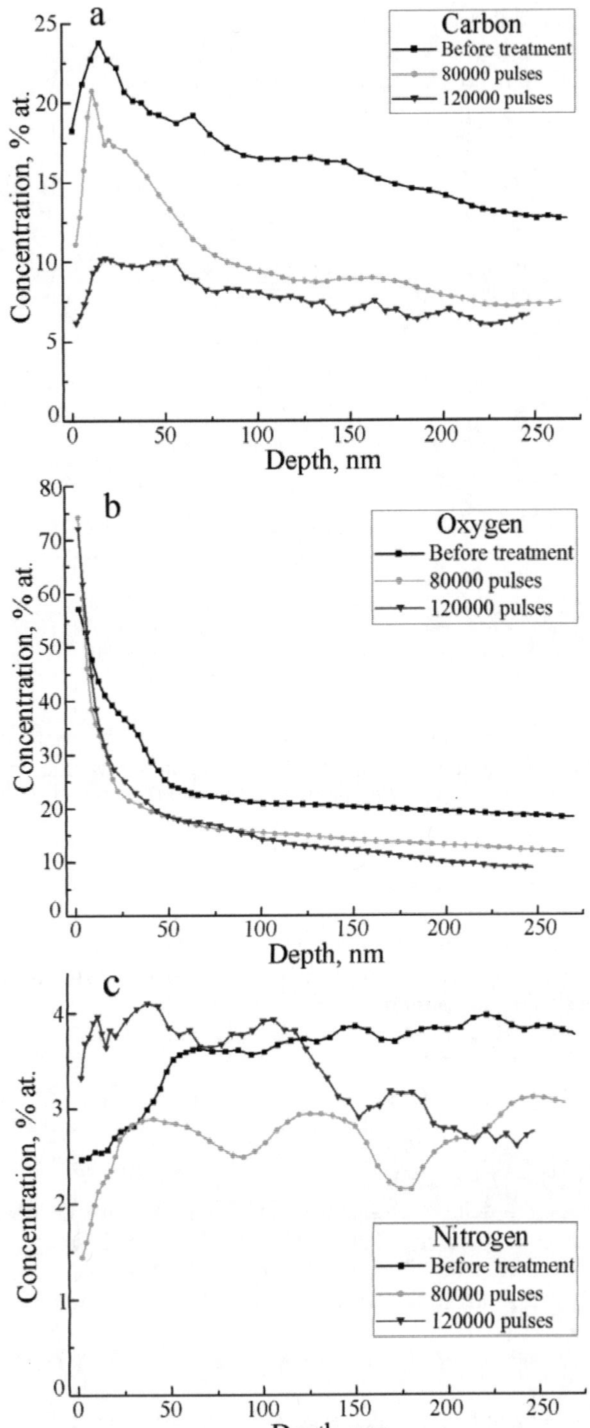

Figure 1. Carbon (a), oxygen (b) and nitrogen (c) concentrations at a depth of the niobium surface before and after REP DD treatment in nitrogen flow.

As shown in figure 2, the surface free energy increased 2.8 times after treatment by 80000 discharge pulses. Moreover, the growth of the surface free energy is also due to formation of high polarity chemical groups with oxygen and hydroxide bonds after plasma treatment.

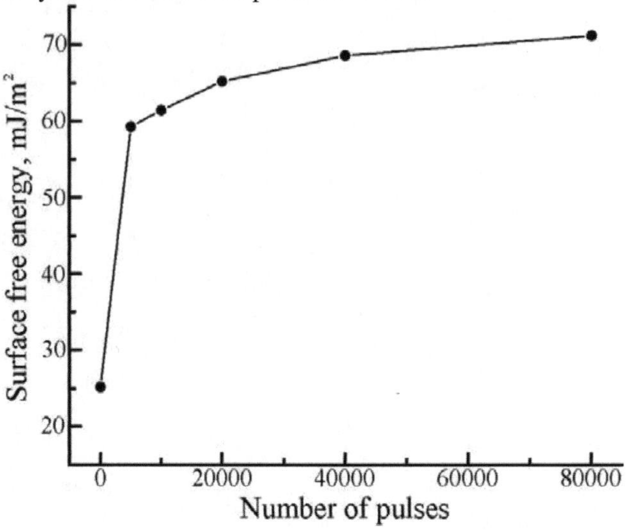

Figure 2. Surface free energy of niobium specimen versus number of REP DD pulses.

The surface profile measurements have shown, that plasma treatment with REP DD in nitrogen flow has polishing effect. Thus roughness parameter R_a (roughness average, nm) of the untreated niobium specimen was equal to 60 nm and after treatment with 80000 REP DD pulses it was 51.2 nm.

4. Conclusion

Niobium surface treatment was performed using a nitrogen plasma of runaway electron preionized diffuse discharge generated at atmospheric pressure. From the observation of Auger electron spectra and surface free energy measurements, it is oblivious that REP DD plasma treatment provides high grade of the surface cleanness along with the intense surface free energy growth, that is important for further bonding, adhesive coating and printing.

Acknowledgement

The work is performed in the framework of the State task for HCEI SB RAS, the project #13.1.3.

References

[1] Rie K-T 1999 *Surf. Coat. Technol.* **112** 56
[2] Leylavergne M, Vardelle A, Dussoubs B and Goubot N 1998 *J. Therm. Spray Tech.* **7** 527
[3] Proskurovsky D I, Rotshtein V P, Ozur G E, Markov A B, Nazarov D S, Shulov V A, Ivanov Yu F and Buchheit R G 1998 *J. Vac. Sci. Technol.* A **16** 2480
[4] Tarasenko V F 2014 *Runaway electrons preionized diffuse discharges* (New York: Nova Science Publishers, Inc.)
[5] Erofeev M V, Shulepov M A, Oskomov K V and Tarasenko V F 2015 *J. Phys.: Conf. Series* **652** 012039
[6] Erofeev M V, Baksht E Kh, Burachenko A G and Tarasenko V F 2015 *Tech. Phys.* **60** 1316
[7] Owens D K and Wendt R C 1969 *J. Appl. Polymer Sci.* **13** 1741

IOP Conf. Series: Journal of Physics: Conf. Series **869** (2017) 012041 doi:10.1088/1742-6596/869/1/012041

Adsorption of CO_2 on Fe-doped graphene nano-ribbons: Investigation of transport properties

W Othman[1], M Fahed[1], S Hatim[2], A Sherazi[3], G Berdiyorov[4] and N Tit[1]

[1] Department of Physics, UAE University, P. O. Box 15551, Al-Ain, U.A.E.
[2] Department of Chemical and Petroleum Engineering, UAE University, P. O .Box 15551, Al-Ain, U.A.E.
[3] Department of Mechanical Engineering, P. O. Box 15551, Al-Ain, U.A.E.
[4] QEERI, Hamad bin Khalifa University, Qatar Foundation, Doha, Qatar

Email: ntit@uaeu.ac.ae

Abstract. Density functional theory combined with the non-equilibrium Green's function formalism is used to study the conductance response of Fe-doped graphene nano-ribbons (GNRs) to CO_2 gas adsorption. A single Fe atom is either adsorbed on GNR's surface (aFe-graphene) or it substitutes the carbon atom (sFe-graphene). Metal atom doping reduces the electronic transmission of pristine graphene due to the localization of electronic states near the impurity site. Moreover, the aFe-graphene is found to be less sensitive to the CO_2 molecule attachment as compared to the sFe-graphene system. These behaviours are not only consolidated but rather confirmed by calculating the IV characteristics from which both surface resistance and its sensitivity to the gas are estimated. Since the change in the conductivity is one of the main outputs of sensors, our findings will be useful in developing efficient graphene-based solid-state gas sensors.

1. Introduction

Graphene-based sensors have advantages over solid-state gas sensors in terms of sensitivity, response and recovery times, low power consumption and low cost [1]. Moreover, they can operate at room temperature and under ambient conditions. In addition, gas sensing and catalytic properties of graphene can be further enhanced by decorating it with metal and metal-oxide nanoparticles [2,3]. Although the choice of non-noble metal as a dopant is mainly motivated by its low cost, Fe atoms can perform as good as noble metal atoms (such as Pt atom) in terms of improving the sensitivity and selectivity of graphene [4]. Transition metals such as Fe, Co, Ni, and Cu are also known to increase the sensitivity and selectivity of graphene-based sensors for gas detection [4,5]. Since the changes in the resistivity after the gas molecule adsorption is the main output of solid-state sensors, a fundamental understanding of the electronic transport properties of graphene under these conditions enables the utilization of the full potential of functionalized graphene for practical applications.

In the present work, we use density functional theory (DFT) in combination with the non-equilibrium Green's function formalism to study the electronic transport response of zig-zag graphene nano-ribbons [6] doped with Fe atom and its effect on the adsorption of carbon dioxide (CO_2) molecule, which is one of the major greenhouse gases. Capturing, storing and converting CO_2 has become a major problem confronting environmental challenges such as the issue of climate change [7,8]. We consider two different cases, either the Fe atom substitutes the carbon atom of graphene (sFe-graphene) or it is adsorbed on graphene's surface (aFe-graphene). The main purpose of this study is to fully explore the effect of the CO_2 molecule adsorption on the electronic transport properties of pristine and functionalized graphene. The obtained results are explained in terms of electron localization in the system. Our findings will be useful to design more sensitive graphene-based gas sensors.

2. Computational model and method

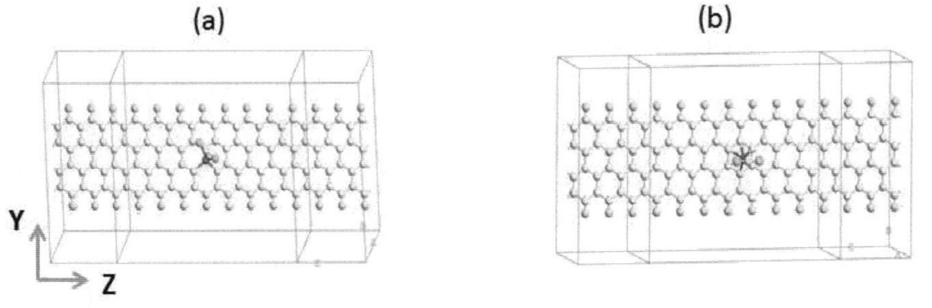

Figure 1: Device geometries: Relaxed structures of hydrogen-passivated GNRs with a single CO_2 adsorbed on Fe atom: (a) Fe as substitutional dopant [sFe]; (b) Fe as ad-atom [aFe]. The conduction/transmission is along z-direction, whereas vacuum space of 20 Å is left along x-direction.

Two device geometries are considered. Each contains an active layer of length 22.149 Å and two electrodes each of size 7.383 Å. The active region is long enough to study the effect of a single impurity. Each GNR has a width of 7.10 Å and contains about 64 carbon atoms. Defects and dopants play important role in the electric properties of materials. The adsorption of CO_2 on carbon vacancy has already been studied in our group [9]. Here, we wanted to study the adsorption of CO_2 on graphene nano-ribbons in assistance of Fe-catalyst. Meanwhile assess the effect of binding of Fe on the adsorption properties. So, we considered two realizations: (a) Fe atom strongly attached to the surface in a substitution of carbon atom, and (b) Fe atom just relaxed on a hollow site above the GNR.

In the computational method, we first optimize the atomic structures using DFT within the general gradient approximation (GGA) of Perdew-Burke-Ernzerhof (PBE) for the exchange-correlation energy [10]. The Brillouin zone sampling/integration was performed using 1x1x100 Monkhorst-Pack technique [11]. The convergence criteria for total energy and Hellman-Feynman forces were 0.001 eV and 0.01 eV/Å, respectively. The transmission and IV curve were calculated along z-direction between the two electrodes. The non-equilibrium Green's function formalism is used to calculate the transport properties. All simulations were obtained using the first-principles computational package Atomistic toolkit [12].

3. Results and discussions

Figure 2 shows the transmission spectra versus energy for GNR (dotted black line), GNR with Fe atom attached to the surface (green line) and GNR+Fe with adsorption of CO_2 on the GNR-Fe composite. Two different cases for the positions of Fe doping are considered: (a) Fe as ad-atom on the hollow site of GNR with a Fe-C distance of about 2.05 Å; and (b) Fe as a substitutional dopant replacing carbon atom. As a reference, the dotted black line in Figure 2 presents the zero-bias T(E) of pristine graphene nano-ribbon without Fe and without CO_2 as well. This latter transmission curve shows a step-like behaviour with an enhanced transmission at the Fermi level, which are typical for graphene nano-ribbons. These features originate from the edge-localized electronic states with energies close to the Fermi level. Figure 2a shows that the aFe adsorption results in considerably

IOP Conf. Series: Journal of Physics: Conf. Series **869** (2017) 012041 doi:10.1088/1742-6596/869/1/012041

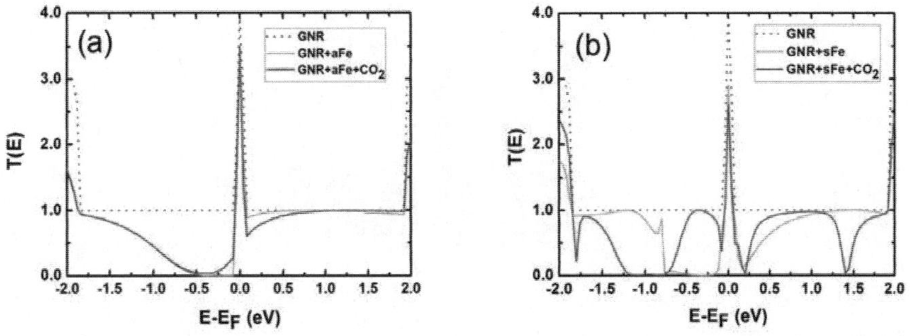

Figure 2: Zero-bias transmission spectra versus energy for GNR (dotted line); GNR with Fe atom (green line); and GNR-Fe with adsorbed CO_2 molecule (red line) for cases: (a) aFe and (b) sFe.

reduction when decreasing the electron energy below Fermi level (at about - 0.5 eV). A bit variation occurs when the CO_2 gets adsorbed on aFe as shown by red curve in Figure 2a. In contrast to these features, in the case of sFe shown in Figure 2b the discrepancy between T(E) before the landing of CO_2 molecule and after on sFe is huge. This fact reveals that sFe should play the role of a better gas sensor to CO_2 than aFe does. This will be confirmed with further studies shown below.

In order to get a better insight about the origin of the obtained changes in the transmission spectra due to the doping, we calculated the projected self-consistent Hamiltonian (PSH) eigen-states. These eigen-states are associated with the poles of the green's function at the given electron energy. In Figure 3, we chose to show the PSH eigenstates of sFe-GNR sample without and with CO_2 adsorbed molecule in order to explain the huge discrepancy in T(E) spectra. The energy is taken to be $E\text{-}E_F = -1.0$ eV. Figure 3 shows three eigen-states corresponding to: (a) GNR, (b) GNR-sFe, and (c) GNR-sFe with CO_2 molecule. Panel (a) shows the eigen-state to be uniformly-distributed as an extended Bloch-like eigen-function; whereas panel (b) shows a strong localization of electrons around Fe atom; and panel (c) shows also still strong and a bit perturbed but confined near the vicinity of Fe site.
Thus, strong electronic localization is the reason causing the huge discrepancy in T(E) behaviours.

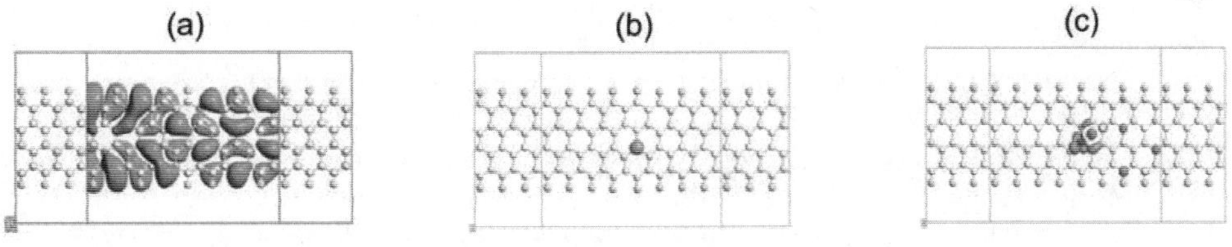

Figure 3: Iso-surface plots of the PSH eigen-states for: (a) pristine GNR; (b) sFe-GNR; and (c) sFe-GNR with adsorbed CO2 molecule. E = -1.0 eV.

Last but not the least, Figure 4 displays the results of IV-curve calculated using DFT with GGA. Figure 4a shows the results of two IV curves for the case of sFe in order to carry on further the preceding analysis. Actually, the IV-curve of GNR-sFe shows higher conductivity than the IV-curve of GNR-sFe with adsorbed CO_2 molecule. From each IV curve, the resistance can be derived, based on it the sensitivity can be estimated using the formula:

$$S = \frac{|R_g - R_a|}{R_a} \times 100\% \qquad (1)$$

where R_g and R_a are the resistances of the sample in exposure and in absence of the gas, respectively. Figure 4b shows the resistance versus bias and Figure 4c shows the sensitivity versus bias.

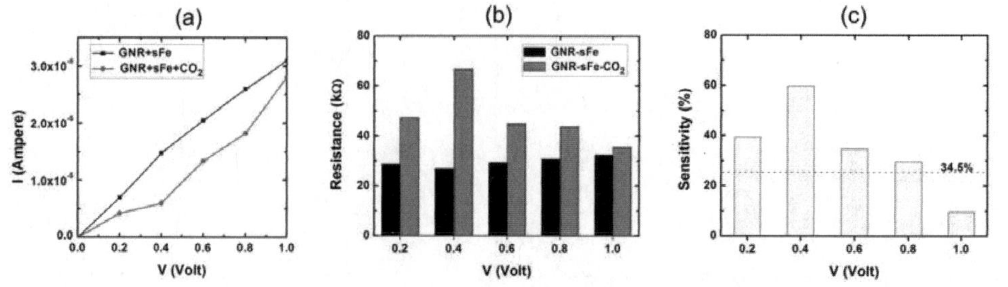

Figure 4: (a) IV-curves calculated for GNR-sFe without and with CO_2 molecules, (b) Resistances corresponding to these two cases and (c) Gas sensitivity are shown versus bias.

These latter should be weakly-dependent or independent of bias. Nonetheless, the average value does confirm a higher gas sensitivity (of about 34.5%) corresponding to the case of sFe.

4. Conclusions:

State-of-the-art computational method, based on DFT combined with non-equilibrium Green's functions formalism, is employed to calculate transmission spectrum and IV-curve of Fe-doped GNRs. Two scenarios are put under the scope: (i) Fe atom used as ad-atom relaxed on hollow site (aFe), and (ii) Fe atom used as a dopant substituting carbon atom (sFe). Tests of adsorption of CO_2 molecule are carried out and the results showed that sFe to have stronger binding to the surface and thus greater effect on the states near Fermi level so it drastically affect the conductivity of the sample. More interestingly is that such behaviours escalated to yield high sensitivity toward the detection of CO_2 molecule.

Acknowledgements

One of us (N.T.) is indebted to the partial financial supports of UAEU-UPAR (Fund # 31S169) and UAEU-Research-Center (Fund # 31R068).

References

[1] Varghese SS, Lonkar S, Singh KK, Swaminathan S and Abdala A 2015 *Sensors Actuat.* B **218** 160
[2] Yin PT, Shah S, Chowalla M and Lee KB 2015 *Chem. Rev.* **115** 2483
[3] Mishra AK and Ramaprabhu S 2011 *Energy Environ. Sci.* **4** 889
[4] Zhang HP, Luo XG, Song HT, Lin XY, Lu X and Tang Y 2014 *Appl. Surf. Sci.* **317** 511
[5] Tang Y, Ma D, Chen W and Dai X 2015 *Sensors Actuat.* B **211** 227
[6] Son Y W, Cohen M L and Louie S G 2006 *Phys. Rev. Lett.* **97** 216803
[7] Keith DW 2009 *Science* **325** 1647
[8] Chu S and Majumdar A 2012 *Nature (London)* **488** 294
[9] Tit N, Said K, Mahmoud N, Kouser S and Yamani ZH, 2016 *Appl. Surf. Sci.* **394** 219
[10] Perdew J P, Burke K, Ernzerhof M 1996 *Phys Rev. Lett.* **77** 3865
[11] Monkhorst H J and Pack J D 1976 *Phys. Rev.* B **13** 5188
[12] Distributed by Quantum-Wise company, Copenhagen, Denmark: http://www.quantumwise.com

Mechanical properties of the iron-based superconductor $SmFeAsO_{1-x}F_x$

W Malaeb[1], I Kazah[1], R Awad[1] and M Fujioka[2]

[1]Physics Department, Faculty of Science, Beirut Arab University, Beirut 11-5020, Lebanon
[2]Research Institute for Electronic Science, Hokkaido University, Sapporo, Hokkaido 001-0020, Japan
Email: w.malaeb@bau.edu.lb

Abstract. We have investigated the mechanical properties of the iron-based superconductor (FeSC) $SmFeAsO_{1-x}F_x$ (Sm1111) with a wide doping range $0.09 \leq x \leq 0.3$ by performing the Vickers microhardness test. The estimated Vickers microhardness number (H_v) was analysed using Meyer's law, Hays-Kendall approach, elastic/plastic deformation (EPD) model and proportional specimen resistance (PSR) model, where the PSR model showed the best match with our experimental data. H_v values showed an increase as x increases, indicating that fluorine doping results in an improvement in the mechanical properties of the Sm1111 compound and makes it more convenient for applications.

1. Introduction

There has been much interest in iron-based superconductors (FeSCs) [1] which have been extensively studied recently [2-5]. $SmFeAsO_{1-x}F_x$ (Sm1111) belongs to what is known as the "1111" FeSC family, and still keeps the highest bulk T_c record of around 58 K [6, 7]. It was reported that with F doping in Sm1111, the crystallographic transition from the tetragonal to the orthorhombic structure disappears and superconductivity (SC) simultaneously appears at x≈0.06-0.08 and T_c^{onset} rapidly increases above x=0.08 [6]. Also XRD data show that the impurity phases start to increase from x=0.16 [6]. Given the high T_c value in addition to its large upper critical field and high magnetic critical current density reaching 1.5-2×10^4 A/cm^2 [8], Sm1111 has been considered a promising candidate for applications such as wires and tapes [6]. For this reason, many of its properties have been studied [6, 7], yet some other properties have not been fully studied like the mechanical properties which are very critical to judge the convenience of a compound for applications. An important mechanical property is the microhardness, the measure of the material's resistance to localized plastic deformation like small dents or scratches [9]. A popular microhardness measuring technique is the Vickers indentation method [9] which we have implemented here to determine the Vickers microhardness number (H_v) of a wide doping range $0.09 \leq x \leq 0.3$ of Sm1111. H_v results showed a normal indentation size effect (ISE) similar to several cuprate systems [10-13]. In addition, H_v values showed an increase as x increases indicating that fluorine doping results in an improvement in the mechanical properties of the Sm1111 compound and makes it more convenient for applications.

2. Experimental technique

The Vickers microhardness measurements of Sm1111 samples with $0.09 \leq x \leq 0.3$ were carried out in atmospheric air at room temperature using a digital microhardness tester (MHVD-1000IS). The loads applied on the sample surface by the pyramidal indenter varied between 0.49 and 9.8 N with different loading times varying between 10 and 60 s in steps of 10 s. To guarantee the surface position independence of the collected data, the indentations were made at three different regions of the surface from which an average H_v value was evaluated for each load. The Vickers microhardness number H_v is defined as the ratio of the applied load to the pyramidal contact area of indentation, and is given as:

$$H_v = 1854.4\frac{F}{d^2}, \tag{1}$$

where F is the applied load in Newton and d is the diagonal length in micrometers.

3. Results and discussion

The variation of the Vickers microhardness number H_v calculated from equation (1), with the applied load F for Sm1111 samples with $0.09 \leq x \leq 0.3$ is displayed in figure 1. The first observation is that for all doping levels, H_v decreases as F increases reaching saturation at high loads. This behavior, known as the indentation size effect (ISE), has been observed in several cuprate systems [10-13]. However, it is sharper in cuprates compared to what is observed here for the FeSCs case which probably indicates better grain connectivity in FeSCs compared to cuprates. Second, we notice that H_v increases as the doping content x increases up to x=0.22 beyond which no considerable increase is observed. Similar to the case of cuprates, the increase of H_v values as x increases may be attributed to the reduction of porosity, the resistance to crack propagation among the grains and the increase of grain connectivity [10-13] resulting from replacing oxygen with fluorine. Therefore, a general trend of the effect of fluorine doping in Sm1111 compound is an improvement in the mechanical properties which makes it more convenient for applications.

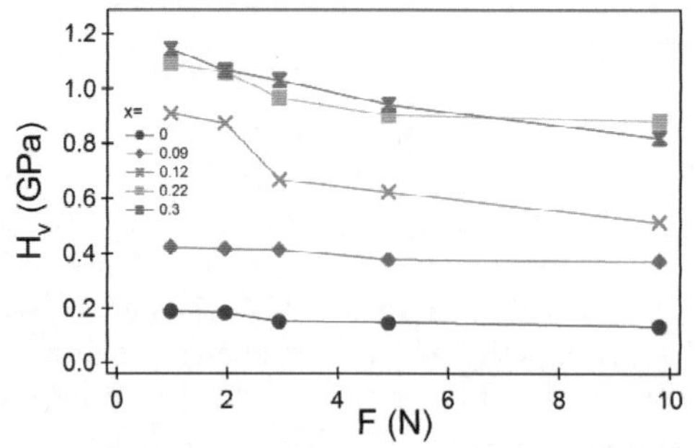

Figure 1. Variation of H_v with F at a dwell time of 40 s for $SmFeAsO_{1-x}F_x$ with x=0, 0.09, 0.12, 0.22 and 0.3.

Next we present in figure 2 our data analysis according to Meyer's law [14], Hays and Kendall (HK) approach [15], the elastic/plastic deformation (EPD) model [16, 17] and the proportional specimen resistance (PSR) model [10-13]. To check Meyer's law given by equation (2), lnF is plotted as a function of lnd in figure 2 (a). For all doping levels, the exponent n known as the Meyer number is less than 2 which indicates a normal ISE in agreement with other high-T_c superconductors like cuprates [10-13] and MgB_2 compound [18].

$$F = Ad^n \tag{2}$$

The values of W_{HK}, the limiting applied force value and the load-independent microhardness constant A_1 according to the HK model given by Eq. (3) were determined from F vs. d^2 plot displayed in figure 2 (b) and the obtained values are listed in table 1.

$$F - W_{HK} = A_1 d^2 \tag{3}$$

HK load-independent microhardness values (H_{HK}), listed in table 1, were obtained from Eq. (4):

$$H_{HK} = 1854.4\frac{F-W_{HK}}{d^2} \tag{4}$$

IOP Conf. Series: Journal of Physics: Conf. Series **869** (2017) 012042 doi:10.1088/1742-6596/869/1/012042

The constants A_2 and d_0 values, related to the EPD model expressed by equation (5), were determined from figure 2 (c) where $F^{0.5}$ is plotted versus d and listed in table 1.

$$F = A_2(d + d_0)^2 \tag{5}$$

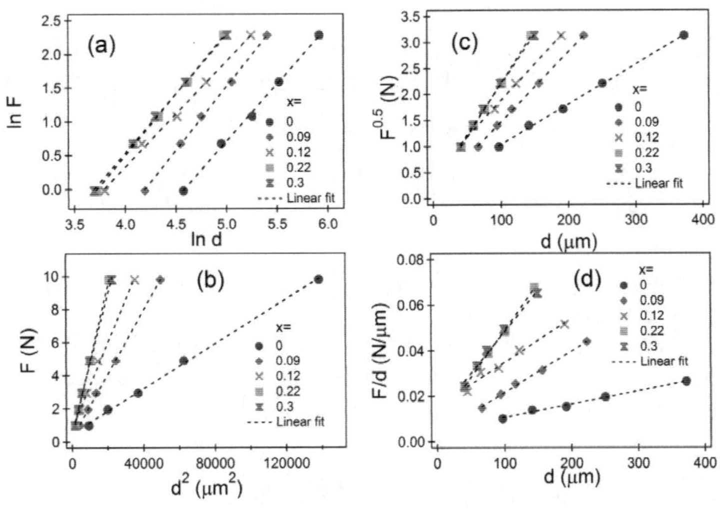

Figure 2. Vicker's microhardness data of $SmFeAsO_{1-x}F_x$ with x=0, 0.09, 0.12, 0.22 and 0.3 analyzed according to (a) Meyer's law: variation of *ln(F)* with *ln(d)*, (b) HK approach: variation of the applied force *(F)* with the square of the semi-diagonal length *(d)*, (c) EPD model: variation of $(F^{0.5})$ with *(d)*, and (d) PSR model: the ratio *(F/d)* versus *(d)*. In each plot, the dotted lines represent the linear fitting of the data corresponding to each doping level from which the fitting parameters of table 1 were determined (refer to text).

Then the elastic/plastic deformation load-independent microhardness H_{EPD} is evaluated from Eq. (6) and also listed in table 1:

$$H_{EPD} = 1854.4 \times \frac{F}{(d+d_0)^2} \tag{6}$$

Plotting F/d versus d in figure 2 (d), we determined the constants α_1 and β related to the PSR model given by equation (7), and listed them in table 1.

$$F = \alpha_1 d + \beta d^2 \tag{7}$$

We then evaluate the load-independent microhardness H_{PSR} values from the following relation and list them in table 1:

$$H_{PSR} = 1854.4 \times \frac{\alpha_1 d+\beta d^2}{d^2}. \tag{8}$$

Table 1 summarizes the H_v values obtained from the models analysis discussed above compared with the experimental H_v values at the saturation/plateau region. While the other two models show a reasonable agreement with our experimental results, the PSR model shows the best match. Therefore, it is noticed that in Sm1111 compound, although T_c saturates around x=0.12 [6, 7], the mechanical properties keep improving as x increases up to x=0.22.

Table 1. Fitting parameters determined from the models discussed in the text for $SmFeAsO_{1-x}F_x$ with x=0, 0.09, 0.12, 0.22 and 0.3 together with the H_v values determined from the models compared with experimental H_v values at the plateau region.

x	H_v (GPa) (plateau region)	Hays and Kendall (HK)			Elastic Plastic Deformation (EPD)			Proportional Speciman Resistance (PSR)		
		$A_1\times10^{-4}$ ($N/\mu m^2$)	W_{HK} (N)	H_{HK} (GPa)	$A_2\times10^{-4}$ ($N/\mu m^2$)	d_0 (μm)	H_{EPD} (GPa)	$\alpha\times10^{-2}$ (N/μm)	$\beta\times10^{-4}$ ($N/\mu m^2$)	H_{PSR} (GPa)
0.0	0.1443	0.7	0.5	0.1258	0.59	34.8	0.1092	0.49	0.6	0.1472
0.09	0.3878	2.0	0.2	0.3616	1.82	8.51	0.3413	0.33	2.0	0.4106
0.12	0.6011	3.0	0.8	0.4708	145	28	0.3594	1.54	2.0	0.795
0.22	0.9179	5.0	0.3	0.8405	4.16	8.1	0.7698	0.75	4.0	0.8806
0.3	0.9313	4.0	0.6	0.7992	3.72	13.4	0.7062	1.16	4.0	0.9594

4. Conclusion

Our Vickers microhardness results on a wide doping range of the FeSC Sm1111 have showed some common features with cuprates especially the ISE. Analyzing these results according to the available models, we deduced that the PSR model showed the best match with our experimental data. We confirmed that increasing the doping level in Sm1111 improves the mechanical properties of this compound and makes it more convenient for applications.

References

[1] Kamihara Y *et al* 2008 *J. Am. Chem. Soc.* **130** 3296.
[2] Si Q *et al* 2016 *Nature Reviews Materials* **1** 16017.
[3] Johnson P D *et al* 2015 *Iron-based Superconductivity* (Springer).
[4] Scalapino D J 2012 *Rev. Mod. Phys.* **84** 1383.
[5] Johnston D C 2010 *Adv. Phys.* **59** 803.
[6] Fujioka M *et al* 2013 *Supercond. Sci. Technol.* **26** 085023.
[7] Fujioka M *et al* 2013 *J. Phys. Soc. Jpn.* **82** 094707.
[8] Fujioka M *et al* 2013 *J. Phys. Soc. Jpn.* **82** 024705.
[9] Callister W D and Rethwisch D G 2014 *Material Science and Engineering* (Wiley).
[10] Rahal H T *et al* 2016 *J. Supercond. Nov. Magn.* DOI 10.1007/s10948-016-3654-4.
[11] Barakat M M *et al* 2015 *J Alloy Compd.* **652** 158.
[12] Awad R *et al* 2014 *J. Alloy. Compd.* **610** 614.
[13] Awad R *et al* 2014 *J. Supercond. Nov. Magn.* **27** 1757.
[14] ElMustafa A A and Stone D S 2003 *J. Mech. Phys. Solids* **51** 357.
[15] Hays C and Kendall E G 1973 Metallography **6** 275.
[16] Weiss H J 1987 *Phys. Status Solidi (a)* **99** 491.
[17] Bull S J *et al* 1989 *Philos. Mag. Lett.* **59** 281.
[18] Kölemen U 2006 *J. Alloys Compd.* **425** 429.

Advanced plasmonic interfaces for optimized light trapping in photovoltaics

Z M Saleh[*,1,2], H Nasser[2], E Özkol[2], A Bek[2,3] and R Turan[2,3]

[1]Department of Physics, Arab American University, Jenin, Palestine
[2]Center for Solar Energy Research and Applications, METU, Ankara, Turkey
[3]Department of Physics, Middle East Technical University, Ankara, Turkey
Email: zaki.Saleh@aauj.edu

Abstract. Plasmonic interfaces are integrated to photovoltaic devices to enhance light trapping and improve efficiency. The optimum thickness of the spacer layer used to passivate the absorber layer and adjust its distance from the metal nanoparticles remains unclear. We integrate plasmonic interfaces consisting of Ag nanoparticles and silicon nitride spacers of different thicknesses to the back of a-Si:H absorber to investigate the optimum thickness of the spacer layer and use the photocurrent in a-Si:H to indicate the enhancement in light-trapping. For integration to the back or front of the device, the localized surface plasmon resonance (LSPR) is shifted and broadened into the red with increased spacer layer thickness and the effect is more pronounced for integration to the back. An appreciable enhancement of photocurrent in a-Si:H is consistent with this broadening of LSPR and has a critical dependence on spacer layer thickness.

Keywords— plasmonic interfaces, photocurrent, light trapping, spacer layer.

1. Introduction

The performance of photovoltaic devices depends on efficient light absorption and photocarrier collection mechanisms. Even if the most efficient carrier collection is achieved, significant portion of the solar spectrum remains unused. Advanced light management schemes to optimize light absorption include surface texturing, up and down conversion, plasmonics and other surface nanostructures [1,2]. Plasmonic interfaces consisting of metal nanoparticles integrated to PV devices with a spacer layer have shown a promising potential for enhancing light-trapping and improving solar cell efficiency [2 – 4]. Extinction of light as characterized by a minimum in transmittance occurs at the localized surface plasmon resonance (LSPR) when the frequency of incident photons matches the collective frequency of oscillating electrons at the surfaces of metal nanoparticles [5,6].

Optical extinction is red-shifted and broadened into the 500 – 700 nm region with increasing nanoparticle size and also with spacer layer thickness [2]. Theoretical calculations predict that the scattering cross-section increases while preferential scattering decreases with increasing spacer layer thickness, suggesting that the optimum photocurrent is a compromise between the increased scattering cross-section and decreased preferential scattering at a given spacer layer thickness [3]. Scattering can be preferentially directed towards the a-Si:H absorber by the appropriate selection of refractive indices.

In this study, we integrate silver nanoparticles to the back surface of a-Si:H serving as the absorber layer (figure 1a) using SiNx spacer layers of 0.30 and 60 nm in thickness and use the enhancement in the spectral dependence of photocurrent normalized to incident photon power as a means of measuring enhanced light trapping dependence on spacer layer thickness.

IOP Conf. Series: Journal of Physics: Conf. Series **869** (2017) 012043 doi:10.1088/1742-6596/869/1/012043

2. Samples and experiment

The self-assembled Ag nanoparticles were fabricated by dewetting 15-nm-thick sputtered Ag films on glass substrates in N_2 at 420 °C for one hour as was explained previously [7]. The SiNx spacer layers of 0, 30, and 60 nm thicknesses followed by a 150-nm-thick a-Si:H layer were deposited using plasma-enhanced chemical vapor deposition (PE-CVD). Another sample with the plasmonic interface on top of the a-Si:H layer using a 30-nm thick spacer was fabricated for comparison.

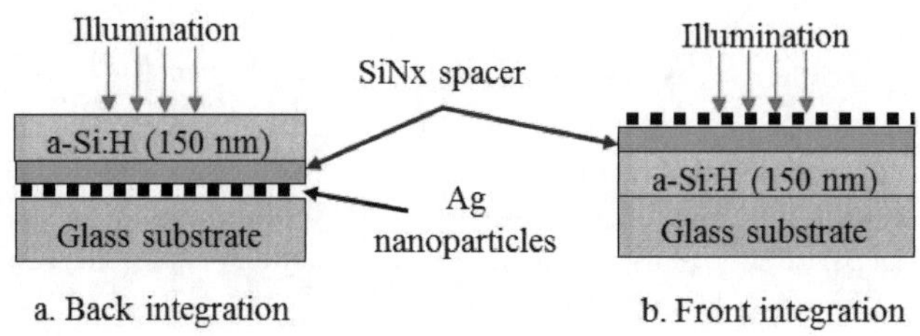

Figure 1. Schematic diagrams for the back (a), and front (b), integrations of plasmonic interfaces dewetted at 420 °C (front interface).

Scanning electron microscope (SEM) images and optical transmittance and reflectance over the spectral range of 400 – 1100 nm were performed using an optical system as described previously [7]. Aluminum electrodes of standard coplanar geometry were evaporated on the a-Si:H side using a shadow mask. The photocurrent in a-Si:H was measured versus wavelength under illumination from a halogen lamp through a monochromator with 10 nm wavelength increments over the spectral range of 400 – 1100 nm.

3. Results

In figure 2, we show an SEM image of Ag nanoparticles dewetted at 420 °C on a glass substrate, the Ag nanoparticle size distribution and the transmittance versus wave length. The SEM image shows round and completely separated particles while the particle size distribution shows a normal distribution with a mean particle size of about 110 nm. Round particles in the range of this size are essential in promoting scattering and minimizing losses. Previous studies show that smaller size particles exhibit large parasitic losses and high absorption [9]. The transmittance curve shows a pronounced plasmonic resonance at about 520 nm wavelength. These measurements were repeated for the two other interfaces to make sure that all interfaces were identical before the subsequent steps are added.

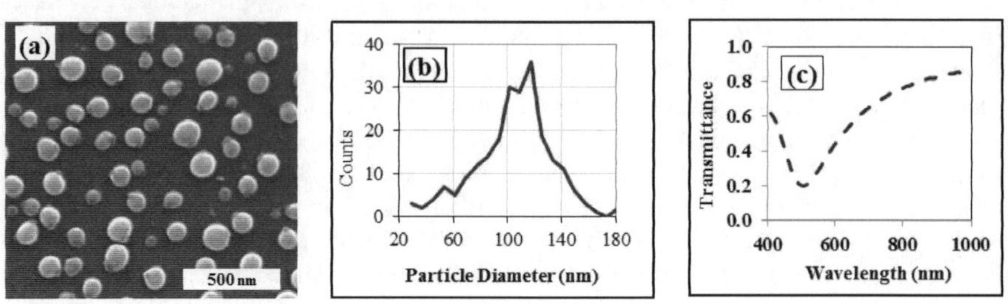

Figure 2. (a) SEM micrograph of Ag nanoparticles dewetted on glass substrate at 420 C, (b) the corresponding particle size distribution, and (c) the transmittance showing a minimum at 520 nm.

Figure 3a shows the extinction spectra calculated as $E = 1 - T - R$, where T is transmittance and R is reflectance, for the three interfaces with 0 (no spacer), 30-nm and 60-nm thick SiNx spacers. The extinction maxima are clearly shifted and broadened into the red with increased spacer layer thickness. The effect of these plasmonic interfaces on the extinction spectrum of a-Si;H are plotted in figure 3b which shows the extinction spectra for the flat a-Si;H reference (no interface) and two others with interfaces integrated to the back using spacer layer thicknesses of 30 and 60 nm. The clean spectra in figure 3a, are complicated by the addition of the a-Si:H due to interference with reflections from the various interfaces in the stack. Despite the wiggles in the extinction spectra, the shift and broadening are still clear in figure 3b when compared with the flat a-Si:H reference although the effect of the spacer thickness is obscured.

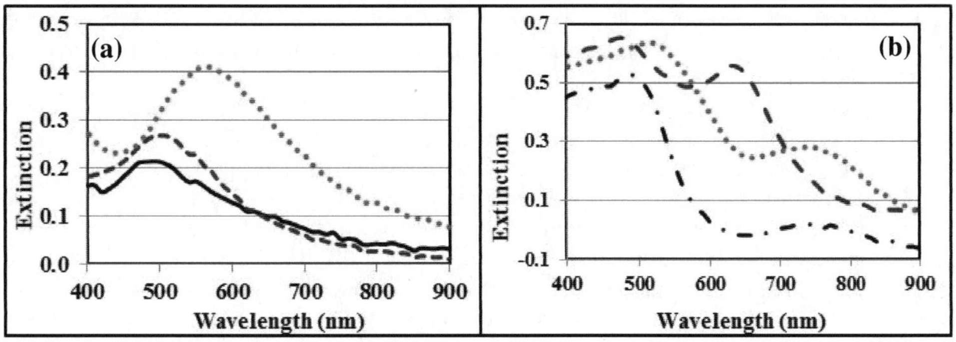

Figure 3. (a) Extinction spectra for three plasmonic interfaces with identical Ag-nanoparticles size distributions using spacer layer thicknesses of zero (solid line), 30 nm (dashed) and 60 nm (dotted). (b) the extinction spectra of a flat 150 nm thick a-Si:H reference with no interface (dashed-dotted) and with two plasmonic interfaces with spacer layer thicknesses of 30 nm (dashed line) and 60 nm (dotted line).

Figure 4 shows the photocurrent versus wavelength for the two samples with plasmonic interfaces compared to the flat a-Si:H reference. The photocurrent shows appreciable enhancement in the region of 500 – 700 nm for both samples and the effect is more pronounced for the sample with the thinner spacer layer, which is consistent with their extinction spectra. The difference in their photocurrent enhancements below 500 nm will be discussed below.

The sample with zero spacer (not shown here) exhibits anomalous and non-repeatable results which we attribute to the direct contact of the a-Si:H with the conductive Ag nanoparticles acting as defects limiting the current [8,9].

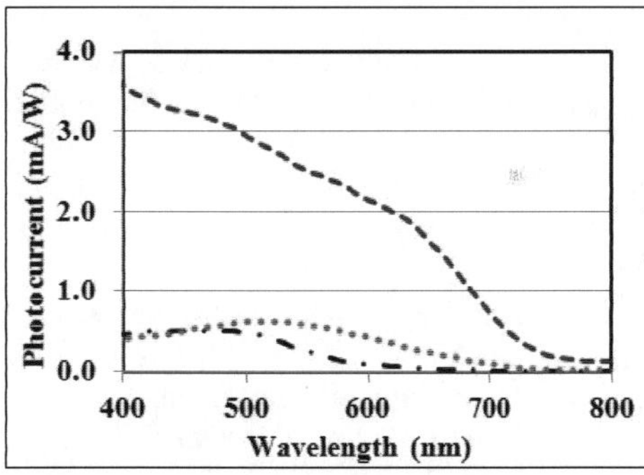

Figure 4. Photocurrent versus wavelength in a-Si:H with plasmonic interfaces using: 30-nm (dashed line) and 60-nm (dotted line) spacer layers and flat a-Si:H reference (dashed-dotted line: multiplied by 50.

4. Discussion

Extinction represents light that is either absorbed or scattered by the interface. Scattering can be made preferentially into the a-Si:H due to its high refractive index. To determine whether scattering is taking place, we measured the photocurrent in a-Si:H versus wavelength and used the enhancement in photocurrent as an indication of enhanced light trapping and preferential scattering into the absorber layer. The broadened Extinctions observed in figure 3 for samples with plasmonic interfaces compared to the flat a-Si:H reference represent either enhanced absorption by the interface (which decreases photocurrent) or enhanced preferential scattering into a-Si:H (which increases photocurrent). The clear enhancement in photocurrent in the spectral range of 500 – 700 correlate well with the shift and broadening into the red of the extinction spectra. This correlation strongly suggests that preferential scattering is taking place.

The lower photocurrent enhancement for the region below 500 nm may be attributed to the higher absorption in the thicker SiNx layer although rough calculations suggest that this alone may not account for the large difference between 30 nm and 60 nm spacer layer thickness. For front integration, light is already travelling towards the absorber layer and the plasmonic interface may improve photocurrent by preferential (forward) scattering the otherwise reflected light but may at the same time block or absorb part of the incident light which offsets the enhancement effect. We therefore see no clear enhancement for integration to the front [9]. For the back integration, on the other hand, the interface can only backscatter the otherwise transmitted light with no possibility of blocking it. These results suggest that the proper position for the plasmonic interface is the back of the absorber layer in agreement with previous results [10].

To achieve optimum light trapping and thus maximum device efficiency, it is therefore recommended that plasmonic interfaces are integrated to the back of the device (opposite side of light exposure). Nanoparticle shape should be round and the size should be approximately100 nm in diameter to minimize parasitic loss and maximize preferential scattering. The remaining critical parameter of spacer layer thickness is not as straightforward. The spacer layer should be thin enough to promote preferential scattering but sufficiently thick to keep the absorber passivated. As the above results indicate, increasing the thickness to higher than 30 nm reduces the photocurrent because preferential scattering decreases. Our internal measurements (not shown here) show that SiNx spacers less than 20 nm in thickness may not be continuous which exposes the Ag to the absorber.

Acknowledgements The authors acknowledge the support by The Scientific and Technological Research Council of Turkey (TÜBİTAK) grants 113F239 (A.B.) and 113F375 (A.B.), TUBITAK-BIDEB-2221 (Z.M.S.) and TUBITAK BIDEB-2215 (H.N.).

References

[1] Atre A C, García-Etxarri A, Alaeian H and Dionne J A 2012 *J. Opt.* **14** 024008
[2] Pillai S and Green M A 2010 *Solar Energy Materials & Solar Cells* **94** 1481
[3] Catchpole K R and Polman A 2008 *Optics Express* **16** 21799
[4] Eminian C, Haug F J, Cubero O, Niquille X and Ballif C 2011 *Progr. Photovolt.* **19** 260
[5] Temple T L, Mahanama G D K, Reehal H S and Bagnall D M 2009 *Solar Energy Materials & Solar Cells* **93** 1978
[6] Kim S K, Cho C H, Kim B H, Choi Y S, Park S J and Lee K 2009 *Appl. Phys. Lett* **94** 183108
[7] Saleh Z M, Nasser H, Özkol E, Günöven M, Bek A and Turan R 2013 *Plasmonics* **9** 357
[8] Lükermanna B, Heinzmanna U and Stiebig U 2012 *Proc. SPIE* **8471** 84710S
[9] Saleh Z M, Nasser H, Özkol E, Günöven M, Abak K, Canli S, Bek A and Turan R 2015 *J Nanopart Res* **17** 419
[10] Beck F J, Mokkapati S, Polman A and Catchpole K R 2010 *Applied Phys. Lett* **96** 033113

IOP Conf. Series: Journal of Physics: Conf. Series **869** (2017) 012044 doi:10.1088/1742-6596/869/1/012044

On the phase diagram of $CrAs_{1-x}Sb_x$

Kh. A. Ziq, M Abu baker, M Faiz and E Al-Nahari

Physics Department, King Fahd University of Petroleum and Minerals, Dhahran, Saudi Arabia

Email: kaziq@kfupm.edu.sa

Abstract. Transport, magnetic, XRD and XPS studies have been performed on $CrAs_{1-x}Sb_x$ in an effort obtain the phase diagram in the intermediate range ($x \sim 0.5$) in the hexagonal phase. We found that $CrAs$ undergoes antiferromagnetic and structure transition at $T_N = 240K$. Upon Sb substitution ($x > 0.3$), the Neel temperature is sharply suppressed with slope ($-10.7\frac{K}{x}\%$) in much similar way to the suppression caused by pressure in $CrAs$ single crystal. The chemical compositions obtained using XPS are in good agreement with nominal values. Moreover, XPS reveals no changes in the valance states for the constituent elements of $CrAs_{1-x}Sb_x$. Unexpectedly, we found that arsenic at the surface has higher atomic ratio than the nominal composition. This may be due to segregation of arsenic to the surface of sample.

1. Introduction

Recent discovery of superconductivity in $CrAs$ under high pressure (10 kbar) opened new possibility in discovering superconductivity in other transition metal based pnictide [1,2]. At ambient pressure, CrAs is non-superconductor. CrAs belongs to a large family of MX compounds where M is a transition metal and X is P, As or Sb, and has "commonly" orthorhombic structure [3,4]. Its magnetic, transport and structural properties were studied extensively in the seventies and eighties of twentieth century in an effort to probe the magnetic and semiconducting properties. It is well established that $CrAs$ undergoes first order antiferromagnetic transition at Neel temperature (T_N). Most of the previous magnetic, structural and transport properties were investigated above liquid nitrogen temperature [3,4]. The Neel and the structure transition temperatures were continually suppressed with increasing pressure [1,2].

In this work, we investigated the phase diagram of $CrAs_{1-x}Sb_x$, in the intermediate region of the phase diagram (x~0.5) where the Neel temperature reaches its minimum value. We use transport and XPS analyses to investigate the variation of T_N with Sb substitution and the electronic states of Cr, As and Sb.

2. Experimental method

Solid state reaction was used to prepare $CrAs_{1-x}Sb_x$ with x = 0.00, 0.45, 0.50, 0.60 and 0.65. Stoichiometric ratios of high purity (99.99%) powders of Cr, As and Sb were mixed, ground and pressed into pellets under ~3 Ton/cm^2 then sealed in quartz tubes under partial argon pressure. To reduce the volatility of As and Sb during annealing, all samples were heated slowly (~60°C/h) to 600°C and annealed for 10 hours. The temperature was raised slowly (~60°C/h) to 1100°C and annealed for 24 hours. The temperature was reduced to 900°C and was kept for 72 hours. Finally, the samples were annealed at 300 ^{o}C for another 72 hours, and then quenched to room temperature.

IOP Conf. Series: Journal of Physics: Conf. Series **869** (2017) 012044 doi:10.1088/1742-6596/869/1/012044

The resistance was measured in a temperature range 4-300K using four probes ac resistance bridge (LR-750) operating at $16Hz$. The temperature was monitored using Lake Shore-336 temperature controller along with closed cycle refrigerator (Model SHI-4T). LABVIEW software was used to control all measuring systems and collect the data. The XPS analysis was performed using a Thermo-Scientific (Escalab-250 Xi). The elemental composition and the oxidation state of all elements (Cr, As and Sb) in each sample were obtained by this technique.

3. Results and discussions

3.1. XRD analysis

patterns for all samples were analyzed using MAUD software. The CrAs structure is orthorhombic with lattice parameter a= 6.2082 Å, b=5.6519 Å and c=3.4691 Å. For $x \geq 0.45$; the structure becomes hexagonal, with traces of the orthorhombic phase. In the hexagonal phase both lattice parameters a and b shrink dramatically reduced to 3.6932 Å, while the lattice parameter c; elongates considerably from 3.4793 to 5.7651 Å. Our results are consistent with earlier studies reported [3-4].

3.2. XPS analysis

The electronic state and chemical composition of the $\mathrm{CrAs_{1-x}Sb_x}$ samples were obtained using XPS. The spectra of all elements in each sample along with deconvoluted peaks are analyzed. A representative data for CrAs sample is shown in figure 1. Chromium $2p$ core level spectrum shows multiplet splitting due to unpaired d- electrons, in addition to spin-orbit splitting ($2p3/2$ and $2p1/2$). The binding energy for Cr $2p3/2$ is assigned to the alloy and the valence state is Cr^{+3}. Similarly, arsenic and antimony $3d$ core level spectra also consist of spin-orbit doublets. The binding energy of As $3d5/2$ is assigned to alloy and the valence state is $\mathrm{As^{-3}}$. Unexpectedly, arsenic composition is found higher than the nominal value, indicating that arsenic segregates to the surface of the sample. However, using depth profile analysis, we found that the overall concentration of As and Sb agrees within 5-10% of the nominal concentration.

3.3. Transport measurements

The measured resistance for $\mathrm{CrAs_{1-x}Sb_x}$ are presented in figure 2. The measurements were performed during cooling range 300-4K. For x = 0, the resistivity curves clearly shows a drop at about 263K followed by a maximum near 230K. For Sb concentrations (x~0.05), the maximum becomes more pronounced. The resistance of *CrSb* exhibits a maxima as the temperature decrease. The drop in the resistance coincides with the antiferromagnetic transition. The transition temperatures are determined from the inflection point in the derivative of the resistivity with respect to temperature. These results are in agreement with transport measurement carried out on single crystal of CrAs by H. Kotegawa *et al* [2]. The antiferromagnetic transition drops to about 150K for x=0.5. The variations of the antiferromagnetic transition temperature with Sb concentration is presented in figure 4. The figure shows a sharp linear drop in T_N near x = 0.45 with a slope (~ -10.67 K/x). This is similar to drop in T_N observed in the variations of T_N with applied pressure reported in [1]. The in antiferromagnetic transition suppressing is striking and is clearly shown in figure 3. It is worth mentioning that superconducting transition in these doped polycrystalline samples have not been observed down to 4.0K, the lowest temperature available in our system.

IOP Conf. Series: Journal of Physics: Conf. Series **869** (2017) 012044 doi:10.1088/1742-6596/869/1/012044

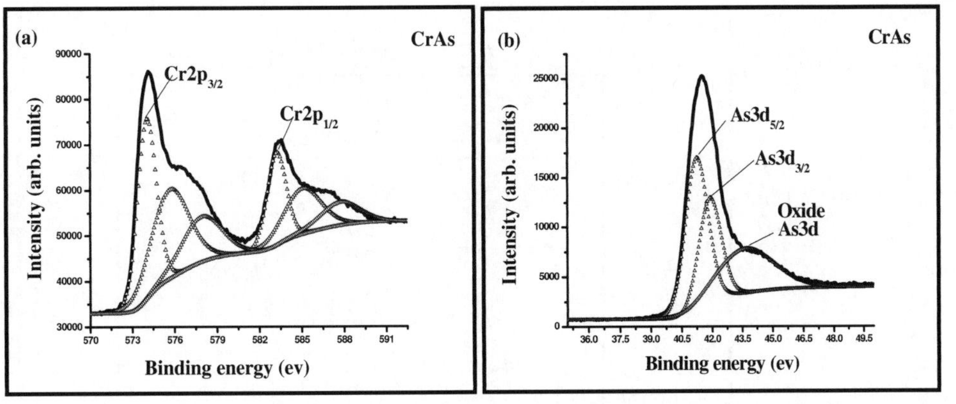

Figure 1: XPS spectra of CrAs sample: (a) Cr2p core level spectrum, (b) As3d core level spectrum. The experimental spectra are represented by solid line. The fitted peaks are represented by open triangle.

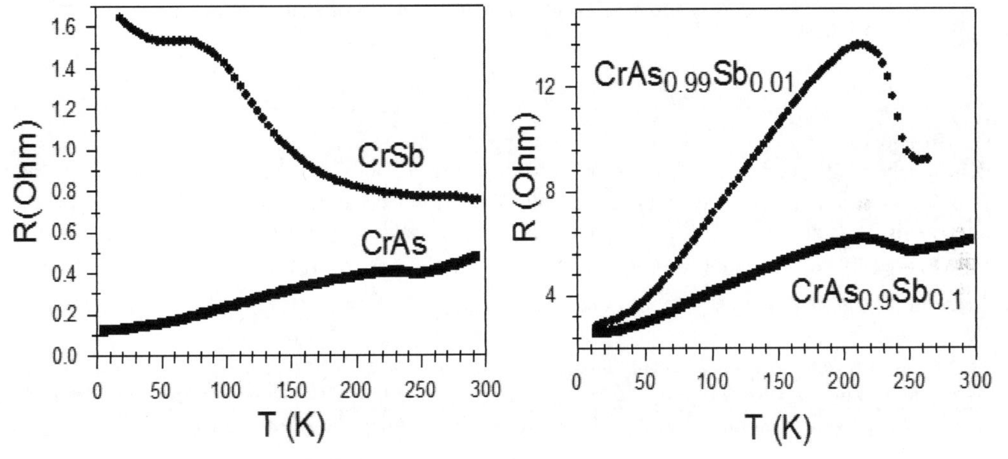

Figure 2: Variations of the resistance with temperature for $CrAs_{1-x}Sb_x$.

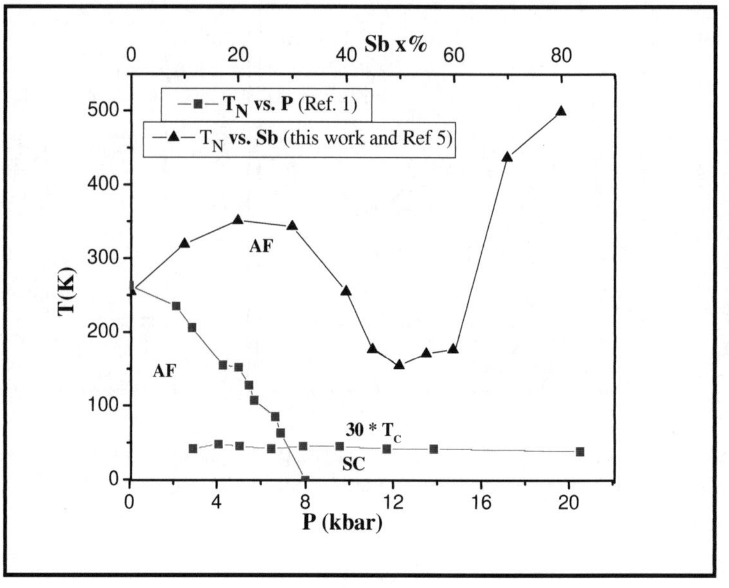

Figure 3: Combined phase diagram of $CrAs_{1-x}Sb_x$.

4. Conclusion

XPS analysis revealed that valence states for the elements are Cr^{+3} and As^{-3} did not change with Sb^{-3} substitutions. Transport measurements revealed the antiferromagnetic phase transition that drops with increasing Sb near *0.5*, then increase gradually. We observed a liner drop in T_N near $x=0.45$ with linear slope ($-10.7\frac{K}{x}\%$) in a much similar way to the suppression of T_N for CrAs under the effects of external pressure.

Acknowledgments

We would like to acknowledge the support of King Fahd University of Petroleum & Minerals for this work. This project was funded by NSTIP under project number **11-ADV1631-04**.

References

[1] Wu W, Cheng J, Matsubayashi K, Kong P, Lin F, Jin C *et al* 2014 Nature communications **5** 5508.
[2] Kotegawa H, Nakahara S, Tou H, and Sugawara H 2014 J. Physical Society of Japan **83** 093702.
[3] Selte A K K, Jamison W E, Andressen A F, Engebretsen J E 1971 Acta Chemica Scandinavica **25** 1703.
[4] Boller H and Kallel A 1971 Solid State Communications **9** 1699.
[5] Suzuki Tand Ido H 1986 J. Mag. and Mag. Mat. **54** 935.

IOP Conf. Series: Journal of Physics: Conf. Series **869** (2017) 012045 doi:10.1088/1742-6596/869/1/012045

The effect of PVP on morphology, optical properties and electron paramagnetic resonance of $Zn_{0.5}Co_{0.5}Fe_{2-x}Pr_xO_4$ nanoparticles

Z Bitar[1], D El-Said Bakeer[2] and R Awad[1]

[1]Physics Department, Faculty of Science, Beirut Arab University, Lebanon
[2]Physics Department, Faculty of Science, Damanhur University, Damanhur, Egypt
E-mail: zouheirbitar@hotmail.com

Abstract. Zinc Cobalt nano ferrite doped with Praseodymium, $Zn_{0.5}Co_{0.5}Fe_{2-x}Pr_xO_4$ ($0 \leq x \leq 0.2$), were prepared by co-precipitation method from an aqueous solution containing metal chlorides and two concentrations of poly(vinylpyrrolidone) (PVP) 0 and 30g/L as capping agent. The samples were characterized using X-ray powder diffraction (XRD), Transmission Electron Microscope (TEM), UV-visible optical spectroscopy, Fourier transform infrared (FTIR) and Electron Paramagnetic Resonance (EPR). XRD results display the formation of cubic spinel structure with space group Fd3m and the lattice parameter (a) is slightly decreased for PVP capping samples. The particle size that determined by TEM, decreases for PVP capping samples. The optical band energy E_g increases for PVP capping samples, confirming the variation of energy gap with the particle size. The FTIR results indicate that the metal oxide bands were shifted for the PVP capping samples. EPR data shows that the PVP addition increases the magnetic resonance field and hence decreases the g-factor.

1. Introduction

The interest in research of metal spinel ferrite nanoparticle, with the formula MFe_2O_4, M = metal(s), has increased significantly in the last 10 years due to their potential applications in spintronics, biomedical and magnetic refrigeration [1-4]. MFe_2O_4 can be described as a cubic, closely packed arrangement of oxygen atoms, and M^{2+} and Fe^{2+} ions can occupy either tetrahedral (A) or octahedral (B) sites. Among all methods, chemical co-precipitation is most suitable because of its simplicity, better control on particle size and other properties of materials. To control the growth of the spinel ferrite nanoparticles, poly(vinyl pyrrolidone) (PVP) is added during the synthesis for capping the surface of the particles [5-6]. The results indicate a decrease in the particle size for PVP capping samples.

The aim of the work is to study the effect of PVP on zinc cobalt nano ferrite doped with a rare earth element through structural, optical and magnetic measurements.

2. Experimental

Metal chlorides reagents, poly (vinyl pyrrolidon) (PVP), and deionized water were used as precursors. A separate solution of PVP was prepared by dissolving 3g of PVP in 100 ml before mixing the suitable stoichiometric amounts of metal chloride reagents. For co-precipitation of such prepared solution, solution of NaOH (4M) was added in drop wise with constant stirring produced by a magnetic stirrer. The pH of the solution was maintained at 12.5-13, in order to transform the hydroxides of metals into nanoparticles. The solution was maintained at 333K with constant stirring for 2 hours. This duration was sufficient for the transformation of hydroxides into spinel ferrite. The precipitated particles were then washed several times with distilled water to remove the impurities. The precipitated particles were then dried in an oven at 373K for 24 hours. The resulting solid was grounded into powder and calcined at temperature 823K for 6 hours. The nanosized ($Zn_{0.5}Co_{0.5}Fe_{2-x}Pr_xO_4$) were characterized by XRD using Bruker D8 advance powder diffractometer with Cu-Kα radiation (λ = 1.54056 Å) in the range $10^{\circ} \leq 2\theta \leq 80^{\circ}$. The surface morphology was examined using Jeol transmission electron microscope JEM-

100CX, operated at 80 kV. The FTIR analysis was carried by FTIR 8400S Shimadzu in the range 350-700 cm^{-1}. UV-visible absorbance spectrum of sample has been taken in the wavelength range from 200 nm to 600 nm by using Jasco-V-670 - UV-Vis spectrometer. EPR spectrum of samples was recorded on X- band EPR Spectrometer Model: ER0485 Make: Bruker, for determining g-value and resonance magnetic field.

3. Results and discussion

The XRD results of the ($Zn_{0.5}Co_{0.5}Fe_{2-x}Pr_xO_4$) nanoparticles with x= 0.04 and 0.08, free and PVP capping are shown in Figure 1. The XRD results confirm the presence of nanoferrite with a face-centered cubic structure [7]. The XRD patterns also show that the peaks become broader for PVP capping samples and their intensity decrease contributing to the decrease in the crystallite size.

Table1. Various parameters of Zn-Co nano ferrite measured by XRD and TEM with and without PVP capping.

	Free from PVP				**Capped with PVP**			
	x=0	x=0.04	x=0.08	x=0.2	x=0	x=0.04	x=0.08	x=0.2
Crystallite size XRD(nm)	28.3	19.4	18.7	14.7	13.3	12.24	8.51	14.63
Particle size TEM(nm)	24.69	17.24	15.9	12.06	13.08	10.34	7.58	7.04
Lattice parameter a (A^0)	8.4404	8.4349	8.4291	8.4057	8.3948	8.4020	8.4022	8.4092
ν_1 (cm^{-1})	570.74	576.74	586.38	632.67	577.02	591.341	583.141	571.458
ν_2 (cm^{-1})	434	449.43	430.14	441.71	403.957	419.444	397.853	433.333
Band gap energy (eV)	3.05	3.06	3.07	3.10	3.064	3.075	3.081	3.085
Hr (G)	2171	2500	2518	2826	2399	2663	2678	3116
g-factor	3.041	2.641	2.622	2.337	2.752	2.479	2.465	2.119

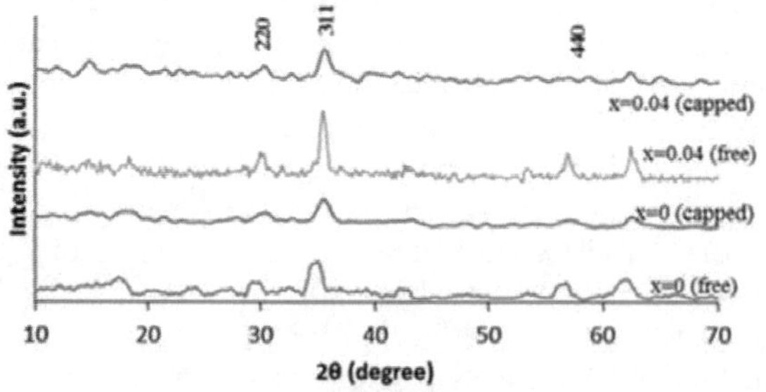

Figure 1. XRD patterns of $Zn_{0.5}Co_{0.5}Fe_{2-x}Pr_xO_4$ nanoparticles with x=0 and 0.04 for free and PVP capping samples.

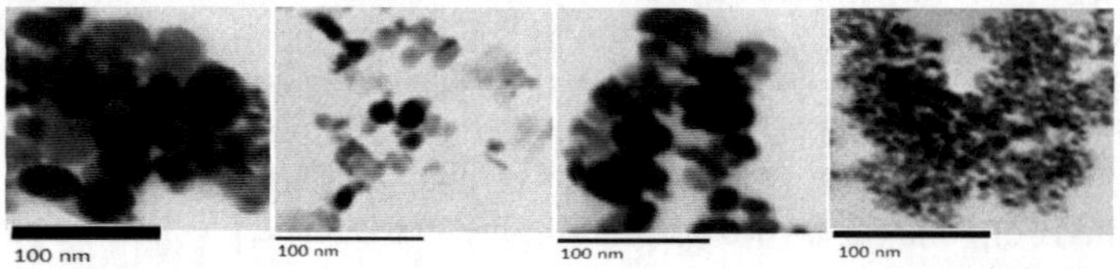

(a) x=0 (free) (b) x=0 (capped) (c) x=0.08 (free) (d) x=0.08(capped)

Figure 2. TEM images of $Zn_{0.5}Co_{0.5}Fe_{2-x}Pr_xO_4$ nanoparticles.

The average crystallite size (D) was calculated from the full width at the half maximum (FWHM) of the XRD patterns and the lattice parameter, by using Debye-Scherer formula and Bragg's equation respectively [8].

IOP Conf. Series: Journal of Physics: Conf. Series **869** (2017) 012045 doi:10.1088/1742-6596/869/1/012045

Table 1 shows that both the lattice parameter and the average crystallite size decrease for PVP capping samples. This means that PVP reduces the agglomeration of the nanoparticles [9]. Figure 2 (a, b, c and d)) shows the TEM images for free and PVP capping samples of $Zn_{0.5}Co_{0.5}Fe_{2-x}Pr_xO_4$ nanoparticles with x=0 and 0.08. It is clear that the presence of PVP significantly affects the crystallite size, distributions and dispersion of the resulting nanoparticles. Uniform nanoparticles with narrow size distribution and sphere shapes with weak agglomeration were obtained for PVP capping samples [10]. PVP results in significant size reduction of nanoparticles. The crystallite size obtained from TEM, is quiet similar to that obtained from XRD with approximate deviation from 2 to 15% as listed in the table 1.

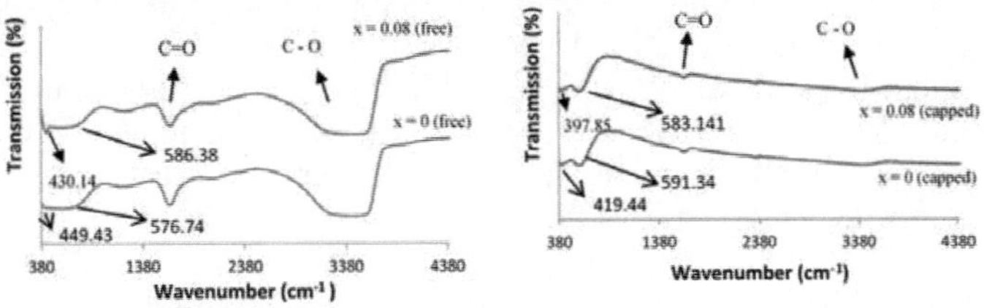

Figure 3. FTIR spectra of $Zn_{0.5}Co_{0.5}Fe_{2-x}Pr_xO_4$ with x=0 and 0.08 for free and PVP capping

Figure 3 shows the FTIR spectra for $Zn_{0.5}Co_{0.5}Fe_{2-x}Pr_xO_4$ with x=0 and 0.08. The FTIR spectra indicate the two principle absorption bands in the range 350-700 cm^{-1}, which correspond to intrinsic stretching vibrations of the metal at the octahedral site, Moct↔O (observed from ν_2= 382 to 456cm^{-1}) and at the octahedral site, Mtet↔O (observed from ν_1=570 to 633cm^{-1}). For PVP capping samples, a shift in these two vibration bands is observed, as listed in table 1. For PVP capping samples, the broad band between 3000 and 3600 cm^{-1} related to stretching hydroxyl (O-H) group is replaced by sharper peak [11]; Also a shift in the peaks between 1620 and 1650 cm^{-1} towards lower wave number region is observed which indicates that C=O bond is getting weakened, and there exists an interaction between metal ions and PVP through oxygen of C=O group of the polymer [12].

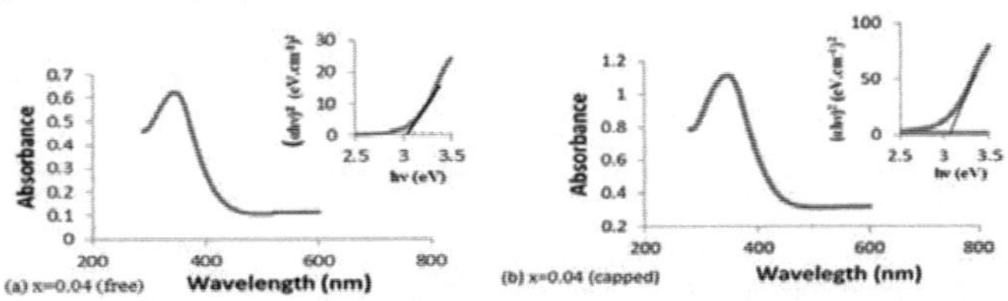

Figure 4. Plot of Absorbance as a function of wavelength $Zn_{0.5}Co_{0.5}Fe_{2-x}Pr_xO_4$ with x=0.04 for (a) free and (b) PVP capping samples; The inset figures are for Tauc plot.

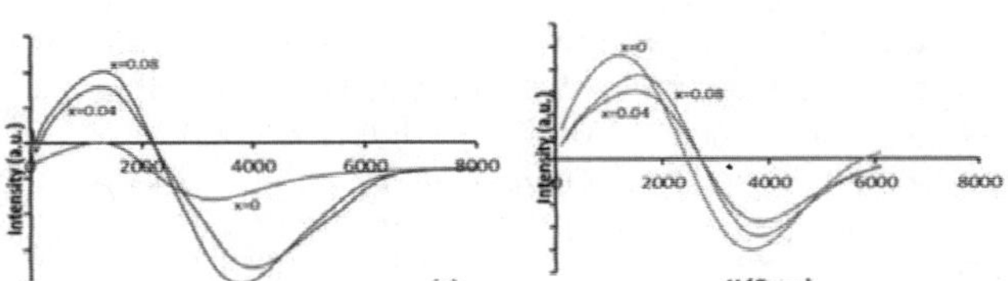

Figure 5. EPR spectra for $Co_{0.5}Zn_{0.5}Fe_{2-x}Pr_xO_4$: (a) without (b) with PVP capping

Table 1 shows the values of bandgap energy of free and PVP capping samples. The energy bandgap E_g was obtained from the Tauc plot, using the intercept of the linear portion of the curve $(\alpha h\nu)^2$ versus hv to y=0 as shown in figure 4. For the obtained energy gap of PVP-capped nanoparticles was found to increase which confirms the decrease of particle size when PVP was added. Such increase in the band gap may be attributed to the quantum confinement effect created by PVP surfactant [13].

X-band EPR spectra (9.24 GHz) recorded at room temperature for free and PVP capping $Co_{0.5}Zn_{0.5}Fe_{2-x}Pr_xO_4$ are shown in figure 5. A single and strong symmetric EPR signal was observed for all samples with the resonance magnetic field H_r. For PVP capping samples, the value of H_r was increased as listed in table 1.

The spectroscopic Landé factor (g-value) was determined from the magnetic resonance field H_r by:

$$g = hv(\beta\ H_r)^{-1}, \quad (1)$$

where h is Planck's constant, v is the microwave frequency (9.24×10^9 Hz), β is the Bohr magnetron. All the spectra were analysed and fitted to single EPR absorption line to obtain the values of magnetic resonance fields H_r and g-values as listed in table 1.

The increase of resonance field, as shown in figure 5, could be contributed to the decrease in super exchange interactions for the PVP capping samples. The g-values decreased for PVP capping samples, as listed in table 1; this is probably due to the enhanced exchange interaction between ferromagnetic core and antiferromagnetic surface layer [14].

4. Conclusion

The nanoparticles of $Zn_{0.5}Co_{0.5}Fe_{2-x}Pr_xO_4$ were successfully fabricated by the co-precipitation method with PVP as a capping agent with concentrations 0g/L and 30g/L. This work focused on the effect of PVP on Zn-Co nano ferrite doped with rare earth element. The average particle size was determined by TEM is in agreement with that calculated from XRD. For PVP capping samples, the particle size and the band gap energy was decreased. The FTIR confirmed the presence of metal oxide bands where their corresponding wavenumbers were shifted when PVP was added. The EPR spectra show that the resonance magnetic field was increased and the g-factor was decreased for the PVP capping samples. The effect of PVP in reducing the size and improving the EPR spectra of nanoparticles can have benefit in biomedical applications such as Magnetic Resonance Imaging MRI.

Furthermore, this work can be extended by adding different concentrations of PVP, in order to find the optimum quantity.

Acknowledgment

This work was performed in the material science lab, Physics department, Faculty of Science, Beirut Arab University, Beirut, Lebanon; in cooperation with physics department in Alexandria University, Alexandria, Egypt.

References

[1] Luders U, Barthelemy A and Bibes M 2006 *Adv. Mater.* **18** 1733–1736
[2] Pankhurst Q A, Connolly J and Jones S K 2003 *J. Phys.* **36** R167–R181
[3] Berry C C 2005 *J. Mater. Chem.* **15** 543–547
[4] Estelirch J, Sanchez-Martin M and Busquets M 2015 *Int. J. of Nan.* **10** 1727-1741
[5] Verma R, Mantri B, Ramphal and Srivastava A 2015 *Adv.Mater.Lett.* **6**(4) 324-333
[6] Kamari H, Naseri M and Saion E 2014 *Int. Nan. Lett.* **4** 118-129
[7] Amiri S and Shokrollahi H 2013 *J. Magn. and Magn. Mat.* **345** 18–23
[8] Suryanarayana C and Norton M 1996 *X-Ray Diffraction: A Practical Approach* (New York: Plenum) 207-221
[9] Sivakumar P, Ramesh R, Ramanand A, Ponnusamy S and Muthamizhchelvan C 2011 *Mat.Let.* **65(3)** 483-485
[10] Naseri M, Saion E, Abbastabar Ahangar H, Hashim M and Shaari A 2011 *P.tech.* **212** 80-88
[11] Gökdemir F, Yüzbaşıoğlu V, Keskin B, Özdemir O and Kutlu K 2014 *Adv. Mat. Lett.* **5(7)** 367
[12] Anasuya K, Veeraiah M and Hemalatha P 2015 *Res. J. Chem. Sci.* **5(2)** 64-69
[13] Sharma M, Kumar S, and Pandey O 2010 *J. Nano. Res.* **12(7)** 2655-2666
[14] Venkataraju C and Paulsingh R 2015 *Int. J. Nano Dimens. Nat. Conf. 384 Recent Innov. Appl. Sci. Humanit.* **4** 183

IOP Conf. Series: Journal of Physics: Conf. Series **869** (2017) 012046 doi:10.1088/1742-6596/869/1/012046

A search for beyond the Standard Model physics using a final state with light and boosted muon pairs at the CMS experiment

A Castaneda[1]
On behalf of the CMS collaboration
[1]Texas A&M University at Qatar, Education City, P.O.Box 23874, Doha, Qatar
E-mail: castaned@cern.ch

Abstract. A search for physics beyond the Standard Model (SM) is performed using a final state with multi-muons and collision data collected by the Compact Muon Solenoid (CMS) experiment at the CERN Large Hadron Collider (LHC). The topology under study considers events with pairs of oppositely charged muons (dimuons) with a low invariant mass and a common origin. This selection offers an extremely low contribution from SM processes and fits into several beyond the SM signatures. Results are interpreted in a model-independent fashion and in the context of two scenarios: Supersymmetry including "dark" sectors (dark-SUSY) and Next-to-Minimal Supersymmetric Standard Model (NMSSM). Both scenarios postulate a non-SM decay of the Higgs boson to a pair of new light bosons which subsequently decay to a pair of muons. In the case of the dark-SUSY scenario, the hypothetical production of dark photons could give an insight on the origin of dark matter.

1. Motivation

On July 2012, the CMS and ATLAS collaborations at the CERN LHC announced the observation of a new particle with a mass of 125 GeV/c^2 and properties consistent with those of the predicted Higgs boson [1, 2]. This discovery constituted one of the greatest achievements made by the collaborations thus far and it marked the beginning of the LHC research program. Since then, extensive maintenance and upgrade work has been conducted on the accelerator and its different subsystems in order to reach the desired collision energy and particle rate. This effort will continue in the coming years. In the case of Higgs physics, there are two well-identified research programs at the LHC. The first one aims to collect enough data to perform precision measurements of its properties in order to determine if any of those deviate from SM expectations. The second program focuses on searches for new physics considering non-SM decays of the Higgs boson. Some of those new physics scenarios could be accessible with the available data. In this letter we explore one of the non-SM decays of the Higgs boson to a pair of new light bosons that subsequently decay to a pair of oppositely charged muons. One of the strengths of this analysis is the model-independent design; this allows for the results to be reinterpreted by other models with similar topologies. In addition, two benchmark scenarios are considered. The first one is in the context of the NMSSM that effectively extends the Higgs sector and allows for the production of new light bosons [3–5]. The second one is in the context of supersymmetric models with additional "hidden" or "dark" sectors (dark-SUSY) [6–8] which postulate the production of dark photons (γ_D) that, unlike SM photons, are massive and could decay to SM particles via a kinetic mixing parameter (ϵ). Previous searches for pair production of new light bosons that decay into dimuons have been performed at the LHC [9–13]. The two latest results, from ATLAS and CMS, consider models in which the γ_D can be long-lived.

2. CMS detector and data acquisition

The CMS detector was designed and built with several subsystems that provide an accurate measurement of muon properties. The central feature of the CMS apparatus is a superconducting solenoid of 6m internal diameter which provides a strong magnetic field of 3.8T that is used to measure muon momentum with high precision. Muons are identified and their trajectories are reconstructed using information from three different detector technologies: Drift Tubes (DTs), Cathode Strip Chambers (CSCs), and Resistive Plate Chambers (RPCs). A more detailed description of the CMS detector and its subsystems can be found in [14]. The data was collected by the CMS experiment in 2015 using 13 TeV proton-proton collision energy and it corresponds to a total integrated luminosity of 2.8fb^{-1}. A dedicated tri-muon trigger running online and refined with complex algorithms offline was used to select the data for this analysis.

3. Analysis strategy and SM backgrounds

Events are selected if they contain at least four high quality reconstructed muons with a minimum transverse momentum (p_{T}) of 8 GeV/c and a pseudo-rapidity (η) range below 2.4. In addition, the leading muon is required to have a $p_{\mathrm{T}} > 17\,\mathrm{GeV}/c$ and to be located in the central region ($|\eta| < 0.9$) to ensure a high and flat trigger efficiency. Muons are paired based on a common vertex criteria (Kalman filter method) and low invariant mass selection of $m(\mu^+\mu^-) < 9\,\mathrm{GeV}/c^2$. Once all of the dimuons are formed, only events with exactly two dimuons are selected. Furthermore, each dimuon is required to have at least one hit in one of the first three pixel barrel layers or in one of the two pixel endcap layers. This constraint defines the fiducial volume used for this search. To suppress backgrounds with dimuons coming from jets, dimuons are required to be isolated from all other activity in the event. Finally, the invariant masses of the two reconstructed dimuons should be compatible with each other to within the detector resolution. This selection defines our signal region; it can be visualized as a corridor in the 2D dimuon invariant mass space, as shown in figure 1. The complete offline selection criteria can be found in [15]. Monte Carlo simulated samples for both the NMSSM and dark-SUSY scenarios were produced in order to optimize the selection criteria and quantify the sensitivity for our benchmark scenarios. Table 1 shows the total event selection efficiency $\epsilon_{\mathrm{full}}^{MC}$ using one dark-SUSY sample with specific parameters for the dark photon. To provide a simple recipe for future reinterpretations of the results in the contexts of other models, we separately determine the geometric and kinematic acceptance of this analysis calculated using generator level information only. We denote this acceptance as α_{gen}. The model independence of the ratio $\epsilon_{\mathrm{full}}^{MC}/\alpha_{\mathrm{gen}}$ (which holds for variations on NMSSM and dark-SUSY parameters) permits an estimate of the full event selection efficiency of this analysis for any arbitrary new physics model predicting the final state signature with a pair of new light bosons.

Three main SM processes contribute to the background in the signal region. The dominant contribution comes from events with two b-quarks ($b\bar{b}$), in which each b-quark decays to a pair of muons via double semi-leptonic decay; this background is estimated directly from the data by examining a control region where this process is enhanced and then extrapolating the contribution to the signal region. The second contribution is the prompt production of two J/ψ particles coming from two different mechanisms (double and single parton scattering) and is estimated with a combination of data and Monte Carlo simulation. The third (and smallest) contribution comes from electroweak production of four muons where the dominant process is the production of two Z bosons that further decay into muons; this background is estimated purely

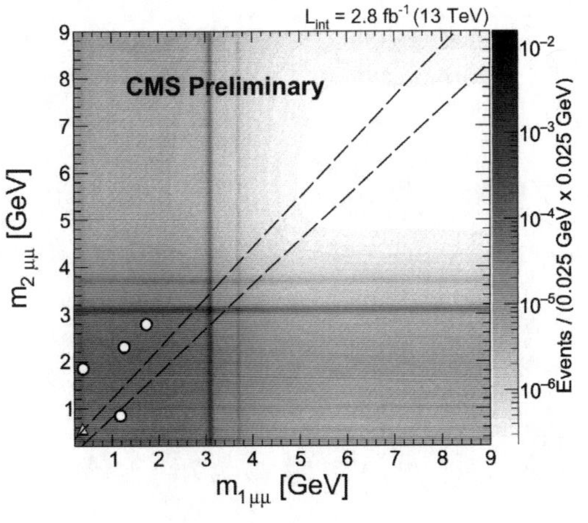

Figure 1. Distribution of dimuon masses ($m_{\mu\mu_1}$ vs. $m_{\mu\mu_2}$), with the single event observed (triangle) and background expectation (shading).

Table 1. Event selection efficiencies $\epsilon_{\text{full}}^{MC}(m_h, m_{\gamma_D})$ and kinematic acceptances $\alpha_{\text{gen}}(m_h, m_{\gamma_D})$ for a dark SUSY benchmark model with a dark photon lifetime of 2 mm.

$m_h[GeV]$	125	125	125	125
$m_{\gamma_D}[GeV]$	0.25	0.7	1.5	8.5
$\epsilon_{\text{full}}^{MC}$	0.00631 ± 0.00021	0.01733 ± 0.00042	0.03325 ± 0.00056	0.13099 ± 0.00137
α_{gen}	0.00982 ± 0.00026	0.02710 ± 0.00053	0.05096 ± 0.00070	0.20032 ± 0.00 175
$\epsilon_{\text{full}}^{MC}/\alpha_{\text{gen}}$	0.64232 ± 0.01258	0.63956 ± 0.00919	0.65242 ± 0.00637	0.65393 ± 0.00380

from Monte Carlo simulation. The total contribution to the background from SM processes into the signal region is estimated to be $0.74^{+0.54}_{-0.32}$ (stat.)±0.15 (syst.) events.

4. Results and interpretation

The full event selection criteria is applied in data; it results in one observed event inside the signal region. This observation is consistent with the expectation from SM, therefore results are interpreted as exclusion limits. The model independent limit on $\sigma(\text{pp} \to 2\text{a} + \text{X}) \times \mathcal{B}^2(\text{a} \to 2\mu) \times \alpha_{\text{gen}}$ is 1.7fb and it is constant in the entire m_a range as a consequence of having a constant $\epsilon_{\text{full}}/\alpha_{\text{gen}}$ for each signal mass. For the NMSSM scenario, the 95% CL upper limit is derived for $\sigma\,(\text{pp} \to \text{h}_{1,2} \to 2\text{a}_1) \times \mathcal{B}^2(\text{a}_1 \to 2\mu)$ as a function of m_{h_1} for three choices of m_{a_1} (where a_1 is the new light boson), as shown in figure 2. In the case of the dark-SUSY scenario, a 95% CL limit on the product of the Higgs boson production cross section and the branching fractions of the Higgs boson decay to a pair of dark photons is determined. The limit is set in the $(m_{\gamma_\text{D}}, \varepsilon)$ plane and it is shown in figure 3, along with limits from other experimental searches. The lifetime is directly related to the kinetic mixing parameter ε and the mass of the dark photon m_{γ_D} via $\tau_{\gamma_\text{D}}(\varepsilon, m_{\gamma_\text{D}}) = \varepsilon^{-2} f(m_{\gamma_\text{D}})$, where $f(m_{\gamma_\text{D}})$ is a function that depends only on the mass of the dark photon [16].

5. Conclusions and future perspectives

A search for physics beyond the SM is proposed. The models under study postulate the production of new light bosons from a non-SM decay of the Higgs boson. After a detailed study of analysis selection criteria and estimation of background contributions, there is no

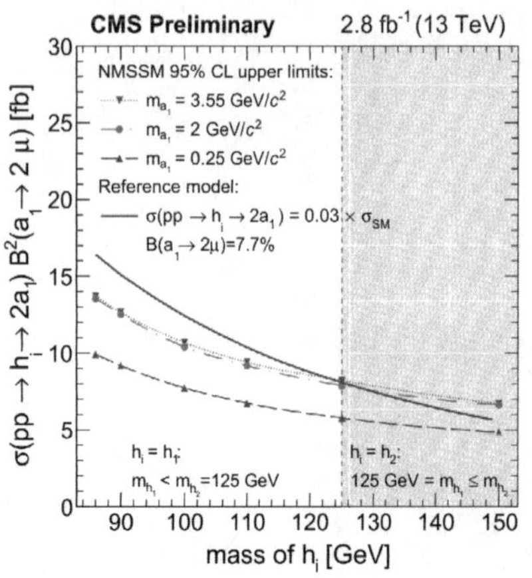

Figure 2. The 95% CL upper limits on the production of new light bosons (a_1) as a functions of m_{h_1}, for the NMSSM scenario.

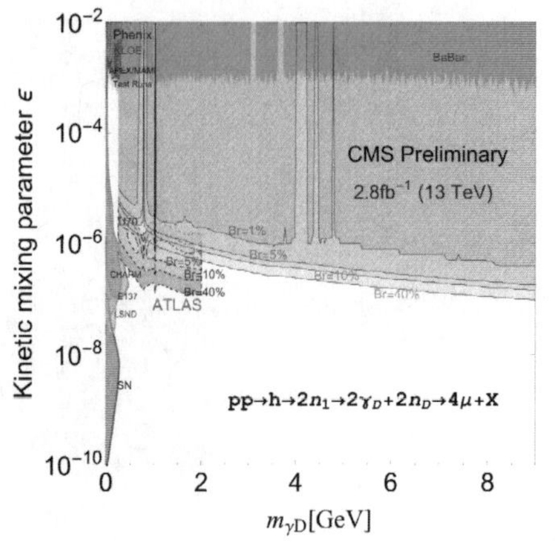

Figure 3. 95% CL upper limits on the production of dark photons (black solid curves) in the $(m_{\gamma_D}, \varepsilon)$ plane.

evidence for deviations from SM expectations. Results are interpreted as exclusion limits in a model-independent fashion and as a function of NMSSM and dark-SUSY parameters. An updated analysis is in preparation in which the data collected by the CMS experiment in 2016 will be analyzed. This data corresponds to about ten times the luminosity used here; it can potentially increase the probability of observing new physics phenomena and also improve the current analysis methods.

Acknowledgments
This work was made possible by the NPRP award [NPRP 9-328-1-066] from the Qatar National Research Fund. The statements made herein are solely the responsibility of the author[s].

References
[1] Aad G *et al.* (ATLAS) 2012 *Phys. Lett. B* **716** 1–29
[2] Chatrchyan S *et al.* (CMS) 2012 *Phys. Lett. B* **716** 30–61
[3] Drees M 1989 *Int. J. Mod. Phys. A* **4** 3635
[4] Maniatis M 2010 *Int. J. Mod. Phys. A* **25** 3505–3602
[5] Ellwanger U, Hugonie C and Teixeira A M 2010 *Phys. Rept.* **496** 1–77
[6] Arkani-Hamed N, Finkbeiner D P, Slatyer T R and Weiner N 2009 *Phys. Rev. D* **79** 015014
[7] Baumgart M, Cheung C, Ruderman J T, Wang L T and Yavin I 2009 *JHEP* **04** 014
[8] Falkowski A, Ruderman J T, Volansky T and Zupan J 2010 *JHEP* **05** 077
[9] Chatrchyan S *et al.* (CMS) 2011 *JHEP* **07** 098
[10] Aad G *et al.* (ATLAS) 2012 *Phys. Lett. B* **721** 32–50
[11] Chatrchyan S *et al.* (CMS) 2013 *Phys. Lett. B* **726** 564–586
[12] Aad G *et al.* (ATLAS) 2014 *JHEP* **11** 088
[13] Khachatryan V *et al.* (CMS) 2016 *Phys. Lett. B* **752** 146–168
[14] Chatrchyan S *et al.* (CMS) 2008 *JINST* **3** S08004
[15] 2016 A Search for Beyond Standard Model Light Bosons Decaying into Muon Pairs Tech. Rep. CMS-PAS-HIG-16-035 CERN Geneva URL `https://cds.cern.ch/record/2232052`
[16] Batell B, Pospelov M and Ritz A 2009 *Phys .Rev. D* **79** 115008

IOP Conf. Series: Journal of Physics: Conf. Series **869** (2017) 012047 doi:10.1088/1742-6596/869/1/012047

Search for dark matter and supersymmetry in the vector boson fusion topology in proton-proton collisions at CMS

A Celik[1,2] and A M C Hernandez[3] On Behalf of the CMS Collaboration

[1] Texas A&M University, College Station, TX, USA
[2] Fermi National Accelerator Laboratory, IL, USA
[3] Texas A&M University at Qatar, Doha, Qatar
Email: alicelik@physics.tamu.edu

Abstract: A search for pair production of dark matter candidates and supersymmetry (SUSY) production with two jets in vector-boson fusion (VBF) topology is presented using data collected by the Compact Muon Solenoid (CMS) detector in proton-proton collisions at the Large Hadron Collider (LHC). Final states with no leptons are expected in pair production of dark matter particles or scalar quarks in SUSY compressed mass-spectra scenarios. Final states with low-energy leptons are expected in the production of charginos and neutralinos in SUSY compressed mass-spectra scenarios. Results for both zero and two lepton final states at 8 TeV are presented with brief prospects at 13 TeV.

1. Introduction

Understanding the nature of dark matter (DM) in the Universe would rank among the biggest leaps forward in the progress of fundamental physics. Supersymmetry remains perhaps the best motivated of possible theories to simultaneously describe the nature of DM and solve the gauge hierarchy problem in the Standard Model (SM). The lightest neutralino ($\tilde{\chi}_1^0$) in SUSY could be the lightest supersymmetric particle (LSP) and stable as the DM candidate. However, for all of its attractive features, there are no experimental signs yet of its existence. Current SUSY limits now reach as high as around 1.7 TeV for gluinos in certain scenarios [1]. It could be because SUSY has a more challenging nature. SUSY scenarios with compressed mass-spectra are of great interest. This motivates to conduct searches with vector-boson fusion (VBF) type events. In this paper, results from VBF searches in Run-1 along with prospects in Run-2 are presented.

The search strategy using events with dijet in the VBF topology at the LHC has been suggested in Refs. [2-4]. Two analyses reported here are the first searches for SUSY and DM with two jets in the VBF topology at the LHC: (i) two leptons with VBF dijet to search for charginos and neutralinos in the Minimal Supersymmetric Standard Model (MSSM) [5]; (ii) large momentum imbalance with VBF dijet to test DM-gauge boson couplings in a framework of an effective field theory (EFT) and also test SUSY in compressed mass-spectrum scenario [6]. Below are brief summaries of each analysis.

2. Search for charginos/neutralinos using the VBF processes

Figure 1 shows one of representative Feynman diagrams for pair production of charginos and neutralinos in VBF processes. Figure 2 (left) shows the invariant mass distributions of two jets (m_{jj}) in VBF topology for events with two leptons, comparing between the data and the SM expectation [5]. Upper limits are set on the cross sections for pair production of charginos and neutralinos with two associated jets, assuming the supersymmetric partner of the τ lepton to be the lightest slepton and to be lighter than the lighter chargino. The data sample corresponds to an integrated luminosity of 19.7 fb^{-1} at 8 TeV recorded by the CMS detector [7]. Figure 2 (right) shows the observed limit on cross sections for two scenarios. For a compressed mass-spectrum scenario in which the mass difference between the $\tilde{\chi}_1^0$ and the next lightest, mass-degenerate, gaugino particles $\tilde{\chi}_2^0$ and $\tilde{\chi}_1^{\pm}$ is 50 GeV, lower mass limit of 170 GeV is set for $\tilde{\chi}_2^0$ and $\tilde{\chi}_1^{\pm}$. This scenario was not excluded by the latest CMS analyses in the trilepton channel [8]. The VBF topology has been demonstrated to be complementary to a standard SUSY search in trilepton channel at the LHC.

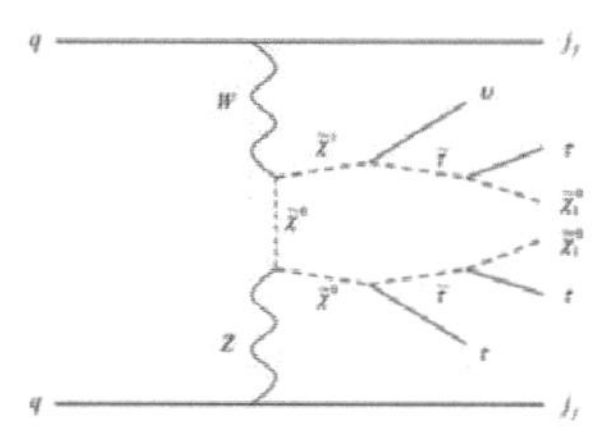

Figure 1. A representative Feynman diagram of pair production of charginos and neutralinos in the VBF processes [5].

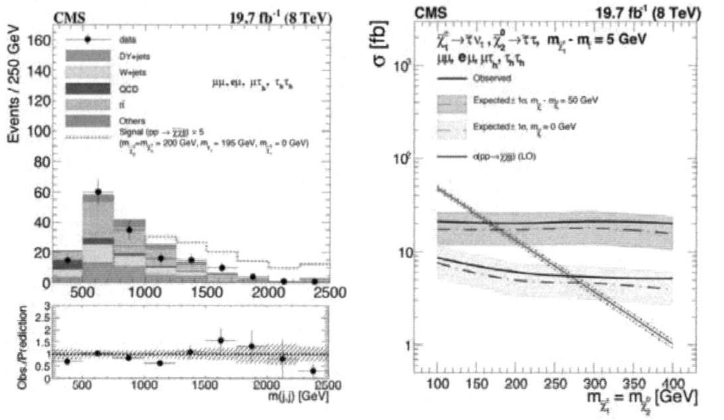

Figure 2. (Left) Dijet invariant mass distribution for the combination of all dilepton channels in missing transverse energy + VBF events, comparing data and the SM expectation [5]. (Right) 95% C.L. upper limits on the cross section as a function of $m(\tilde{\chi}_1^{\pm}) = m(\tilde{\chi}_2^0)$ for the scenario where $m(\tilde{\tau}) = m(\tilde{\chi}_1^{\pm}) -$ 5 GeV in two different $m(\tilde{\chi}_1^0)$ definitions: $m(\tilde{\chi}_1^0) = m(\tilde{\chi}_1^{\pm}) -$ 50 GeV (compressed mass spectra) and $m(\tilde{\chi}_1^0) = 0$ GeV (large mass gap) [5].

3. DM and Compressed Mass-Spectra SUSY

Events with transverse momentum imbalance and two jets in VBF topology is used for the first time to search for pair production of (i) DM candidates (figure 3 top) and (ii) compressed mass-spectra squarks (figure 3 bottom). The data sample corresponds to an integrated luminosity of 18.5 fb^{-1} of pp collision at 8 TeV recorded by the CMS detector. The observed dijet mass spectrum is consistent with the SM expectation (figure 4).

The results are used to test DM-gauge boson coupling (g_{eff}) in an EFT framework. DM masses (m_χ) are explored as a function of contact interaction strength (Λ) (figure 5 left). The EFT approach is defined to be valid for each parameter point of m_χ and Λ if the fraction of MC signal events (R_Λ) classified as not having large momentum transfer is 80% or more. Here the MC signal event is considered as having large momentum transfer if the center-of-mass energy of the DM pair ($E_{\text{cm}}^{\chi\chi}$) is larger than the mediator mass parameter of the EFT, M = $\Lambda\, g_{\text{eff}}$. Truncated limits are calculated by adding the requirement $E_{\text{cm}}^{\chi\chi} < \Lambda\, g_{\text{eff}}$ to the signal acceptance, following Refs. [9, 10]. More signal events are removed in higher DM mass regions where R_Λ curves tend to go up and truncated limits go down. The curve of the DM relic abundance Ωh^2= 0.12 is calculated using the MadDM program [11], assuming that DM pairs annihilate to electroweak gauge boson pairs. The DM is more abundant than observed in the regions above and left from the Ωh^2= 0.12 line.

The results are also used to set the most stringent limit on third-generation squark production mass below 315 GeV is found (figure 5 right), assuming a 5 GeV mass difference with respect to the lightest neutralino. The result can be compared to the "monojet" searches with a mass limit of 250 GeV in a compressed scenario of bottom squark and top squark [12]. It is demonstrated as a powerful tool for DM and SUSY searches at the LHC.

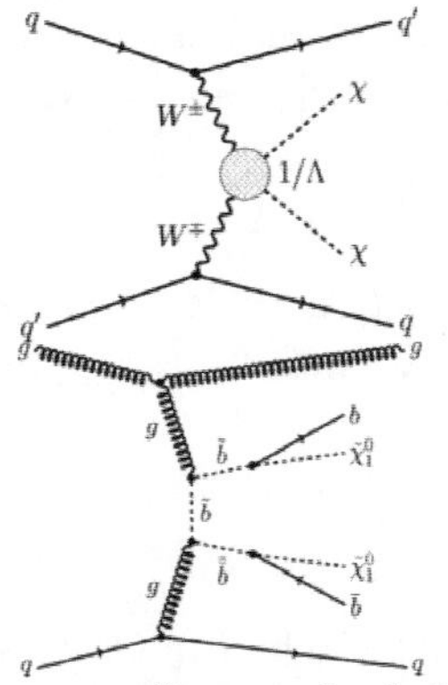

Figure 3. Feynman diagrams for dark matter pair production in a VBF process (top) and for bottom squark pair production (bottom) [6]. Given a nearly degenerate bottom squark and the lightest neutralino (LSP), the final state b-quarks are too soft to be observed.

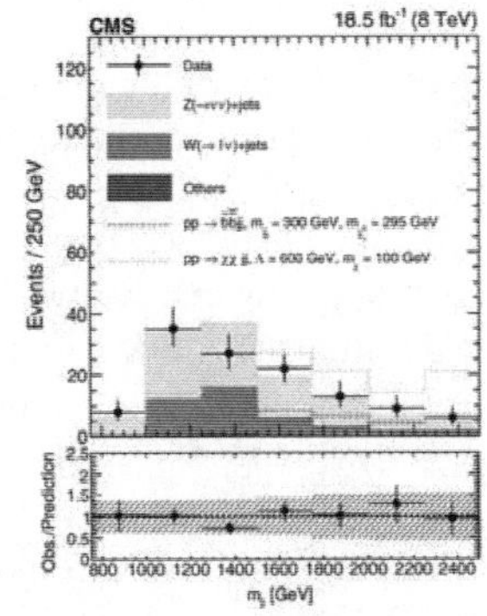

Figure 4. Dijet mass distribution of the data (dots), estimated background (stacked histograms), and signal samples (dashed lines) after the event selection. The last bin includes all events above 2250 GeV. The ratio plot (below) shows the yields in data divided by predicted yields for each bin. The shaded band includes systematic and statistical uncertainties in the background prediction [6].

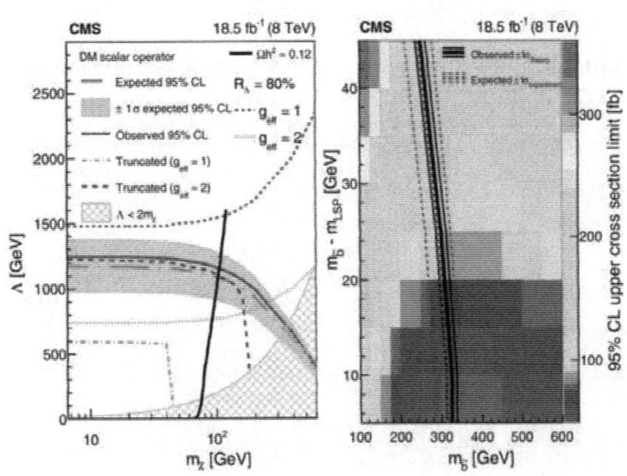

Figure 5. (left) Contact interaction scale (Λ) limit at 95% C.L. as a function of the DM mass (m_χ). Values of (m_χ, Λ) below the curve are excluded. The validity of the effective field theory is quantified by (i) R_Λ = 80% contours and (ii) truncated limits for different values of the effective coupling (g_{eff}). The DM relic abundance Ωh^2= 0.12 is calculated as described in the text [6]. (right) Bottom squark pair production 95% C.L. upper cross section limit as a function of the bottom squark mass and the mass difference between the bottom squark and the LSP. The observed (expected) cross section limit includes one standard deviation bands for the theoretical (experimental) uncertainty [6].

4. Summary

The searches for events with 0 and 2 leptons plus two jets in the VBF topology at 8 TeV are presented, using the data of 8-TeV *pp* collisions recorded by the CMS detector. The dijet mass distributions are consistent with the SM expectations, testing the DM-gauge boson coupling for the first time and setting the stringent limits on squark mass in compressed mass-spectra scenarios. It demonstrates a new approach using VBF for SUSY and DM searches at the LHC.

In Run 2 of the LHC, the CMS experiment recorded so far (2015, 2016) a data sample correspond to an integrated luminosity of about 40 fb^{-1}. This dataset will allow to substantially extend the sensitivity of analyses with events containing 0, 1 or 2 leptons and two jets in the VBF topology in compressed mass-spectra scenario.

Acknowledgements
A.C. was supported by Fermilab LPC Graduate Scholar program, Republic of Turkey Ministry of National Education and US Department of Energy (grant DE-SC0010813); and A.M.C.H. by the Qatar National Research Fund (grant NPRP 9-328-1-066).

References

[1] CMS Collaboration 2014 *Phys. Lett.* B. **733** 328.
[2] Dutta B, Gurrola A, Johns W, Kamon T, Sheldon P and Sinha K 2013 *Phys. Rev.* D **87** 035029.
[3] Delannoy A G, Dutta B, Gurrola A, Johns W, Kamon T, Luiggi E, Melo A, Sheldon P, Sinha K, Wang K, and Wu S 2013 *Phys. Rev. Lett.* **111** 061801.
[4] Dutta B, Gurrola A, Hatakeyama K, Johns W, Kamon T, Sheldon P, Sinha K, Wu S, and Wu Z 2015 *Phys. Rev.* D **92** 095009.
[5] CMS Collaboration 2015 *J. High Energy Physics* **11** 189.
[6] CMS Collaboration 2017 *Phys. Rev. Lett.* **118** 021802.
[7] CMS Collaboration 2014 The CMS experiment at the CERN LHC *J. Instrum.* **3**, S08004.
[8] CMS Collaboration 2014 *Eur. Phys. J.* C **74** 3036.
[9] Abercrombie D et al. 2015 Dark Matter Benchmark Models for Early LHC Run-2 Searches: Report of the ATLAS/CMS Dark Matter Forum [arXiv:1507.00966].
[10] CMS Collaboration 2016 *Phys. Rev.* D **93** 052011.
[11] Backović M, Kong K, and McCaskey M, 2014 MadDM v.1.0: Computation of dark matter relic abundance using MadGraph 5 *Physics of the Dark Universe* 5-6 18.
[12] CMS collaboration 2015 *J. High Energy Physics* **06** 116.

IOP Conf. Series: Journal of Physics: Conf. Series **869** (2017) 012048 doi:10.1088/1742-6596/869/1/012048

Discovery potential for directional dark matter detection with nuclear emulsions

A M Guler, *for the NEWSdm Collaboration*

Middle East Technical University, 06800 Ankara, Turkey
Email: ali.murat.guler@cern.ch

Abstract. Direct Dark Matter searches are nowadays one of the most exciting research topics. Several Experimental efforts are concentrated on the development, construction, and operation of detectors looking for the scattering of target nuclei with Weakly Interactive Massive Particles (WIMPs). In this field a new frontier can be opened by directional detectors able to reconstruct the direction of the WIMP-recoiled nucleus thus allowing to extend dark matter searches beyond the neutrino floor. Exploiting directionality would also give a proof of the galactic origin of dark matter making it possible to have a clear and unambiguous signal to background separation. The angular distribution of WIPM-scattered nuclei is indeed expected to be peaked in the direction of the motion of the Solar System in the Galaxy, i.e. toward the Cygnus constellation, while the background distribution is expected to be isotropic. Current directional experiments are based on the use of gas TPC whose sensitivity is limited by the small achievable detector mass. In this paper we show the potentiality in terms of exclusion limit of a directional experiment based on the use of a solid target made by newly developed nuclear emulsions and read-out systems reaching sub-micrometric resolution.

1. Introduction

The most (~26%) of the matter in the Universe is in the form of non-baryonic cold Dark Matter [1]. The evidence for the existence Dark Matter came from the observation that various luminous objects move faster than one would expect if they only felt the gravitational attraction of other visible objects. The existence of such matter requires new particles beyond the standard model of particle physics. Among the several possibility the most discussed one are weakly interacting massive particles (WIMPs). They are neutral stable particles that were in thermal equilibrium in the early universe and have to be non-baryonic, cold, and nearly non-interacting. A large, and growing, experimental effort are devoted to detect and characterize WIMPs.

The most promising experimental method for the dection of WIMPs is the direct detection of the recoil energy deposited by elastic scattering of WIMPs on the target nuclei. Although new generation detectors are capable of measuring the direction of a nuclear recoil track, resulting from the elastic scattering of a target nucleus by an incoming WIMP, they are commonly limited in sensitivity by background. Neutrino interactions are an irreducible source of background since no detector can be shielded from the flux of incident neutrinos. However, even in the presence of irreducible backgrounds, directional detection can still provide an unambiguous observation of dark matter interactions. The motion of our solar system around the galactic center should produce an apparent dark matter wind which is detectable as diurnal modulations of the magnitude and direction of a dark matter interaction signal in a terrestrial detector. In contrast, backgrounds are expected to be relatively isotropic.

Since the track length of the recoil nucleus is too short, direct searches requires a very high spatial resolution. The lenght of a WIMP-induced nuclear recoil depends on the recoil energy and density of the target material. For example, a nucleus recoiling with energy 100 keV travels 100 nm in a solid. For the direct detection of Dark Matter there are several approaches [2] based on different technologies like Multi-Wire Proportional Chambers(MWPCs), Micro Pattern Gaseous Detectors (MPGDs), optical readouts, and nuclear emulsions. Each technology has its own advantages and disadvantages. For example, experiments based on low pressure gaseous Time Projection Chambers (TPCs) is hardly scalable to very large detectors masses needed to reach a good sensitivity to the Spin-Independent (SI) case. They are only sensitive to the Spin-Dependent (SD) coupling of WIMPs with less than 1 kg mass detectors. However, the use of a solid target for directional searches would overcome the mass limitation of gaseous detector thus allowing to reach an high sensitivity in the low cross section sectors of the SI case. Nevertheless, in a solid medium, the track of the WIMP-scattered nuclear recoil will have the length of a nuclear recoil of the order of a few hundred nanometers, much shorter than in the case of a gaseous target where the recoil length is expected to be of the order of a few millimeters. The nuclear emulsion based techniques proposed by NEWSdm Collaboration [3] fullfils both conditions i) high spatial resolution and ii) massive target. Emulsions consist of silver-halide crystals dispersed in a polymer layer, typically gelatin. Silver-halide crystals are semiconducting with a band gap of 2.6 eV, and work as a sensor to detect charged particles. The size of a silver grain after development is typically around 50 nm. A nuclear recoil can be reconstructed by imaging sets of these silver grains.

In this paper, we report the discovery potential of WIMPS using an emulsion-based detector.

2. Experimental concept

The detector exploits new generation nuclear emulsions with nanometric grains. An R&D conducted by the Nagoya University in collaboration with the Fujifilm Company has established the production of films with nanometric grains. The so-called Nano Imaging Trackers (NIT) and Ultra-Nano Imaging Trackers (U-NIT), have grains of 45 and 18 nm diameter respectively (see figure 1) [4,5]. The detector is conceived as a bulk of NIT surrounded by a shield to reduce the external background. The detector is then placed on an equatorial telescope in order to absorb the earth rotation, thus keeping fixed the detector orientation with respect to the incoming apparent WIMP flux, i.e. toward the Cygnus constellation. The angular distribution of the WIMP-scattered nuclei is therefore expected to be strongly anisotropic with a peak centered in the forward direction.

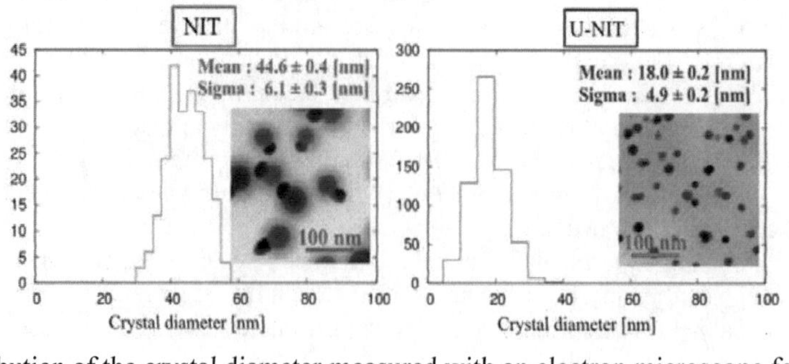

Figure 1: Distribution of the crystal diameter measured with an electron microscope for NIT (left) and U-NIT (right) emulsions [3].

The presence of both light and heavy nuclei in the emulsion gel results in a good sensitivity to WIMPs with both light and heavy masses. Figure 2 shows the correlation between the track length of the recoiled nucleus and its kinetic energy for the different target nuclei. Although Ag and Br are the most effective targets for WIMP masses in this range, the detection capability is reduced since their ranges are shorter than lighter elements at the same energy. Instead, for a WIMP with a mass around 10 GeV/c^2, the

IOP Conf. Series: Journal of Physics: Conf. Series **869** (2017) 012048 doi:10.1088/1742-6596/869/1/012048

kinematics favours lighter nuclei that, for a given kinetic energy, have a longer range. Therefore, the contribution of the C, N and O ions is essential for WIMP masses around 10 GeV/c^2.

The estimated WIMP rates are of the order of 1 event/kg/year, much lower than the usual radioactive backgrounds. For this reason, the detector has to be placed underground to be protected from cosmic-ray induced background. Moreover, a careful control of the radioactive contamination of the materials used for the detector construction and a precise estimation of the corresponding induced background are needed.

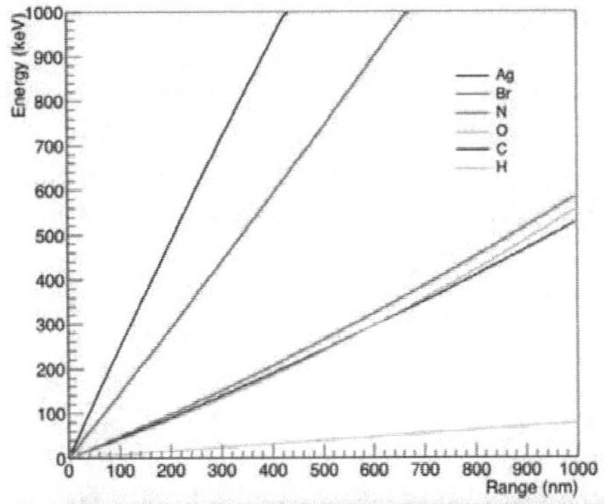

Figure 2: Correlation between the track length of the recoiled nuclei and their kinetic energy, for different target nuclei in NIT emulsions [3].

2.1. Read-out technique

Fast automatic optical microscopes are used to read out nuclear emulsions. After the exposure, the search for signal candidates requires the scanning of the whole emulsion volume. The analysis of NIT emulsions is performed with a two-step approach: a fast scanning, based on techniques developed for the OPERA experiment [6] for the signal preselection: an ellipse is fitted to the optically reconstructed track, and candidate events are selected by applying a cut on the ratio of the lengths of the major and minor axes. After elliptical shape recognition with optical microscopy candidate events are confirmed by super-high resolution microscopy techniques.

In order to simulate the effect of a WIMP-induced nuclear recoil and to measure the efficiency and the resolution of the optical microscope, a Kr ion beam at different energies was exposed NIT films When analysed with the optical microscope, submicrometric tracks produced by Kr ions appear as shown in figure 3.

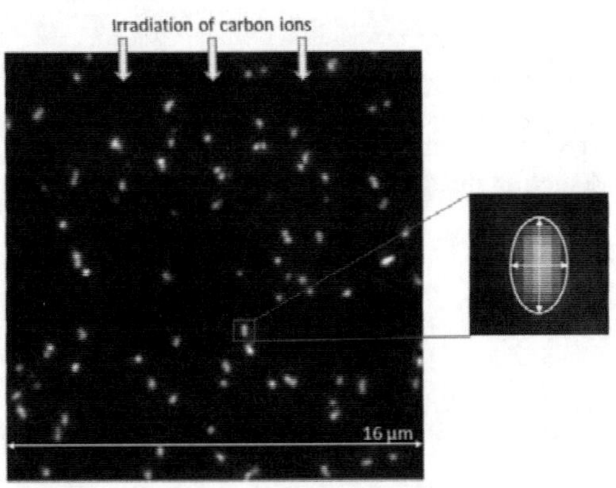

Figure 3: Kr ions implanted on NIT films. The image is taken with an optical microscope. The selection of candidate tracks is based on the elliptic fit of the clusters [3].

IOP Conf. Series: Journal of Physics: Conf. Series **869** (2017) 012048 doi:10.1088/1742-6596/869/1/012048

3. Discovery Potential

The final sensitivity of low-energy rare event searches is strongly limited by the background induced by radioactivity. Two main categories have to be taken into account: the environmental or external background and the intrinsic one. The main background sources α, β, γ-rays and and neutron induced recoils, while NIT are essentially not sensitive to minimum ionizing particles (MIP). α originating from U and Th isotopes having energies of the order of MeV, can be identified by measuring the track length. Moreover, γ-rays and β can be rejected by properly regulating the emulsion response, in terms of number of sensitized crystals per unit path length (i.e. the sensitivity), through a chemical treatment of the emulsion itself, therefore they are considered as reducible background. However, the contribution from neutrons is usually considered as irreducible, since they induce nuclear recoils as WIMPs do. The directional detection has the unique capability of distinguishing the WIMP signal from the background by exploiting the feature of the signal, expected to be peaked in the direction of motion of the Sun. On the contrary, there is no reason for the neutron-induced nuclear recoil spectrum to present the same feature. In particular, the neutron background is expected to be isotropically distributed. Neutrons in underground laboratories originate either from cosmic muon interactions, or from environmental radioactivity, or from spontaneous fissions and (α,n) reactions. The first two sources can be reduced by an appropriate shielding; the latter one, coming from the intrinsic radioactivity of the target materials, is associated with an isotropic distribution in the laboratory frame.

The 90% C.L. upper limit in case of null observation is shown in figure 4 for an exposure of 1 kg.year of NIT emulsions, with a minimum detectable track length ranging from 200 nm down to 50 nm and in the hypothesis of zero background [3]. Even not including the directionality discrimination of the signal and assuming to reach a negligible background level, such an experiment would cover a large part of the parameter space indicated by the DAMA/LIBRA results with a small (1 kg) detector mass, using a powerful and complementary approach.

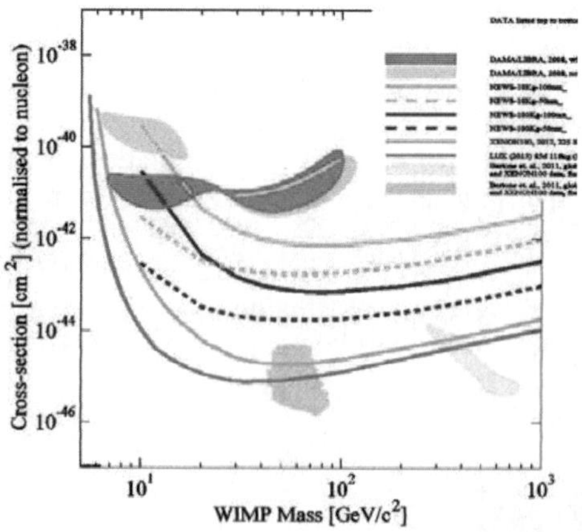

Figure 4: Sensitivity at 90% C.L, in the zero background hypothesis for an experiment with a mass of 10 kg (green) and 100 kg (blue) for two value of detection threshold: 100 nm (dashed lines) and 50 nm (solid line) [3].

References

[1] Jungman G, Kamionkowski M, Griest K 1996 *Phys. Rep.* **267** 195.
[2] Battat J B R *et al* 2016 *Phys. Rep.* **662** 1.
[3] Letter of Intent NEWS Collaboration arXiv:1604.04199, LNGS-LOI 48/15.
[4] Natsume M *et al* 2007 *Nucl. Instr. Meth.* A575 439.
[5] Naka T *et al* 2013 *Nucl. Instr. Meth.* **A718** 519.
[6] OPERA Collaboration 2009 *JINST* **4** P04018.

IOP Conf. Series: Journal of Physics: Conf. Series **869** (2017) 012049 doi:10.1088/1742-6596/869/1/012049

Studying total proton–proton cross section collision at large hadron collider using gene expression programming

A Radi[1,2]

[1]Department of Physics, College of Sciences, Sultan Qaboos University, Al Khoudh, Muscat 123, Oman
[2]Department of Physics, Faculty of Sciences, Ain Shams University, Abbassia, Cairo 11566, Egypt
E-mail: Amr.radi@cern.ch

Abstract. New technique is presented for modeling total cross section of proton-proton (p-p) collision from low to ultra-high energy regions using gene expression programming (GEP). GEP, as a machine learning technique is usually used for modeling physical phenomena by discovering a new function $\sigma_T(\sqrt{s})$. In case of modeling the p-p interactions at the Large Hadron Collider (LHC), GEP is used to simulate and predict the total cross-section which is a function of total center-of-mass from low to high energy $\sqrt{s}$. The discovered function shows a good match as compared with the other models. The predicted values of total cross section are in good agreement with Particle Data Group (PDG).

1. Introduction

The study of the interactions between elementary particles requires high energy collisions as in LHC [1-3]. LHC possibly will yield the way in the direction of our awareness of particle physics beyond the Standard Model. The proton-proton interaction is one of the fundamental interactions in high-energy physics. In order to fully exploit the enormous physics potential, it is important to have a complete understanding of the reaction mechanism.

Recently, different modeling methods, based on soft computing systems, include the application of evolution algorithms [4-7]. GEP [8] can be used as an alternative tool for the simulation of these interactions [5-8]. GEP is a global optimization algorithm and an automatic programming technique that had been applied in particle physics [9-10]. It is a recently developed evolutionary computation method for function discovery and data analysis. GEP uses population of individuals, selects them according to fitness, and produces genetic variation using one or more genetic operators [11].

In this paper, we have discovered the functions of the total cross section $\sigma_T(\sqrt{s})$ of p-p collision from low to ultra-high energy regions using the GEP technique. This paper has five sections. Section 2, gives a review to the basics of the GEP technique. Section 3 explains how GEP is used to model the p-p collision. Finally, the results and conclusion are provided in sections 4 and 5 respectively.

2. An overview of gene expression algorithms

GEP stores the individuals as symbolic strings (linear) which are then presented as an expression tree for evaluation [9]. GEP, provides the possible solutions to a given optimization problem, using the Darwinian principle of survival of the fittest. It uses biologically inspired operations like replication, recombination and mutation. In GEP, the chromosome or genome composed of a linear, symbolic string of fixed length, each consists of one or more genes. Each gene can be represented by an algebraic expression and are composed of a head and a tail. The head contains symbols that represent both functions (F={sqr, -,+ , *,/, or other operators}) and terminals (T={x,y,z, random constants}), whereas the tail contains only terminals. As an example, the algebraic expression: sqr((x-y)+x*z) can be represented as an Expression Tree (ET) as shown in figure 1.

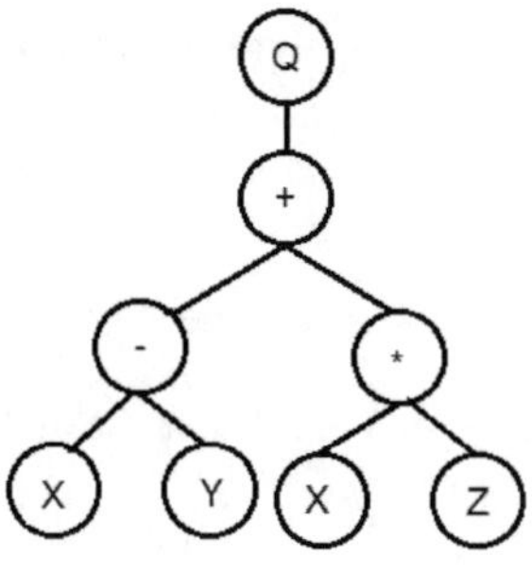

Figure 1. An example of gene ET.

Here {Q, -,+ , *} represents the head (*h*), "Q" represents the square root function, and {x, y, x, z} represents the tail t.

Each gene individual in the population is assigned a fitness value, which quantifies how well it performs in solving the problem. The fitness value is computed by a problem dependent fitness function. The fitness function that can be applied for evaluating performance of generated solution are measured with the summed square of residuals (SSE), R-square is the ratio of the sum of squares of the regression (SSR) and MSE is the mean square error or the residual mean square [12].

3. The proposed GEP for the p-p collision

Total cross sections are proposed with two individual GEP models. The model is trained/predicated using experimental data to simulate the p-p collision that calculates the total cross sections $\sigma_T(\sqrt{s})$.

GEP has the potential to discover a new model, to show that the data sets are subdivided into two sets (training and predication). GEP, as configured in table 1, runs until the fitness function is reduced to an acceptable level (MSE = 0.0005). GEP discovers a new model from the training set. The predicated set is used to examine the generalization capabilities of the model.

Table 1. GEP optimal parameters that were obtained.

GEP Parameters	σ_T ($\sqrt{s}$)
Generations	800
Populations	30000
Function set	{*,/,-,+,ln,sqrt}
Terminal Set	{constant, $\sqrt{s}$ }
Fitness function	MSE
Selection method	Elits and rank
Mutation rate	0.01
Crossover rate	0.9

Frontiers in Theoretical and Applied Physics/UAE 2017 (FTAPS 2017) IOP Publishing
IOP Conf. Series: Journal of Physics: Conf. Series **869** (2017) 012049 doi:10.1088/1742-6596/869/1/012049

4. Results

The discovered function is tested using the experimental data of the total cross sections σ_T with the center-of-mass energy $\sqrt{s}$. The training data considers both the experimental observations at the LHC in the range (10 GeV – 8 TeV) and the Particle Data Group for p-p collision experiments [13-17].

The discovered function has been tested to associate the input patterns to the target output patterns using the error function. The final discovered function for describing the total cross sections σ_T with the center-of-mass energy is given by

$$\boldsymbol{\sigma_T (\sqrt{s}) = ((\mathrm{sqrt}((\sqrt{s}+\exp(7)))+7.5)+\sin((\mathrm{sqrt}((\sqrt{s}))+(1/(\sqrt{s}/2))))} \quad (1)$$

The model is in excellent agreement with the extrapolation from low to ultra-high energy regions as in figure 2. In figure 3, the values of the total cross-sections $\sigma_T (\sqrt{s})$ is compared with results of the center-of-mass energy $\sqrt{s}$ between 10 GeV and the higher energies 50000 GeV from cosmic rays [17].

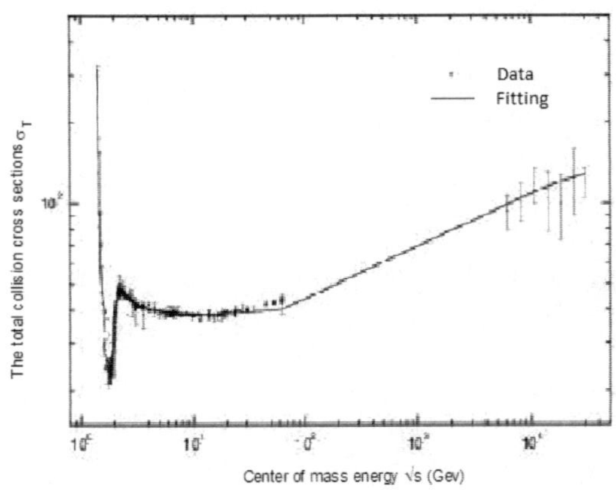

Figure 2. The progress of the total cross sections σ_T function of the centre-of-mass energy $\sqrt{s}$ from low to ultra-high energy regions.

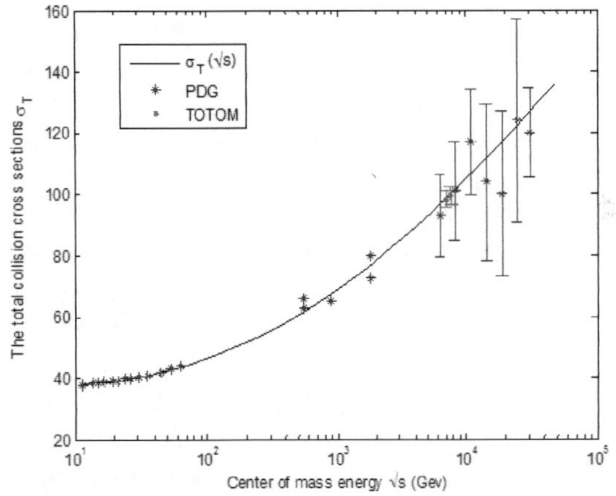

Figure 3. The progress of the total cross sections σ_T function of the centre-of-mass energy $\sqrt{s}$ between 10-50000 GeV.

In table 2, the elastic scattering cross-section was subtracted from the total cross section to obtain a value for the inelastic cross-section which can then be compared/verified with the measurements of the CMS [13], ATLAS [14], ALICE [15] and TOTAM [16] experiments.

Table 2. Results of the TOTEM measurements at the LHC energy of $\sqrt{s}$ = 7 & 8 TeV [16].

Experiment	Inelastic cross-section (mb)	Elastic cross-section (mb)	Total cross-section calculated (mb)	Total cross-section predicted (mb) using $GEP_T (\sqrt{s})$
σ (TOTOM)	73.5 (- 0.6 +1.1)	24.8 (± 1.4)	98.3 (± 3.0)	98.1
σ (CMS)	68.0 (± 8.4)			
σ (ATLAS)	69.4 (± 9.1)			
σ (ALICE)	72.7 (± 6.2)			
$\sqrt{s}$ = 8 TeV			102.9(± 3.0)	103.1

Moreover, the discovered function σ_T ($\sqrt{s}$) predicts the total cross sections σ_T at $\sqrt{s}$ = 7 TeV which is 98.1 mb. This values is the same as experimental of TOTOM (98.3 mb) [16]. The predicted values of σ_T ($\sqrt{s}$) at LHC are 103.1 mb and 127.3 mb at $\sqrt{s}$ = 8 TeV, and $\sqrt{s}$ = 14 TeV, respectively. Those results similar to Nakamura and Block [17, 19,20].

5. Conclusions

The current work presents GEP as a new technique for constructing functions of the total cross sections, σ_T ($\sqrt{s}$) of p-p collision. The discovered functions show good match to the experimental data. Moreover, they are capable of predicting experimental data for σ_T($\sqrt{s}$) that are not used in the training session. Consequently, the predicted values of σ_T ($\sqrt{s}$) at the LHC energy in terms of the same parameters are in good agreement with the cosmic-ray experimental data from Particle Data Group. GEP has become one of important research areas in the field of high energy physics.

References

[1] Khachatryan V *et al* 2014 CMS Collaboration *Eur. Phys. J.* C **74** 3053.
[2] Aaboud M *et al* 2016 ATLAS Collaboration *Phys. Lett.* B **761** 158-178.
[3] TOTAM Collaboration and ATLAS Collaboration 2016 arXiv:1609.08847v3 [hep-ph] .
[4] Eiben A, Smith A, Introduction to Evolutionary Algorithms (Springer, Berlin, 2003).
[5] El-Bakry S and Radi A 2007 *Int. J. Mod. Phys.* C **18** 351.
[6] El-dahshan E, Radi A, El-Bakry M 2009 *Int. J. Mod. Phys.* C **20** 1817.
[7] Whiteson S, Whiteson D 2009 *Eng. Appl. Artif. Intel.* **22** 1203.
[8] Ferreira C 2006 Gene Expression Programming: Mathematical Modeling by an Artificial Intelligence, 2nd Edition, Springer-Verlag Germany.
[9] Teodorescu L, High energy physics data analysis with gene expression programming. In 2005 IEEE Nuclear Science Symposium Conference Record, Vol. **1**, pp. 143-147, (2005).
[10] Teodorescu L and Sherwood D 2008 Computer Physics Communications **178 (6)** 409–419.
[11] Ferreira C 2001 "Gene expression programming in problem solving" WSC6 tutorial.
[12] Cohen J, Cohen P, West S and Aiken L 2003 applied multiple regression/correlation analysis for the behavioral sciences. Mahwah, NJ: Erlbaum.
[13] CMS-Collaboration 2011 Performance Analysis Note, CMS-PAS-FWD-**11**-001.
[14] Aad G *et al* 2011 arXiv: 1104. 0326 [hep-ex] (2011).
[15] Poghosyan M G *et al* 2011 arXiv:1109.4510 [hep-ex].
[16] TOTEM Collaboration2011 *EPL* **96** 21002.
[17] Nakamura K *et al* 2010 *J. Phys.* G **37** 075021.
[18] Cudell J *et al* 2002 *Phys. Rev. Lett.* **89** 201801.
[19] Block M M 2011 arXiv: 1109, 2940 [hep-ph].
[20] Block M M and Halzen F 2005 hep-ph/0506031.

IOP Conf. Series: Journal of Physics: Conf. Series **869** (2017) 012050 doi:10.1088/1742-6596/869/1/012050

Analytical study of fractional equations describing anomalous diffusion of energetic particles

A M Tawfik[1,2], H Fichtner[1], R Schlickeiser[1] and A Elhanbaly[2]

[1] Institut für Theoretische Physik IV, Ruhr-Universität Bochum, Universitätsstrasse 150, D-44780 Bochum, Germany

[2] Theoretical Physics Research Group, Mansoura University, Mansoura 35516, Egypt

E-mail: Ashraf.Attia@ruhr-uni-bochum.de

Abstract. To present the main influence of anomalous diffusion on the energetic particle propagation, the fractional derivative model of transport is developed by deriving the fractional modified Telegraph and Rayleigh equations. Analytical solutions of the fractional modified Telegraph and the fractional Rayleigh equations, which are defined in terms of Caputo fractional derivatives, are obtained by using the Laplace transform and the Mittag-Leffler function method. The solutions of these fractional equations are given in terms of special functions like Fox's H, Mittag-Leffler, Hermite and Hyper-geometric functions. The predicted travelling pulse solutions are discussed in each case for different values of fractional order.

1. Introduction

Anomalous diffusion is an ubiquitous phenomenon and one of the fundamental processes for transport of matter in different systems [1-2]. The anomalous diffusion, based on the mean square displacement of the diffusing species $< z^2(t) \sim t^\alpha >$ for a slow process with $0 < \alpha < 1$ is called sub-diffusive. The sub-diffusion process has been observed in many physical systems such as spatially disordered or fractal media and in the temporal fluctuations of the medium [3]. It is well- known that sub-diffusion equations in terms of fractional derivatives can be obtained from Continuous Time Random Walk models [4]. In general, the anomalous regimes are characterized by non-Gaussian statistics like Levy statistics, which encompasses probability distributions with power-law tails. One of the methods describing the anomalous process is to replace the ordinary time and space derivatives in a standard kinetic equation by fractional time and space derivatives. In this paper, we introduce the fractional Klein-Kramer equation which can be considered as an extension of the fractional diffusion equation by using the Caputo fractional derivative [5]. Caputo's definition the derivative of any continuous function $\phi(t)$ is as follows

$$^{c}D_t^\alpha \phi(t) = \frac{1}{\Gamma(n-\alpha)} \int_0^t (t-\tau)^{n-\alpha-1} \frac{d^n}{d\tau^n}\phi(\tau)\, d\tau,\ n-1 < \alpha < n, n \geq 0,\ n \in N \tag{1}$$

Some endeavors to understand the energetic particle transport in the cosmos [6] consider the Langevin equation [7] for the coordinate $z(t)$

$$\frac{dv}{dt} = -\eta v + \frac{F(z)}{m} + \Gamma(t), \quad \frac{dz}{dt} = v, \tag{2}$$

where m is the mass of a diffusing particle, η denotes a friction coefficient, $F(z)$ is the external force field and $\Gamma(t)$ is a Gaussian white noise and represent the source of the anomalous behavior. Below we introduce the approximations of the fractional Klein-Kramer equation to describe the sub-diffusion process. Two limiting cases can be distinguished, namely the fractional modified Telegraph from which the probability distribution function can be derived and the fractional Rayleigh equation controlling the velocity distribution in the force-free limit. The paper is organized as follows. In section 2, the fractional modified Telegraph and the fractional Rayleigh equations are derived from the fractional Klein-Kramer equation and by using the Laplace transform with the Mittag-Leffler function method, we obtain the analytical solutions of the two fractional equations. Finally, the predicted traveling pulse solutions for each case at different values of fractional order are illustrated in section 3.

2. The Klein-Kramer equation and its approximations

From the Langevin equation (2), one can derive the corresponding fluctuation-averaged phase space dynamics governed by the Klein-Kramer equation, which describes the Brownian motion of particles with the presence of an external force $F(z)$

$$\frac{\partial W(z,v,t)}{\partial t} = \left[-v\frac{\partial}{\partial z} + \frac{\partial}{\partial v}\left(\eta v - \frac{F(z)}{m}\right) + \frac{\eta k_B T}{m}\frac{\partial^2}{\partial v^2}\right] W(z,v,t), \tag{3}$$

where $W(z,v,t)$ is the distribution function of energetic particles, k_B is Boltzmann constant, and T is the absolute temperature. The generalization to anomalous diffusion behavior leads to the fractional Klein-Kramer equation

$$D_t^{\alpha} W(z,v,t) = \left[-vD_z^{\beta} + \frac{\partial}{\partial v}\left(\eta v - \frac{F(z)}{m}\right) + \frac{\eta k_B T}{m}\frac{\partial^2}{\partial v^2}\right] W(z,v,t), \tag{4}$$

where D_t^{α} and D_z^{β} are the Caputo fractional derivative operator defined in equation (1) for time and space respectively.

2.1. Analytical solution of the fractional modified Telegraph equation

The first approximation of the fractional Klein-Kramer equation is obtained by using the Davies approach [8] and by integrating equation (4) over both $\int .dv$ and $\int .vdv$. A combination of both resulting equations leads to the fractional modified Telegraph equation with an external force

$$\tau^{\alpha} D_t^{\alpha} D_t^{\alpha} w(z,t) + D_t^{\alpha} w(z,t) = BD_z^{\beta} w(z,t) + AD_z^{\beta} D_z^{\beta} w(z,t), \tag{5}$$

where $w(z,t) = \displaystyle\int_{-\infty}^{\infty} W(z,v,t)\, dv$, $\frac{1}{\eta} = \tau$, $A = \frac{k_B T}{m\eta}$, $B = \frac{-F(z)}{m\eta}$ and F is a constant force.The values of all parameters depend on the comparison with the modified Telegraph equation in Ref. [9] which describes the diffusion of energetic particles for isotropic density case. We introduce a separation ansatz of the form $w(z,t) = \varphi(z)\ T(t)$ together with a choice of initial conditions that is compatible with [9]:

$$T(0) = 1, \quad \partial_t T(0) = 0. \tag{6}$$

We use the Mittag-Leffler function method [10] to solve the spatial part $\varphi(z)$ and Laplace transform to solve $T(t)$. Finally, the general solution of the space time fractional modified Telegraph equation is given by

$$w(z,t) = \sum_{n=0}^{\infty}\sum_{m=0}^{\infty} \frac{(-1)^n \left(\frac{-B\pm\sqrt{B^2+4A\lambda}}{2A}\right)^m z^{m\beta}}{n!\ \Gamma(m\beta+1)} \frac{t^{\alpha n}}{\tau^{\alpha n}} \tag{7}$$
$$\left[\frac{t^\alpha}{\tau^\alpha} H^{1,1}_{1,2}\left[\lambda\tau^{-\alpha}t^{2\alpha}\middle| \begin{matrix}(-n,1)\\(0,1),(-n\alpha-\alpha,2\alpha)\end{matrix}\right] + H^{1,1}_{1,2}\left[\lambda\tau^{-\alpha}t^{2\alpha}\middle| \begin{matrix}(-n,1)\\(0,1),(-n\alpha,2\alpha)\end{matrix}\right]\right],$$

where λ is the separation constant, which we take as minus unity.

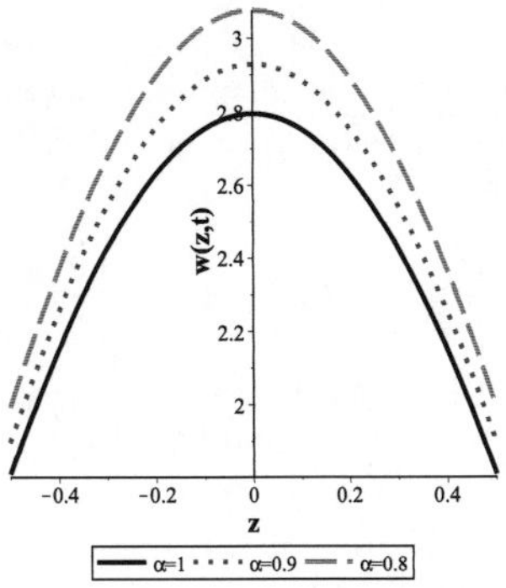

Figure 1. Predicted travelling pulse solutions of the fractional modified Telegraph equation (7) for different values of α at $\beta = 1$.

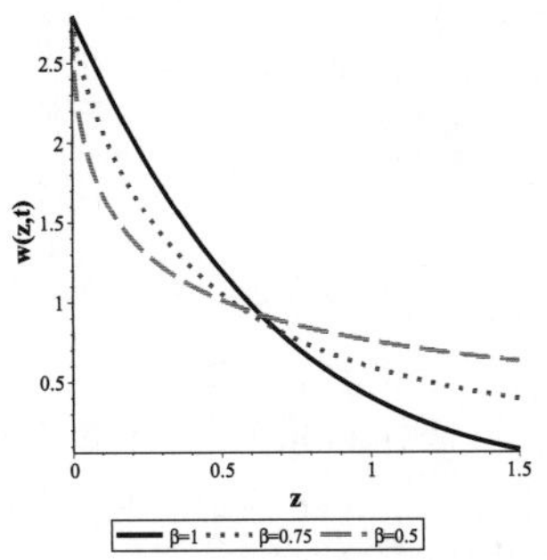

Figure 2. Predicted travelling pulse solutions of the fractional modified Telegraph equation (7) for different values of β at $\alpha = 1$.

2.2. Analytical solution of the fractional Rayleigh equation

To describe the distribution in velocity space averaging out the particle position in the absence of an external force with integration by parts the fractional Klein-Kramer equation (4) will be reduced to the fractional Rayleigh equation

$$D_t^\alpha \psi(v,t) = \eta\left[\frac{\partial}{\partial v}v + A\eta\frac{\partial^2}{\partial v^2}\right]\psi(v,t) \tag{8}$$

where $\psi(v,t) = \int_{-\infty}^{\infty} W(z,v,t)\ dz$. The Rayleigh describes the relaxation of the probability distribution function $\psi(v,t)$ toward the stationary Maxwell distribution [11]. To solve the fractional Rayleigh equation (8) consider the function $\psi(v,t)$ to be separated into two parts $\psi(v,t) = F(v)\ \omega(t)$, with the initial value given by

$$\omega(0) = 1, \quad \partial_t\omega(0) = 0. \tag{9}$$

Then, the general solution of the fractional Rayleigh equation reads:

$$\psi(v,t) = \frac{e^{\frac{v^2}{2A\eta}}}{2} E_{\alpha,1}(-\lambda t^\alpha)\left[H\left(\frac{\lambda}{\eta}, \frac{v}{\sqrt{2A\eta}}\right) + F_1\left(\frac{-\lambda}{2\eta}, \frac{1}{2}, \frac{v^2}{2A\eta}\right)\right], \tag{10}$$

where λ is the separation constant, which we take as minus unity, $E_{\alpha,1}$ is the Mittag-Leffler function [12], H is the Hermite function [13] and F_1 is the Hyper-geometric function [14].

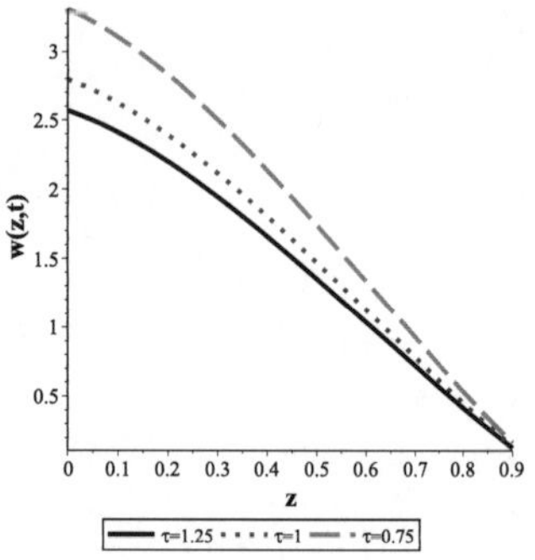

Figure 3. Predicted travelling pulse solutions of the fractional modified Telegraph equation (7) for different values of τ at $\alpha = \beta = 1$.

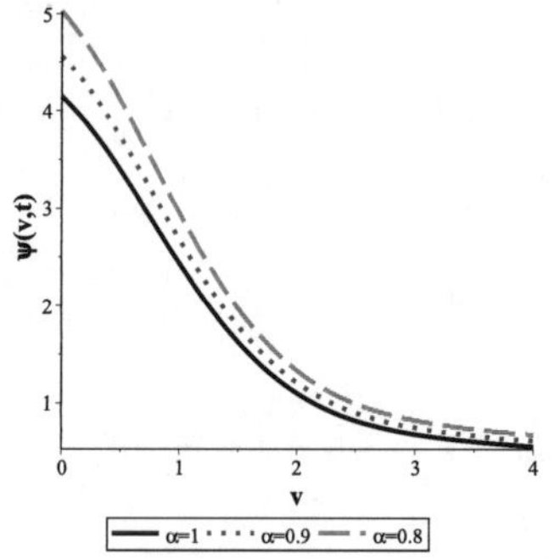

Figure 4. Predicted travelling pulse solutions of the fractional Rayleigh equation (10) for different values of α.

3. Discussion and conclusion

Litvinenko *et al* [9] derived a modified Telegraph equation for isotropic density case, which describes the probability distribution function for the energetic particle transport in the heliosphere. We choose the physical parameters in equation (5) to construct a generalization of the modified telegraph equation to describe the sub-diffusion process of the energetic particles. In figure 1, the solution of the fractional modified telegraph equation illustrates the probability distribution function for both normal diffusion and sub-diffusion processes for different values of time fractional order and attempts to show the accurate prediction on a shorter time scale. Also, the sub-diffusion of the energetic particle is characterized by the non-Gaussian distribution such as in figure 2. Furthermore, in figure 3, we show the change of the probability distribution function for different values of relaxation time. Hence, depending on the parameters, the relaxation time can show either weak or strong z dependence. Finally, in the velocity space, the velocity distribution of the energetic particles in the sub-diffusion limit deviate from the Maxwellian as illustrated in the solution of the fractional Rayleigh equation figure 4.

Acknowledgments

A M Tawfik would like to thank the Egyptian Ministry of Higher Education for supporting his research activities.

References

[1] Crank J 1975 *The mathematics of diffusion* (Oxford University Press Clarendon)
[2] Compte A 1997 *Phys. Rev.* E **55** 6821
[3] Li Y, Farrher G and Kimmich R 2006 *Phys. Rev* E **74** 066309-1
[4] Goychuk I 2012 *Phys. Rev* E **86** 021113-1
[5] Caputo M 1967 *Geophys. J. Roy. Astr. Soc* **13** 529
[6] Metzler R and Klafter 2000 *J. AIP Conference Proceedings* **502** 375
[7] Metzler R and Klafter 2000 *Phys. Rev* E **61** 132
[8] Davies R.W 1954 *Phys. Rev.* E **93** 1169
[9] Litvinenko Y, Effenberger F and Schlickeiser R 2015 *ApJ* **806** 217
[10] Rida S Z and Arafa A A M 2011 *International Journal of Differential Equations* **2011** 814132
[11] Metzler R and Sokolov I M 2002 *Europhys. Lett* **58** 482
[12] Gorenflo R, Kilbas A A, Mainardi F and Rogosin S V 2014 *Mittag-Leffler Functions, Related Topics and Applications* (Springer:Berlin)
[13] Nico T 1996 *Special Functions: An Introduction to the Classical Functions of Mathematical Physics* (Wiley-New York)
[14] Bailey W N G 1935 *Generalized Hypergeometric Series* (Cambridge University Press)

Fission barrier of actinides and superheavy nuclei: effect of pairing interaction

H Abusara

Department of Physics, BirZeit University, Ramallah, Palestine

E-mail: habusara@birzeit.edu

Abstract. Systematic calculations of fission barriers for axial deformation are performed for even-even nuclei in actinides region of the nuclear chart and superheavy nuclei. These calculations were performed using relativistic Hartree-Bogoliubov (RHB) formalism with separable pairing.

1. Introduction

In the last couple of decades the value of of the height of the inner fission barrier in even-even nuclei has gained a great interest by the nuclear physics community due to its importance to several physics phenomena. For example the formation and stability of super heavy nuclei is determined by the width and height of its fission barrier [1, 2].

Fission barrier, has been studied using covariant density functional theory (CDFT) with different pairing schemes. Inner fission barriers in many nuclei have been calculated in the axially symmetric relativistic mean field (RMF) + BCS approach [3–7]. The authors of [8] showed that the usage of constant gap approximation leads to unphysical results for the fission barriers and suggested that treatment of pairing interaction using effective density-dependent zero-range force is far better approximation of pairing. The authors of [9, 10] used Constant pairing strength in the BCS part, and explored the role of triaxilaity.

In this manuscript a systematic calculations of the fission barriers in the regions of actinides and superheavy of the nuclear chart using relativistic Hartree-Bogoliubov (RHB) theory with sparable pairing formalism. The manuscript is organized as follows; The axial RHB theory with separable pairing and its details related to the calculations of fission barriers are discussed in section 2. The results of the calculation of the fission barriers, and the comparison with experimental data are discussed in section 3. Finally, section 4 summarizes the results of this work.

2. Theoretical framework and the details of numerical calculations

The starting point of Covariant Density Functional Theory (CDFT) is a standard Lagrangian density [11],

$$\begin{aligned}\mathcal{L} \quad &= \bar{\psi}\left(\gamma(i\partial - g_\omega\omega - g_\rho\vec{\rho}\vec{\tau} - eA) - m - g_\sigma\sigma\right)\psi \\ &+\tfrac{1}{2}(\partial\sigma)^2 - \tfrac{1}{2}m_\sigma^2\sigma^2 - \tfrac{1}{3}g_2\sigma^3 - \tfrac{1}{4}g_3\sigma^4 - \tfrac{1}{4}\Omega_{\mu\nu}\Omega^{\mu\nu} + \tfrac{1}{2}m_\omega^2\omega^2 \\ &-\tfrac{1}{4}\vec{R}_{\mu\nu}\vec{R}^{\mu\nu} + \tfrac{1}{2}m_\rho^2\vec{\rho}^{\,2} - \tfrac{1}{4}F_{\mu\nu}F^{\mu\nu}\end{aligned} \tag{1}$$

where the nucleus consists of point-like nucleons, described by Dirac spinors ψ, interacting via the exchange of mesons with finite masses and leading to the interactions of finite range. The mesons create effective fields in a Dirac equation, which corresponds to the Kohn-Sham equation [12] in the non-relativistic case.

The Lagrangian (1) contains as parameters the meson masses m_σ, m_ω, and m_ρ and the coupling constants g_σ, g_ω, and g_ρ and e is the charge of the protons. The calculations are performed with the NL3* parameterization [13].

$$\begin{aligned}\hat{H}(r) &= \sum_{i=1}^{A} \psi_i(r)^\dagger[\alpha p + \beta m]\psi_i(r) + \frac{1}{2}[(\nabla\sigma)^2 + m_\sigma^2\sigma^2 + \frac{2}{3}g_2\sigma^3 + \frac{1}{2}g_3\sigma^4] - \frac{1}{2}[(\nabla\omega)^2 \\ &+ m_\omega^2\omega^2] - \frac{1}{2}[(\nabla\rho)^2 + m_\rho^2\rho^2] + [g_\sigma \rho_s \sigma + g_\omega \gamma_\mu \omega^\mu + g_\rho \vec{j}_\mu \cdot \vec{\rho}_\mu + e j_{p\mu} A^\mu] - \frac{1}{2}[(\nabla A)^2]\end{aligned} \quad (2)$$

Equation (2) represents the Hamiltonian of the nucleus in terms of the mass of the nucleons, mass and the field of the mesons, and the nucleonic density. For details on the theoretical formalism, see [14].

The RHB energy density functional E_{RHB} is given by:

$$E_{RHB}[\hat{\rho}, \hat{k}] = E_{RMF}[\hat{\rho}] + E_{pair}[\hat{k}] \quad (3)$$

$E_{RMF}[\hat{\rho}]$ is given by:

$$E_{RMF}[\psi, \bar{\psi}, \sigma, \omega^\mu, \vec{\rho^\mu}, A^\mu] - \int d^3r II \quad (4)$$

and the $E_{pair}[\hat{k}]$ is given by:

$$E_{pair}[\hat{k}] = \frac{1}{4}\sum_{n_1 n_1'}\sum_{n_2 n_2'} k^*_{n_1 n_1'} < n_1 n_1' | V^{PP} | n_2 n_2' > k_{n_2 n_2'} \quad (5)$$

$< n_1 n_1' | V^{PP} | n_2 n_2' >$ are the matrix elements of the two-body pairing interaction.

$$\begin{aligned} V^{PP}(r_1, r_2, r_1', r_2') &= -G\delta(R - R')P(r)P(r') \\ R &= \frac{1}{\sqrt{2}}(r_1 + r_2) \qquad r = \frac{1}{\sqrt{2}}(r_1 - r_2) \end{aligned} \quad (6)$$

$$P(r) = \frac{1}{(4\pi a^2)^{3/2}} e^{-r^2/2a^2} \quad (7)$$

The calculations are performed by imposing constraints on the axial mass quadrupole moment. The method of quadratic constraints uses a variation of the function

$$\left\langle \hat{H} \right\rangle + C_{20}(\langle \hat{Q}_{20} \rangle - q_{20})^2 \quad (8)$$

where $\left\langle \hat{H} \right\rangle$ is the total energy and $\langle \hat{Q}_{20} \rangle$ denotes the expectation value of the mass quadrupole operator

$$\hat{Q}_{20} = 2z^2 - x^2 - y^2 \quad (9)$$

In these equations, q_{20} is the constrained value of the multipole moment, and C_{20} the corresponding stiffness constant [15].

IOP Conf. Series: Journal of Physics: Conf. Series **869** (2017) 012051 doi:10.1088/1742-6596/869/1/012051

3. Results and discussion

3.1. Actinides

A systematic calculations has been performed for even-even nuclei in the actinides region of the nuclear chart, where experimental data exists, allowing only axial deformation. It has been shown in [9] that triaxiality lowers the height of the barrier by $1-4$ MeV, and that the average deviation from experimental data, was around 1 MeV. The axial calculations provide an upper limit on the height of the inner fission barrier. If one looks at the results shown in figure 7 of [9] and extract the values of the inner fission barrier for the axial results, the average deviation between theoretical and experimental results can then be computed. The average deviation would increase by an extra 1.5 MeV and becomes around 2.5 MeV. The average deviation

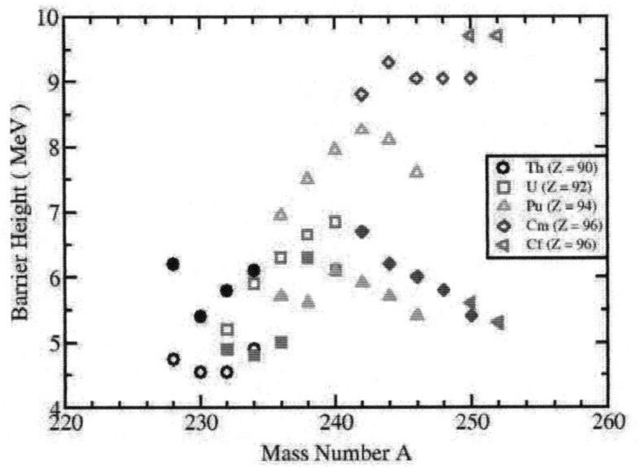

Figure 1. (Color online) Open symbols are the theoretical values of the height of the fission barrier, while filled symbols are the experimental values taken from [16].

between the theoretical and experimental data shown in figure 1 is 1.8 MeV, which is around 30% better than what was obtained in [9]. This result points that the usage of RHB theory with separable paring formalism enhances the description of the fission barrier.

It can be noticed that with the exception of Th isotopes, our calculations over estimate the experimental values of the fission barrier. Our values of the Th isotopes barrier heights is around 4.5 MeV and the experimental data are around 5.5 MeV. However, the trend in the change in the barrier heights as a function of the mass number is similar in both cases. For Pu, Cm and Cf isotopes, our values of the barrier heights is shifted by a constant value from the experimental data. We expect that the inclusion of triaxiality in the calculations would bring our results closer to the experimental data. The major difference comes in the U isotopes, the experimental values are constant in the beginning of the chain then there is a sudden jump. Our results show a steady increases along the chain.

3.2. Superheavy

As mentioned in Sec.3.1 separable pairing perform better than BCS approximation in the pairing channel. It was shown in [10] that triaxiality doesn't affect the height of the inner fission barrier in superheavy region, Z=112-120.

In figure 2, the height of fission barrier for the superheavy nuclei is plotted as a function of the mass number. For $Z = 112$, one can see that the height of the barrier decreases with the number of neutrons until it reaches a minimum value of 3.5 MeV at $N = 168$, then it starts to increase to a maximum value of 4.5 MeV. However, this trend is different from the one for $Z =$

114 and 116 isotopes, where the barrier height increases along the isotopic chain. It starts from 4 MeV up to 5 MeV for Z = 114, and from 4 to 5.5 MeV in Z = 116 For Z = 118, the results

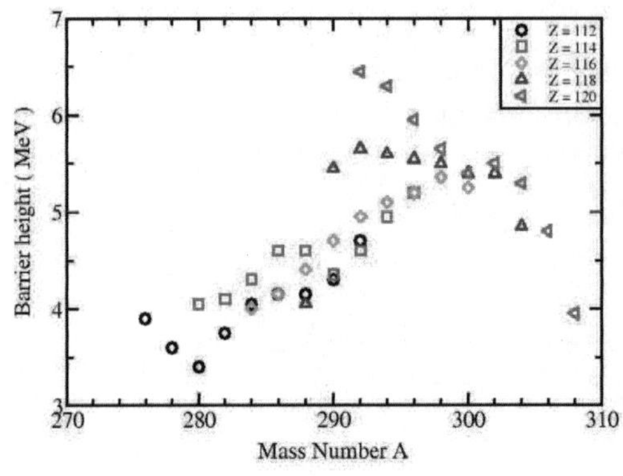

Figure 2. (Color online) Height of inner fission barrier for superheavy nuclei.

show a constant value of the fission barrier, and then it starts to decrease around A = 300. In Z = 120 chain, the highest value of the barrier is obtained for A = 292, which corresponds to N = 172. $^{292}120$ (292 is the mass number and 120 is the proton number) is known to be spherically doubly magic, and thus it is expected to have the highest fission barriers among the nuclei in that region. The difference between the largest and smallest value is around 3 MeV.

4. Conclusion

Systematic calculation of the fission barrier in two regions; actinides and superheavy, using RHB theory with separable pairing formalism has been performed. It is been demonstrated, when only axial deformation is allowed, that RHB theory with separable pairing perform better than RMF+BCS. The average deviation of the height of the fission barrier is reduced by 30% as compared with the results obtained in [9]. The height of the inner fission barrier in superheavy nuclei was also calculated.

References

[1] Sobiczewski A and Pomorski K 2007 *Prog. Part. Nucl. Phys.* **58** 292.
[2] Itkis M G, Oganessian Y T, and Zagrebaev V I 2002 *Phys. Rev.* C **65** 044602.
[3] Blum V, Maruhn J, Reinhard P-G and Greiner W 1994 *Phys. Lett.* B **323** 262.
[4] Rutz K, Maruhn J A, Reinhard P-G and Greiner W 1995 *Nucl. Phys.* A **590** 680.
[5] Ren Z, Tai F and Chen D-H 2002 *Phys. Rev.* C **66** 064306.
[6] Zhang W, Zhang S-S, Zhang S-Q and Meng J 2003 *Chin. Phys. Lett.* **20** 1694.
[7] Long W H, Van Giai N and J. Meng J 2006 *Phys. Lett.* B **640** 150(2006).
[8] Karatzikos S, Afanasjev A V, Lalazissis G A and Ring P 2010 *Phys. Lett.* B **689** 72.
[9] Abusara H, Afanasjev A V and Ring P 2010 *Phys. Rev.* C **82** 044303
[10] Abusara H, Afanasjev A V and Ring P 2012 *Phys. Rev.* C **85** 024314
[11] Gambhir Y K, Ring P and Thimet A 1990 *Ann. Phys. (N.Y.)* **198** 132.
[12] Kohn W and Sham L J 1965 *Phys. Rev.* A **137** 1697.
[13] Lalazissis G A, Karatzikos S, Fossion R, Peña Arteaga D, Afanasjev A V and Ring P 2009 *Phys. Lett.* B **671** 36.
[14] Niksic T, Paar N, Vretenar D and Ring P 2014 *Comp. Phys. Comm.* **185** 1808
[15] Ring P and Schuck P 1980 *The Nuclear Many-Body Problem* (Springer-Verlag, Berlin).
[16] Samyn M, Goriely S and Pearson J M 2005 *Phys. Rev.* C **72** 044316.

New interpretation of data of the Earth's solid core

H H Guliyev

Department of Tectonophysics and Geomechanics, Institute of Geology and Geophysics of Azerbaijan National Academy of Sciences, Ave. H. Javid 119, Baku, AZ 1143, Azerbaijan
E-mail: hatamguliyev@gmail.com

Abstract. The commonly accepted scientific opinions on the inner core as the deformable solid globe are based on the solution of the problem on the distribution of elastic parameters in the inner structures of the Earth. The given solution is obtained within the necessary integral conditions on its self-weight, moment of inertia concerning the axes of rotation and periods of free oscillations of the Earth. It is shown that this solution does not satisfy the mechanics of the deformable solid body with sufficient local conditions following from basic principles concerning the strength, stability and actuality of velocities of propagation of elastic waves. The violation of local conditions shows that the inner core cannot exist in the form of the deformable solid body within the commonly accepted elastic parameters.

1. Introduction

The results of geomechanical analysis of geophysical data are given within the non-classical linearized approach (NLA). At the same time numerical data of Preliminary Reference Earth Model (PREM) [1] are used considering the fact that the parameters of the inner core provided in this paper taken as a basis in all pre-proposed theoretical models of the medium [2].

The purpose of geomechanical analysis is to determine the conditions for pressure and deformation that ensures the accuracy of calculations of physical and mechanical parameters of the model of the Earth's solid core on the basis of geophysical dataset.

2. Achievements of theoretical limit of strength

Let's consider the case when the medium is evenly and uniformly deformed prior to the beginning of fracture. In this case, if the pressure value does not exceed the theoretical limit of strength, all calculations on the physico-mechanical parameters are correct. It is shown in the NLA [3] that the value of the theoretical limit of the strength of the medium is determined from $P = \mu$ under the conditions of compression (we are interested in this variant of deformation) for a perfect elastic isotropic material. This is found from the condition of the loss of ellipticity of the basic equation of motion [3].

3. Instability of equilibrium state

The uniform distribution of deformation in the medium can also be violated as a result of the instability (in various forms) without fracture.

NLA allows defining the limits of change of deformation within the framework of which the equilibrium of uniformly deformed states is stable [4, 5, 6]. Deformation in the globe is nonuniformly distributed long before the limits of strength of material were reached in case of realization of instability. Stability of solid globe is considered to concretize the discussion [5]. It is considered that: a) elastic isotropic body with potential of harmonic type within the theory of large initial deformations and conditions of stability are obtained in the form

$$0<\lambda_1<1;\ \left(\lambda+\frac{2}{3}\mu\right)\left(\lambda+\frac{4}{3}\mu\right)^{-1}<\lambda_1<1; \tag{1}$$

b) elastic body with quadratic potential within the theory of the second variant of small initial deformations and conditions of stability are obtained in the form

$$(2-\lambda_1)\left(\lambda+\frac{2}{3}\mu\right)>0;\ \mu+3(\lambda_1-1)\left(\lambda+\frac{2}{3}\mu\right)>0; \tag{2}$$

c) elasto-plastic body (deformation theory) within the theory of the second variant of small initial deformations and condition of stability are obtained in the form

$$P<\mu; \tag{3}$$

d) elasto-plastic body (Prandtl-Reuss theory of plasticity) within the theory of the second variant of small initial deformations is obtained in the form

$$P<\lambda_1^2\mu. \tag{4}$$

It is shown for all the considered models of the medium that the equilibrium state is stable in case of setting follower loads on the surface of an isotropic globe under the fulfillment of conditions (1)-(4).

The stability of body in the form of the globe is considered in case of influence of "dead" loads on its surface. It is shown that there is a critical load $P_{cr.}$ (according to the value this load is less than the value μ) in reaching of which uniform deformed state of equilibrium of the globe is unstable. The critical forces and deformation leading to the loss of stability of the equilibrium state of the globe are determined within the theory of the second variant of small initial deformations

$$P_{cr.}=\frac{\mu}{4(1-2\nu)}\left(5-4\nu-\left(16\nu^2-8\nu+9\right)^{\frac{1}{2}}\right). \tag{5}$$

In case of the theory of large initial deformations and the use of harmonic elastic potential, critical value of shortening is defined in the form

$$(\lambda_1)_*=\frac{(3-2\nu)(1+\nu)}{(3-2\nu)(1+\nu)+(1-2\nu)}. \tag{6}$$

In equations (1)-(6), λ and μ are Lame's moduli, ν is Poisson's ratio and P is pressure parameter.

4. Internal instability

Critical values of stress and deformations leading to violation of conditions of ellipticity of basic equations cause the phenomenon in the body which is called the "internal" instability in the theory [5, 6].

In case of modelling the deformation process using harmonic elastic potential within the theory of large initial deformations, the limit value of elongation (shortening) coefficient λ_1^* is defined in the form

$$\lambda_1^*=\frac{1+\nu}{2-\nu},\quad \varepsilon_0^*=\frac{3}{2}\frac{2\nu-1}{(2-\nu)^2}, \tag{7}$$

and in the following form in case of quadratic elastic potential within the theory of large initial deformations

$$\lambda_1^*=\left(\frac{1+\nu}{2-\nu}\right)^{\frac{1}{2}},\quad \varepsilon_0^*=\frac{1}{2}\frac{2\nu-1}{2-\nu}. \tag{8}$$

In case of linear elastic isotropic material within the theory of the second variant of small initial deformations, we get

$$P_{cr.} = \mu, \qquad \varepsilon_0^* = \frac{1}{2}\frac{2\nu - 1}{1+\nu}. \tag{9}$$

ε_0 is a parameter of overall deformation in equations (7)-(9). It follows from the aforementioned equations (1)-(4) and (7)-(9) that the "internal" instability occurs within the NLA in overall uniform deformation (compression) of isotropic globe at a level of pressure comparable in value with shear moduli.

5. Elastic wave propagation in the deformed medium

Such problems are intensively studied in [4, 6,7]. The fulfilment of condition of ellipticity of basic equations also provides an actuality (not negative value) of velocities of propagation of small perturbations.

In case of overall uniform pre-compression of isotropic medium, "true" velocities of propagation of elastic waves in it are defined by expressions [6]

$$\rho C_l^2 = \lambda + 2\mu - PK_p^R; \qquad \rho C_S^2 = \mu - PK_S^R. \tag{10}$$

Where C_l, C_S are "true" velocities of quasi-pressure and quasi-shear elastic waves; K_p^R, K_S^R are coefficients of nonlinear action of isotropic medium [7].

In cases of the theory of the second variant of small and large initial deformations, pressure elastic wave with true velocity cannot be propagated in the quadratic elastic potential under the fulfillment of condition

$$\frac{P}{\mu} \geq \frac{2(1-\nu^2)}{(1-2\nu)(3-\nu)}; \qquad \frac{P}{\mu} \geq \frac{2(1-\nu^2)}{(1-2\nu)(5-3\nu)} \tag{11}$$

in the stressed isotropic medium. This condition takes the following form for shear elastic waves

$$\frac{P}{\mu} \geq \frac{1+\nu}{2-\nu}; \qquad \frac{P}{\mu} \geq \frac{1+\nu}{3(1-\nu)}. \tag{12}$$

6. Numerical results and discussions

The calculated values of the critical forces and elongations leading to the loss of stability of the equilibrium state of the deformed globe are given in table 1. The values of P^*/μ are calculated according to equation (9). "Internal" instability is realized in reaching these values of surface loads in the globe in case of small initial deformations.

Table 1. Critical parameters of instability.

ν	**0**	**0.1**	**0.2**	**0.3**	**0.4**	**0.41**	**0.45**
P^*/μ	1	1	1	1	1	1	1
P_*/μ	0.5	0.53	0.57	0.60	0.64	0.64	0.65
λ_1^*	0.5	0.58	0.67	0.76	0.88	0.89	0.94
$(\lambda_1)_*$	0.75	0.79	0.84	0.89	0.94	0.94	0.97

The values of λ_1^* calculated according to the equation (7) shows that similar process also occurs in the case of large initial deformations. The values of P_*/μ and $(\lambda_1)_*$ which causes the loss of stability in the globe on the geometric form change are calculated according to equations (5) and (6) respectively.

Comparisons of the values P^*/μ with P_*/μ and λ_1^* with $(\lambda_1)_*$ shows that the process of loss of stability of the globe on the geometric form change precedes the process of the "internal" instability.

Calculations are also performed according to the equations (11) and (12). Elastic pressure waves with true velocity cannot propagate in the globe in reaching the numerical values of P_l/μ given in table 2. Similar results for P_S/μ are also obtained in the case of shear elastic waves. The data obtained for small initial deformations are given in the numerators and the data obtained for large initial deformations are given in the denominators.

Table 2. Critical values P_l/μ and P_S/μ for different Poisson's ratio.

ν	**0**	**0.1**	**0.2**	**0.3**	**0.4**	**0.41**	**0.45**
P_l/μ	$\frac{0.6667}{0.4}$	$\frac{0.8534}{0.5266}$	$\frac{1.1429}{0.7273}$	$\frac{1.6852}{1.1098}$	$\frac{3.2308}{2.2105}$	$\frac{3.5689}{2.4518}$	$\frac{6.2549}{4.3699}$
P_S/μ	$\frac{0.5}{0.3333}$	$\frac{0.5789}{0.4074}$	$\frac{0.6667}{0.5}$	$\frac{0.7647}{0.6190}$	$\frac{0.875}{0.7778}$	$\frac{0.8868}{0.7966}$	$\frac{0.9355}{0.8788}$

Contrary to the results of table 2, data given in PREM [1] shows that elastic waves propagate in the solid core of the Earth even under higher baric conditions.

7. Conclusions

The establishment of features of deformation process in the baric conditions of the internal solid core related to reaching the ultimate strength of the medium, forms of instability of the isotropic globe and the impossibility of propagation of elastic waves in the medium is new and not taken into account in the existing theoretical models of the Earth [1, 2].

References

[1] Dziewonski A M and Anderson D L 1981 Preliminary reference Earth model *Physics of the Earth and Planetary Interiors* **25 (4)** pp 297–356

[2] Sumita I and Bergman M I 2007 Inner-Core Dynamics *In Book: Treatise on Geophysics, 1st Edition (Editor-in-Chief: Gerald Schubert)*, **8** pp 299–318

[3] Kuliev G G 1988 *Strength of Materials* **20 (5)** 623–629

[4] Akbarov S D 2015 *Dynamics of pre-strained bi-material elastic systems: Linearized three-dimensional approach* (Switzerland: Springer) p 1004

[5] Guz A N 1986 *Fundamentals of three-dimensional theory of stability of deformable bodies* (Kyiv: Vishcha shkola Head Publishing House) p 511

[6] Guz A N 1986 *Elastic waves in bodies with initial stresses, propagation patterns* (2nd Volume) (Kyiv: Naukova Dumka) p 536

[7] Guliyev H H 2009 Nonlinear actions of elastic medium and their effect on the propagation velocity of elastic waves *Proceedings of the NAS of Azerbaijan Earth Sciences* **2** pp 31–39.

IOP Conf. Series: Journal of Physics: Conf. Series **869** (2017) 012053 doi:10.1088/1742-6596/869/1/012053

Up- and down-quark contributions to the nucleon electromagnetic form factors at low Q^2

I A Qattan[1] and J Arrington[2]

[1] Department of Physics, Khalifa University of Science and Technology, P.O. Box 127788, Abu Dhabi, United Arab Emirates

[2] Physics Division, Argonne National Laboratory, Argonne, Illinois, 60439, USA

E-mail: issam.qattan@kustar.ac.ae

Abstract. The spatial distribution of charge and magnetization within the nucleon (proton and neutron) is encoded in the elastic electromagnetic form factors $G_E^{(p,n)}$ and $G_M^{(p,n)}$. These form factors have been precisely measured utilizing elastic electron scattering, and the combination of proton and neutron form factors allows for the separation of the up- and down-quark contributions to the nucleon form factors. We expand on our original analyses and extract the up- and down-quark contributions to the nucleon electromagnetic form factors from worldwide data with an emphasis on precise new data covering the low-momentum region, which is sensitive to the large-scale structure of the nucleon. From these, we construct the flavor-separated Dirac and Pauli form factors and their ratios, and compare the results to recent extractions and theoretical calculations and models.

1. Introduction

The nucleon electromagnetic form factors $G_E^{(p,n)}(Q^2)$ and $G_M^{(p,n)}(Q^2)$ are fundamental quantities in nuclear and elementary particle physics as they provide information on the spatial distributions of charge and magnetization within the nucleon. They are a function of the four-momentum transfer squared of the virtual photon, Q^2. In the nonrelativistic limit, they are the Fourier transform of the charge and magnetization distributions. Therefore, isolating the up- and down-quark contributions to the nucleon form factors is essential to examine spatial asymmetries in the quark distributions.

There are primarily two methods used to extract the proton form factors. The first is the Rosenbluth or longitudinal-transverse (LT) separation method [1] which uses measurements of unpolarized cross section, and the second is the polarization transfer/polarized target (PT) method [2] which requires measurements of the spin-dependent cross section. In the one-photon exchange (OPE) approximation or the Born value, the unpolarized cross section is proportional to the "reduced" cross section, $\sigma_R = G_M^2 + (\varepsilon/\tau)G_E^2$, where $\tau = Q^2/4M_N^2$, M_N is the nucleon mass, and ε is the virtual photon longitudinal polarization parameter defined as $\varepsilon^{-1} = [1+2(1+\tau)\tan^2(\theta_e/2)]$, where θ_e is the scattering angle of the electron. Measuring σ_R at several ε points for a fixed Q^2 value, one can separate G_E^p and G_M^p. However, for cases where ε/τ is extremely small (large), it is difficult to extract G_E^p (G_M^p) with precision. On the other hand, the polarization measurements are sensitive only to the ratio G_E^p/G_M^p. Therefore, by taking ratios of polarization components, many of the systematic uncertainties in the polarization measurements cancel, allowing for precise measurements of the ratio $R_p = \mu_p G_E^p/G_M^p$ [3], where

μ_p is the proton magnetic dipole moment. The two methods yield strikingly different results for the ratio R_p in the region $Q^2 > 2$ (GeV/c)^2 [4], where the Rosenbluth extractions show approximate scaling, $\mu_p G_E^p / G_M^p \approx 1$, while the recoil polarization data indicate a nearly linear decrease in R_p with Q^2 with some hint of flattening out above 5 (GeV/c)^2. Recent studies suggest that hard two-photon exchange (TPE) corrections to the unpolarized cross section may resolve the discrepancy [5–9].

2. Extraction of the flavor-separated form factors

Recent precise measurements of the neutron's electric to magnetic form factor ratio $R_n = \mu_n G_E^n / G_M^n$ up to 3.4 (GeV/c)^2 [10], combined with existing $R_p = \mu_p G_E^p / G_M^p$ measurements in the same Q^2 range allowed for a separation of the up- and down-quark contributions to the nucleon form factors at large Q^2 [11]. This first analysis, referred to as "CJRW" in this work, examined the scaling behavior of the up- and down-quark contributions at large Q^2, supporting the idea that diquark correlations play an important role [12]. Recently, we extended the flavor separation analysis [13–15] using combined cross section and polarization measurements of elastic electron-proton scattering with an emphasis on precise new data from Ref. [16] covering the low-momentum region, which is sensitive to the large-scale structure of the nucleon. In our work, we account for effects neglected in the original work where we apply TPE corrections in the extraction of the proton form factors based on the approach of Ref. [17]. The TPE correction applied in our work, based on the parametrization from Ref. [18], is linear in ε [19] and vanishes in the limit $\varepsilon \to 1$ [20, 21]. We also compare our results to a parametrization of the proton form factors extracted [5] after applying the hadronic calculation for TPE from Ref. [22]. We also include additional new G_M^n data from CLAS [23] and performed a new global fit to G_M^n which we used, along with the parametrization of R_n [10], to construct G_E^n, as well as account for the uncertainties associated with all of the form factors measurements as the CJRW analysis accounted only for uncertainties on R_n which was the dominant uncertainty for their flavor-separated results. Finally, we use our results of the flavor-separated form factors to construct the flavor-separated Dirac, $F_1^{(u,d)}$, and Pauli, $F_2^{(u,d)}$, form factors and their ratios.

3. Results and discussion

In this section, we present our results of the flavor-separated form factors $F_1^{(u,d)}$ and $F_2^{(u,d)}$. We then compare our results to the CJRW extractions which allows for examination of the effect of the TPE corrections applied, additional uncertainties, as well as updated form factor data set used. We also compare our results to the Venkat *et al.* [24] ("VAMZ"), and Arrington *et al.* [5] ("AMT") extractions which use improved proton form factor parametrization obtained assuming different treatment of TPE corrections at lower Q^2 values. In addition, we used the Venkat plus the G_M^n and R_n fits mentioned above, and looked at the impact of our updated G_M^n fit by replacing this with the Kelly [25] fit for G_M^n ("VAMZ-Kelly"). Finally, we compare the results to recent theoretical calculations and fits to the flavor-separated form factors: a Dyson-Schwinger equation ("DSE") calculation [12], a pion-cloud relativistic constituent quark model ("PC-RCQM") [26], a relativistic constituent quark model whose hyperfine interaction is derived from Goldstone-boson exchange ("GBE-RCQM") [27], and a generalized parton distribution (GPD) calculations [28].

Figure 1 shows the flavor-separated contributions of F_1 and F_2. It was reported in the CJRW analysis that while both the up-quark form factors, F_1^u and F_2^u, continued to rise compared to the down-quark, the down-quark contributions, F_1^d and F_2^d, strikingly exhibited $1/Q^4$ scaling behavior above $Q^2 = 1.0$ (GeV/c)^2 in agreement with the moments of the generalized parton distributions predictions [29]. However, these predictions are based on fits which included nucleon form factors data, except for the most recent G_E^n data. Our results for F_1^u and F_2^u suggest as well that both form factors continue to rise with increasing Q^2 but with F_2^u values

IOP Conf. Series: Journal of Physics: Conf. Series **869** (2017) 012053 doi:10.1088/1742-6596/869/1/012053

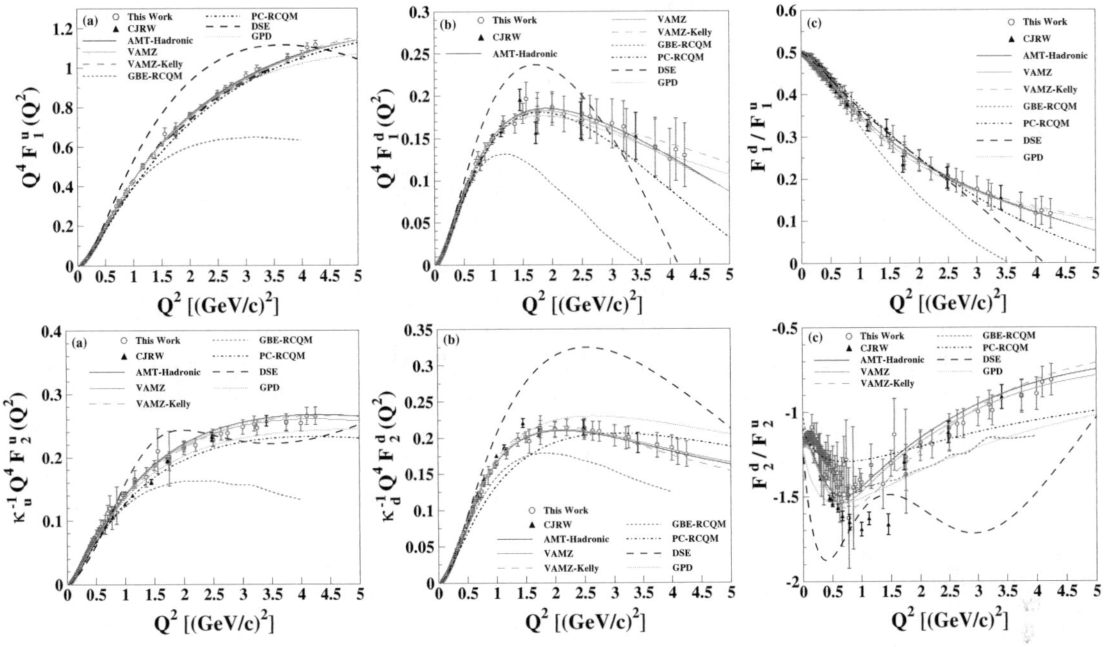

Figure 1. The flavor-separated form factors $F_1^{(u,d)}$ (top) and $F_2^{(u,d)}$ (bottom) and their ratios from our analysis [15] and the CJRW extractions [11]. Also shown are the AMT [5] and VAMZ fits [24], and the values from the GBE-RCQM [27], PC-RCQM [26], the DSE [12], and the GPD [28] models.

somewhat higher than the CJRW extractions at low Q^2. This can be seen clearly in the ratio F_2^d/F_2^u. On the other hand, our results suggest that the down-quark contributions are falling more rapidly than the up-quark contributions at high Q^2 with respect to the $1/Q^4$ behavior which is more apparent for F_2^d and the ratio F_2^d/F_2^u in agreement with global parametrizations and theoretical calculations which predict faster falloff than the apparent $1/Q^4$ scaling behavior. However, such behavior is sensitive to the parametrization of G_E^n at $Q^2 > 2$ (GeV/c)^2 [10]. The difference in the results obtained are attributed mainly to the TPE corrections applied to the proton form factors and to lesser extent to the use of the updated G_M^n parametrization. In addition, while the G_E^n uncertainties have the largest impact, the additional uncertainties from the proton and G_M^n yield a non-negligible increase in the total uncertainties.

Figure 2 shows the ratios $\kappa_u^{-1}F_2^u/F_1^u$ and $\kappa_d^{-1}F_2^d/F_1^d$. Here $\kappa_{u,d}$ are the $Q^2 = 0$ limits of $F_2^{u,d}$ with $\kappa_u = \mu_u - 2 = 1.67$ and $\kappa_d = \mu_d - 1 = -2.03$. The ratios are scaled by $\kappa_{u,d}^{-1}$ and normalized to $1/F_1^{u,d}$ yielding 0.5(1) for the up(down)-quark contribution. At low Q^2, the ratio F_2^u/F_1^u falls rapidly up to $Q^2 \simeq 1$ (GeV/c)^2 where it starts to decrease slowly. While our values are somewhat larger than those obtained in the CJRW extractions for $Q^2 < 1.5$ (GeV/c)^2 due to the difference in the F_2^u values, they are in good qualitative agreement with form factor fits and calculations. The ratio $\kappa_d^{-1}F_2^d/F_1^d$ increases slowly at low Q^2 and becomes almost constant for $Q^2 > 1$ (GeV/c)^2. Our results are consistent with both the CJRW results and extractions based on form factors fits. However, our results differ from all calculations as they predict a rapid falloff of F_1^d at large Q^2 compared to F_2^d which clearly leads to a rapid rise in the ratio. The high-Q^2 measurements of G_E^n planned at Jefferson Lab after the 12 GeV upgrade [30] are clearly critical to pin down the behavior of this ratio as well as to examine the theoretical calculations and models.

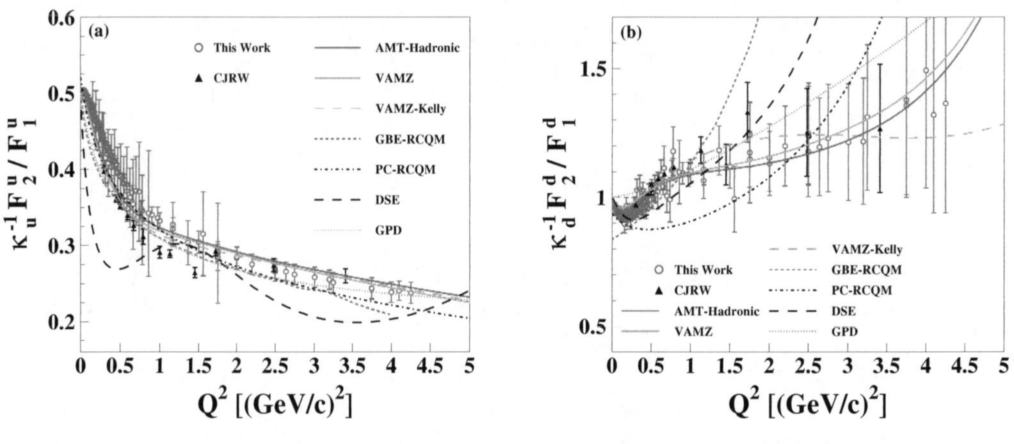

Figure 2. The ratios $\kappa_u^{-1}F_2^u/F_1^u$ and $\kappa_d^{-1}F_2^d/F_1^d$. Points and curves as in Fig. 1.

Acknowledgments

This work was supported by Khalifa University of Science and Technology and by the U.S. Department of Energy, Office of Nuclear Physics, under contract DE-AC02-06CH11357.

References

[1] Rosenbluth M N 1950 *Phys. Rev.* **79** 615
[2] Dombey N 1969 *Rev. Mod. Phys.* **41** 236
[3] Punjabi V *et al.* 2005 *Phys. Rev.* **C71** 055202; Puckett A J R *et al.* 2012 *Phys. Rev.* **C85** 045203; Puckett A J R *et al.* 2010 *Phys. Rev. Lett.* **104** 242301; Zhan X *et al.* 2011 *Phys. Lett.* **B705** 59; Ron G *et al.* 2011 *Phys. Rev.* **C84** 055204; Arrington J, Perdrisat C, and de Jager C W 2011 *J. Phys. Conf. Ser.* **299** 012002
[4] Qattan I A *et al.* 2005 *Phys. Rev. Lett.* **94** 142301; Arrington J 2003 *Phys. Rev.* **C68** 034325
[5] Arrington J, Melnitchouk W, and Tjon J A 2007 *Phys. Rev.* **C76** 035205
[6] Carlson C E and Vanderhaeghen M 2007 *Ann. Rev. Nucl. Part. Sci.* **57** 171
[7] Arrington J, Blunden P, and Melnitchouk W 2011 *Prog. Part. Nucl. Phys.* **66** 782
[8] Qattan I A and Alsaad A 2011 *Phys. Rev.* **C84** 054317 [Erratum: 2011 *Phys. Rev.* **C84** 029905]
[9] Adikaram D *et al.* (CLAS Collaboration) 2015 *Phys. Rev. Lett.* **114** 062003; Rachek I A *et al.* 2015 *Phys. Rev. Lett.* **114** 062005; Rimal D *et al.* (CLAS Collaboration) 2016 *Preprint* arXiv:1603.00315 [nucl-ex]
[10] Riordan S *et al.* 2010 *Phys. Rev. Lett.* **105** 262302
[11] Cates G, de Jager C, Riordan S, and Wojtsekhowski B 2011 *Phys. Rev. Lett.* **106** 252003
[12] Cloet I *et al.* 2009 *Few-Body Syst.* **46** 1
[13] Qattan I A and Arrington J 2012 *Phys. Rev.* **C86** 065210
[14] Qattan I A and Arrington J 2014 *Eur. Phys. J. WOC* **66** 06020
[15] Qattan I A, Arrington J, and Alsaad A 2015 *Phys. Rev.* **C91** 065203
[16] Bernauer J C *et al.* (A1 Collaboration) 2014 *Phys. Rev.* **C90** 015206
[17] Qattan I A, Alsaad A, and Arrington J 2011 *Phys. Rev.* **C84** 054317
[18] Borisyuk D and Kobushkin A 2011 *Phys. Rev.* **D83** 057501
[19] Tvaskis V *et al.* 2006 *Phys. Rev.* **C73** 025206
[20] Arrington J, Blunden P G, and Melnitchouk W 2011 *Prog. Part. Nucl. Phys.* **66** 782; Chen Y -C, Kao C -W, and Yang S -N 2007 *Phys. Lett.* **B652** 269
[21] Arrington J 2004 *Phys. Rev.* **C69** 032201
[22] Blunden P G, Melnitchouk W, and Tjon J A 2005 *Phys. Rev.* **C72** 034612
[23] Lachniet J *et al.* (CLAS Collaboration) 2009 *Phys. Rev. Lett.* **102** 192001
[24] Venkat S, Arrington J, Miller G A, and Zhan X 2011 *Phys. Rev.* **C83** 015203
[25] Kelly J J 2004 *Phys. Rev.* **C70** 068202
[26] Cloet I C and Miller G A 2012 *Phys. Rev.* **C86** 015208
[27] Rohrmoser M, Choi K -S, and Plessas W 2011 *Preprint* arXiv:1110.3665 [hep-ph]
[28] Gonzalez-Hernandez J O, Liuti S, Goldstein G R, and Kathuria K 2013 *Phys. Rev.* **C88** 065206
[29] Diehl M, Feldmann T, Jakob R, and Kroll P 2005 *Eur. Phys. J.* **C39** 1
[30] Dudek J *et al.* 2012 *Eur. Phys. J.* **A48** 187

IOP Conf. Series: Journal of Physics: Conf. Series **869** (2017) 012054 doi:10.1088/1742-6596/869/1/012054

The optimum cut-off radius in Monte Carlo simulation of Yukawa potential point particles

I Suwan[1], H Hussein[1], A Hussein[2] and M Daragmeh[1]
[1] Department of Mathematics and Statistics, The Arab American University, Jenin, Palestine
[2] Department of Physics, The Arab American University, Jenin, Palestine

E-mail: iyad.suwan@aauj.edu

Abstract. In this paper, infinite systems of point particles with Yukawa potential and periodic boundary conditions are simulated using Monte Carlo technique in three dimensions. Because of the short range nature of the Yukawa potential, cut-off radius r_{cut} is considered in calculations (i.e, for each particle *i*, the effect of the other particles on it inside a sphere of radius r_{cut} is taken into account). The cut-off radius used in Monte Carlo simulation affects the physical behavior of the system being simulated. A sequence of r_{cut} values are used. When the change in the total potential energy becomes negligible, the optimum value of the cut-off radius is determined. This value is found to be independent of density and temperature in the *NVT*-ensemble case.

1. Introduction

Computer simulations of three-dimensional Yukawa fluids play a great role in many applications in Physics and Engineering. The short range nuclear forces produced from a meson interaction theory were proposed by Hideki Yukawa in 1935 [1]. According to this theory, each nucleon emits and reabsorbs pions in continuous behavior. The emitted pion may interact with new nucleon instead of returning to its parent nucleon. The resulting momentum, due to the shifting pions, is equivalent to the action of the nuclear forces. These forces are repulsive at very short ranges and attractive at large distances between nucleons. The static Yukawa potential (also called a screened Coulomb potential) is a spherically symmetric solution of the Klein-Gordon equation that takes the form [1]

$$V(r) = g\frac{e^{-\lambda r}}{r}, \tag{1}$$

where r is the distance between particles, λ and g are constants.

The Yukawa potential can be used to model the dynamical interactions between particles in plasmas and atomic physics [2]. The Monte Carlo simulation determined the phase diagram of a system consisting of hard core attractive Yukawa potential [3]. The fluid-solid phase transition was studied, and a transition curve was obtained for Yukawa systems near the one component plasma limit [4]. The location of the phase equilibrium and interfacial properties of two-dimensional Yukawa fluids were determined [5]. The linear and quadratic static density response functions of three-dimensional Yukawa liquid were computed by applying an external potential in molecular dynamics simulation [6]. The two-nucleon potential from the cut-off Yukawa theory was studied, and the two-nucleon interaction was

IOP Conf. Series: Journal of Physics: Conf. Series **869** (2017) 012054 doi:10.1088/1742-6596/869/1/012054

calculated up to the fourth order [7]. The melting line of the Yukawa system was located by determining the free energy of both fluid and solid phases by computer simulations [8]. In fact, many researches have been arising in the last and in the present centuries in the field of systems with Yukawa potential. The Yukawa potential is a short range one; hence, a cut-off radius of interaction can be used in simulations. The main goal of this research is obtaining the lowest cut-off radius that gives the total potential energy of the system. In order to achieve this purpose, Monte Carlo simulations are performed for the system at different values of temperature and density in an *NVT*-ensemble.

2. The cut-off radius

In the course of Monte Carlo simulation, the total potential energy is calculated before and after each Monte Carlo step. A straightforward simplification is truncating the interaction potential beyond a cut-off radius. The total Yukawa potential energy in this case is approximated by

$$U(r_{ij}) = \sum_{i<j} \bar{V}(r_{ij}), \tag{2}$$

where

$$\bar{V}(r) = \begin{Bmatrix} V(r), & r \le r_{cut} \\ 0, & r > r_{cut} \end{Bmatrix}. \tag{3}$$

The accuracy of this approximation depends on the value of r_{cut}. In the case of Yukawa potential, r_{cut} can be typically chosen large enough such that the perturbation introduced into the simulation by the use of $\bar{V}(r)$ instead of $V(r)$ is negligible.

The potential truncation is generally possible if the particles effectively interact via short range pair potential. An important factor for the performance of Monte Carlo simulation code is its scalability with the number of particles in the system. Calculating all atomic interactions in the system during simulation is practically impossible. But, in systems with Yukawa potential, the inter atomic forces decreases rapidly by increasing the distance between particles; hence, considering the interactions within r_{cut} is reasonable. The optimal choice of r_{cut} is accomplished according to the fact that the difference between energy values approaches *zero* by increasing r_{cut}.

3. Numerical results

The simulation of Yukawa systems is performed by implementing periodic boundary conditions. The cubic box is replicated to form an infinite lattice. When a particle moves in the original box, its periodic images in other boxes move in the same way. In this work, ***N*** point-particles in the original box are simulated at fixed temperature ***T*** and fixed volume ***V***. The number of particles used in calculations which is 125 is traditionally sufficient to get the desired results. The system under consideration is simulated in a cubic box using different values of r_{cut}, and the average value of the potential energy is calculated. For this purpose, the potential function (1) is reduced to

$$V(r^*_{ij}) = \frac{e^{-ar^*_{ij}}}{r^*_{ij}}, \tag{4}$$

where $a = g\lambda$, and $r^*_{ij} = \frac{r_{ij}}{g}$. Theoretically, using the reduced parameter $a = 1$ in simulations will give enough information about the optimum reduced value of the cut-off radius r^*_{cut}. The simulation is done at different values of temperature T^* and density ρ^* in the ranges [0.5, 3] and [0.25, 2], respectively. The convergence of Yukawa potential as a function of Monte Carlo sweep is tested. The reduced cut-off radius r^*_{cut} associated with large convergence speed and high accuracy is found. For example, figure 1 shows the average energy of the simulated system for $T^* = 2.5$ at different values of ρ^* and r^*_{cut}. The

IOP Conf. Series: Journal of Physics: Conf. Series **869** (2017) 012054 doi:10.1088/1742-6596/869/1/012054

main result that can be clearly noticed from figure 1 is that r^*_{cut} affects the average energy of the simulated system. By increasing the value of r^*_{cut} , the accuracy of the average potential energy becomes better until reaching a specific limit. Using cut-off radius greater than $r^*_{cut} = 3.5$ will give no significant change in the potential energy.

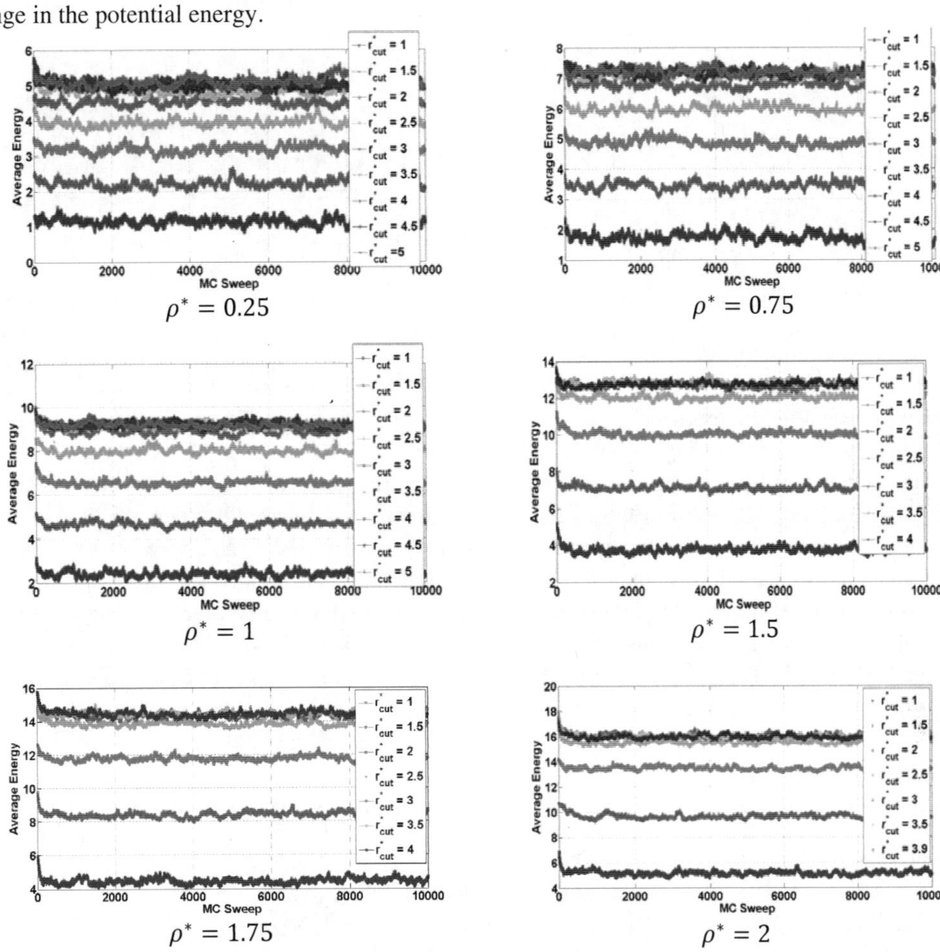

Figure 1. Average energy Versus Monte-Carlo sweeps for $T^* = 2.5$ at different values of ρ^* and r^*_{cut}.

The above procedure is repeated at many different values of ρ^* and T^*. In all cases, no remarkable change in the average potential energy happens at values of r^*_{cut} above 3.5. In addition, the fitting tool in MATLAB is used in order to obtain an equation that approximates the average energy in terms of density and Yukawa potential. The determination coefficient obtained is 0.9893. This approximation is formulated by

$$E(\rho^*, V) = 3.689 + 4.367\rho^* - 24.83V + 1.026(\rho^*)^2 - 11.58\rho^*V + 48.07V^2 \qquad (5)$$

At $T^* = 0.5$ and $T^* = 1.5$, figure 2 shows the value of the reduced average energy difference $E_{r^i_{cut}} = U_{r^i_{cut}} - U_{r^{i-1}_{cut}}$, ($r^i_{cut} > r^{i-1}_{cut}$). The behavior shows that by increasing the cut-off radius, the difference in the total potential energy using any two successive values of r^*_{cut} approaches *zero*. In both cases, $r^*_{cut} = 3.5$ gives very accurate average energy.

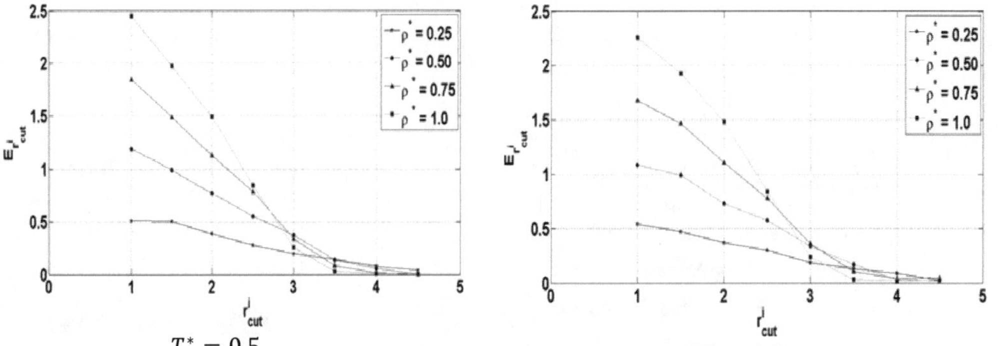

$T^* = 0.5$ $T^* = 1.5$

Figure 2. Energy difference at successive values of r_{cut}^* for $T^* = 0.5$ and 1.5 at different values of ρ^*.

4. Conclusions

One of the main goals of statistical physics is understanding the structure of liquids and gases from given inter particle interaction potentials. Major obstacles in this field are the limited range and the finite accuracy in measuring any observation. The Monte Carlo program used involves the calculation of the potential energy of a particular configuration. In principle, calculating the potential at any particle must include interactions of infinite number of terms which is not applicable in practice. Hence, in the course of simulation, an approximation using cut-off radius is used. The convergence of Monte Carlo simulation of *N*-particle system to the equilibrium state of Yukawa potential energy is affected by the cut-off radius. Increasing this radius increases the accuracy but causes a very slow convergence. The results of simulation at different values of temperature and density show that the reduced cut-off radius greater than 3.5 will not improve the accuracy. At all choices of temperature and density, the value 3.5 of the cut-off radius is optimum and using it in simulation makes it remarkably efficient.

Acknowledgments

This work was supported by the faculty of scientific research in the Arab American University, Jenin.

References

[1] Yukawa H 1935 *Proc. Phys. Math. Soc. Japan* **17** 48
[2] Jiao Li G and Ho N K 2013 *Int. J. Quantum Chem.* **113** 2569
[3] Hagen M H J and Frenkel D 1994 *J. Chem .Phys.* **101** 4093
[4] Farouki R Tand Hamaguchi S 1994 *J. Chem. Phys.* **101** 9885
[5] Mendez-Maldonado G A, Gonzalez-Melchor M and Alejandre J 2012 *Condens. Matter Phys.* **15** 23002
[6] Magyar P, Donko Z, Kalman G J and Golden K I 2014 *Phys. Rev.* E **90** 023102
[7] Gartenhaus S 1955 *Phys. Rev.* **100** 900
[8] Meijer E J and Frenkel D 1991 *J.Chem. Phys.* **94** 2269

DNA compaction by poly (amido amine) dendrimers of ammonia cored and ethylene diamine cored

K Qamhieh[1] and J Al-Shawwa[1]

[1] Physics Department, College of science and Technology, Al-quds University, Jerusalem, Palestine.
Email: khawlaq@gmail.com

Abstract. The complexes build–up of DNA and soft particles poly amidoamine (PAMAM) dendrimers of ammonia cored of generations (G1-G6) and ethylenediamine cored of generations (G1-G10) have been studied, using a new theoretical model developed by *Qamhieh* and coworkers. The model describes the interaction between linear polyelectrolyte (LPE) chain and ion-penetrable spheres. Many factors affecting LPE/dendrimer complex have been investigated such as dendrimer generation, the Bjerrum length, salt concentration, and rigidity of the LPE chain represented by the persistence length. It is found that the wrapping chain length around dendrimer increases by increasing dendrimer`s generation, Bjerrum length, and salt concentration, while decreases by increasing the persistence length of the LPE chain. Also we can conclude that the wrapping length of LPE chain around ethylenediamine cored dendrimers is larger than its length around ammonia cored dendrimers.

1. Introduction

Gene therapy derives its name from the idea that genetic materials can be used to supplement or alter gene expression within a specific cell. The challenge here is to transport the DNA through the cell walls, which can be achieved by condensing DNA with a cationic particles, such as poly (amido amine) (PAMAM) dendrimers. Dendrimers are branched polymeric molecules, which have a specific size and shape, affording a narrow weight distribution. With its unique molecular architecture and related properties, dendrimers have a great potential as delivery vectors for gene transfection. Transfection efficiency and functionality of DNA complexed with dendrimers depend on the structure, size, and the charge density of these polymers. The properties of the complexes were studied by various experimental and theoretical modeling [1-5]. The objective of our work is to study the complexation of DNA with dendrimers of different generations for ammonia cored dendrimers and for ethylinediamine cored dendrimers. Through our study different factors have been investigated, like salt concentration of the solution, persistence length (l_p) represents the flexibility of LPE chain, dendrimer generation, and the strength of electrostatic interaction on the complexation represented by Bjerum lengh (l_B).

2. Analytical model

A theoretical model for the complexation of one sphere with one LPE chain, developed by Qamhieh et al. was adopted [5]. The complex was considered to be formed between soft (ion penetrable) sphere of radius R and charge Ze represents the dendrimer, where Z is the valency and e is the elementary charge, and the DNA as a LPE chain of l_p persistence length, radius $r = 1$ nm, length $L >> R$ and the charge density of $\lambda = -e / b$, where b is the axial spacing between negative charges on the chain. The sphere and the LPE chain are placed in 1:1 salt solution in a system characterized by Bjerrum length $l_B = e^2/\varepsilon k_B T$, where ε is the dielectric constant of the medium, k_B is Boltzmann constant, T is the thermal energy, and Debye screening lengthis $\kappa^{-1} = (8c_s\pi l_B)^{-1/2}$ where c_s is the salt concentration of the solution. The total free energy of a system consisting of one dendrimer and one LPE chain is given as the following:

$$F(l) = F_{compl}(l) + F_{chain}(L-l) + F_{compl-chain}(l) + F_{elastic}(l) \quad (1)$$

where l is the length of the LPE chain wrapped around the sphere, and the remaining chain is of length $(L - l)$. The first term is the electrostatic charging free energy of a spherical complex of charge $Z(l)e$ given by:

$$F_{compl}(l) \cong \frac{3Z^2(l) l_B k_B T}{8\pi(\kappa R)^2}\left[\cosh(\kappa R) - \frac{\sinh(\kappa R)}{\kappa R}\right]\frac{e^{-\kappa R}}{R} \quad (2)$$

$Z(l) = (Z - l/b)$ represents the charge of the complex consists of the sphere and the corresponding wrapped chain. The second term is the total entropic electrostatic free energy of the remaining chain $(L-l)$ given by:

$$F_{chain}(L-l) \cong \frac{k_B T}{b}\cdot\Omega(a)\cdot(L-l)\cdot(1-\xi^{-1}) \quad (3)$$

Where $\Omega(a)$, is the entropic cost describing the condensed DNA counterion, $\xi = l_B / b$ is the so-called Manning parameter [6]. The third term is the electrostatic free energy of the interaction between the complex and the rest of the chain and is given by

$$F_{compl-chain}(l) \cong \frac{3Z(l) l_B k_B T}{4\pi(\kappa R)^2}\left[\cosh(\kappa R) - \frac{\sinh(\kappa R)}{\kappa R}\right] \times \left[\ln(r) - \sum_{n=0}^{\infty}\frac{(-1)^n}{(n+1)!\cdot(n+1)}\cdot(\kappa r)^{n+1}\right]\Bigg|_R^{L-l} \quad (4)$$

The final term in the free energy is the elastic (bending) free energy required to bend l of the chain around a sphere of radius R and is given by

$$F_{elastic}(l) \cong \frac{k_B T l_p}{2R^2} l \quad (5)$$

3. Results and disscution

In order to estimate the optimal wrapping length l_{opt} of the LPE chain around the dendrimer, equation 1 should be solved. The model is applied to obtain a mechanistic understanding of how the complexation of 2000 bp DNA of L = 680 nm, and b = o.17 nm, is affected when it is combined with one dendrimer of ammonia cored (G1- G6), or methylenediamine cored (G1-G10), in aqueous solutions, at room temperature. The l_{opt} is found to be larger with ethylenediamine cored dendrimers than it is with ammonia cored dendrimers as shown in figures 1-4. The difference is clearer with small generations.

3.1 Effect of dendrimer generation on optimal chain length condensed on PAMAM dendrimer

The complexation of one LPE chain represents the DNA of Bjerum length l_B= 0.71 nm, and a persistence length l_p = 3 nm, with one dendrimer at 1:1 salt concentration corresponding to Debye screening length of 6 nm, have been studied. Figure 1 shows that the optimal length is increased by increasing the generation (size) of the dendrimer, as the higher generation dendrimers have lower curvature and higher surface charge density, as a result they cost less free energy for bending than lower generations.

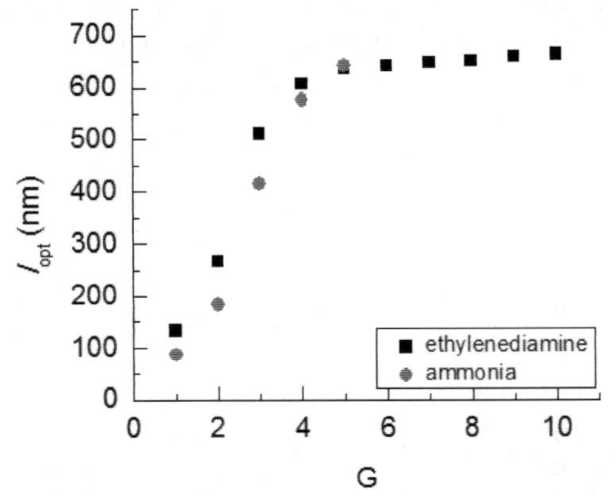

Figure 1. Optimal LPE chain condensed on (a) ammonia cored dendrimer, (b) ethylenediamine cored as a function of dendrimer generation. The persistence length l_p = 3nm, the Bjerum length l_B = 0.71 nm, and the chain length L= 680 nm, at 1:1 salt concentration corresponding to Debye screening length of 6 nm.

3.2 Effect of Bjerrum length on optimal chain length condensed on PAMAM dendrimer

The l_{opt} for complexation of flexible LPE chain of persistence length l_p = 3 nm, with dendrimers is shown as a function of l_B in Figure 2. From the figure we can conclude that the l_{opt} increases as the l_B increases, as long as l_B is less than b, when l_B exceeds the value of b the dendrimer is saturated and the l_{opt} is fixed by increasing l_B. With large generations larger than G5, the whole LPE chain is almost condensed around the dendrimer. At low l_B and less than b the complexation is very sensitive, a small change in l_B gives a large change in optimal length.

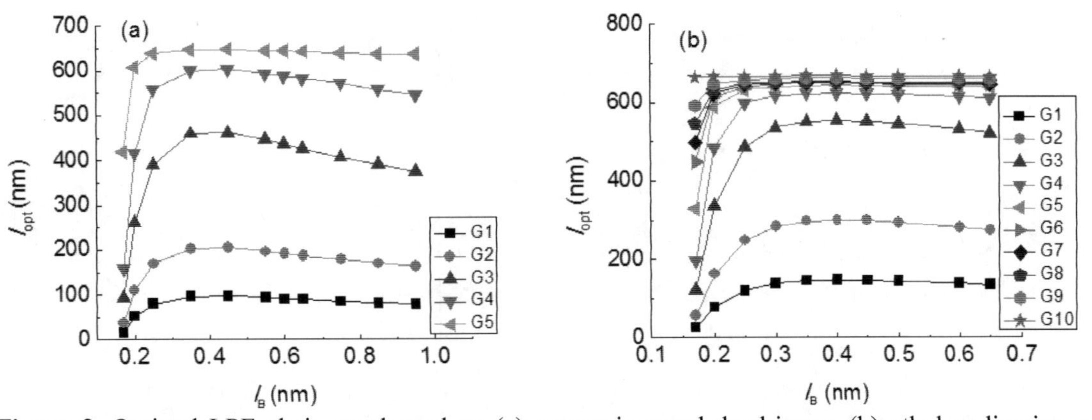

Figure 2. Optimal LPE chain condensed on (a) ammonia cored dendrimers, (b) ethylenediamine cored dendrimers as a function of l_B . The persistence length l_p = 3nm.

3.3 Effect of salt concentration on optimal chain length condensed on PAMAM dendrimer

Figure 3 shows the l_{opt} as a function of 1:1 salt concentration for the complexation of flexible LPE chain and l_B = 0.71 nm, with PAMAM dendrimers. The figure shows that the l_{opt} is increased by increasing the salt concentration with small generations of dendrimers, while with large generations the effect is less and the l_{opt} will not be affected by increasing the salt concentration.

3.4 Effect of Persistence length on optimal chain length condensed on PAMAM dendrimer

Figure 4 shows the optimal chain length as a function of the persistence length for the complexation of LPE chain of Bjerum length l_B = 0.71 nm, and at 1:1 salt concentration corresponding to Debye screening length of 6 nm, with PAMAM dendrimers of different generations. From which we can conclude that the optimal chain length is decreased slightly by increasing the persistence length as the bending energy is decreased by increasing stiffness of the chain.

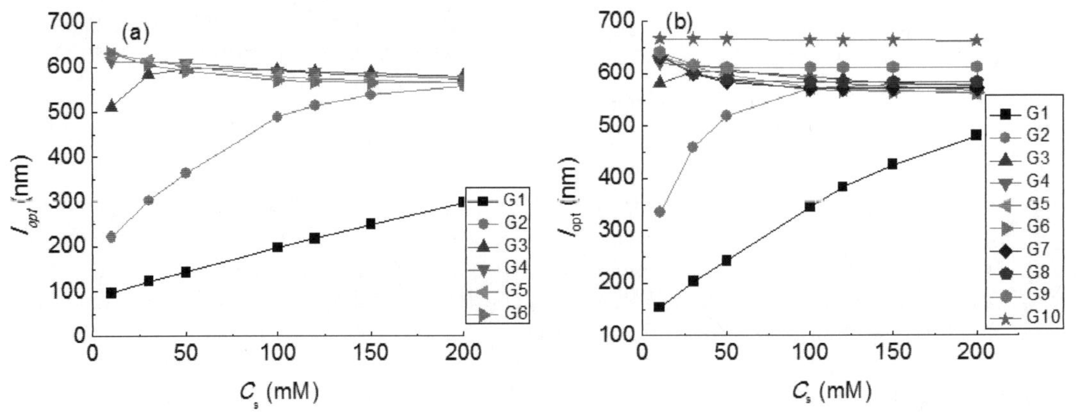

Figure 3. Optimal LPE chain condensed on (a) ammonia cored dendrimers, and (b) ethylenediamine cored dendrimers as a function of 1:1 salt concentration, and l_B = 0.71 nm.

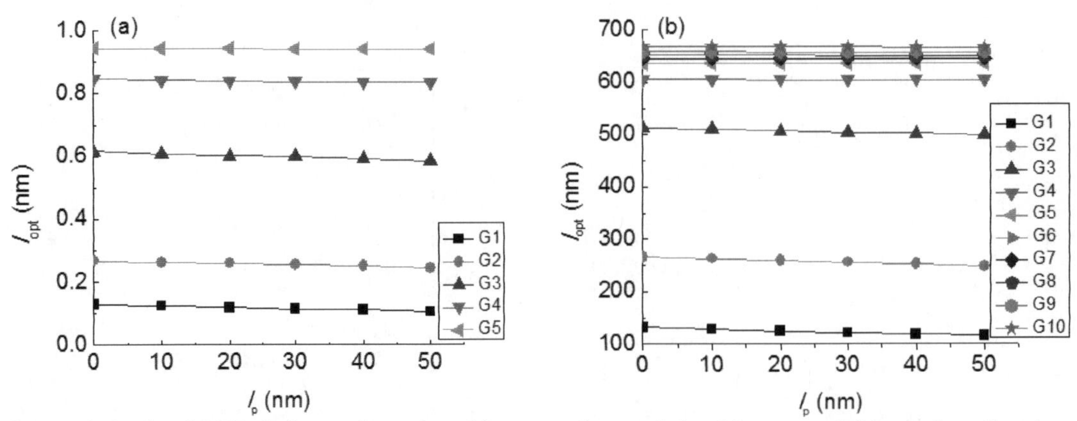

Figure 4. Optimal LPE chain condensed on (a) ammonia cored dendrimers, and (b) ethylenediamine cored dendrimers as a function of persistence length l_p, the Bjerum length l_B = 0.71 nm.

IOP Conf. Series: Journal of Physics: Conf. Series **869** (2017) 012055 doi:10.1088/1742-6596/869/1/012055

4. Conclusions

From the study we can conclude that the complexation of DNA with PAMAM dendrimers can be affected by different factors, by which we can control the compaction of the DNA with dendrimers. The effect is larger on the complexation with ethylenedimamine cored dendrimers than with ammonia cored dendrimers.

References

[1] Ainalem M -L and Nylander T 2011 *Soft Matter* **7** 4577.
[2] Schiessel H, Bruinsma R F and Gelbart W M 2001 *J. Chem. Phys.* **115** 7245.
[3] Qamhieh K, Nylander T and Ainalem M -L 2009 *Biomacro molecules* **10** 1720.
[4] Qamhieh K, Nylander T, Black C F, Attard G S, Dias R S and Ainalem, M -L 2014 *Physical Chemistry Chemical Physics* **16** 13112.
[5] Qamhieh K, and Khaleel A 2014 *Colloids and Surfaces A: Physicochemical and Engineering Aspects* **442** 191.
[6] Manning Q 1978 *Revs. Biophys.* **11** 179.

Frontiers in Theoretical and Applied Physics/UAE 2017 (FTAPS 2017) IOP Publishing
IOP Conf. Series: Journal of Physics: Conf. Series **869** (2017) 012056 doi:10.1088/1742-6596/869/1/012056

Reconstructing the distortion function for non-local gravity

K Y Roobiat[1] and R Pazhouhesh[1]

[1]Physics department, University of Birjand, Birjand, Iran
Email:rpazhouhesh@gmail.com

Abstract. We develop Maggiore and Mancarella non-local model by using new function, $f(m^2\frac{R}{\Box^2})$ and we obtained analytical hyperbolic tangent function. This new model has expansion history exactly as same as ΛCDM (Lambda Cold Dark Matter) with same matter content, but without cosmological constant or dark energy. However, background evolution in our model and ΛCDM are the same, but these models may be distinguishable in structure formation or observation that will contain additional information more than background level.

1. Introduction

Universe expansion was discovered by Edwin Hubble [1] in 1929. Multiple evidences such as cosmic microwave background [2], supernova observations [3] and galaxy distribution [4] lead to the discovery of accelerating expansion of the Universe. The standard interpretation of this acceleration, in the form of a cosmological constant [5] raises some fundamental problems and questions. Why the cosmological constant should be much smaller than any other scale in physics and why it should have the right magnitude have all recently came into question [6]. Modified gravity theories have been also considered for Universe accelerating expansion [7]. For example, $f(R)$ theories [8] which generalize the Einstein-Hilbert Lagrangian from R to $f(R)$ can be used to create a model which exactly reproduces the ΛCDM expansion history just by choosing $f(R) = R - 2\Lambda$ [9]. More generally, $f(R)$ theories can deviate from observations even with no perturbation consideration.

In 2007 Deser & Woodard [10], proposed a new class of modified gravity models, which add some inverse differential operators of the curvature invariants terms to the Einstein-Hilbert action. The simplest choice is the inverse of the d'Alembert operator operator acting on the Ricci scalar, $f\left(\frac{1}{\Box}R\right)$. The function $f\left(\frac{1}{\Box}R\right)$ is dimensionless obviously, therefore this model does not have explicit mass scale in contrast to the non-local models that we will discuss later. In the literature, more activity is dedicated for identifying the form of the function $f\left(\frac{1}{\Box}R\right)$ in order to obtain background evolution of ΛCDM. The final result for, $f(\frac{1}{\Box}R) = a_1[\tanh(a_2Y + a_3Y^2 + a_4Y^3) - 1]$ (where $Y = \frac{1}{\Box}R + a_5$, and $a_1, \dots a_5$ are coefficients obtained by fitting $f(X)$ to the observed expansion history) [11] which does not agree with observations. A good review about this model can be found [12].

Another choice is adding directly a term involving $\left(R\frac{1}{\Box^2}R\right)$ to the Einstein-Hilbert Lagrangian and it will reproduce dark energy dynamically again. This modification has been argued as a consistent infrared modification of General Relativity (GR) theory and also it is ghost–free [13]. This scenario has been greatly studied in the field of cosmological perturbations and other observable imprints [14].

Therefore, we focus in particular on the generalized $R\frac{1}{\Box^2}R$ model, and the paper is organized as follows: We introduce a generalized non-local gravity model in section 2, and in section 3 equations of motion (EOM) are derived. We find distortion function by reconstructing the ΛCDM expansion history in section 4. Finally, in section 5 conclusions and discussion are given.

2. The model

A non-local modified GR model proposed by Maggiore and Mancarella was based on the following action [13]

$$S = \frac{1}{16\pi G}\int d^4x\sqrt{-g}\left[R - \frac{1}{6}R\frac{m^2}{\Box^2}R\right], \tag{1}$$

in order to reproduce the observed amount of dark energy. Where $\frac{1}{\Box}$ is formal inverse of d'Alembertian operator ,$\Box$, in the scalar representation which can be expressed as the convolution with a bi-scalar Green's function $G(x, y)$:

$$\left(\frac{1}{\Box}R\right)(x) \equiv \int d^4y\sqrt{-g(y)}G(x,y)R(y) \quad , \quad \Box_x G(x,y) = \frac{1}{\sqrt{-g(x)}}\delta^4(x-y) \tag{2}$$

This model seems to have advantages to other non-local models that have already been confronted with observations [10-12,15]. The non-local term in equation (1) is controlled by a mass scale, m, in contrast to the non-local models of Deser & Woodard [10] and Barvinsky [16], which does not have such mass scale. The parameter m is approximately of the order of H_0. As a result, this modification will not affect the solar system tests and therefore in small scale, we cannot observe any deviation from the GR. This model will generate dark energy dynamically and by choosing $m \approx 0.28H_0$, we can reproduce the present observed value of cosmological constant [13]. Additional features of $\frac{1}{\Box}$R and $\frac{1}{\Box^2}$R models, is that by choosing t_s () to lie inside the radiation-dominated era one obtains a natural onset for the appearance of dark energy. Indeed, since $R = 0$ during radiation dominated era, the deviation from GR starts in matter domination era and the precise value of t_s (start time of the accelerating expansion of the universe) is not important.

We develop the Maggiore and Mancarella model by replacing $\left(-\frac{1}{6}R\frac{m^2}{\Box^2}R\right)$ term with $f\left(m^2\frac{R}{\Box^2}\right)$ in equation (1). As a consequence, we use the following action

$$S = \int\left[\frac{R}{16\pi G}\left[1 + f\left(m^2\frac{R}{\Box^2}\right)\right] + \mathcal{L}_{matter}\right]\sqrt{-g}d^4x. \tag{3}$$

The free parameter $f(X)$ in equation (3) is known as the nonlocal distortion function and it is function of its dimensionless argument which will be determined by matching with expansion history of ΛCDM model.

3. Equations of motion

In all models field equations are obtained by varying the actions with respect to the metric, $g_{\mu\nu}$. For flat Friedmann–Lemaitre–Robertson–Walker (FLRW) metric, the EOM will take the following forms

$$3H^2 + \Delta G_{tt} = 8\pi G\rho \tag{4}$$

$$-2\dot{H} - 3H^2 + \frac{1}{3a^2}\delta^{ij}\Delta G_{ij} = 8\pi GP \tag{5}$$

where ρ and P are the total energy density and pressure of matter respectively and the non-local terms, ΔG_{tt} & $\frac{1}{3a^2}\delta^{ij}\Delta G_{ij}$, are

$$\begin{aligned}\Delta G_{tt} &= (3H^2 + 3H\partial_t)\left[f\left(m^2\frac{R}{\Box^2}\right) + \frac{1}{\Box^2}\left[Rf'\left(m^2\frac{R}{\Box^2}\right)\right]\right] \\ &+ \frac{1}{2}m^2\left[\partial_t\left(\frac{R}{\Box^2}\right)\partial_t\left(\frac{1}{\Box}\left[Rf'\left(m^2\frac{R}{\Box^2}\right)\right]\right) + \partial_t\left(\frac{R}{\Box}\right)\partial_t\left(\frac{1}{\Box^2}\left[Rf'\left(m^2\frac{R}{\Box^2}\right)\right]\right)\right] \\ &+ \frac{1}{2}m^2\left(\frac{R}{\Box}\right)\left[\frac{1}{\Box}\left[Rf'\left(m^2\frac{R}{\Box^2}\right)\right]\right]\end{aligned} \tag{6}$$

IOP Conf. Series: Journal of Physics: Conf. Series **869** (2017) 012056 doi:10.1088/1742-6596/869/1/012056

$$\frac{1}{3a^2}\delta^{ij}\Delta G_{ij}$$
$$= \left(-2\dot{H} - 3H^2 - \partial_t^2 - 2H\partial_t\right)\left[f\left(m^2\frac{R}{\Box^2}\right) + m^2\frac{1}{\Box^2}\left[Rf'\left(m^2\frac{R}{\Box^2}\right)\right]\right]$$
$$+\frac{1}{2}m^2\left[\partial_t\left(\frac{R}{\Box^2}\right)\partial_t\left(\frac{1}{\Box}\left[Rf'\left(m^2\frac{R}{\Box^2}\right)\right]\right) + \partial_t\left(\frac{R}{\Box}\right)\partial_t\left(\frac{1}{\Box^2}\left[Rf'\left(m^2\frac{R}{\Box^2}\right)\right]\right)\right]$$
$$-\frac{1}{2}m^2\left(\frac{R}{\Box}\right)\left[\frac{1}{\Box}\left[Rf'\left(m^2\frac{R}{\Box^2}\right)\right]\right], \quad (7)$$

where prime denotes the derivative respect to argument of function.

4. Reconstruction function

We will use Woodard and Deffayet method [11] in order to find the reconstruction function. By subtracting equation (5) from equation (4), we obtain following second order differential equation

$$6H^2 + 2\dot{H} + \left\{\left(6H^2 + 2\dot{H} + 5H\partial_t + \partial_t^2\right)\left[f\left(m^2\frac{R}{\Box^2}\right) + m^2\frac{1}{\Box^2}\left[Rf'\left(m^2\frac{R}{\Box^2}\right)\right]\right]\right.$$
$$\left. + m^2\left(\frac{R}{\Box}\right)\left[\frac{1}{\Box}\left[Rf'\left(m^2\frac{R}{\Box^2}\right)\right]\right]\right\} = 8\pi G(\rho - P) \quad (8)$$

Because we want to reproduce ΛCDM cosmology with the same matter content, but without any cosmological constant, therefore the Hubble parameter, H, appearing in equation (8), is a solution of the famous standard Friedmann's equations. We can compute $8\pi G(\rho - P)$ term from Friedmann's equations and it can be plugged in to equation (8)

$$\left(6H^2 + 2\dot{H} + 5H\partial_t + \partial_t^2\right)\left[f(X) + m^2\frac{1}{\Box^2}[Rf'(X)]\right] + m^2\left(\frac{R}{\Box}\right)\left[\frac{1}{\Box}[Rf'(X)]\right] = -6H_0^2\Omega_\Lambda \quad (9)$$

Where, $\Omega_\Lambda = \frac{\Lambda}{3H_0^2}$ and $X = m^2\frac{R}{\Box^2}$.

If we define a new function $Y(t) = \frac{1}{\Box^2}[Rf'(X)]$, we obtain an ODE for $Y(t)$

$$C_6(t)Y^{(6)} + C_5(t)Y^{(5)} + C_4(t)Y^{(4)} + C_3(t)Y^{(3)} + C_2(t)Y^{(2)} + C_1(t)Y^{(1)} + C_0(t) = 0 \quad (10)$$

Where $= \frac{1}{\Box}R$, $Y^{(n)} = \frac{d^n}{dt^n}Y$ and C_n coefficients are function of (H, b, X) which we don't show them here.

For solving equation (10) numerically, we have to calculate first the C_n coefficients. Therefore, we wrote a program in Fortran language for computing C_n coefficients, and then equation (10) can be solved with the Runge-Kutta method. In this step we need to know the initial condition of $Y^{(n)}\big|_{t_0}, n = 1, 2, \ldots, 5$, where t_0 is the initial starting point. We can choose all initial conditions be zero, because $Rf\left(m^2\frac{R}{\Box^2}\right)$ as well as $f\left(m^2\frac{R}{\Box^2}\right)$ must tend to zero in the initial time. After finding $Y^{(1)}(t)$, we can obtain $f(t)$ with the condition $f(t_0) = 0$. Finally, by computing $t(X)$ we can find distortion function $f(X)$. The numerical solution for distortion function $f(X)$ is shown in figure 1. For finding a suitable function that will be fitted with the numerical results of figure 1, we tried to use many functions and finally, we found the following best analytic function

$$f(X) = -0.48999 \times \tanh(1.32489 \times X^{0.8} - 1.24632 \times X + 0.35843 \times X^{1.2}) \quad (11)$$

Moreover, the fitted curve and regular residuals are demonstrated in figure 2.

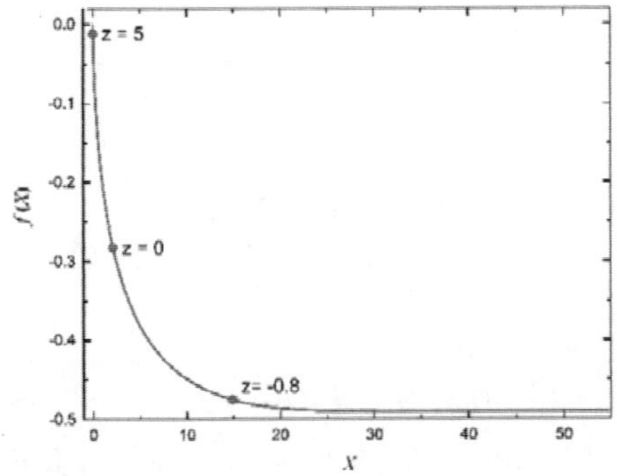

Figure 1. Numerical result for distortion function $f(X)$.

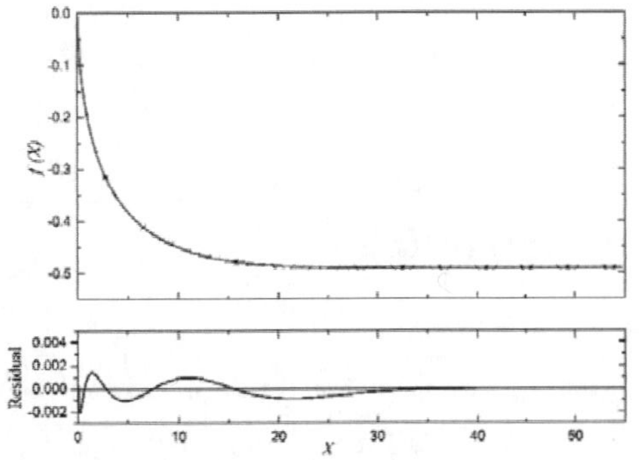

Figure 2. Fitted curve (red line) and regular residuals (blue line) for distortion function $f(X)$.

5. Results and discussion

We chose Maggiore and Mancarella model and develop it in the form of equation (3) to reproduce ΛCDM expansion history. For this purpose, we used Woodard and Deffayet method to find distortion function $f(X)$ numerically, and found the analytic function in the form of equation (11).

This model has expansion history exactly as same as ΛCDM with same matter content, but without cosmological constant or dark energy. However, background evolution in our model and ΛCDM are the same, but these models may be distinguishable in structure formation or observation that will contain additional information more than background level. Derivative of function $f(X)$ at $X = 0$, is infinite. As a result, this term, can make significant corrections to GR theory when it expands around flat space. In equation (4) we have only $Rf'(X)$, however, $f'(X)$ will be very large at $X \approx 0$, but R is very small and consequently $Rf'(X)$ will be finite. It would be extremely interesting to apply those tests for close massive binary, solar system and gravitational lensing.

References

[1] Hubble E 1929 *PNAS* **15** 3
[2] Perlmutter S *et al* 1999 *ApJ* **517** 565
[3] Engelen A V *et al* 2012 *ApJ* **756** 142
[4] Sanchez A G *et al* 2012 *MNRAS* **425** 1
[5] Alam U, Sahni V and Starobinsky A A 2007 *JCAP* **0702** 011
[6] Ratra B and Peebles P J E 2003 *Rev. Mod. Phys.* **75** 559
[7] Carroll S M *et al* 2004 *Phys. Rev.* D **70** 043528
[8] Hu W and Sawicki I 2007 *Phys. Rev.* D **76** 064004
[9] Dunsby P K S *et al* 2010 *Phys. Rev.* D **82** 023519
[10] Deser S and Woodard R P 2007 *Phys. Rev. Lett.* **99** 111301
[11] Deffayet C and Woodard R 2009 *JCAP* **0908** 023
[12] Woodard R P 2014 *Found.Phys.* **44** 213
[13] Maggiore M and Mancarella M 2014 *Phys. Rev.* D **90** 023005
[14] Codello A and Jain R K 2016 [raXiv:1603.00028 [gr-qc]].
[15] Maggiore M 2014 *Phys. Rev.* D **89** 043008
[16] Barvinsky A O 2012 *Phys. Rev.* D **85** 104018
[17] Tsamis N C and Woodard R P 2014 *JCAP* **1409** 008

Van der Waals dispersion energy between atoms and nanoparticles

M Boustimi[1], M Loulou[1], S Natto[1], A Belafhal[2] and J Baudon[3]

[1] Umm Al-Qura University, Physics Dept, PO Box 715, 21955-Makkah, Saudi Arabia
[2]Laboratory of Nuclear, Atomic and Molecular Physics, Department of Physics, Chouaïb Doukkali University, P. B 20, 24000 El Jadida, Morocco
[3] Laboratoire de Physique des Lasers, CNRS-UMR 7538, Université Paris 13,99 av. J.B. Clément, Villetaneuse, France

E-mail: mohamed.boustimi@gmail.com

Abstract. In this work, we focus on the atom-surface interaction where the geometry of the surface is highly symmetric (i.e. sphere, cylinder and plane) and the atom is in ground state. We first present the main features of our model, based on the susceptibility tensors of the two partners in interaction, to determine a general expression of the dispersive energy of van der Waals interaction. Some results are given as applications of this model which addresses recent nanophysical problems, for example, when atoms are in the vicinity of metallic nanoshells, nanospheres or nanowires.

1. Introduction

A great deal of effort has been devoted to the problem of microsystems such as an atom or a molecule interacting, at short or large distance, with the surface of a solid. It has been shown, both theoretically and experimentally, that the vicinity of the surface strongly modifies the dynamical properties of the atomic system, giving rise to a wide variety of 'surface effects', such as van der Waals (vdW) interaction, desorption energy, alteration of radiative properties, etc. More recently, the interest in this field has been renewed by the realization of micro-, meso- and nanostructures [1]. These are of various kinds but, for our purpose, they can be classified into two classes: (i) those leading to a positive curvature of the surface, such as microspheres [2] or tips of near-field microscopes (STM, AFM, etc.) [3–5]; (ii) those in which the curvature of the surface is negative, such as microcavities and porous materials [6–11], fullerenes [9], carbon nanotubes [12]. The latter class deals with the important general problem of the physics of confined atoms or molecules. Theoretical methods have been developed to calculate dispersive and inductive interactions between a microsystem and a macroscopic one treated as a continuous and homogeneous medium. A very concise and general formalism, using generalized susceptibilities [13], is particularly well-suited to treat the problem of ground state atom or molecule in the vicinity of a sphere. In this formalism, the van der Waals energies are readily calculated once the surface field gradient susceptibility tensor, is determined. The paper is organized as follows: in section2 the problem of an atom inside a nanoshell of inner radius a and outer radius b is considered (see figure 1). The response potentials and the propagators inside the nanoshell are used to evaluate the van der Waals energy. The empty cavity is obtained by making the outer radius infinite and the case of an atom in the vicinity of a filled nanosphere is obtained by taking the limit when the inner radius goes to zero. In section3 similar quantities as before, i.e., the response potential, the propagators outside the nanowire of radius a (see figure 2) are introduced to determine the van der Waals for an atom near a metallic nanowire.

2. van der Waals energy between atoms and nanoparticles

In this work, we only present the essentials which enables us to determine the dispersion van der Waals energy (for more details one can see for example Refs.13, 16).

2.1. Atom near a metallic nanoshell

Consider an atom in interaction with a spherical nanoshell of inner radius *a* and external radius *b* (see figure1). The metallic nanoshell is assumed as a continuum medium with dielectric function taking into account the mobility of the electrons. For more realistic description we use a simplified Lindhard dielectric function, in the hydrodynamic model, involving the spatial dispersion effect in the metal [14].

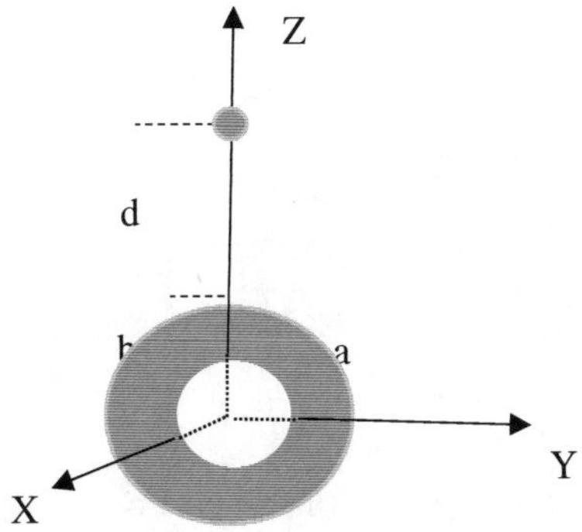

Figure 1. Atom at distance *d* from a nanoshell of inner and outer radii *a* and *b*.

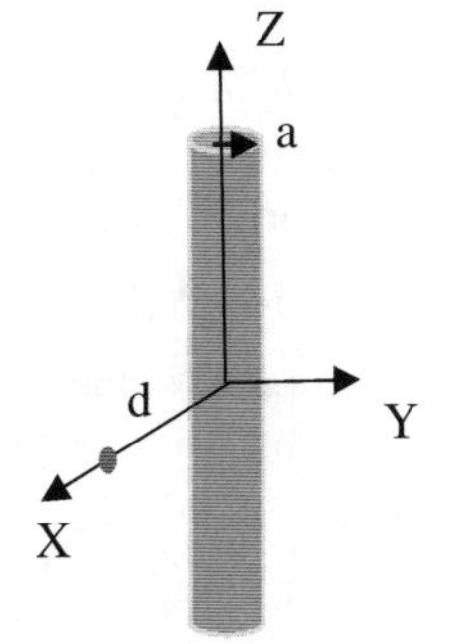

Figure 2. Atom on x axis in the vicinity of a cylindrical nanowire.

At intermediate distances, avoiding both chemisorption and retarded effects, the nanoshell is considered as continuum medium and dispersion van der Waals energy is given by [15]

$$U_d(\vec{R}) = \frac{\hbar}{2\pi}\sum_{m,n=1}\frac{1}{(2m-1)!!(2n-1)!!}\int_0^\infty d\xi \quad {}^{(m)}\chi^{(n)}(\omega)[m+n] \quad {}^{(m)}S^{(n)}(\vec{R},\vec{R},i\xi) \tag{1}$$

${}^{(m)}S^{(n)}(\vec{R},\vec{R},i\xi)$and ${}^{(m)}\chi^{(n)}(\omega)$are respectively the susceptibility tensor of the field gradient of the nanoshell and the multipolar polarizability tensor of the atom at frequency $\omega = i\xi$. The tensors are of order (n+m) and the symbol [m+n] represents their contraction. When one considers only the dipolar contribution, this expression simplifies to

$$U_d(\vec{R}) = \frac{\hbar}{2\pi}\int_0^\infty d\xi \quad \alpha(i\xi)\{S_{xx}(\vec{R},\vec{R},i\xi) + S_{yy}(\vec{R},\vec{R},i\xi) + S_{zz}(\vec{R},\vec{R},i\xi)\} \tag{2}$$

$\alpha(\omega)$is the dynamic atomic polarizability which could be written, following Drude Model [13], as

$$\alpha(\omega) = \frac{\alpha(0)\omega_0^2}{\omega_0^2-\omega^2} \tag{3}$$

where $\alpha(0)$ is the static polarizability, ω_0 the angular frequency of the vibrating electrons and the susceptibility tensor is defined as [16]

$${}^1S^1(\boldsymbol{r},\boldsymbol{R},\omega) = -\sum_{\ell,m} T^1_{\ell m}\frac{a^{2\ell+1}}{r^{2\ell+1}}\Delta_\ell(a,b,\omega)O^1Y_\ell^m(\Omega) \tag{4}$$

$T_{\ell m}$ involves Clebsch-Gordan coefficients, $\Delta_\ell(a,b,\omega)$is the reflection factor of the nanoshell and $O^1$is a tensor operating on the spherical harmonic Y_ℓ^m:

$$O^1Y_\ell^m(\Omega) = -r^{\ell+2}\nabla\left[\frac{1}{r^{\ell+1}}Y_\ell^m(\Omega)\right] \tag{5}$$

The reflection factor of a nanoshell, with radii a<b, is written then

$$\Delta_\ell(a,b,\omega) = \frac{\ell\{aG_\ell(b,\omega)+\ell[F_\ell(a,\omega)G_\ell(b,\omega)-F_\ell(b,\omega)G_\ell(a,\omega)]\}-b[a+\ell F_\ell(a,\omega)]}{(\ell+1)\{aG_\ell(b,\omega)+\ell[F_\ell(a,\omega)G_\ell(b,\omega)-F_\ell(b,\omega)G_\ell(a,\omega)]\}-b[a+\ell F_\ell(a,\omega)]} \tag{6}$$

with $G_\ell(r,\omega) = b^2 \sum_{k,k'} N_{k\ell} N_{k'\ell} j_\ell(kr) j_\ell(k'r)\, \Xi_\ell^{-1}(k,k',\omega)$,
and $F_\ell(r,\omega) = a^2 \sum_{k,k'} N_{k\ell} N_{k'\ell} j_\ell(kr) j_\ell(k'a)\, \Xi_\ell^{-1}(k,k',\omega)$,
Following the work done by C. Girard et al [17], we neglect the off-diagonal elements of the matrix $\Xi_\ell(k,k',\omega)$ which will be simplified and related to Lindhard dielectric function $\varepsilon(k,\omega)$ by

$$\Xi_\ell(k,k',\omega) = k^2 \varepsilon(k,\omega)\delta_{k,k'} \tag{7}$$

The reflection coefficient of a nanosphere is easily deducible from eq.(6) when the radius a goes to zero

$$\Delta_\ell(b,\omega) = \frac{\ell G_\ell(b,\omega) - b}{(\ell+1)G_\ell(b,\omega) + b} \tag{8}$$

By taking into account this, the van der Waals energy reads to

$$U_d(R) = \sum_\ell C_\ell(a,b) \frac{a^{2\ell+1}}{R^{2\ell+1}} \tag{9}$$

where $C_\ell(a,b)$ are called the dispersion coefficients which carry the nonlocal behavior of the metal

$$C_\ell(a,b) = \frac{\hbar}{2\pi}(\ell+1)(2\ell+1)\int_0^\infty d\xi\, \Delta_\ell(a,b,i\xi)\, \alpha(i\xi) \tag{10}$$

2.2. Atom near a metallic nanowire
In a similar manner, except that the cylindrical symmetry involve a new set of eigenmodes, we have a relation describing the dispersive interaction between an atom and a metallic nanowire is

$$U_d(R) = \frac{\hbar}{\pi^2}\alpha(0)\omega_0^2 \sum_{n=1} \int_0^\infty dk_\parallel \left[k_\parallel^2 K'^2_n(k_\parallel R) + \left(k_\parallel^2 + \frac{n^2}{R^2}\right) K_n^2(k_\parallel R)\right] \int_0^\infty d\xi\, \frac{\Delta_n(k_\parallel, a, i\xi)}{\omega_0^2 + \xi^2} \tag{11}$$

Where $k_\parallel$ is the component of the wave vector **k** along the wire z-axis. The reflection factors of the nanowire are given by the following relation [18]

$$\Delta_n(k_\parallel, a, i\xi) = -\frac{I_n(k_\parallel a) - k_\parallel F_n(k_\parallel, a, i\xi) I_n'(k_\parallel a)}{K_n(k_\parallel a) - k_\parallel F_n(k_\parallel, a, i\xi) K_n'(k_\parallel a)} \frac{K_n'(k_\parallel a)}{I_n'(k_\parallel a)} \tag{12}$$

I_n, K_n, I_n' and K_n' respectively are the modified Bessel functions and their derivatives. The functions F_n are given by the following relation [19]

$$F_n(k_\parallel, a, i\xi) = \frac{\xi^2}{\xi^2 + \omega_P^2} \frac{I_n(k_\parallel a)}{k_\parallel I_n'(k_\parallel a)} + \frac{\omega_P^2}{\xi^2 + \omega_P^2} \frac{I_n(\beta a)}{\beta I_n'(\beta a)} \tag{13}$$

These functions contain all the dynamic information about the nanowire.

3. Numerical results and conclusions

Numerical calculations have been made for a Kr atom in the vicinity of a gold nanoshell, a nanosphereand a nanowire. Note that, only in a local treatment, the dispersive energy U_d depends on

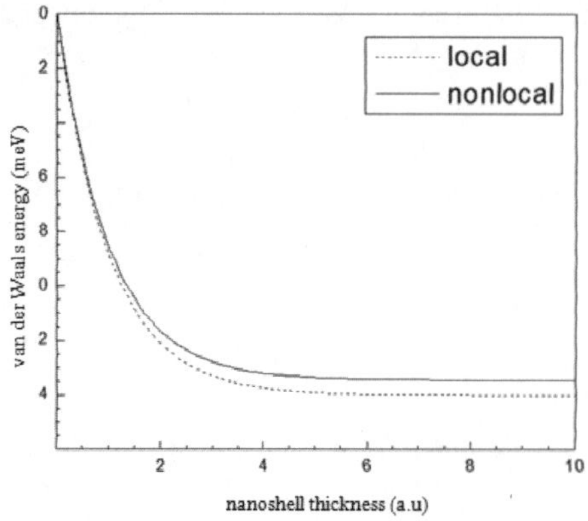

Figure 3. Van der Waals energy, in terms of the thickness, for Kr atom-Au Nanoshell (inner radius=5nm and outer radius=5.5nm). Full line, nonlocal treatment; dotted line, local treatment.

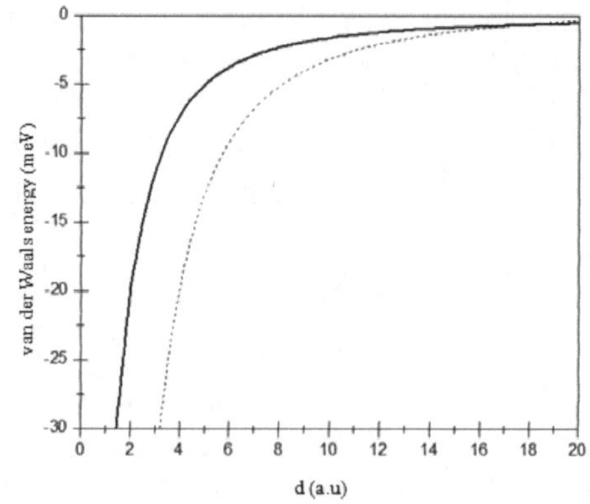

Figure 4. Van der Waals energy in terms of d the approach distance Kr atom-Au nanowire (radius 1.5 nm). Full line, nonlocal treatment; dotted line, local treatment.

IOP Conf. Series: Journal of Physics: Conf. Series **869** (2017) 012057 doi:10.1088/1742-6596/869/1/012057

the distance d as Cd^{-3} (where the coefficient C is geometry dependent). In figure 3, the energy is plotted, when the atom is near a nanoshell as a function of the thickness $e=b-a$, the distance d being fixed at 6 a.u and a difference between local and nonlocal treatments namely for small distances is observed. The asymptotic value of U_d at large e coincides with the energy of the atom in front of an infinite metallic medium limited by a plane while case of a nanosphere is simply deduced by cancelling the radius a from the equations given above. Figure 4 represents the energy values when a Krypton atom is absorbed by a gold nanowire which confirms that the local energy is larger than the nonlocal one.

One can conclude that with this method, using susceptibilities of the two partners in interaction, one gets van der Waals energies (namely the quantum part of energy due to fluctuating charges) between atoms and nanoparticles with spherical or cylindrical symmetry. The method is readily extendable to molecules provided that one adds higher multipolar contributions. Spatial dispersion effect becomes very important when the atom is very close to the particle (few angstroms).

Acknowledgments
M. Boustimi thanks the King Abdulaziz City for Science and Technology for supporting him within the framework of the project number 11-NAN2198-10.

References
[1] Whiters F*et al* 2015 *Nature Mat.* **14** 301
[2] Labani B, Girard C and Maghezi S 1990 *Mol. Phys* **69** 85
[3] Abraham F F, Batra I P and Ciraci S 1988 *Phys. Rev. Lett* **60** 1314
[4] Van Labeke D, Labani B and Girard C 1989 *Chem. Phys. Lett* **162** 399
[5] Hartmann U 1990 *Phys. Rev.* B **42** 1541
[6] An K, Childs J J, Dasari R R and Feld M S 1994 *Phys. Lett* **73** 3375
[7] Kinoshita T, Takahashi Y and Yabuzaki T 1994 *Phys. Rev.* B **49** 3648
[8] Derycke I, Vigneron J P, Lambin Ph, Lucas A A and Derouane E G 1991 *J. Chem. Phys* **94** 4620
[9] Lucas A, Gensterblum G, Pireaux J J, Thiry P A, Caudano R, Vigneron J P, Lambin Ph and Kratschmer W 1992 *Phys. Rev.* B **45** 13694
[10] Klimov V V, Ducloy M and Letokhov V S 1996 *J. Mod. Opt* **43** 549
[11] Klimov V V, Ducloy M and Letokhov V S 1996 *J. Mod. Opt* **43** 2251
[12] Ebbessen T W and Takada T 1995 *Carbon* **33** 973
[13] Girard C, Maghezi S and Hache F 1989 *J.* Chem. *Phys* **91** 5509
[14] Fuchs R and Claro F 1987 *Phys. Rev.* B **35** 3722
[15] Girard C, Girardet C and Silvi B 1988 *Chem. Phys* **125** 261
[16] Boustimi M 1999 *Ph.D. Thesis, El Jadida*, Morocco
[17] Girard C and Hache F 1987 *Chem. Phys* **118** 249
[18] Boustimi M, Baudon J, Candori P and Robert J 2002 *Phys. Rev.* B **65** 155402
[19] Lindhard J and Vidensk K D 1954 *Selsk. Mat. Fys. Medd* **28** 8.

Dependence of η-mode and anomalous transport on entropy drift in magnetoplasma

M Y Khan and F Rasheed

Department of Mathematics, Riphah International University Islamabad Pakistan

Email: myaqubsultani@gmail.com

Abstract. More than 90 years passed when name plasma was coined and more than 60 years in controlling plasma. It seems to be missing something, may be entropy that depends upon maximum constituents of plasma. Inculcating entropy in theoretical and experimental research may lead to the easy confinement. In this paper, Yaqub's work is extended, transport phenomenon is studied with drift approximation inculcating entropy and η-mode depending upon entropy gradient drift is observed. New dispersion relation is derived by using Branginskii transport equations. Dependence of Anomalous transport on entropy is also observed. This work will be helpful in understanding different phenomena of space plasmas and tokamak specifically the pedestal H-mode.

1. Introduction

Although the concept of entropy plays an important role in diverse fields ranging from classical thermodynamics of Celsius, dissipation theorems, statistical thermodynamics of Boltzmann and Gibbs, Stochastic thermodynamics, Quantum mechanics and other fields of science, like plasma [1], there is no single definition of entropy, therefore, an attempt is made to define distribution function of entropy and entropy drift as well [2]. The reason is, the research on plasma is diverging continuously. Though, it is helpful to understand the different aspects of plasma in Tokamak and space plasmas. This trend is enhancing the problem of confining plasma. Therefore, it is better to converge the research to reduce the number of variables to single main variable. For this purpose, entropy can be taken as single variable. Because, maximum variables are connected to entropy. In this way, a minute change in plasma status can be observed on the canvas of entropy. Thus, confinement of plasma may be easier only by controlling entropy. Although, at this stage, it is not a simple task but, in the long run, entropy based devices to confine plasma will be made. Nuclear Fusion is a process in which light nuclei fuse to form a heavier nuclei and release energy.

Thus, Nuclear Fusion process is a large source of energy but, the control of this process is still not achieved. For this purpose, different schemes have been proposed such as, Magnetic Confinement Fusion and Inertial Confinement Fusion schemes etc. In magnetic confinement problems arise due to anomalous transport of particles and energy. In Classical, Neoclassical processes and Anomalous transport [3-6] triggered by Coulomb collisions have been studied extensively. Thus, Anomalous transport is the main problem to the confinement of plasma due to temperature-gradient driven drift modes [7-13]. Entropy fluctuations were observed experimentally first time in 1989 by Zhang et al.[14], two of the three peaks of spectrum were ion acoustic waves propagating oppositely, while, the third peak named as entropy

IOP Conf. Series: Journal of Physics: Conf. Series **869** (2017) 012058 doi:10.1088/1742-6596/869/1/012058

fluctuations at $\omega = 0$. They used Braginskii's model to calculate different coefficients of transport in low frequency, long-wavelength range. Muller et al. [15] investigated the ratio of shear viscosity η to entropy density S and suggested the lower bound of this ratio $\eta / s = (1/4\pi)(\hbar / k_B)$, for a quantum fluid. Kobayashi et al. [16] suggested that entropy mode were able to generate robust, radially inward, particle transport comparable to the level observed in Levitated dipole experiment [17]. Brantov et al. [18] solved kinetic equations and found the entropy mode in entire range of frequencies with the help of Landau electron-electron, electron-ion and ion-ion collision integrals in fully ionized plasma. Therefore, it is important to investigate the role of entropy drift on electron-temperature-gradient driven plasma.

In this paper, entropy drift is added to drift approximation, a new dispersion relation is derived in which η_e mode is observed depending on entropy and anomalous transport is also investigated. This investigation will help to understand the electrostatic turbulence in linear regime and electron energy transport in inhomogeneous tokamak as well as in space plasmas.

2. Basic Model Equations and Dispersion Relation

Taking the gradients of equilibrium quantities along the x-axis, our model equations which consists of electron continuity equation, momentum equation and energy balance equation along with entropy equation can be written as:

$$m_e n_e (\frac{\partial}{\partial t} + \mathbf{v}_e \cdot \nabla)\mathbf{v}_e = -e[\mathbf{E} + \mathbf{v} \times \mathbf{B}] - \nabla P \tag{1}$$

$$\frac{\partial}{\partial t} n_e + \nabla \cdot (\mathbf{n}_e \mathbf{v}_e) = \mathbf{0} \tag{2}$$

$$\left(\partial_t + \tfrac{5}{3} v_B \cdot \nabla\right) T + n_e T_e (\nabla . v_n) = -\nabla . \left[\frac{5 c n_e T_e}{2 e B_0} \bar{z} \times \nabla T_e \right], \tag{3}$$

In the case of low-frequency $\omega \ll \omega_{ce}$ where the cyclotron frequency [13] is $\omega_{ce} (= eB_0 / m_e c$, charge of electron is e, electron mass is m_e and speed of light is c) and using total drift-approximation:

$$\mathbf{v}_e = \mathbf{v}_{EB} + \mathbf{v}_{De} + \mathbf{v}_{pe} + \mathbf{v}_S + v_{ez} \hat{z}, \tag{4}$$

and equations (1)-(3), we get the reduced forms:

$$\left(\partial_t + \mathbf{v}_E \cdot \nabla + \mathbf{v}_B \cdot \nabla + \mathbf{v}_S \cdot \nabla\right) N - \left(\mathbf{v}_B - \mathbf{v}_n\right) \cdot \nabla \Phi + \mathbf{v}_B \cdot \nabla T + \rho_e^2 [\nabla(\mathbf{v}_{D1} \cdot \nabla)] \nabla_\perp \Phi + m_e S_0 [(1+\eta_e)\mathbf{v}_n - \mathbf{v}_B] \cdot \nabla S - \rho_e^2 \left(\partial_t + \mathbf{v}_E \cdot \nabla + \mathbf{v}_{De0} \cdot \nabla + \mathbf{v}_{ez} \partial_z\right) \nabla_\perp^2 \Phi + \partial_z [v_{ez}(1+N)] = 0, \tag{5}$$

$$\left(\partial_t + \mathbf{v}_E \cdot \nabla + \mathbf{v}_S \cdot \nabla + \mathbf{v}_S \cdot \nabla + v_{ez} \partial_z\right) v_{ez} = v_{te}^2 [\Phi - (N+T)], \tag{6}$$

$$\left(\partial_t + \mathbf{v}_E \cdot \nabla + \tfrac{5}{3} \mathbf{v}_B \cdot \nabla + \mathbf{v}_S \cdot \nabla\right) T - \tfrac{2}{3} \left(\partial_t + \mathbf{v}_E \cdot \nabla + \mathbf{v}_S \cdot \nabla\right) N + \tau \left(\eta_e - \tfrac{2}{3}\right) \mathbf{v}_n \cdot \nabla \Phi = 0, \tag{7}$$

In equilibrium, the external magnetic field is $B_0 = B_0 \hat{z}$, $n_{e0}(x)$ is the electron number density, where the different drifts are, entropy drift $\mathbf{v}_S = (c m_e / e B_0) \hat{z} \times \nabla S_e$, diamagnetic drift $v_{De} = (c / e B_0 n_e) \hat{z} \times \nabla (n_e T_e)$, polarization drift $\mathbf{v}_{pe} = -(c / B_0 \omega_{ce})[\partial_t + (\mathbf{v}_{\mathbf{E}} + \mathbf{v}_{De}) \cdot \nabla + v_{ez} \partial_z \nabla_\perp \phi]$, ∇n_{e0} drift $v_n = (c T_{e0} / e B_0) \hat{z} \times \nabla \ln n_{e0}$, ∇B_0 drift is $\mathbf{v}_B = (c T_{e0} / e B_0) \hat{z} \times \nabla \ln B_0$, $E \times B$ drift $v_E = \frac{E \times B}{B^2}$ and v_{ez} is electron-fluid velocity along z-axis. The different notations are, T_e the electron temperature and n_e the electron number density, φ the wave potential, $N = n_{e1} / n_{e0}$,

$S = S_1/S_0, \Phi = e\phi/T_{e0}$ and $\rho_e = cT_{e0}/eB_0\omega_{ce}$. The first order $\mathbf{v}_{D1} = (cT_{e0}/eB_0)\hat{z}\times\nabla(N+T)$, zeroth order $v_{De0} = (1+\eta_e)v_n$ of electron-diamagnetic drift, with $\eta_e = L_T^{-1}/L_n^{-1} = (\partial \ln T_{e0}/\partial x)/(\partial \ln n_0/\partial x)$, $\mathbf{v}_T = (cT_{e0}/eB_0)\hat{z}\times\nabla \ln T_{e0}$ and $v_{te} = \sqrt{T_0/m_e}$ is the electron-thermal speed, $\gamma = \frac{5}{3}$ for three dimensional case and T_{e0} is the electron temperature in equilibrium. It is assumed that perturbations are proportional to $\exp[i(\mathbf{k}\cdot\mathbf{r} - \omega t)]$, where $\mathbf{k}$ the wave vector and ω is the perturbation frequency. Using Fourier transformation, we get the dispersion relation,

$$(\omega-\omega_D+\omega_S)(\omega-\omega_S)\left(\omega-\frac{5}{3}\omega_D-\omega_S\right)-\frac{k_zT_0}{m_e}\left(\omega-\frac{5}{3}\omega_D-\omega_S\right)$$

$$-\frac{2}{3}(\omega-\omega_S)\left\{\omega_D(\omega-\omega_S)+\frac{k_z^2T_0}{m_e}\right\}-k_\perp^2\rho_e^2(\omega-\omega_S)[\omega-\eta(1+_e)\omega_n](\omega-\omega_D-\omega_S)$$

$$-m_eS_0(\omega-\omega_S)\left(\omega-\frac{5}{3}\omega_D-\omega_S\right)[(1+\eta_e)\omega_n-\omega_D]+\frac{k_zT_0}{m_e}\left(\omega-\frac{5}{3}\omega_D-\omega_S\right)$$

$$-\tau(\omega-\omega_S)(\omega_D-\omega_n)\left(\omega-\frac{5}{3}\omega_D-\omega_S\right)-\tau(\eta-\frac{2}{3})\omega_n\left\{\omega_D(\omega-\omega_S)+\frac{k_z^2T_0}{m_e}\right\}=0 \quad (8)'$$

The notations used in above equation are, $\omega_S = \mathbf{k}\cdot\mathbf{v}_S$, $\omega_D = \mathbf{k}\cdot\mathbf{v}_B$, $\mathbf{k}\cdot\mathbf{v}_{De0} = (1+\eta_e)\omega_n$ and $\omega_n = \mathbf{k}\cdot\mathbf{v}_n$ .We now discuss the case for $\omega_D = \omega_n = 0$, then dispersion relation reduces to $\omega = \omega_S$ and

$$(\omega-\omega_S)^2 + k_\perp^2\rho_e^2\omega(\omega-\omega_S) - \frac{2}{3}\frac{k_z^2T_0}{m_e} = 0, \quad (9)$$

$$\omega = \omega_S \pm \left[\frac{\frac{2}{3}v_{te}^2k_z^2 - k_\perp^2\rho_e^2\omega_S}{1+k_\perp^2\rho_e^2}\right]^{\frac{1}{2}}. \quad (10)$$

For $\frac{2}{3}v_{te}^2k_z^2 < k_\perp^2\rho_e^2\omega_S$, entropy instability exists which can play important role in destabilizing plasma to understand anomalous transport.

3. Anomalous Transport

Now, it will be shown that Anomalous transport due to ETG-mode across the confining magnetic field is expressed as: $\mathbf{\Gamma} = \langle v_{EB}T_{e1}^*\rangle + \text{c.c.}$, where T_{i1}^* is the complex conjugate of the perturbed electron temperature and the angular brackets represent the ensemble average, thus, we have:

$$T = \frac{1}{\left(\omega-\frac{5}{3}\omega_D-\omega_S\right)}\left[\frac{2}{3}(\omega-\omega_S)N - \tau(\eta-\frac{2}{3})\omega_n\Phi\right].$$

Here, $\omega = \omega_r + i\gamma$, where ω_r is real while γ is imaginary part of ω . We find the electron energy flux $\mathbf{\Gamma}_x$ as:

$$\mathbf{\Gamma}_x = -32\sum_k \frac{cT_{i0}\gamma\pi^2\omega_r^2\omega_{ce}^2}{eB_0k_\perp^2k_yv_{th}^4}\frac{\tau\eta_e\omega_n + \frac{10}{9}\omega_D - \frac{2}{3}\tau\omega_n}{\left(\omega-\frac{5}{3}\omega_D-\omega_S\right)^2+\gamma^2}. \quad (11)$$

The well-known Fick's law is, $\Gamma_x = -\chi_e \frac{dT_{e0}}{dx}$. Thermal conductivity of electron is expressed as:

$$\chi_e = 32\sum_k \frac{cT_{e0}\gamma\pi^2\omega_r^2\omega_{ce}^2}{eB_0k_\perp^2k_yv_{th}^4}\frac{\frac{\tau}{L_T}+\frac{10}{9}\frac{1}{L_B}-\frac{2}{3}\frac{\tau}{L_n}}{\left(\omega-\frac{5}{3}\omega_D-\omega_S\right)^2+\gamma^2}.$$

With the help of Bohm diffusion coefficient, the above equation is, $\chi_e = \Delta D_B$, where Bohm diffusion coefficient is $D_B = cT_{e0}/16eB_0$ and enhancement factor Δ is $\Delta = -\frac{384}{5}\frac{eB_0S_0\omega_{im}\omega_r^2\omega_cL_T}{k_\perp^2k_yv_{th}^2}$.

4. Conclusions

In this paper, the effect of entropy drift on electron- temperature- gradient mode is investigated. It is found that entropy-drift plays an important role in energy and mass transport. In the linear limit, new dispersion relation is derived and specific cases are also discussed, the presence of entropy instability is seen and if density and magnetic field gradients are ignored, one root is $\omega = \omega_S$. Dependence of η_e term on entropy is also observed. Anomalous transport coefficient has been rederived and its dependence on entropy has been found. This result is not only helpful to understand entropy in a better way and to understand anomalous transport of plasma as well. On the basis of entropy drift, new horizon will be opened for research. This work has relevance in magnetically confined plasma in the presence of temperature and density gradients for tokamak and space plasmas.

References

[1] Purushottam D and Gujrati 2015 *Entropy* **17** 710.
[2] Yaqub k M, Iqbal J and Ul Haq A 2014 *Phys. Plasmas* **21** 052308.
[3] Hinton F L and Hazeltine R D 1976 *Rev. Mod. Phys.* **42** 239.
[4] Hirshman S P and Sigmar D J 1981 *Nucl. Fusion* **21** 9.
[5] Balescu R 1985 Transport Processes in Plasmas. North-Holland,Amsterdam Press **2**
[6] Liewer P C 1985 *Nucl. Fusion* **25** 543.
[7] Sugama H, Okamoto M, Horton W and Wakatani M 1996 *Phys. Plasmas* **3** 2379.
[8] Coppi B, Rosenbluth M N and Sagadeev R Z 1967 *Phys. Fluids* **10** 582
[9] Liu C S 1971 *Phys. Rev. Lett.* **27** 1637.
[10] Bharuthram R and Shukla P K 1987 *Plasma Phys. Control Fusion* **29** 1695.
[11] Hassam A B, Antonsen T M Jr, Drake J F and Guzdar P N *Phys. Fluids* B **2** 8.
[12] Jarmén A, Andersson P and Weiland J 1987 *modes. Nucl. Fusion* **27** 941.
[13] Shukla P K 1990 *Phys. Scr.* **42** 725.
[14] Zhang Y Q, DeSilva A W and Mostovych A N 1989 *Phys. Rev. Lett.* **62** 1848.
[15] Muller M, Schmalian J and Frtz L 2009 *Phys. Rev. Lett.* **103** 025301.
[16] Kobayashi S and Rogers B N and Dorland W 2010 *Phys. Rev. Lett.* **105** 235004.
[17] Kesner J 1997 *Plasma Phys. Rep.* **23** 742.
[18] Brantov A V, Bychenkov V Y and Rozmus W 2012 *Phys. Rev. Lett.* **108** 205001.

Effect of electromagnetic fields on some biomechanical and biochemical properties of rat's blood

M A Mohaseb[1,2] , F A Shahin[1], F M Ali[3] and H A Baieth[4]

[1] Beni-Suef University, Faculty of Science, Department of Physics, Beni-Suef, Egypt
[2] Umm-Al-Qura University, Faculty of Applied Science, Department of Physics, Mecca, Saud Arabia.
[3] Cairo University, Faculty of Science, Department of Biophysics, Giza, Egypt
[4] Benha University, Faculty of Science, Department of Physics, Benha, Egypt
E-mail: marefaie@uqu.edu.sa

Abstract: In order to study the effect of electromagnetic fields (0.3 mT, 50 Hz) on some biomechanical and biochemical properties of rats' blood, healthy thirty male albino rats of 150 ± 10 g were divided into three equal groups namely A, B1, B2. Group A used as a control group, group B1 was continuously exposed to a magnetic field of (0.3 mT, 50 Hz) for a period of 21 days for direct effect studies. Group B2 was continuously exposed to the same magnetic field for the same period of time, then was housed away from the magnetic field for a period of 45 days for delayed effects studies. After examination, the results indicated that the apparent viscosity and the consistency index increased significantly and very high significantly for groub B1 and B2 compared to control at $P<0.05$. Red blood cell counts (RBCs) membrane elasticity had significantly and very high significantly decreased for groups B1 and B2. Moreover, delayed effects studies indicated that there is deterioration in the bone marrow functions. These results are supported by the blood film image, where irregularities and deformations in the RBCs membranes had been occurred. We conclude that the cell membrane properties are highly affected by the extremely low frequency (ELF) magnetic fields, which proved to be biologically toxic.

1. Introduction

There are numerous sources of environmental extremely low frequency (ELF) magnetic fields, both naturally occurring and man–made, to which humans are exposed [1]. Large intermittent, time varying magnetic fields are produced in the atmosphere as a result of intense solar activity and thunderstorms, and can reach levels up to 0.5 μT (T=Tesla =10^4 gauss) during magnetic storm [1].

Numerous man made sources of ELF magnetic fields are present in at home, work place and public places such as urban transit systems. These sources include electric power transmission and distribution lines, long-range ELF military communication systems and a wide variety of appliances, tools, and machines used in offices and various industries [1-3]. Recently, there has been a considerable research effort on biological effects and potential hazards of low-frequency electric and magnetic fields.

The aim of the present work is to study the effect of electromagnetic fields (0.3 mT, 50 Hz) on some biophysical and biochemical properties of rat's blood.

IOP Conf. Series: Journal of Physics: Conf. Series **869** (2017) 012059 doi:10.1088/1742-6596/869/1/012059

2. Materials and Methods

2.1. Experimental animals
Thirty male albino rates, each of average weight of 150±10 g were divided into three groups as follows group A as control group, group B1 is directly exposed continuously to magnetic field of 0.3 mT (50Hz) for 21 days, group B2 was used for late effect studies after 45 days.

2.2.Determination of the rheological properties of blood
Viscosity is an expression of the degree of slipperiness between two layers of fluid. Shear stress is expressed as force per unit area (F/A). The shear stress, required to produce a shear rate, is defined as the viscosity [4]. Rheological properties of blood are determined by the Brookfield DV-III programmable rheometer. Brookfield DV-III programmable rheometer measures fluid's parameters of shear stress and viscosity at given shear rates according this relation, shear stress = (shear rate)^n, where n is the zero shear Newtonian viscosity [5]. The exponent n, often called flow index.

2.3.Osmotic fragility
A stock solution of buffered sodium chloride, osmotically equivalent to 100 g/L (10%). The percent of hemolysis (H%) can be plotted as a function of the percentage of sodium chloride concentration [6].

2.4.Histology
The blood film was made by a manual method for making wedge blood smears.The RBCs were examined by using image analyzer system called SAMICA (Scanning and Measuring with Automatic Image Contour Analysis) with magnification of 1000 times.

2.5. Statistical analysis
Analysis of data was performed using student ANOVA test and comparing between means using Least Significant Difference (LSD) at P < 0.05 as out lined in [7].

3.Results

3.1.Rheological Properties of Blood
Figure 1(a) illustrates the variation of the blood viscosity (cP) as a function of the shear rate (s^{-1}) from all groups. As shown, the viscosity of the blood from groups B1and B2 has higher values at all shear rates as compared with control group A. Moreover, this increase in the value of the blood's viscosity is maximum at shear rate (7.50 s^{-1}) for samples from group B2 as compared with control group A. The variation of the shear stress (dyne/cm^2) as a function of the shear rate (s^{-1}) for the blood samples from groups A, B1 and B2 are shown in figure 1(b).

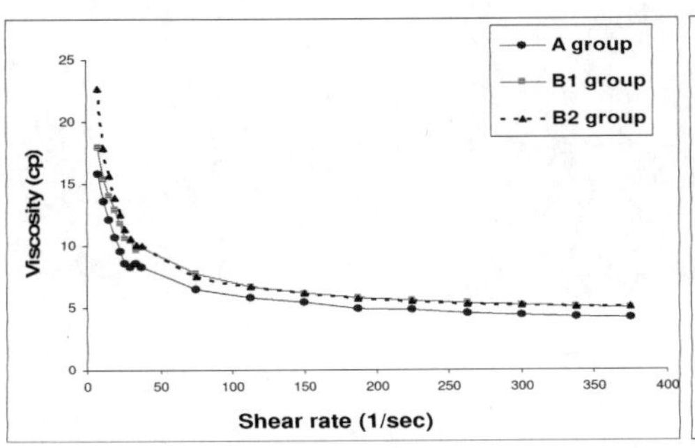

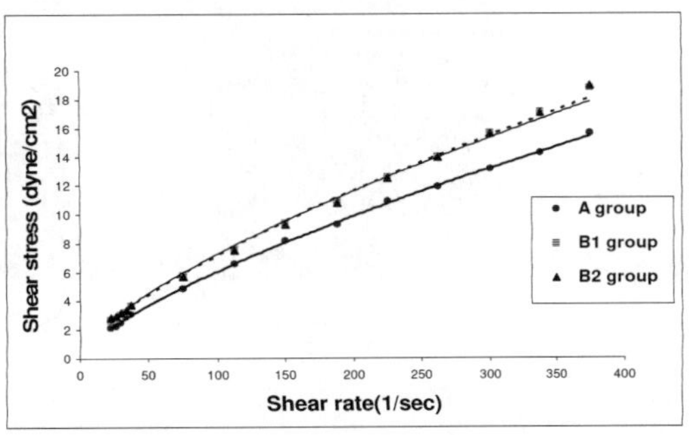

Figure 1: Variation of the blood viscosity (left) and shear stress (right) versus shear rate for groups A, B1, and B2.

3.2. Osmotic Fragility Test

Figure 2(a,b) shows the results of osmotic fragility measurements for the RBCs collected from animals of different groups of A, B1 and B2, where the percentage of the hemolysed cells are plotted as a function of the percentage of NaCl concentration. The differentiations of these results were plotted against the NaCl% and shown in figure for the three groups. The differential curves results indicate the behavior of the curves pass through a maximum (C %) at which maximum hemolysis rate occurs.

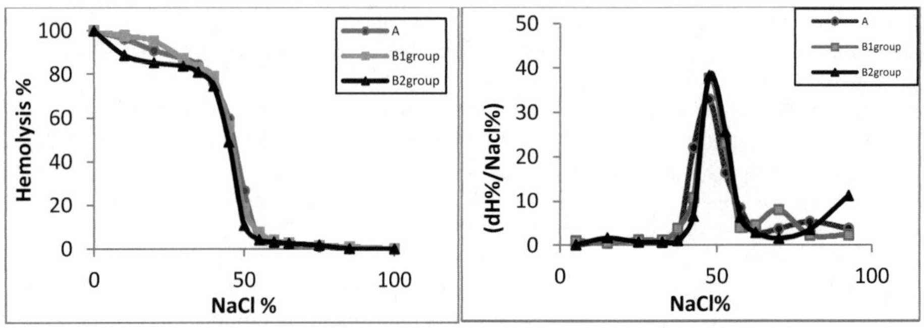

Figure 2: Osmotic fragility (left) and differentiation curves(right) for group A, B1, B2.

3.3. Histology

Figure 3 shows an image for a blood film from all groups as scanned and magnified 1200 times by the image analyzer. The photo 3-a (left) indicates regular traditional forms of the RBCs. Alos, we note that the compression occurred on some adjacent cell membrane (marked by arrows) due to the repulsive coulomb forces originated from the positive surface charges on the RBCs, membrane. It is clear from the figure 3-b (middle), that changes had occurred in the morphology of RBCs membrane of the exposed group. These changes, in the cell membrane, can be easily noticed from the formed aggregates of several adjacent blood cells and some cells diffused in each other forming common membrane (marked by arrows). Figure 3-c (right) illustrates the scanned blood film image from the exposed group 45 days post exposure to magnetic field. The image shows that the newly generated RBCs' membrane had many disorders as compared with control group. It can be also noticed that some cells diffused in others, all cell membranes are irregular and the back ground of the slide is unclear because of hemolysis.

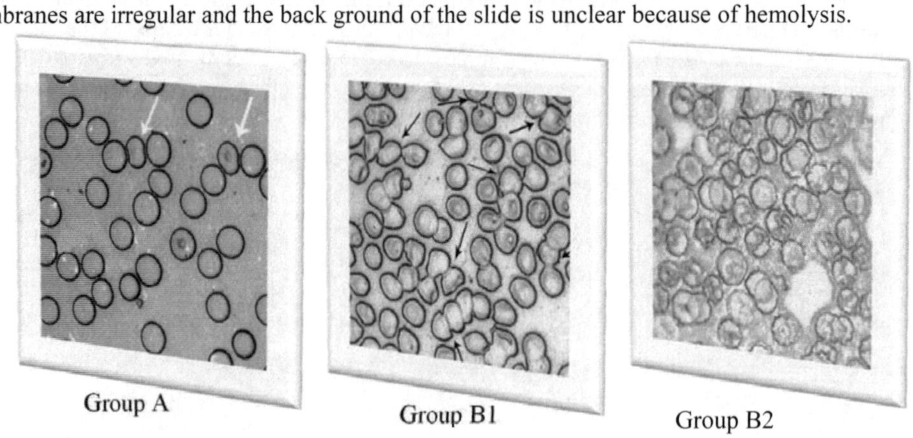

Figure 3. Blood film for groups A,B1, and B2 (x1200).

4. Discussion

The studies focused on the mechanical properties of the RBC' membrane, which plays an important role on the RBCs metabolic activity and it was used as an example to the possible injury that may occur in other cells from different organs. This was achieved through viscosity, osmofragility studies and blood film images.

From the present data, one may easily find that the apparent viscosity of the blood increases significantly and the half maximum width of the osmo fragility curves suffered highly significantly decrease for blood collected from group B1. Moreover, from the blood film for group B1 one may find the formation of aggregates and diffusion of adjacent cells. Therefore, one may presume that the direct effect of the 0.3 mT, 50Hz magnetic field on the cellular membrane is the disturbance of the molecular free motion of the bilayer macromolecules forming the cellular membrane. These magnetic forces will cause some changes in the packing properties of these macromolecules that may lead to changes of the cell membrane mechanical properties such as the elastic range and permeability.

One may easily notice, from the blood film for normal samples from group A, that there are coulomb forces between adjacent RBCs that prevent formation of aggregates or diffusion. These electric forces are strong enough to cause bending to the cell membranes when a cell is in close proximity between two cells (marked by arrows in figure 3(a)(left). Alos, the highly significantly decrease of W_{hmax} for the RBCs from group B2. The decrease of W_{hmax} means the decrease of the elastic range of the cell membrane and hence folding of the cells to pass through the blood capillaries (which have much smaller diameter than the RBCs diameter) will be reduced and hence the RBCs will fail to reach the target cells to carry on their metabolism. One may state here that the direct effect of exposures to the 50 Hz, 0.3 mT magnetic field is the change of the cell membrane properties. Delayed effect studies did not show any sort of repair to different organs especially to the RBCS. The newly generated RBCs have irregular membrane form as the blood was full of aggregates of RBCS, in additional to the pronounced hemolysis occurred. This finding indicates that the bone marrow (the blood generating system), was heavily injured, either directly by the influence of the magnetic field or indirectly as a feedback to the defects of the RBCs, which failed to do their metabolic activities with the bone.

5. Conclusions

Exposure to the 50 Hz, 0.3 mT magnetic fields is biologically toxic. When investigating the hazards of non ionizing radiation, it is necessary to study the physical properties of the RBCs and to have blood film magnified to 1200 times. Also, It is necessary to encourage and to support research work in the radiation hazards for non ionizing radiation; since it is distributed everywhere and there is no dosimetry system that exists till now.

References

[1] Tenforde T S 1992 Annu. Rev. public Health **13** 173
[2] Stuchly M A *1986 Health Phys.*, **51** 215
[3] Bernhardt J H and Matthes R 1992 *ELF and RF electromagnetic sources, in Non-Ionizing Radiation.* Greene, M.W., Ed., UBC press. Vancouver, British Columbia, Canada, **59**
[4] Boron W and Boulpaep E. 2003 *Medical Physiology: A Cellular and Molecular Approach. Saunders, Philadelphia* Pp: 429-435
[5] Lenk R S 1978 *The characterization of viscous flow, viscosity, shear rate and shear stress. In 'polymer rheology'* Applied science publishers LTD. London.
[6] Brown A and Hunter R C 1993 *Hematology: Principles and Procedures.* Lea & Febiger, Philadelphia
[7] PC-STAT 1995 One way analysis of variance procedure. Georigis University

Development of Micro-Pattern Gas Detectors for the Upgrade of the Muon System of the CMS Experiment at the Large Hadron Collider

Othmane Bouhali[1]
on behalf of the CMS Muon group
E-mail: othmane.bouhali@qatar.tamu.edu

Abstract.
After the discovery of the long awaited Higgs boson in 2012, the Large Hadron Collider (LHC) at the European Organization for Nuclear Research (CERN) and its two general purpose experiments (ATLAS and CMS) are preparing to break new grounds in High Energy Physics (HEP). The international HEP collaboration has established a rigorous research program of exploring new physics at the high energy frontiers. The program includes substantial increase in the luminosity of the LHC putting detectors into a completely new and unprecedented harsh environment. In order to maintain their excellent performance, an upgrade of the existing detectors is mandatory. In this work we will describe ongoing efforts for the upgrade of the CMS muon detection system, in particular the addition of detection layers based on the Gas Electron Multiplier (GEM) technology. We will summarize the past 5-year R&D program and the future installation and operation plans.

1. Introduction

The Large Hadron Collider has been built to shed light on several fundamental questions in particle physics. The Compact Muon Solenoid (CMS) experiment [1] is one of its two general purpose experiments designed and built to detect and reconstruct particles produced during LHC collisions. With the discovery of the Higgs boson [2, 3], the international collaboration has established a rigorous research program involving precise measurements of the Higgs properties and searches for new physics. This requires a large increase in the LHC luminosity which puts stringent requirements on the detectors. In order to maintain its excellent performance, CMS is undergoing a series of upgrades of its components, among them its muon system. This latter is a critical component of CMS due to the strong role of new physics channels involving muon particles in the final state. The CMS muon system is composed of three detector technologies: Resistive Plate Chambers (RPC), Drift Tubes (DT) and Cathode Strip Chambers (CSC). Until now the redundancy of the CMS muon system could not yet be established in the pseudorapidity region $\eta > 1.6$ (figure 1). In the past 5 years the muon collaboration has conducted an extensive R&D program to recover this redundancy, extend the capabilities of the forward region to cope with the future increase in the LHC luminosity and enhance the trigger and reconstruction performance.

[1] Present address: Texas A&M University at Qatar, Engineering Building, Education City, Qatar Foundation, P.O.Box 23874, Doha, Qatar.

IOP Conf. Series: Journal of Physics: Conf. Series **869** (2017) 012060 doi:10.1088/1742-6596/869/1/012060

2. The forward muon system upgrade

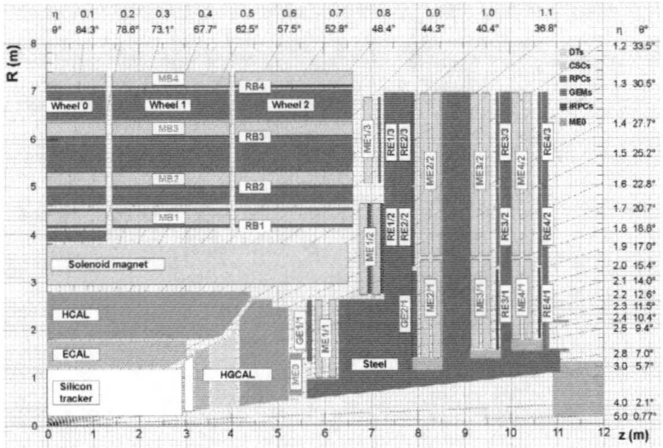

Figure 1. Illustration of the transverse section of CMS showing, in red, the planned GEM stations: GE1/1, GE2/1 and in yellow the ME0 [5].

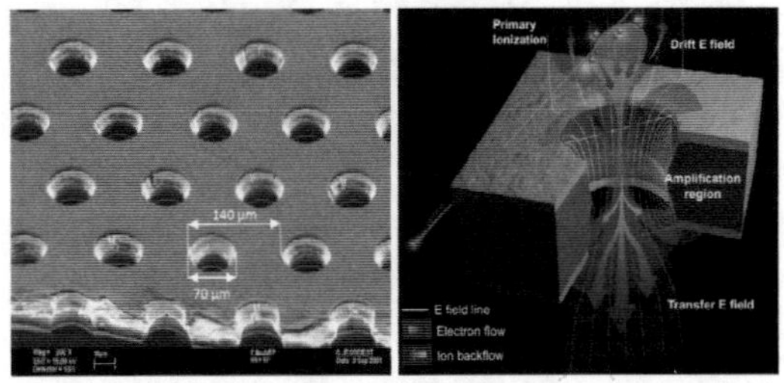

Figure 2. Left: Scanning Electron Microscope picture of a GEM foil showing the hole pattern. Right: Illustration of the avalanche formation inside a GEM hole [5].

The forward region will be instrumented with additional layers of a new technology based on GEM [4] detectors. The GEM is an amplification stage consisting of a 50 μm thick polymer foil, metal coated on each side, and perforated regularly by a high density of holes (figure 2 left). In our application, the hole diameter is 70 μm, and the hole spacing is 140 μm. With a suitable voltage of a few hundreds of volts, applied between the two metal coated sides of the GEM, a high electric field of up to 100 kV/cm is reached inside the holes (figure 2 right). When a charged particle crosses the detector, it ionizes gas molecules inside the detector active volume, creating pairs of electrons and positive ions. Thanks to the electric field inside the detector, electrons will drift towards the positive anode creating a signal that can be measured by appropriate readout electronics. The detector gain can be further enhanced by inserting three GEM foils inside the same gas volume known as triple-GEM detector (figure 3). The triple-GEM was proposed by the collaboration to equip the region of $1.55 < \eta < 2.18$. These detectors have to fulfill several requirements: uniform response, high detection efficiency, good spatial and time resolution [5].

The collaboration has designed, built and operated large-area of triple-GEM chambers [5]. These trapezoidal chambers, denoted as GE1/1, have an active area of 990 $\times$ (220-455) mm^2 each, and are sectorized in η partitions to cover 10° each in the azimuthal direction and provide radial readout with strips pointing towards the beam pipe (figure 4). Two chambers are mounted face-to-face to form a double detection layer referred to as "superchamber". The standard gas mixture consists of either Ar/CO_2 or $Ar/CO_2/CF_4$. The readout electronics are based on a binary technology VFAT [5, 6].

3. Performance results

In the past five years several GE1/1 prototypes were built in laboratories and tested with charged particles at CERN [7] and Fermilab [8]. Below we summarize the results from these efforts.

3.1. Detection efficiency and spatial resolution

Figure 5 shows the detection efficiency as a function of the electric field applied to the GE1/1 chamber. Three regions of the chamber are reported. An efficiency of 97-98% per chamber is achieved. With 97% efficiency per chamber, a superchamber will have 99% efficiency if the signals of its two face-to-face chambers are combined as a logical "OR". Figure 6 shows the

IOP Conf. Series: Journal of Physics: Conf. Series **869** (2017) 012060 doi:10.1088/1742-6596/869/1/012060

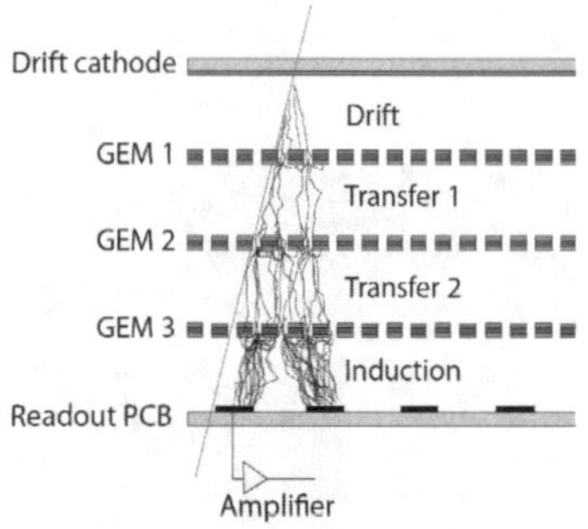

Figure 3. A schematic view of a triple-GEM detector [5].

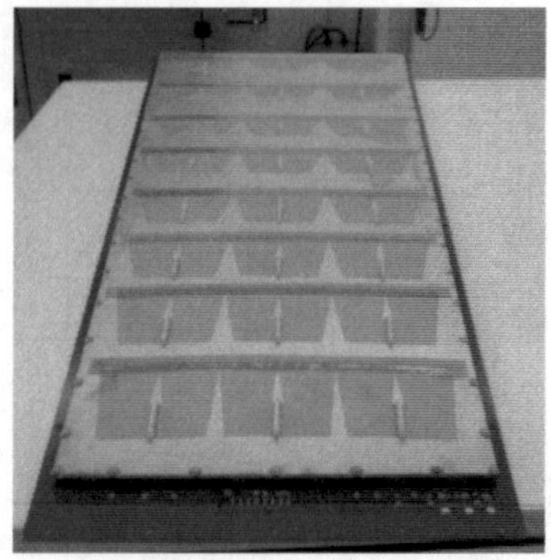

Figure 4. Picture of GE1/1 prototype chamber [5].

distribution of the residual between the particle coordinate measured by the GE1/1 chamber and that extrapolated from an external reference detector. The width of this distribution, after adequate corrections, gives the spatial resolution of the GE1/1 detector. A spatial resolution of 268 μm is achieved which is well within the required accuracy [5].

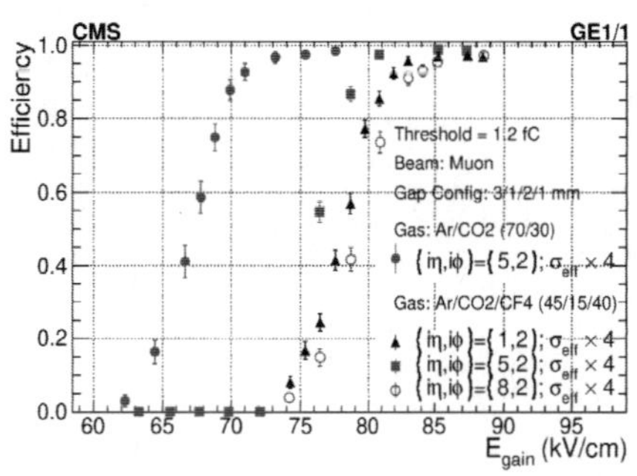

Figure 5. Detection efficiency as a function of the electric field in the GE1/1 chamber [5]. Three different areas of the chamber are reported.

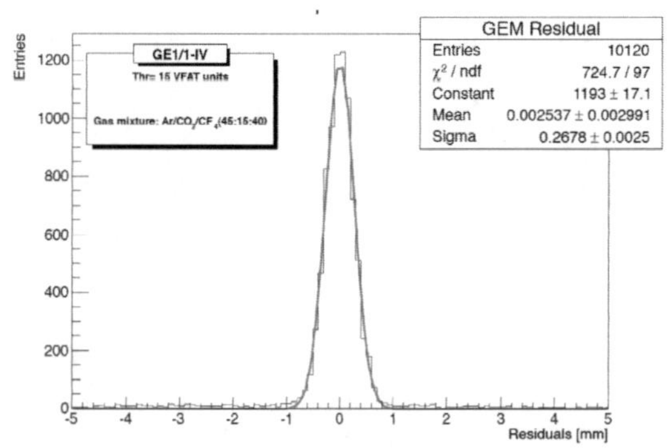

Figure 6. Residual distribution of the GE1/1 chamber as obtained from a test with a beam of charged particles [5].

3.2. Time resolution

A key parameter to match the CMS trigger requirement is the time resolution of the detector. Figure 7 shows the time resolution as a function of the electric field for both Ar/CO_2 and $Ar/CO_2/CF_4$. The difference in the time resolution between both gas mixtures is due to the difference in electron drift velocity between both gases. Nevertheless, in each case at least 8 ns time resolution is obtained which largely matches the CMS trigger requirements in this region for an LHC bunch crossing of 25 ns.

3.3. Uniformity reponse

Due to the unprecedented large area of GEM foils used in GE1/1, a uniformity of response across the whole chamber needs to be ensured. The collaboration has optimized the assembly procedure to fulfill this requirement and prevent any deformation of the GE1/1 or its components. Figure 8

IOP Conf. Series: Journal of Physics: Conf. Series **869** (2017) 012060 doi:10.1088/1742-6596/869/1/012060

shows the relative gain obtained from a uniformity study across the full chamber area. The observed gain variations are within the range set by the collaboration.

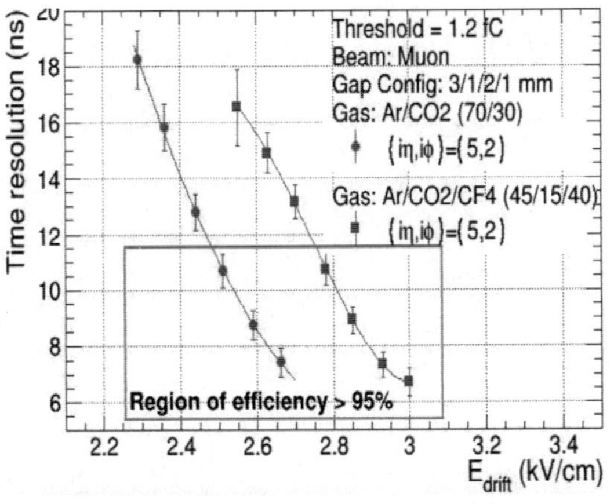

Figure 7. Time resolution as a function of the electric field in a GE1/1 for two different gas mixtures.

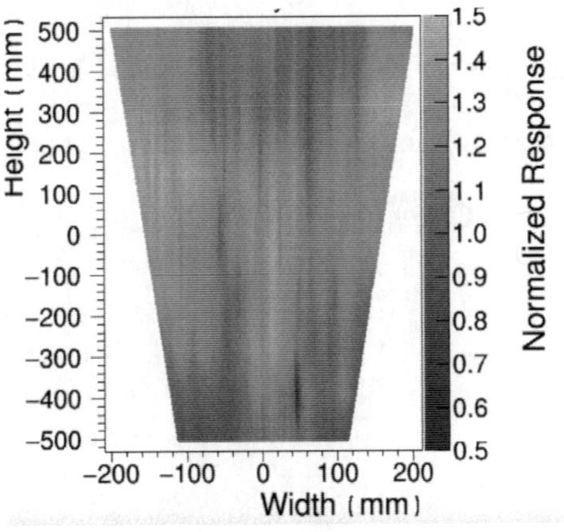

Figure 8. Gain uniformity scan across a full GE1/1 chamber. The gain variations are within the acceptable limit set by the collabroation.

4. Summary and future upgrades

Thanks to the excellent performance achieved during its R&D program, the GE1/1 project has been approved by the collaboration. In total 36 superchambers will be installed in the LHC Long Shutdown (LS2) in 2018. As a preparation for this milestone, 5 GE1/1 chambers have been recently installed successfully in CMS.

Moving forward, the collaboration has embarked in an R&D program for two additional upgrades. The first one, referred to as GE2/1, consists of installing two rings of double layered triple-GEM chambers covering the region of $1.6 < \eta < 2.45$ (figure 1) and planned for the LHC Long Shutdown (LS3) in 2024. The second one, called ME0, consists of additional layers to extend the coverage to $2.0 < \eta < 3.0$. (figure 1).

Acknowledgments

The author is supported by the Qatar National Research Fund under grant NPRP award [NPRP 9-328-1-066].

References

[1] CMS Collaboration, *J. Inst.***3** (2008) S08004.
[2] ATLAS Collaboration, *Phys. Lett.* B**716** (2012) 1
[3] CMS Collaboration, *Phys. Lett.* B **716** (2012) 30
[4] F. Sauli, *Nucl. Instrum. Meth.* A**386** (1997) 531.
[5] CMS GEM Collaboration, CMS technical design report for the muon endcap GEM Upgrade , CERN-LHCC-2015-012, CMS-TDR-013, Jun 2015,
[6] P. Aspell et al., proceedings of the Topical Workshop on Electronics for Particle Physics, 4612 Praque, Czech Republic, 03-07 Sept., pp. 292296. 2007. 4613 doi:10.5170/CERN-2007-007.292.
[7] D. Abbaneo et al., IEEE Nucl. Sci. Symp. Med. Imag. Conf. Rec. (2011) 18061810.
[8] D. Abbaneo et al., IEEE Nucl. Sci. Symp. Med. Imag. Conf. Rec. (2014).

Microscopic description of the competition between spontaneous fission and α-decay in neutron-rich Ra, U and Pu nuclei

R Rodríguez-Guzmán[1] and L M Robledo[2]

[1] Physics Department, Kuwait University, Kuwait 13060, Kuwait

[2] Departamento de Física Teórica, Universidad Autónoma de Madrid, 28049-Madrid, Spain

E-mail: raynerrobertorodriguez@gmail.com

Abstract. Constrained mean-field calculations, based on the Gogny-D1M energy density functional, have been carried out to describe fission in Ra, U and Pu nuclei with neutron number $144 \leq \mathrm{N} \leq 176$. Fission paths, collective masses and zero-point quantum vibrational and rotational corrections are used to compute the spontaneous fission half-lives. We also pay attention to isomeric states along the considered fission paths. Alpha decay half-lives have also been computed using a parametrization of the Viola-Seaborg formula. Though there exists a strong variance of the predicted fission rates with respect to the details involved in their computation a robust trend is obtained indicating, that with increasing neutron number fission dominates over α-decay. Our results also suggest that a dynamical treatment of pairing correlations is required within the microscopic studies of the fission process in heavy nuclear systems.

1. Introduction

Fission is a large amplitude collective motion the theoretical description of which, still remains a challenge in modern nuclear structure physics with a major impact on several basic research and technological areas [1, 2]. A detailed knowledge of fission in heavy and superheavy nuclei is required, to better understand the very limits of the nuclear stability. The fission properties of neutron-rich nuclei are relevant in the nucleosynthesis and abundances of elements with mass number $\mathrm{A} > 120$ due to the fission recycling in the r-process [3]. Furthermore, microscopic studies of the fission paths and related properties are very useful to deepen our knowledge of the different decay channels (fission, α-decay, ...) in heavy nuclear systems. In this contribution, we report on our results regarding the competition between spontaneous fission and α-decay in neutron-rich Ra, U and Pu nuclei [4–6].

In this section, we briefly outline the theoretical framework used in our calculations [4–6]. We have resorted to the Hartree-Fock-Bogoliubov (HFB) approximation [7] based on the parametrization D1M [8] of the Gogny [9] energy density functional (EDF). We have used as constraining operators the axially symmetric quadrupole $\hat{Q}_{20}$ and octupole $\hat{Q}_{30}$ operators as well as the necking $\hat{Q}_{Neck}(z_0, C_0)$ operator to reach two-fragment (2F) solutions starting from the one-fragment (1F) ones. Aside from the constraints already mentioned, as well as the usual HFB ones on both the proton and neutron numbers, a constraint on the operator $\hat{Q}_{10}$ is used to prevent spurious center of mass effects. The HFB quasiparticle operators [7] have been

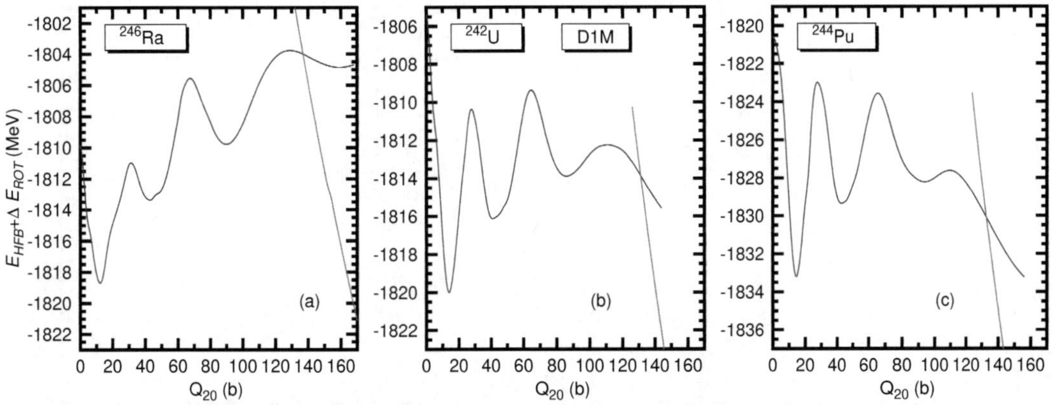

Figure 1. (Color online) The HFB plus the zero point rotational energies obtained with the Gogny-D1M EDF are plotted in panels (a), (b) and (c) as functions of the quadrupole moment Q_{20} for the nuclei ^{246}Ra, ^{242}U and ^{244}Pu. Both the 1F (blue) and 2F (red) solutions are included in the plot. For more details, see the main text.

expanded in an axially symmetric (deformed) harmonic oscillator (HO) basis containing states with J_z quantum numbers up to 31/2 and up to 26 quanta in the z direction. The lengths of the HO basis have been optimized so as to minimize the total HFB energy. For the solution of the HFB equations, an approximate second order gradient method [10] has been used.

We have computed the spontaneous fission half-life t_{SF} within the WKB formalism

$$t_{SF} = 2.86 \times 10^{-21} \times \left(1 + e^{2S}\right) \tag{1}$$

where the action S along the (one-dimensional projected) quadrupole constrained fission path reads

$$S = \int_a^b dQ_{20}\sqrt{2B(Q_{20})\left(V(Q_{20}) - (E_{GS} + E_0)\right)} \tag{2}$$

with integration limits a and b corresponding to the classical turning points for the energy $E_{GS} + E_0$. The potential $V(Q_{20})$ is given by the HFB energy corrected by the rotational $\Delta E_{ROT}(Q_{20})$ and vibrational $\Delta E_{vib}(Q_{20})$ energies. The collective masses $B(Q_{20})$ have been computed using both the Generator Coordinate Method (GCM) and the Adiabatic Time-Dependent HFB (ATDHFB) schemes. For the parameter E_0 we have assumed a typical value of 1.0 MeV. In our calculations, we take the 2F curves as intersecting the 1F ones [4–6]. We have computed the α-decay half-lives using a recent parametrization

$$log_{10}t_\alpha = \frac{AZ + B}{\sqrt{Q_\alpha}} + CZ + D \tag{3}$$

of the Viola-Seaborg formula [4, 11].

2. Results

We have computed the fission paths (i.e., the 1F and 2F curves) for Ra, U and Pu nuclei with the neutron numbers $144 \leq \mathrm{N} \leq 176$. A typical outcome of our calculations is shown in figure 1

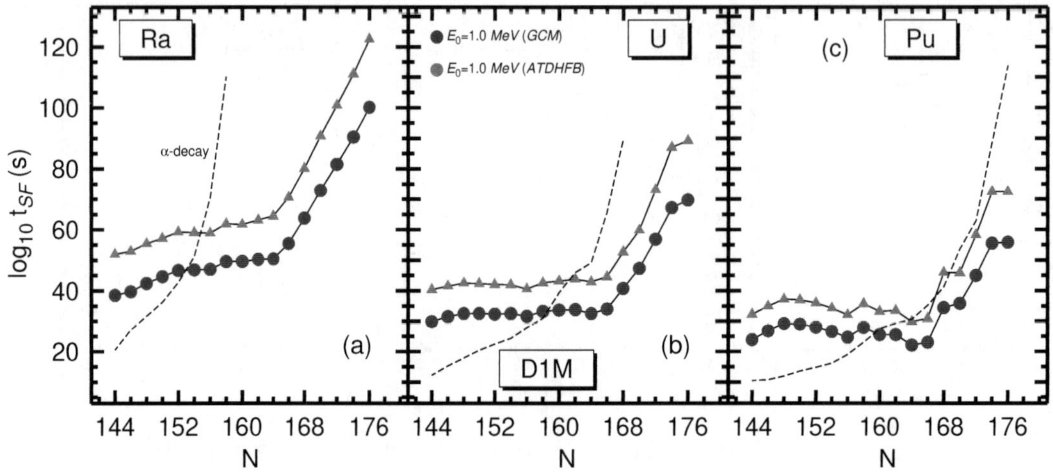

Figure 2. (Color online) The spontaneous fission half-lives t_{SF}, predicted within the GCM and ATDHFB schemes, for the considered Ra, U and Pu nuclei, are depicted as functions of the neutron number in panels (a), (b) and (c). Results have The α-decay half-lives are plotted with dashed lines. For more details, see the main text.

where the HFB plus the zero point rotational energies are plotted, as functions of the quadrupole moment, for ^{246}Ra, ^{242}U and ^{244}Pu, taken as illustrative examples. A similar analysis has been carried out for all the other studied nuclei. The zero-point vibrational energies $\Delta E_{vib}(Q_{20})$ are not included in figure 1, as they are rather constant as functions of Q_{20}. As can be seen, the ground state is located around $Q_{20} = 12b$ while the first fission isomer appears around $Q_{20} = 42b$ with an excitation energy of 2.33, 3.91 and 3.83 MeV in ^{246}Ra, ^{242}U and ^{244}Pu. This first fission isomer, is separated from the ground state by an inner barrier, the top of which is located around $Q_{20} = 30b$. The height of the inner barrier amounts to 7.70, 9.66 and 10.16 MeV, respectively. Note, that a reduction of the inner barrier heights, typically by a few MeV, takes place once triaxial degrees of freedom are taken into account. However, such a reduction comes together with an increase of the collective inertia and, as a result, its influence on the predicted t_{SF} values equation (1) is quite limited (see, [4] and references therein). From figure 1, we observe the second and third fission barriers as well as a (parity-breaking) second fission isomer ($Q_{20} \approx 90 - 94b$) in between them with excitation energy of 8.92, 6.13 and 4.95 MeV for ^{246}Ra, ^{242}U and ^{244}Pu. Octupole correlations (i.e. $Q_{30} \neq 0$ values) start to play a role for $Q_{20} \geq 60b$ leading to a reduction of the outer barriers as compared with a reflection-symmetric situation. For the three isotopic chains, well defined second fission isomers appear along the 1F paths for the mass numbers $240 \leq A \leq 254$ showing that the shell effects leading to such isomers are systematically present in our calculations. Some additional shallow minima are also found along the 1F paths for heavier nuclei [4–6]. A change in tendency is observed in the 1F fission paths of the studied Ra, U and Pu nuclei around the neutron number N=164 followed, by a steady increase in the height of the inner and second barriers as well as in the excitation energies of the second fission isomers. From the increase in the height of the inner barriers and the widening of the 1F curves one could expect, as it is shown to be the case below, larger spontaneous fission half-lives as we move towards more neutron-rich regions.

The spontaneous fission half-lives are plotted in figure 2 as functions of the neutron number for the studied Ra, U and Pu nuclei. As can be seen, the ATDHFB t_{SF} values are always larger than the GCM ones [4–6]. For example, for ^{242}U they are $\log_{10} t_{SF} = 42.36$ s and 32.58 s, respectively. Increasing the parameter E_0 [see, equation (2)] leads to a reduction of t_{SF} however, the trend observed in the figure remains the same. This is also true if, instead of the parametrization D1M, we resort to any other of the most recent parametrizations (i.e., D1S and/or D1N) of the Gogny-EDF [4]. For the Ra, U and Pu isotopes with N $\geq$ 166 one observes a steady increase in the spontaneous fission half-lives. This behavior agrees well with the change in tendency observed in the corresponding 1F paths and makes the heavier neutron-rich isotopes stable against fission, diminishing the importance of fission recycling in the r-process. In figure 2, we have also depicted the α-decay half-lives equation (3). Here, one should keep in mind that the parameter set D1M is well suited for such calculations as it has been tailored to provide a better description of the nuclear masses [8]. As observed in the figure, the α-decay lifetime, which is initially shorter than the fission one, becomes longer for large enough neutron numbers. In the case of Ra isotopes our calculations predict the crossing point to be N $\approx$ 152-154 while for Pu isotopes it is found to be N $\approx$ 160-164, i.e., around two mass units later than for U nuclei. Finally, let us also mention that pairing correlations have a strong impact on the predicted spontaneous fission half-lives due to the inverse dependence of the collective masses with the square of the pairing gap [12]. We have repeated the calculations changing the strength of the pairing fields by 5 and 10 % [4] and found that the predicted spontaneous fission half-lives change by several orders of magnitude. However, even in such a case, the main conclusions stated before (i.e., there is a steady increase of t_{SF} in the heavier neutron-rich nuclei and fission becomes dominant over α-decay with increasing neutron number) still hold indicating that the trend predicted by our calculations is a robust one. Our results also suggest the need for a dynamical treatment of pairing correlations within microscopic fission studies. The first steps in this direction have already been undertaken in Ref. [13].

R. R.-G. acknowledges valuable discussions with the members of the Department of Physics, Kuwait University as well as the support received to participate in the conference. The work of L. M. R. was partly supported by the spanish MINECO grant nos. FPA2015-65929 and FIS2015-63770-P.

References

[1] Krappe H J and Pomorski K Springer, Berlin, 2012 *Theory of Nuclear Fission*, Lectures Notes in Physics, **838**
[2] Schunck N and Robledo L M 2016 *Rep. Prog. Phys.* **79** 116301
[3] Martínez-Pinedo G *et al* 2007 *Prog. Part. Nucl. Phys.* **59** 199
[4] Rodríguez-Guzmán R and Robledo L M 2014 *Phys. Rev.* C **89** 054310
[5] Rodríguez-Guzmán R and Robledo L M 2014 *Eur. Phys. J.* A **50** 142
[6] Rodríguez-Guzmán R and Robledo L M 2016 *Eur. Phys. J.* A **52** 12
[7] Ring P and Schuck P Springer, Berlin, 1980 *The Nuclear Many-Body Problem*
[8] Goriely S, Hilaire S, Girod M and Péru S 2009 *Phys. Rev. Lett.* **102** 242501
[9] Berger J F, Girod M and Gogny D 1989 *Nucl. Phys.* A **502** 85c
[10] Robledo L M and Berstch G F 2011 *Phys. Rev.* C **84** 014312
[11] Dong T and Ren A 2005 *Eur. Phys. J.* A **26** 69
[12] Brack M *et al* 1972 *Rev. Mod. Phys.* **44** 320; Berstch G F and Flocard H 1991 *Phys. Rev.* C **43** 2200
[13] Giuliani S A, Robledo L M and Rodríguez-Guzmán R 2014 *Phys. Rev.* C **90** 054311

IOP Conf. Series: Journal of Physics: Conf. Series **869** (2017) 012062 doi:10.1088/1742-6596/869/1/012062

Monitoring pulsating giant stars in M33: star formation history and chemical enrichment

A Javadi[1] and J Th van Loon[2]

[1]School of Astronomy, Institute for Research in Fundamental Sciences (IPM), Tehran, 19395-5531, Iran

[2]Lennard-Jones Laboratories, Keele University, ST5 5BG, UK

E-mail: atefeh@ipm.ir

Abstract. We have conducted a near-infrared monitoring campaign at the UK InfraRed Telescope (UKIRT), of the Local Group spiral galaxy M33 (Triangulum). A new method has been developed by us to use pulsating giant stars to reconstruct the star formation history of galaxies over cosmological time as well as using them to map the dust production across their host galaxies. In first Instance the central square kiloparsec of M33 was monitored and long period variable stars (LPVs) were identified. We give evidence of two epochs of a star formation rate enhanced by a factor of a few. These stars are also important dust factories, we measure their dust production rates from a combination of our data with *Spitzer* Space Telescope mid-IR photometry. Then the monitoring survey was expanded to cover a much larger part of M33 including spiral arms. Here we present our methodology and describe results for the central square kiloparsec of M33 [1–4] and disc of M33 [5–8].

1. Introduction

Luminous, cool evolved stars are powerful tracers of the underlying stellar populations, as they stand out above all other stars especially at infrared wavelengths and are thus the first stars that can be resolved in increasingly distant galaxies. Asymptotic Giant Branch (AGB) stars in particular represent stellar populations ranging in age from tens of millions of years (i.e. "the present") to more than ten billion years (formed at redshift ~ 2). The cool molecular atmospheres of AGB stars lead to broad absorption troughs in their optical spectra; while normally these are oxygen-bearing molecules such as TiO and VO, stars in certain mass ranges dredge up carbon – that had been synthesized in their interiors – and replace the absorption bands by entirely different ones due to C_2 and CN. This changes their colours, generally rendering carbon stars redder than their oxygen (M-type) counterparts. In principle, this makes them easy to distinguish photometrically.

Triangulum offers a superb opportunity to study the structure and formation history of a spiral galaxy, and to study various aspects of stellar evolution. The galaxy is sufficiently massive to contain large populations of stars to be used as tracers of star formation, and to sample brief phases of stellar evolution such as those associated with strong mass loss on the AGB. Its disc is seen under a more favourable viewing angle than the larger Andromeda galaxy (M31), and also its more modest size ($\sim 1°$) makes it easier to study it in its entirety. The methodology comprises three stages: firstly, we find stars that vary in brightness with large amplitude (typically a magnitude) and long period (months to years), and identify them on

IOP Conf. Series: Journal of Physics: Conf. Series **869** (2017) 012062 doi:10.1088/1742-6596/869/1/012062

the basis of their colours and luminosity as cool giant stars at the endpoints of their evolution. Secondly, we use the fact that these stars no longer evolve in brightness, to uniquely relate their brightness to their birth mass, and use the birth mass distribution to reconstruct the star formation history (SFH). Lastly, we measure the excess infrared emission from dust produced by these stars, to quantify the amount of matter they return to the interstellar medium in M33.

2. Observations

The project exploits a large observational campaign between 2003-2007, an investment of over 100 hr on the UK InfraRed Telescope (UKIRT). In first instance, only the central square kiloparsec ("bulge") were monitored and analysed, with the UIST camera. Previously, we presented the variable star catalogue and SFH [1,2], where we found a main formation epoch of redshift 1 but with fluctuations up to the present day. As the new wide-field camera (WFCAM) became available, the campaign was expanded to cover two orders of magnitude larger area, comprising the disc of M33 and its spiral arms. Also, we presented the photometric catalogue of WFCAM data [5,6]. This catalogue comprises of 403 734 stars, among which 4643 stars were identified as LPVs– AGB stars, super–AGB stars and RSGs [9].

3. Star formation history

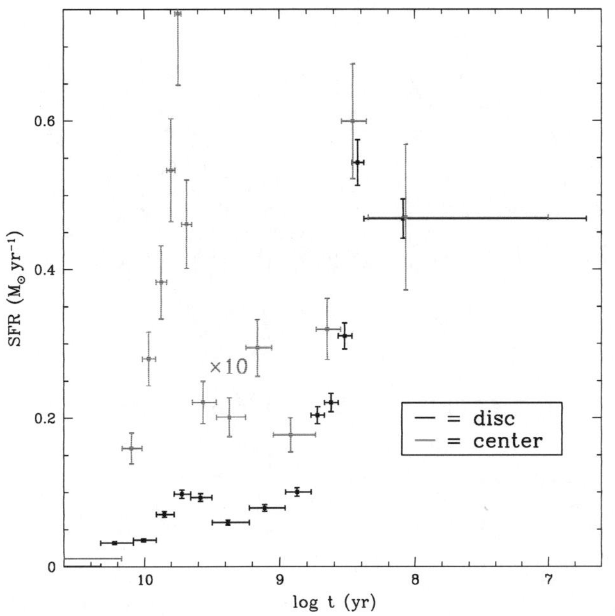

Figure 1. The SFH of M33 in the disc (black) and in the center (red).

Resolved stellar populations within galaxies allow us to derive star formation histories on the basis of colour magnitude diagram modelling, rather than from integrated light. The LPVs are at the end-points of their evolution, and their luminosities directly reflect their birth mass (via the core mass). Stellar evolution models provide this relation. The distribution of LPVs over luminosity can thus be translated into the star formation history, assuming a standard initial mass function. LPVs were formed as recently as < 10 Myr ago and as long ago as > 10 Gyr, so they probe almost all of cosmic star formation. We have successfully used this new technique in

M 33 [2,8], the Magellanic Clouds [10] and NGC 147 and NGC 185 [11,12], using Padova models which also provide the lifetimes of the LPV phase.

In figure 1, we present the SFH for the center and disc of M 33. In the central regions two main epochs of star formation are seen; one that occurred ~ 4–5 Gyr ago, peaking around 4 Gyr ago at a level~ 2.5 times as high as during the subsequent couple of Gyr and the one that occurred ~ 300 Myr–20 Myr ago with a rate ~ 1.5 times higher than the aforementioned peak. Two main epochs of star formation are also seen in the disc of M 33; one that occurred ~ 6 Gyr ago and lasted ~ 3 Gyr and the one that occurred ~ 250 Myr ago and lasted ~ 200 Myr. More that 71 % of stars mass in M 33 were created during the first epoch of star formation and less than 13 % were created during the recent epoch of star formation.

4. Dust production rate of evolved stars

A majority of LPVs have been detected at mid-IR wavelengths with *spitzer* Space Telescope [13]. Therefore, we can estimate the dust production rates across a galactic disc of M 33. We derive mass-loss rates and luminosities in two steps. First, we use DUSTY code [14] to model the spectral energy distributions (SEDs) of LPVs for which mid-IR counterparts have been identified to construct relations between the dust optical depth and bolometric corrections on the one hand, and near-IR colours on the other. Then, by using these relations, we convert the near-IR colours of all red giant variables to luminosities and mass-loss rates.

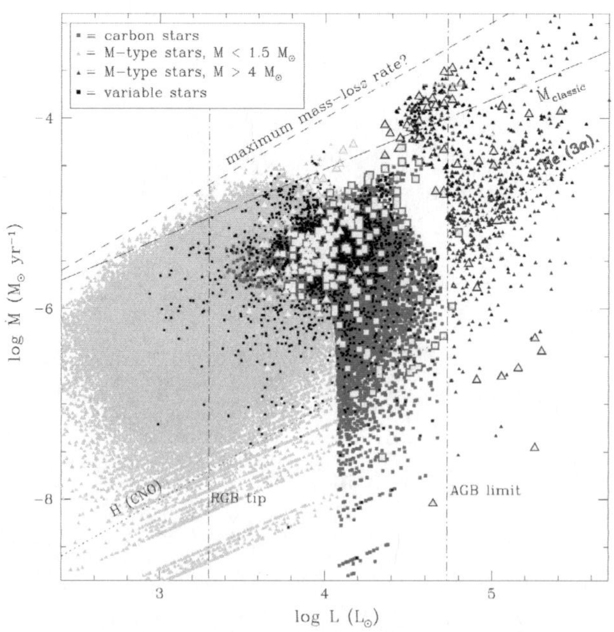

Figure 2. Mass-loss versus luminosity for all stars including non-variable stars. The blue and green triangles show the massive luminous M-type stars and low-mass stars (at lower luminosities), respectively. The red squares show the AGB carbon stars. Large yellow symbols identify the stars modelled with DUSTY [14]; other UKIRT variables are identified by black squares. The vertical dash–dotted lines show the tip luminosity of the RGB and classical limit of the most massive AGB stars (Hot Bottom Burning effects are excluded). The dotted lines show the mass-consumption rates by shell hydrogen burning (CNO cycle) on the AGB and core helium burning (triple-α reaction) in red supergiants. The dashed lines show the limits to the mass-loss rate in dust-driven winds due to single scattering (classic) and multiple scattering.

5. Ongoing-work and conclusions

Currently, we are using the mass-loss rates to investigate the correlations between the dust production rate, luminosity, and amplitude. Also, we are trying to establish a link between the dust return and the formation of stars within the prominent spiral arm pattern. We will show where mass is returned, how this compares to the gas in spiral arms and inter-arm regions, and from this we will estimate gas recycle times and gas depletion in star formation.

In conclusion, using a new technique developed by us, we showed how the SFH varies across M33, e.g. whether star formation has propagated inwards or outwards through the disc. Also, we measured the lag between stars of different ages and the spiral arms in which they were formed.

Acknowledgments

We are grateful for financial support by The Leverhulme Trust under grant No. RF/4/RFG/2007/0297, by the Royal Astronomical Society, and by the Royal Society under grant No. IE130487.

References

[1] Javadi A, van Loon J Th and Mirtorabi M T 2011 *MNRAS* **411** 263
[2] Javadi A, van Loon J Th and Mirtorabi M T 2011 *MNRAS* **414** 3394
[3] Javadi A, van Loon J Th and Mirtorabi M T 2011 *ASPC* **445** 497
[4] Javadi A, van Loon J Th, Khosroshahi H and Mirtorabi M T 2013 *MNRAS* **432** 2824
[5] Javadi A, Saberi M, van Loon J Th, Khosroshahi H, Golabatooni N and Mirtorabi M T 2015 *MNRAS* **447** 3973
[6] Javadi A, van Loon J Th and Khosroshahi H 2015 *PKAS* **30** 355
[7] Javadi A, van Loon J Th and Khosroshahi H 2016 *MmSAI* **87** 278
[8] Javadi A, van Loon J Th, Khosroshahi H, Tabatabaei F and Hamedani Golshan R 2017 *MNRAS* **464** 2103
[9] Saberi M, Javadi A, van Loon J Th and Khosroshahi H 2015 *ASPC* **497** 497
[10] Rezaeikh S, Javadi A, Khosroshahi H and van Loon J Th 2014 *MNRAS* **445** 2214
[11] Golshan R H, Javadi A, van Loon J Th and Khosroshahi H 2015 *PKAS* **30** 169
[12] Golshan R H, Javadi A, van Loon J Th, Khosroshahi H and Saremi E 2017 *MNRAS* **466** 1764
[13] McQuinn K B W *et al* 2007 *ApJ* **664** 850
[14] Ivezić Ž and Elitzur M 1997 *MNRAS* **287** 799

IOP Conf. Series: Journal of Physics: Conf. Series **869** (2017) 012063 doi:10.1088/1742-6596/869/1/012063

Inflationary magnetogensis in R^2-inflation on the light of Planck 2015

A AlMuhammad

President of Qatif Astronomy Society, Qatif, KSA
E-mail: anwar@physics.utexas.edu

Abstract. We study the primordial magnetic field (PMF) generated by the simple inflationary model f^2FF in the context of R^2-inflationary model, the favourite model of Planck 2015 results. The scale invariant PMF is achieved and the backreaction problem is avoided as long as, the rate of inflationary expansion, *H*, is less than the upper bound reported by Planck, 2015, in the observable scales of wave numbers, $k\eta$. Based on these results, we find that the upper limit of present magnetic field, $B_0 < 8.1\times10^{-9}$ G. It is in the same order of the one reported by Planck, 2015.

1. Introduction

Large scale magnetic fields are being observed in all kinds of galaxies and cluster of galaxies at wide range of redshifts. Moreover, some evidences for the presence of the magnetic fields in a very low density intergalactic medium (void) were reported [1]. The generation of such a large scale primordial magnetic fields (PMF) is still an open question in cosmology, see the reviews of this subject [2]. In this paper, the simple inflationary model, f^2FF, of PMF is investigated in the context of R^2-inflation [3], in the same way we did for NI and LFI [4-5]. Further, the reheating parameters are constrained in the same inflationary model by using the reported upper limit of PMF by Planck, 2015 [95]. Finally, the present PMF is constrained based on the scale invariant magnetic field generated during inflationary era.

2. Inflationary Model of PMF in R^2-inflation

In this model, a scalar (*inflaton*) field, ϕ, is coupled to the electromagnetic (vector) field A_μ [7-8] through an unspecified coupling function $f(\phi,t)$. The Lagrangian in this model can be written as,

$$\mathrm{L} = -\sqrt{-g}\left(\frac{1}{2}(\partial_\mu\phi)(\partial^\mu\phi) + V(\phi) + \frac{1}{4}g^{\alpha\beta}g^{\mu\nu}f^2(\phi,t)F_{\mu\alpha}F_{\nu\beta}\right) \quad (1)$$

where, $F_{\nu\beta} = \partial_\nu A_\beta - \partial_\beta A_\nu$ is the electromagnetic field tensor and g is the determinant of the spacetime metric $g_{\mu\nu}$. The first term in the Lagrangian is the standard kinetic part of the scalar field, and the second term, $V(\phi)$, is the potential, which decides the model of inflation.

The main reason behind this coupling is to break the conformal symmetry of electromagnetism and hence prevent the dilution of the seed of magnetic field as it is generated in the inflation era. The potential $V(\phi)$ of R^2-inflation can be written [9] as,

$$V(\phi)=M^4\left[1-\exp\left(-\sqrt{2/3}\phi/M_{Pl}\right)\right]^2, \tag{2}$$

where, M is the amplitude of the potential and it can be determined by the amplitude of CMB anisotropies ($M \simeq 4.0\times10^{-5}M_{Pl}$) [10]. Solving equation (1) for A_μ under slow roll approximations gives,

$$A''(\eta,k)+\left(k^2-\frac{f''}{f}\right)A(\eta,k)=0, \tag{3}$$

where $A(\eta,k)$ is the Fourier transform of $A_\mu(\eta,x)$ and η is the conformal time and k is the comoving wave number. Similarly, in the slow roll limit and by assuming that the relation between couplings function and scale factor is in the power form, $f(\eta)\propto a^\alpha$ where, α is free index to be determined we can find,

$$f(\phi)\propto exp\left[-\frac{\alpha}{3M_{Pl}^2}\int^{\phi}\frac{V(\phi)}{V'(\phi)}d\phi\right]. \tag{4}$$

Hence in a power law expansion, $f(N)=D\left[\left(\frac{4N}{3}\right)^{\frac{\alpha}{4}}e^{-\frac{N\alpha}{3}}\right]$, where, N is the e-folds of inflation.

Therefore, solving (3) by using (2) and (4) gives,

$$A(\eta,k)=(k\eta)^{1/2}\left[C_1(k)J_\chi(k\eta)+C_2(k)J_{-\chi}(k\eta)\right], \tag{5}$$

where, $J_\chi(k\eta)$, is the Bessel function of the first kind. The electric and magnetic spectra can be written respectively as,

$$\frac{d\rho_E}{dlnk}=\frac{f^2}{2\pi^2}\frac{k^3}{a^4}\left|\left[\frac{A(\eta,k)}{f}\right]'\right|^2, \quad \frac{d\rho_B}{dlnk}=\frac{1}{2\pi^2}\left(\frac{k}{a}\right)^4 k\left|A(\eta,k)\right|^2. \tag{6}$$

3. Scale invariant PMF

As magnetic fields are being observed in all scales of the universe, it can be considered as a scale invariant quantity. It implies that the magnetic spectrum can be assumed as constant. Thus, the main task is to investigate the conditions that keep this spectrum constant. One of the main challenges to this model is the backreaction problem, at which the scale of the electromagnetic field can exceed the scale of inflation itself. This problem may spoil the inflation. Hence, any viable model should be free from this problem.

In order to achieve the scale invariance condition in interesting values of N, the values of $\chi(\alpha,N)=5/2$ and $\alpha \simeq \{-7.4, 7.4\}$. Therefore, the scale invariant PMF can be obtained in the long wavelength regime, $k\eta \ll 1$ (outside Hubble radius) around these values of α. It can be depicted along with energy density of the inflation, ρ_{Inf} in figure.1.

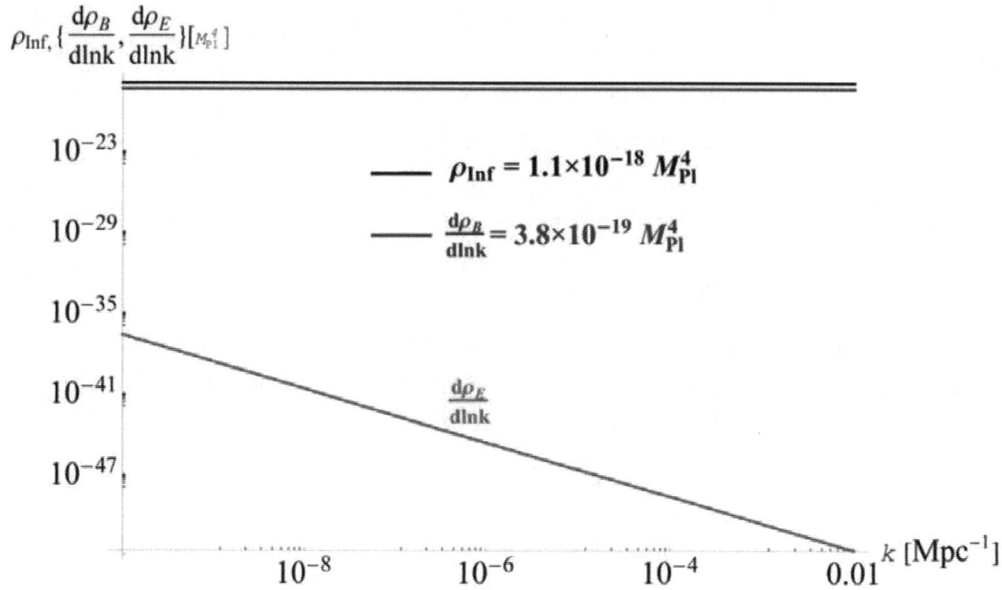

Figure 1. The EM spectra for $\alpha \approx 7.43$, $N = 64$, $\eta = -20$ and the expansion rate (PL) at the pivot scale, $H \approx 3.6\times10^{-5} M_{Pl}$, and $k\eta \ll 1$.

As seen from figure.1 the relation, $\rho_{Inf} > \frac{d\rho_B}{d\ln k} \gg \frac{d\rho_E}{d\ln k}$ is preserved at the rate of expansion, $H \simeq H_* = 3.6\times10^{-5} M_{Pl}$. In fact that value of the Hubble parameter is the upper value during inflation or at the time of pivot scale, at which the space time exits the Hubble radius [11]. Therefore, the problem of the backreaction can be avoided. In order to make sure, the above relation stays valid throughout the inflationary era, we can plot the EM spectra versus η, N, H and α. All of these plots show that the problem of backreaction can be avoided in the interesting cosmological parameters [3].

4. Constraining reheating parameters by means of PMF

The second main result is the constraining of reheating parameters by using R^2-inflation and the present upper limit of PMF reported by Planck, 2015 [6]. In this part, the effect of the scale invariant PMF on the parameters of reheating is investigated. That is basically similar to Ref [12] but by considering R^2-inflationary model instead. The three main values that specify the reheating era are the reheating temperature, $T_{\rm reh}$, the reheating parameter, $R_{\rm rad}$, and the equation of state parameter, $\omega_{\rm reh}$. All of them are constrained by PMF, see how $T_{\rm reh}$ are constrained for different $\omega_{\rm reh}$, in figure.2. It fits well with the range of spectral index reported by Planck 2015 [11]. Also, we calculate the lower limits, $R_{\rm rad} > 6.9$.

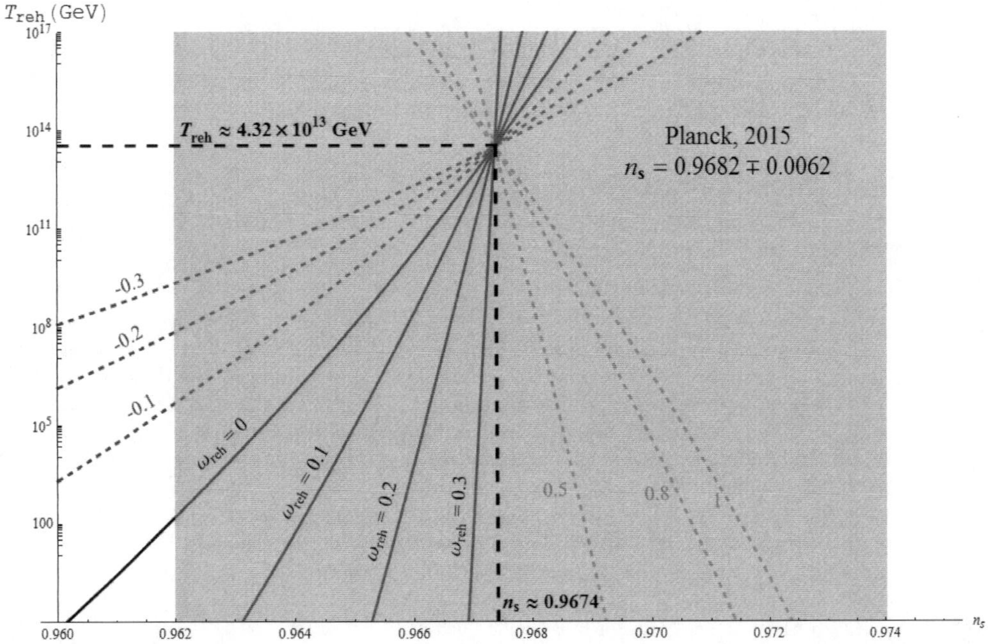

Figure 2. The reheating temperature, $T_{\rm reh}$, versus spectral index, n_s, at the end of R^2-inflation, for some values in $-0.3 < \omega_{\rm reh} < 1$. They all intersect into $T_{\rm reh} \sim 4.32\times10^{13}\,{\rm GeV}$ at $n_s \approx 0.9674$.

5. Constraining the present value of PMF

The main result is the constraining of the value of the present PMF, B_0. Based on the results of the first two parts, we calculate the upper limit of the present PMF as, $B_0 < 8.058\times10^{-9}\,{\rm G}$. It is in the order of ${\rm nG}$, which is the upper bound of PMF reported by Planck 2015 [6].

References

[1] Neronov A and Vovk I 2010 *Sci* **328** 73.
[2] Widrow L *et al* 2012 *Space. Sci. Rev* **166** 37.
[3] AlMuhammad A 2015 Inflationary Magnetogenesis in R^2-Inflation after Planck 2015 *Preprint* astro-ph.CO/1505.05204.
[4] AlMuhammad A and Lopez-Mobilia R 2016, *Astr. Not*, **337** 318
[5] AlMuhammad A and Lopez-Mobilia R 2015 *Gen. Rel. & Grav* **47** 318
[6] Ade P *et al* 2015 Planck 2015 results. XIX. Constraints on primordial magnetic fields *Preprint* astro-ph.CO/1502.01594.
[7] Ratra B 1992 *ApJ.* **391** L1.
[8] Martin J and Yokoyama J 2008 *JCAP* **801** 25
[9] Starobinsky A 1980 *Phys. Lett* **B91** 99.
[10] Martin J, Ringeval C and Vennin V 2014 *Phys. Dark. Univ*, **5** 75
[11] Ade P *et al* 2015 *A & A* **571** A22
[12] Demozzi V and Ringeval C 2012 *JCAP* **1205** 9

Melting behaviour of the candidate minerals for planetary models

S Arafin[1] and R N Singh[1]

[1]Physics department, College of Science, Sultan Qaboos University, Box:36, Al-Khoudh 123, Muscat, Oman

E-mail: sayfin@squ.edu.om

Abstract. The discovery of super-Earths and extra solar giant planets has made high pressure physics relevant for determining the formation and composition of these planets.We have developed a model to determine the pressure dependence of melting temperature, T_m and the Debye temperature, Θ_D of materials utilizing Grüneisen parameter, ξ and the bulk modulus, B as the basic ingredients. The formalism has been applied to estimate the pressure dependence of melting temperature and Debye temperature of one of the candidate minerals such as quartz (stishovite, SiO_2). The pressure range used in the present study is 0-6000GPa commensurate with the interior pressures of giant exoplanets. The results show an excellent agreement with the available experimental and other theoretical results.

1. Introduction

For developing realistic planetary models, knowledge of the melting behavior of the candidate minerals like quartz (stishovite, SiO_2)and MgO at high temperature and pressure is essential. Melting is the most important thermodynamic process that determines the physical and chemical evolution of the interiors of the planets of all sizes. Chemical differentiation and gravitational separation of liquid phase are responsible for the evolution of a terrestrial planet[1]. The melting temperatures, at very high pressures, of the end-members of chemical differentiation such as silicates, iron, iron alloys, and H_2O play an important role with regard to better constraining the internal structure and the physical properties of super-Earths[2]. Silica is not only the most abundant oxide component of the mantles of the terrestrial planets, but also important as a model system to study polymorphic phase changes and interatomic potentials[2,3]. Although free silica may exist locally along the subducted oceanic crust, it may form the bulk composition of super-Earths whose extremely high pressure and temperature conditions allow for greater presence of silica[4-7]. Under these conditions silica and stishovite can have very different properties[8]. The phase diagram of silica becomes, therefore, fundamental to the understanding of the interior of the planets in question.

The pressure variation of melting temperature of silica, at present, is known only up to 500GPa[1] experimentally. Simulations results have reported the melting curve up to 200GPa[9], and recently up to 3400GPa[10]. Since the multimegabar regime corresponds to the conditions found in the core of giant planets and super-Earths[11], it is desirable to explore it for even higher pressures, and from different approaches.

In the present work, we have used a semi-empirical formalism to deal with phase transitions at extremely high pressure and temperature conditions that may exist in the cores of extrasolar planets. Our formalism involves ξ and B and their pressure dependence. Experimentally available values of ξ_0 and B_0 at ambient conditions are the principal ingredients of the present formalism. At higher pressure, we found that the pressure dependence of B and ξ are very sensitive and, change substantially around the structural transformation. We have also incorporated Debye temperature Θ_D in our analysis because of its importance in characterization of rocks and minerals[12]. Any change in the external condition of rocks and other materials formation and transformation can lead to a change in Θ_D.

The layout of the paper is as follows: theoretical formalism is presented in section 2, results and discussions of pressure dependence of the melting temperature, Debye temperature of silica in the pressure range of 0-6000GPa are given in section 3, which is followed by conclusions in section 4.

2. Theoretical formalism

We present here briefly the mathematical formalism leading to semi empirical relation to determine the pressure dependence of melting temperature and the Debye temperature. A detailed formalism to derive the relation of melting temperature as a function of pressure can be found in our earlier works[13-15].

2.1. Melting temperature

The Debye model of solids was incorporated into the present formalism to obtain a relation for the variation of the melting temperature, T_m with pressure, P in terms of bulk modulus B(P), and the Grüneisen parameter $\xi(P)$,

$$(lnT_m)' = (2/B)(\xi - 1/3), \text{where prime denotes differentiation w.r. to } P. \quad (1)$$

Grüneisen parameter, ξ is a measure of the anharmonicity of thermal vibrations in solids. In order to get better information, B and ξ are further expanded in terms of P as

$$B(P) = B_0 + b_1 P + b_2 P^2 + ... \quad (2)$$

$$\xi(P) = \xi_0 + a_1 P + a_2 P^2 + a_3 P^3 + ... \quad (3)$$

We have expanded ξ up to third order to account for its sensitivity on melting at higher pressures where contributions from second order phase transformations may be important. Further expansion of B has been avoided because of complexity in deriving an analytical solution. ξ_0 and B_0 are values at ambient conditions which have been extensively measured for various materials. The coefficients a_n and b_n are the pressure derivatives of ξ and B. Inserting Eqs.(2)-(3) in Eq.(1) and, integrating from pressure 0 to P one can get two solutions for two conditions[13,14].

Solution for condition I: For $b_1^2 < 4B_0b_2$

$$T_m(P) = T_0 \left[\left(1 + \frac{b_1}{b_0}P + \frac{b_2}{B_0}P^2\right)^n exp\left(2DP + \frac{a_3}{b_2}P^2\right)\right] exp\left(\frac{2C}{A}tan^{-1}\frac{AP}{2B_0 + b_1P}\right), \quad (4)$$

$$A = \sqrt{4B_0b_2 - b_1^2}, \quad (5)$$

$$C = \left[2\xi_0 - 2/3 - a_1b_1/b_2 + ((b_1^2 - 2B_0b_2)/b_2^2)\left(a_2 - a_3b_1/b_2^2\right) + a_3B_0b_1/b_2^2\right], \quad (6)$$

$$D = a_2/b_2 - a_3b_1/b_2^2, \quad (7)$$

$$n = a_1/b_2 - a_2b_1/b_2^2 - a_3B_0/b_2^2 + a_3b_1^2/b_2^2. \quad (8)$$

Solution for condition II: For $b_1^2 > 4B_0b_2$

$$T_m(P) = T_0\left[\left(1+\frac{b_1}{b_0}P+\frac{b_2}{B_0}P^2\right)^n exp\left(2DP+\frac{a_3}{b_2}P^2\right)\right]\left\{\frac{\left(b_1+A'\right)P+2B_0}{\left(b_1-A'\right)P+2B_0}\right\}^{C/A'}, \quad (9)$$

$$A' = \sqrt{b_1^2-4B_0b_2};\quad T_0 \text{ is the melting temperature at atmospheric pressure.} \quad (10)$$

2.2. Pressure dependence of Debye temperature, Θ_D

It has been shown[15] earlier that within the Debye-Grüneisen Model the relationship among ξ, B and Θ_D as a function of pressure, P may be written in the form:

$$\frac{1}{\Theta_D(P)}\frac{d\Theta_D(P)}{dP} = \frac{\xi(P)dP}{B_T(P)} \quad (11)$$

Inserting Eqs.(2)-(3) in Eq.(11) and integrating from pressure 0 to P, we get two solutions in terms of Debye temperature Θ_{D0} at ambient condition.

Solution for condition I: For $b_1^2 < 4B_0b_2$

$$\Theta_D(P) = \Theta_{D0}\left[\left(1+\frac{b_1}{b_0}P+\frac{b_2}{B_0}P^2\right)^{n/2} exp\left(DP+\frac{a_3}{2b_2}P^2\right)\right] exp\left(\frac{2C'}{A}tan^{-1}\frac{AP}{2B_0+b_1P}\right) \quad (12)$$

Solution for condition II: For $b_1^2 > 4B_0b_2$

$$\Theta_D(P) = \Theta_{D0}\left[\left(1+\frac{b_1}{b_0}P+\frac{b_2}{B_0}P^2\right)^{n/2} exp\left(DP+\frac{a_3}{b_2}P^2\right)\right]\left\{\frac{\left(b_1+A'\right)P+2B_0}{\left(b_1-A'\right)P+2B_0}\right\}^{C'/A'} \quad (13)$$

$$C' = \left[\xi_0 - a_1b_1/2b_2 + ((b_1^2-2B_0b_2)/2b_2^2)\left(a_2-a_3b_1/b_2^2\right)+a_3B_0b_1/2b_2^2\right] \quad (14)$$

The constants A, D, n and A' are defined earlier in Eqs.(5), (7), (8)and (10) respectively.

3. Results and discussions

The results of our study are presented in Fig.1(A-C). Figure 1A is a plot of low and high pressure melting curves together. It shows an excellent fit of our melting curve (thick solid line) with that of[2] obtained from first principles molecular dynamics simulations for pyrite, cotunnite and Fe_2P phases of silica. Since it is difficult to show discontinuity in the melting curve from theoretical formulations, we have determined the melting curves for the low pressure (0-500GPa) and high pressure (300-6000GPa) regions separately for two sets of parameters (Table 1) to account for the reported phase transition of silica melting curve at about 300GPa. The melting curve obtained from the present formalism shows a minimum slope of 1.34K$\cdot$GPa^{-1} at 3400GPa.

Characteristic points of a melting curve such as maximum and inflexion point are often interpreted to be associated with phase transition. The low pressure region of the present melting curve plotted separately in figure 1B has an excellent fit with that of the experimental curve [1]. Our melting curve shows a minimum slope of 8.09K$\cdot$GPa^{-1} at 350GPa. This value is comparable with the phase transition value of 330GPa found in molecular dynamic simulations studies [2]. Figure 1C which is a plot of Θ_D against pressure, shows a minimum slope of 0.215K$\cdot$GPa^{-1} at 3800GPa. This value of pressure is close to the value, 3400GPa obtained in the case of the melting curve. It can be noticed that the shape of the T_m-P curve is very similar to that of Θ_D-P curve indicating an intimate relationship between the two.

Table 1. Grüneisen parameter, ξ_0 and bulk modulus, B_0(GPa)at ambient conditions. $a_n(\text{GPa}^{-n})$ and $b_n(\text{GPa}^{n-1})$ are the derivatives of ξ and B.

Minerals	ξ_0	a_1	a_2	a_3	B_0	b_1	b_2
α-quartz	1.35	-0.0001035	1.2022x10^{-5}	–	37[8]	6.0[8]	0.1539
stishovite	1.28	-0.00022	5.1044x10^{-10}	2.1969x10^{-11}	295[8]	1.3[8]	0.001882

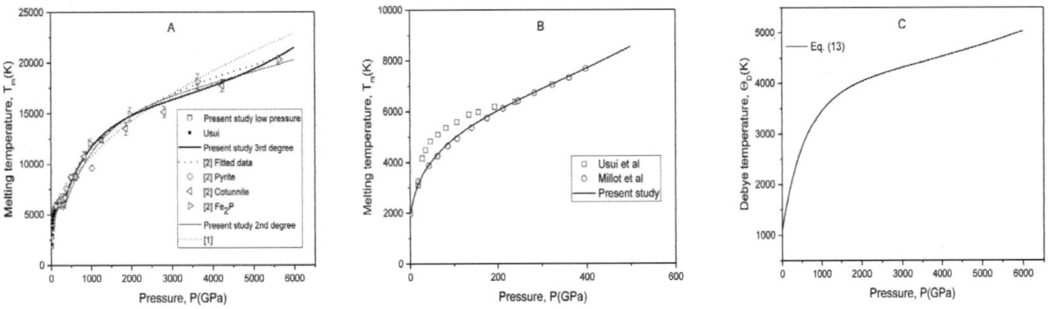

Figure 1. (A) Comparison of the melting curve determined in this study(thick solid line) with those of[1,2,9]. (B) T_m-P curve for low pressure region (0-500GPa). (C) Θ_D -P curve is computed from Eq.(13) using the parameters for high pressure only.

4. Conclusions

An empirical relation to compute T_m-P variation at high pressures is formalised in terms of bulk modulus(B) and Grüneisen parameter(ξ). This is applied to compute the melting temperature of quartz(stishovite, SiO_2) as a function of pressure in the range 0 - 6000GPa. Such a pressure is commensurate to the interior pressure of giant planets. Our computed results are in excellent agreement with available experimental results in the low pressure region. Results exhibit a discontinuity in slope around 350GPa which is comparable with the phase transition value of 330GPa found from molecular dynamic simulation. It has been observed that the value of B and ξ of stishovite at high pressures are quite different from those of α-quartz in the low pressure region.

References

[1] Millot M Dubrovinskaia N, *Černok* A, Blaha S, Dubrovinsky L, Braun D G, Celliers P M, Collins G W, Eggert J H and Jeanloz R 2015 *Science* **347** 418
[2] Gonzalez-Cataldo F, Davis S and Gutierrez G 2016 *Nature* **6** 26537
[3] Hemley R J, Prewitt C T and Kangma K 1994 *Rev. Mineral.* **29** 41
[4] Duffy T, Madhusudhan N and Lee K Mineralogy of Super-Earth Planets 2015 **2** (Elsevier BV)
[5] Valencia D, Ikoma M, Guillot T and Nettelmann N 2010 *Astron. Astrophys.* **516** A20
[6] Elkins-Tanton L T and Seager S 2008 *Astrophys. J.* **685** 1237
[7] Schaefer L and Fegley B 2009 *Astrophys. J.* **703** L113
[8] Lin-gun Liu 1993 *Mech. Mater.* **14**(4) 283
[9] Usui Y and Tsuchiya T 2010 *J. Earth Sci.* **21** 801
[10] Mazevet S, Tsuchiya T, Taniuchi T, Benuzzi-Mounaix A and Guyot F 2015 *Phys. Rev. B* **92** 014105
[11] Umemoto K, Wentzcovitch R M and Allen P B 2006 *Science* **311** 983
[12] Dergachov A L and Starostin V I 1981 *Ore Deposit Geology* **6** 67
[13] Arafin S, Singh R N and George A K 2013 *Physica B* **419** 40
[14] Arafin S and Singh R N 2014 *Appl. Phys. A* **117**(3) 1055
[15] Arafin S and Singh R N 2016 *J. Phys. Chem. Solids* **99** 86

Gamma-Ray Bursts: Characteristics and Prospects

W J Azzam[1], H Zitouni[2] and N Guessoum[3]

[1]Department of Physics, College of Science, University of Bahrain, Sakhir, Bahrain
[2]PTEAM Laboratory, Faculty of Science, University Dr. Yahia Fares, Algeria
[3]Department of Physics, American University of Sharjah, UAE

Email: wjazzam@ucb.edu.bh

Abstract. Gamma-ray bursts (GRBs) are the most powerful explosions in the universe. They have remained the object of intense research ever since their discovery was declassified in the early 1970s. Several space-borne missions have been dedicated to their study, including the Compton Gamma-Ray Burst Observatory (CGRO) in the 1990s and the current *Swift* and *Fermi* satellites. However, despite several decades of focused research, the precise mechanisms behind these enigmatic explosions have not been fully established. In the first part of this paper, we review what is currently known about GRBs. This includes: GRB light-curves and spectra; the different progenitor models, i.e., the "collapsar" and "merger" models; and the afterglow characteristics, including external shocks and the surrounding medium. In the second part of the paper, we present our work, which focuses on utilizing GRBs as cosmological probes. GRBs are ideal cosmological tools, because they have been observed to great distances (redshifts up to $z = 9.4$) and their radiation is unencumbered by any intervening dust. Although GRBs are not standard candles, the discovery of several energy and luminosity correlations, like the Amati relation which correlates the intrinsic spectral peak energy, $E_{p,i}$ to the equivalent isotropic energy, E_{iso}, has ushered in a new era in which GRBs are used to investigate cosmological issues like the star formation rate and the value of the matter-density parameter, Ω_M.

1. Introduction

Gamma-ray bursts (GRBs) are the most powerful explosions in the universe. Their total isotropic energy is in the range of 10^{48}–10^{55} erg [1–6]. A bimodal distribution for GRB durations was detected in the BASTSE data [7], which led to the classification of GRBs into short bursts ($T_{90} < 2$ s) and long bursts ($T_{90} > 2$ s), where T_{90} refers to the time needed for 90% of the fluence to be accumulated [8]. During the 1980s and early part of the 1990s, there was a debate concerning whether GRBs had cosmological or galactic origins. The isotropic spatial distribution of GRBs obtained by BATSE strongly implied a non-galactic origin. The issue was settled in 1997 when the first afterglow observation by the *BeppoSAX* satellite succeeded in pinning down the redshift of GRB970228 [9, 10].

In this paper, we provide a brief review of the main characteristics of GRBs, followed by a discussion of the prospects of using GRBs as cosmological tools. We also briefly present some of our recent results in this area.

IOP Conf. Series: Journal of Physics: Conf. Series **869** (2017) 012065 doi:10.1088/1742-6596/869/1/012065

2. Prompt and Afterglow Emissions

The prompt emission of GRBs consists of high-energy gamma-ray photons as well as hard X-ray photons. The spectrum is non-thermal, and the energy flux has a peak at around a few hundred keV. Although the spectrum varies from one burst to another, it can be fit very well by the Band function, which consists of two power laws joined smoothly at a break energy [11].

The temporal structure of the prompt emission is highly variable. First, the duration of GRBs spans a wide range: from less than 0.1 s to more than 1000 s. Second, within a certain burst, the light curve usually shows dramatic variations in the flux (sometimes by as much as 100%) on very short timescales. In fact, the variability timescale, Δt, is typically a factor of 10^4 shorter than the GRB's duration, T_{90}.

Unlike the prompt emission which is attributed to internal dissipation, the afterglow emission is attributed to external dissipation. As the outflow from some catastrophic event interacts with the surrounding medium, it dissipates energy through external shocks. The afterglow emission ranges from X-rays all the way down to radio waves [12–14].

3. Progenitor Models

Observations indicate that some (but not all) long bursts are associated with supernovae, hence with the violent death of stars. According to the "collapsar" model (the most popular current model for long bursts) when a very massive (M >~ 30 M_{sun}), low-metallicity, rapidly spinning star collapses at the end of its life, the "hypernova" produces a black hole at the core and produces an explosion. The infalling material forms a dense accretion disk that ejects two relativistic jets along the axis of rotation [15, 16].

The "collapsar" model, however, does not explain the formation of short bursts, which typically occur in regions with limited star formation. Several models have been proposed to explain the formation of short GRBs. The most successful models involve the merger of two compact objects, like two neutron stars. As the two compact objects spiral slowly toward one another, they release gravitational energy until the tidal forces rip the progenitors apart and the system collapses into a black hole. The infalling material creates an accretion disk which emits a burst of energy, similar to what is produced in the "collapsar" model but with shorter durations [17, 18].

4. Cosmological Prospects

GRBs hold great potential as cosmological tools for two reasons. First, they have been observed to very high redshifts, up to z = 9.4 so far and with hopes of reaching z = 20 with future satellites [19]. Second, their gamma radiation is unencumbered by any intervening dust. Although GRBs are not standard candles, the discovery of several energy and luminosity correlations (see below) has permitted their use as cosmological probes.

Early attempts to utilize GRB correlations to constrain cosmological models faced several hurdles [19, 20]. First, there was the issue of the paucity of GRBs with measured redshifts. Second, there was the problem of the wide scatter in the correlations. Finally, there was the issue of the circularity problem, which refers to the fact that in order to calibrate the energy and luminosity correlations, one must assume a cosmological model, which defeats the goal of determining cosmological parameters from GRB characteristics. However, as the quality and the abundance of the data improved, there has been a revived interest in GRB cosmology [21].

In a recent study [22], we used a sample of 126 *Swift* bursts to put limits on the cosmological density parameters (Ω_M and Ω_Λ) by employing two GRB correlations: the Amati relation and the Dainotti relation. The Amati relation is a correlation between the intrinsic spectral peak energy, $E_{p,i}$, and the equivalent isotropic energy, E_{iso}, which can be written as

$$E_{p,i}\,(\text{keV}) = K \times [E_{iso} \,/\, 10^{52}\ \text{erg}]^m, \tag{1}$$

where K and m are fitting parameters. The Dainotti relation is a correlation between the X-ray afterglow luminosity, L_X, and the break time, T_a, which is observed in the X-ray flux. We calibrated the Dainotti relation and expressed it as

$$[T_a / (1 + z)] = 10^{25.2} L_X^{-0.46}. \qquad (2)$$

Our analysis involves employing a maximum likelihood method to carry out a "reverse job" in which we search for the cosmological parameters that provide the best fit, which occurs when the likelihood function $-\ln(L)$ is minimized [22]. The fit is not given by specific values but rather by contours that correspond to the same values of $-\ln(L)$, which is achieved by numerically varying Ω_M and Ω_Λ between 0 and 1. We stress that in order for this approach to work, it is important to have high-quality data with little scatter, otherwise the values of Ω_M and Ω_Λ do not converge.

We sum up our results as follows:

- Our analysis indicates that the results obtained using the Amati relation favor a universe dominated by dark energy, whereas those obtained using the Dainotti relation favor a universe dominated by matter.
- The results we obtain are qualitatively in agreement with those obtained using other methods (like Supernovae Ia and the Cosmic Microwave Background).
- For our technique to work well, it is important to obtain large data sets with low scatter.
- We were able to numerically reduce the scatter in the Dainotti relation before utilizing it to place limits on Ω_M and Ω_Λ.

In the future, we hope to pursue this line of analysis by studying other GRB correlations, like the Ghirlanda relation and the Yonetoku relation, and by using larger data sets with the aim of obtaining more stringent limits on the values of the cosmological parameters.

5. Conclusion

Gamma-ray bursts are the most distant and powerful explosions in the universe. They display a wide variety in terms of their light curves and spectra. Although several progenitor models have been proposed to explain the formation of long and short bursts, the precise mechanism behind these enigmatic explosions has yet to be established. In recent years, there has been a revived interest in GRB cosmology, in which correlations between GRB parameters (energies, luminosities, spectra break time, etc.) are utilized to investigate and discriminate between cosmological models. Our study indicates that such an approach is indeed fruitful because it manages to put limits on cosmological parameters.

References

[1] Nakar E 2007 *Phys. Rep.* **442** 166
[2] Zhang B 2011 *Comptes Rendus Physique* **12** 206
[3] Gehrels N and Razzaque S 2013 *Frontiers of Physics* **8** 661
[4] Berger E 2014 *ARA&A* **52** 43
[5] Kumar P and Zhang B 2015 *Phys. Rep.* **561** 1
[6] Mészáros P 2006 *Reports on Progress in Physics* **69** 2259
[7] Meegan C A *et al* 1992 *Nature* **355** 143
[8] Kouveliotou C *et al* 1993 *ApJ* **413** L101
[9] Costa E 1997 *Nature* **387** 783
[10] Van Paradijs J *et al* 1997 *Nature* **386** 686
[11] Band D et al. 1993 *ApJ* **413** 281
[12] Dermer C D 2007 *ApJ* **664** 384

[13] Frail D A *et al* 2000 *ApJ* **534** 559
[14] Frontera F *et al* 2000 *ApJS* **127** 59
[15] MacFadyen A I and Woosley S E 1999 *ApJ* **524** 262
[16] Woosley S E 1993 *ApJ* **405** 273
[17] Vietri M and Stella L 1998 *ApJL* **507** L45
[18] MacFadyen A I 2006 *AIP Conference Proceedings* **836** 48
[19] Azzam W J and Alothman M J 2006a *Adv. Space Research* **38** 1303
[20] Azzam W J and Alothman M J 2006b *Nuovo Cimento B* **121** 1431
[21] Dainotti M G, Del Vecchio R and Tarnopolski M 2016 *Preprint* arXiv:1612.00618v1
[22] Zitouni H, Guessoum N and Azzam W J 2016 *Astrophys. & Space Science J.* **361** 383

IOP Conf. Series: Journal of Physics: Conf. Series **869** (2017) 012066 doi:10.1088/1742-6596/869/1/012066

A Bayesian approach shows no correlation between transit-depth and stellar metallicity for confirmed and candidates *Kepler* gas giants planets

C Nehmé[1,2] and P Sarkis[3]

[1]Department of Physics & Astronomy, Notre Dame University- Louaize, PO Box 72, Zouk Mikael, Lebanon
[2]CEA/DRF/Irfu/SAp, F-91191 Gif-sur-Yvette, France
[3]Universität Bern Space Research & Planetary Sciences Division Bern, Switzerland

E-mail: cnehme@ndu.edu.lb

Abstract. Previous study to investigate the correlation between the transit depth and the stellar metallicity of *Kepler's* (Q1-Q12) gas giant planets (radii of 5-20$R_{\odot}$) has led to a weakly significant negative correlation. We use the cumulative catalog of planets detected by the NASA *Kepler* mission Q1-Q17 catalog, as of April 2015, to perform a solid statistical analysis of this correlation. In the present work, we revise this correlation, within a Bayesian framework, for two large samples: sample A confirmed planets and sample B (confirmed + candidates). We expand a hierarchical method to account for false positives in the studied samples. Our statistical analysis reveals no correlation between the transit depth and the stellar metallicity. This has implications for planet formation theory and interior structure of giant planets.

1. Introduction

NASA's Kepler Mission revolutionized the field of extra solar planets. We now can place statistical constraints on the (observed) properties of planets as well as on theories of planet formation. Clues on the nature of giant planet formation might be revealed from two correlations with stellar metallicity of main sequence stars hosting these planets. The first one is the correlation between the number of giant planets and stellar metallicity revealed by radial velocity surveys[1] and by transit surveys [2]. The second correlation is a positive trend between the mass of heavy-elements in giant planets and the stellar metallicity [3]. As *Kepler* is searching for transiting exoplanets around their host stars. It is primordial to study the correlation between the transit depth and the stellar metallicity of *Kepler's* gas giant candidates. Earlier studies conducted on this correlation between has led to different results. For instance, Dodson-Robinson reported a negative correlation with a weak statistical significance. The author studied the transit depth of 213 giant planets from (Q1-Q12) catalog with estimated radii of 5-20$R_{\oplus}$ with the values of [Fe/H] taken from the *Kepler* Input Catalog (KIC). Dodson-Robinson [4] interpreted the negative correlation as evidence that metal rich planets of a given mass are denser than their metal poor counterparts leading to small radii [6]. Here, we will use the latest available catalog Q1-Q17 [7] to study correlation between the transit depth and stellar metallicity.

IOP Conf. Series: Journal of Physics: Conf. Series **869** (2017) 012066 doi:10.1088/1742-6596/869/1/012066

In the previous study, the author have assumed that the measurement uncertainties are negligible. This assumption is not justified because these uncertainties, especially on radius measurements that depend on the stellar parameters, can be large compared to the scales of interest. Also, a very important component of any study of exoplanet populations is the treatment of detection efficiency. In a transit survey, small planets with long periods are much harder to detect than large planets orbiting close to their stars. Moreover, large-scale surveys, such as *Kepler*, are subject to selection effects and biases. *Kepler* will likely detect a bigger number of planets around bright stars than around faint stars. Thus, a sample of stars with transiting planets may not accurately represent the true intrinsic distribution of the discovered planets. Gaidos E and Mann A [8] reported the importance of including these effects since they can lead to biases in the properties of transiting planets and their host stars. We develop a framework to account for the uncertainties present on the observed parameters and we use Bayesian statistics to study the correlation. Our method is based on hierarchical Bayesian modeling introduced by [9]. We update the method to account for the false positive rate of *Kepler* giant planets.

2. Selection criteria and samples

We use the cumulative catalog of planets detected by the NASA *Kepler* mission which, as of April 2015, consisted of Q1-Q17 catalog [7]. Following [4] and [10] we define gas giant planets as planets having a radius between 5-20$R_\oplus$. The stellar parameters were taken from the *Kepler* stellar Q1-Q16 database [11]. We ended up by having a sample of 390 planets of which 84 are confirmed and 306 are candidates giant planets.

With the goal of performing a robust statistical method, we prepared two different samples. The first sample (sample A) consists of all the 84 confirmed giant planets. The second sample (sample B) contains the 390 confirmed + candidates. We then restrict sample B to stars with a *Kepler* magnitude $K_p < 16$ mag to avoid faint stars and we choose FGK Main sequence stars for *Kepler's* high efficiency of detection giant transiting planets around those stars. In addition to the cuts made on the stellar sample we also require $P < 90$ days because of the low efficiency at detecting long period [12] and SNR > 18.8. After preforming all the cuts (table 1), sample B will be left with 105 confirmed + candidates.

Table 1. Summary of the cuts performed to obtain sample B

Parameter	Value
Stellar effective temperature, T_{eff}	4000 - 7000 K
Stellar gravity, $\log g$ (cm/s^{-2})	4.0 - 5.0
Stellar Radius, $R_\star$	0.7 - 1.4 $R_\odot$
Planetary Radius, R_p	5 - 20 $R_\oplus$
Orbital Period, P	< 90 days
Detection Efficiency, SNR	> 18.8
Kepler magnitude, K_p	< 16 mag

3. Method

Hierarchical Bayesian Modeling (hereafter HBM) allows for intrinsic scatter and heteroscedastic measurement errors i.e the uncertainties for each data points are different. We followed an approach similar to that proposed by Kelly B C [9]. We constructed the likelihood function in a simple way in order to relate the parameters of interest to the observed data, taking into account

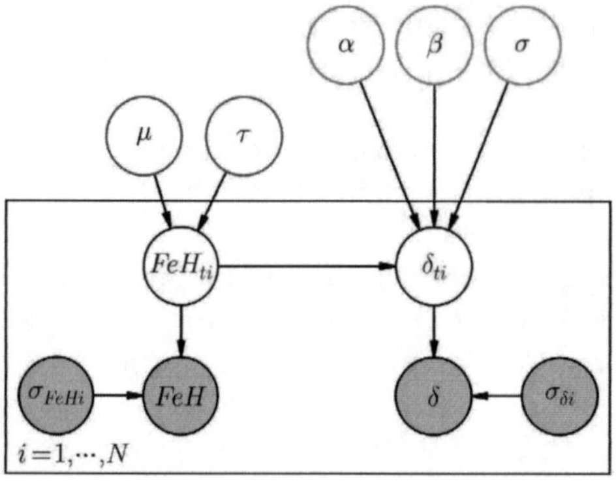

Figure 1. A graphical presentation of our HBM is given in this illustration. The graynodes are the observed parameters. The true missing parameters are in the white nodes. The blue (upper left) nodes are the nuisance parameters and the red nodes(upper right) are the parameters of interest. FeH_i = stellar parameter of the ith planet, $\sigma_{FeH,i}$ = uncertainty on the stellar metallicity of the ith planet, δ_i = transit depth of the ith planet, $\sigma_{\delta,i}$ = uncertainties on the transit depth of the ith planet, FeH_{ti} = true stellar metallicity of the ith planet, δ_{ti} = true transit depth of the ith planet, μ and τ = *nuisance* parameters, α, β and σ = parameters of the linear model

the measurements uncertainties. We extended the method in [9] to account for false positive by Fressin F [13]. This updated statistical method is used to study the correlation between the transit depth (δ) and the metallicity (FeH) of host stars. A graphical illustration of our HBM is given in Figure 1. Markov chain Monte Carlo (hereafter MCMC) was performed using the Python package PySTAN, a package for Bayesian inference. We ran models with 4 Markov Chains, with 5000 iterations for the first sample and 10 000 iterations for the second one. The first 50 per cent of each chain was discarded as "burn-in". This work is the first to perform a full HBM to study correlations in general and the correlation between the transit depth and the metallicity of *Kepler's* giant planets, in particular. Most importantly, the quantification of the intrinsic dispersion which has not been characterized before, is now defined.

4. Results

The equation of the "best-fit" linear models are : $\delta = (0.07 \pm 0.014) + (0.02 \pm 0.14) FeH$ with an intrinsic scatter of $\sigma = 0.03 \pm 0.005$ for the confirmed planets sample (sample A). For sample B the best fit and the intrinsic scatter are $\delta = (0.13 \pm 0.06) + (-0.18 \pm 0.06) FeH$ and $\sigma = 0.05 \pm 0.02$, respectively. The transit depth is plotted against the metallicity of the host star along with their uncertainties, in the right panel of the Figure 2. The "best-fit" models (dotted lines) are shown along with 100 random samples from the MCMC chain. This shows clearly the existence of a large intrinsic scatter. A robust way to demonstrate the absence of correlation between the transit-depth and the stellar metallicity.

5. Discussion

We presented within a Bayesian framework, a study of the correlation between the transit depth of *Kepler's* giant planets and the metallicity of the host star. Data from *Kepler*(Q1-Q17) [7] allowed us to characterize the intrinsic scatter in the relation with a robust statistical analysis. We expand the hierarchical model presentend in [9], to account for the false positive rates. We also considered in the model the relevant selection effects. The use of Markov Chain Monte Carlo (MCMC) to fit models to observations is becoming a standard practice in astronomy. We

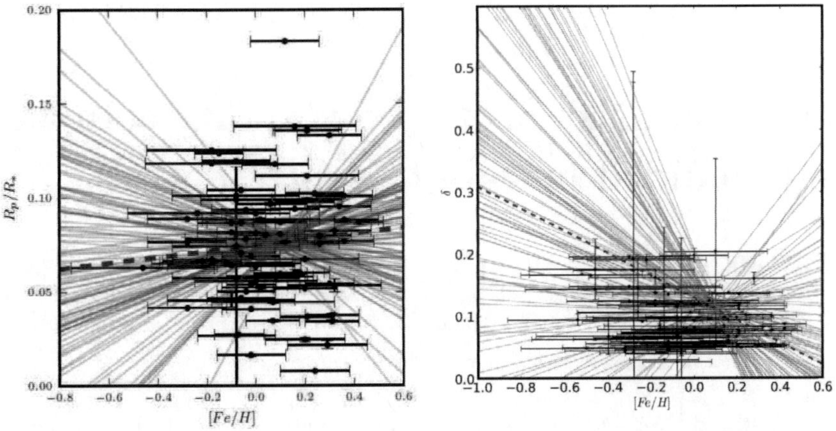

Figure 2. Transit depth (δ) of *Kepler's* giant planets vs. the metallicity of the host star ([Fe/H]).The dashed lines represent the best fit line and the light lines are samples from the MCMC chain. The dashed lines represent the best fit line and the light lines are samples from the MCMC chain.

performed MCMC using the package PyStan to estimate the parameters of the hierarchical linear model. We established that there is no correlation between the transit depth and the stellar metallicity of *Kepler's* gas giant planets. Our result indicate that there is a relatively large intrinsic scatter in the relation. Hence, the previous results could probably be an artifact which shows the importance of accounting for uncertainties and for possible false positives. It also proves the importance of accounting for selection effects and biases within the transit surveys, such as *Kepler* [4] interpreted the negative correlation as evidence that metal-rich planets of a given mass are denser than their metal-poor counterparts, leading to smaller radii [6]. On the contrary, with our robust statistical method, we have proven the independence of transit depth and the stellar metallicity. It certainly warrants further investigations to check what planetary formation model can explain the outcome.

References

[1] Fischer D A and Valenti J **2005** *ApJ* **622** 1102
[2] Wang J and Fischer D A **2013** *arXiv:astro-ph/*1310.7830
[3] Miller N and Fortney J J **2011** *ApJ* **736** L29
[4] Dodson-Robinson S E **2012** *ApJ* **752** 72
[5] Borucki W J, Koch D G, Basri G, Bathalha N *et al* **2011** *ApJ* **736** 19
[6] Fortney J J and Nettelmann N **2010** *Space Sci. Rev.* **152** 423
[7] Mullaly F, Coughlin J L, Thompson S E, Rowe J *et al* **2015** *ApJS* **217** 31
[8] Gaidos E and Mann A **2013** *ApJ* **762** 81
[9] Kelly B C **2007** *ApJ* **665** 1489
[10] Schaufman K C and Laughlin G **2011** *ApJ* **738** 177
[11] Huber D *et al* **2014** *ApJS* **211** 2
[12] Petigura E A, Howard A W and Marcy G W **2013** *PNAS* **110** 19273
[13] Fressin F *et al* **2013** *ApJ* **766** 81

IOP Conf. Series: Journal of Physics: Conf. Series **869** (2017) 012067 doi:10.1088/1742-6596/869/1/012067

Thin accretion disks around a rotating compact object surrounded by quintessence matter

E Larranaga

Observatorio Astronomico Nacional. Universidad Nacional de Colombia. Ciudad Universitaria. Bogota D.C., Colombia

E-mail: ealarranaga@unal.edu.co

Abstract. We study the influence of quintessential matter on the equations describing the properties of a thin accretion disk around a rotating compact object. From these, we show that quintessence matter may play an important role in the accretion process.

1. Introduction

Current astronomical observations suggest that the Universe is undergoing an accelerated expansion. One model used to explain this behavior is the existence of a dynamic scalar field known as quintessence. This field has been incorporated by Kiselev [1] in a static and spherically symmetric solution of the Einstein field equations, which has been generalized to a rotating solution in [2–4]. Some general aspects of spherical accretion onto static Kiselev black holes have been presented in [5]. In this work, we will derive the equations describing the properties of a thin disk [6–8] around a rotating compact object surrounded by quintessence matter to show that this field may play an important role in the accretion.

2. The Rotating Kiselev Solution

Recently, a rotating generalization of the Kiselev's metric [1] was obtained using the Newman-Janis algorithm [2–4] with a line element in Boyer-Lindquist coordinates written as

$$ds^2 = -\left(\frac{\Sigma\Delta}{\Xi} + \frac{\varpi^2\Xi\sin^2\theta}{\Sigma}\right)dt^2 - \frac{2\varpi\Xi\sin^2\theta}{\Sigma}dtd\phi + \frac{\Sigma}{\Delta}dr^2 + \Sigma d\theta^2 + \frac{\Xi}{\Sigma}\sin^2\theta d\phi^2, \qquad (1)$$

where $\Delta = r^2 + a^2 - 2Mr - \alpha r^{1-3\omega}$, $\Sigma = r^2 + a^2\cos^2\theta$, $\Xi = \left(r^2+a^2\right)^2 - a^2\Delta\sin^2\theta$ and $\varpi = \frac{a}{\Xi}\left(2Mr + \alpha r^{1-3\omega}\right)$. This solution describes a compact object characterized by its mass M, angular momentum a and a parameter α which measures the intensity of a quintessence field with equation of state $P_q = \omega\rho_q$, where P_q and ρ_q are its pressure and energy density [1,9,10] and the state parameter takes values in the range $-1 < \omega < -\frac{1}{3}$. In this work, we are interested in the particular value $\omega = -\frac{2}{3}$, which describes an object surrounded by quintessence matter. For black holes, the horizon structure of this solution is defined by the relation $\Delta = 0$, which gives, in general, three horizons: an inner r_-, an outer r_+ and a cosmological horizon r_q. The existence of a black hole spacetime (with horizons) is restricted by the conditions $\alpha \leq \frac{1}{6}$ and $a \leq a_{crit} = \sqrt{\frac{2\left(-1+9\alpha+(1-6\alpha)^{3/2}\right)}{27\alpha^2}}$ (see [2] for the details).

IOP Conf. Series: Journal of Physics: Conf. Series **869** (2017) 012067 doi:10.1088/1742-6596/869/1/012067

3. The accretion disk structure equations

Consider a thin accretion disk with a very small total mass in the equatorial plane around a rotating compact object described by equation (1). In order to study the dynamics of the gas flow, we will consider a general energy-momentum tensor of the form $T^{\mu\nu} = (\rho + \epsilon + P)\, U^{\mu}U^{\nu} + Pg^{\mu\nu} + S^{\mu\nu} + U^{\mu}q^{\nu} + U^{\nu}q^{\mu}$, [7,8] where ρ is the rest-mass density, ϵ is the thermal energy density and P is the pressure, including gas and radiation contributions. U^{μ} represents the 4-velocity of the flow and contains the velocity 3-vector $\vec{u}$. q^{ν} is the energy-flux and $S^{\mu\nu}$ is the viscous stress tensor, written in terms of the bulk viscosity, ξ, and the dynamic viscosity, η, as in [11]: $S^{\mu\nu} = \eta\left(h^{\mu\sigma}\nabla_{\sigma}u^{\nu} + h^{\nu\sigma}\nabla_{\sigma}u^{\mu}\right) + \left(\xi - \frac{2}{3}\eta\right) h^{\mu\nu}\nabla_{\sigma}u^{\sigma}$, with $h^{\mu\nu} = U^{\mu}U^{\nu} + g^{\mu\nu}$. As usual, these contributions satisfy the constrains $S^{\mu\nu}U_{\nu} = 0$ and $q^{\mu}U_{\mu} = 0$.

The structure equations for the accretion disk are the energy momentum balance, $\nabla_{\nu}T^{\mu\nu} = 0$, together with the continuity equation (baryon conservation), $\nabla_{\mu}\left(\rho U^{\mu}\right) = 0$, an equation of state, a viscosity prescription and an equation describing the overall energy transport mechanism. In order to study these equations, we will consider some simplifying assumptions [12,13], in which the structure of the accretion disk is build using cylindrical coordinates (t, r, ϕ, z) and

- The disk is time independent and axisymmetric (derivatives with respect to t and ϕ vanish).
- The disk is geometrically thin, i.e. the proper height satisfies, $H(r) \ll r$ or $\frac{\partial_r P}{\partial_z P} \sim \frac{z}{r} \ll 1$.
- The gas is moving on almost Keplerian orbits with azimuthal velocity $u_{\phi} \approx \sqrt{\frac{GM}{r}}$, but also has a small radial drift velocity, u_r, and a smaller vertical velocity, u_z. From the continuity equation one can obtain the conditions $\frac{u_z}{u_r} \sim \frac{z}{r} \ll 1$ and $\frac{u_r}{u_{\phi}} \sim \frac{H^2}{r^2} \ll 1$ [6].

In order to write the structure equations for the thin disk, we introduce the radial functions

$$A = 1 - \frac{2M}{r} - \frac{\alpha}{r^{3\omega+1}} + \frac{a^2}{r^2}, \tag{2}$$

$$B = 1 - \frac{3M}{r} - \frac{3\alpha}{2r^{3\omega+1}} + \frac{2a}{r}\sqrt{\frac{M}{r} + \frac{\alpha}{2r^{3\omega+1}}}, \tag{3}$$

$$C = 1 - \frac{4a}{r}\sqrt{\frac{M}{r} + \frac{\alpha}{2r^{3\omega+1}}} + \frac{3a^2}{r^2}, \tag{4}$$

$$D = \frac{1}{2\sqrt{r}}\int_{r_0}^{r} \frac{x^2 - 6Mx - 3\alpha x^{1-3\omega} + 8a\sqrt{Mx + \frac{\alpha}{2}x^{1-3\omega}} - 3a^2}{x^2 - 3Mx - \frac{3\alpha}{2}x^{1-3\omega} + 2a\sqrt{Mx + \frac{\alpha}{2}x^{1-3\omega}}}\frac{dx}{\sqrt{x}}, \tag{5}$$

$$F = 1 + \frac{\alpha}{2r^{3\omega+1}}\left(\frac{M}{r}\right)^{-1}, \tag{6}$$

where r_0 is the inner boundary of the disk (marginally stable circular orbit for a black hole [14]).

Considering a non-relativistic gas, $\rho \gg \epsilon$ and $\rho \gg P$ [15], we project $\nabla_{\nu}T^{\mu\nu} = 0$ onto the 4-velocity and onto the 3-space orthogonal to U^{μ} to obtain the equations

$$U^{\nu}\partial_{\nu}\epsilon + (P + \epsilon)\nabla_{\nu}U^{\nu} + S^{\mu\nu}\nabla_{\mu}U_{\nu} + \nabla_{\nu}q^{\nu} + q^{\mu}U^{\nu}\nabla_{\mu}U_{\nu} = 0 \tag{7}$$

and

$$\left[U^{i}U^{\mu} + g^{i\mu}\right]\partial_{\mu}P + \rho U^{\mu}\nabla_{\mu}U^{i} + \nabla_{\mu}S^{i\mu} + U^{i}U_{\nu}\nabla_{\mu}S^{i\mu} + q^{i}\nabla_{\mu}U^{\mu} + q^{\mu}\nabla_{\mu}U^{i} + U^{\mu}\nabla_{\mu}q^{i} - U^{i}U^{\mu}q^{\nu}\nabla_{\mu}U_{\nu} = 0. \tag{8}$$

The velocity U^{μ} defines a local co-rotating reference frame from which it is possible to define an orthonormal tetrad $b^{\mu}_{(\alpha)}$ with $b^{\mu}_{(0)} = U^{\mu}$ and $g_{\mu\nu}b^{\mu}_{(\alpha)}b^{\nu}_{(\beta)} = \eta_{\alpha\beta}$ (details in [7,8,16]). Using the $b^{\mu}_{(\alpha)}$, we define the tetrad components of $S^{\mu\nu}$ and q^{μ}, denoted as $-t_{\alpha\beta}$ and F_{α}, respectively.

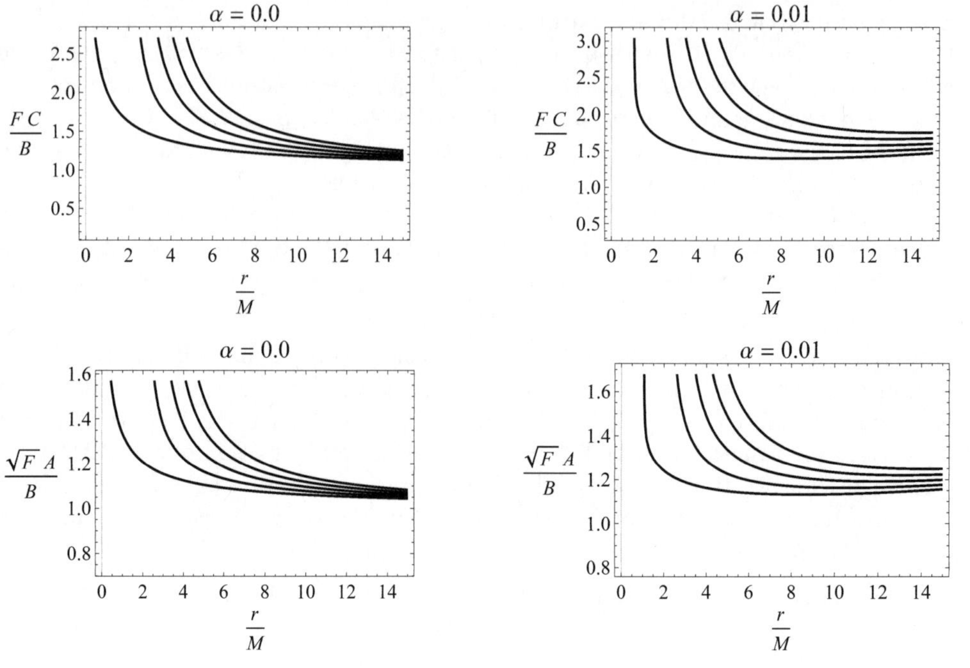

Figure 1. Relativistic correction factors $\frac{FC}{B}$, which affects the hydrostatic equilibrium equation (10) and $\frac{\sqrt{F}A}{B}$ appearing in the energy balance in tetrad components. We plot Kerr, $\alpha = 0$, and the $\alpha = 0.01$ rotating metric. Curves represent $\frac{a}{M} = 0$, 0.25, 0.5, 0.75 and 1.0 from right to left.

The z - component of equation (8) gives the vertical momentum balance,

$$\partial_z P = -\rho z \frac{M}{r^3} \frac{FC}{B}. \tag{9}$$

Introducing the radial component of the 4-velocity, written through the continuity equation as $r \int_{-H(r)}^{H(r)} (\rho U^1)\, dz = r \langle \rho U^1 \rangle = -\frac{\dot{M}}{2\pi}$ with $\dot{M}$ is the total accreted mass flux, the azimuthal component of equation (8) is written in terms of the tetrad components of the stress tensor. After integrating over the height of the disk and using equation (3), it gives

$$\langle t_{r\phi} \rangle = \frac{\dot{M}}{2\pi} \sqrt{\frac{M}{r^3}} \frac{\sqrt{F}D}{A}. \tag{10}$$

Finally, from the energy balance equation (7) in tetrad components we obtain $\frac{\partial F_z}{\partial z} = \frac{3}{4}\sqrt{\frac{M}{r^3}} \frac{\sqrt{F}A}{B} t_{r\phi}$, which integrated over the disk height and using (10) becomes $F_z = \frac{3M\dot{M}}{8\pi r^3} \frac{\sqrt{F}D}{B}$.

When comparing the obtained equations with the non-relativistic results, it is easy to identify that the relativistic corrections are condensed into functions A, B, C, D and F. In Figures 1 and 2 we show the behavior of four of the relativistic correction factors appearing in the structure equations for the particular case of quintessence, $\omega = -\frac{2}{3}$. It is important to note that the existence of quintessence, even for small α, changes the behavior of the corrections importantly.

This work was supported by the Universidad Nacional de Colombia. Hermes Code 38064.

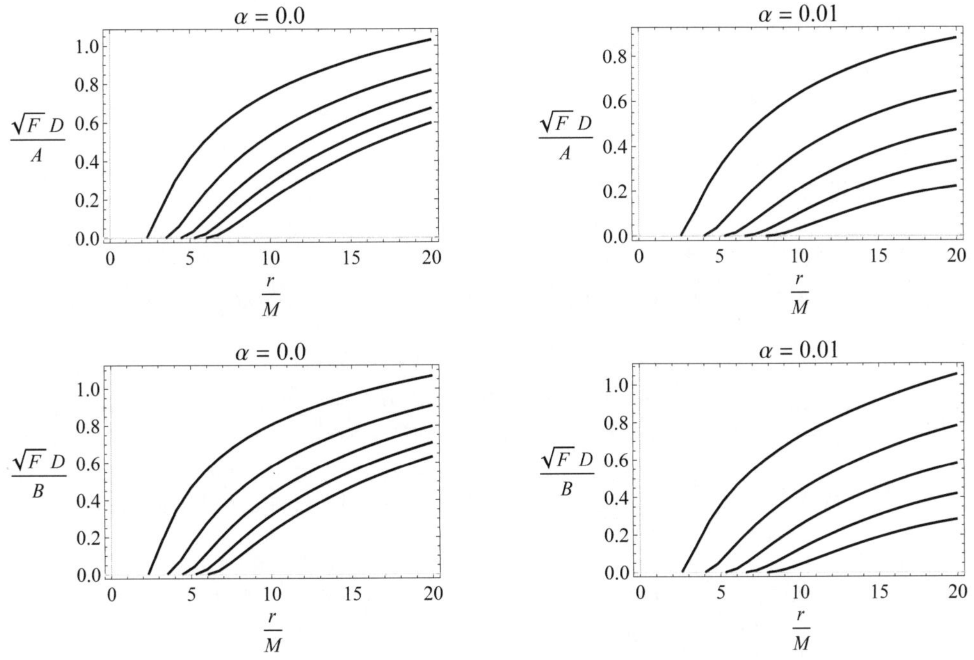

Figure 2. Relativistic correction factors $\frac{\sqrt{F}D}{A}$ affecting the tetrad components of the stress tensor equation (10) and $\frac{\sqrt{F}D}{B}$ characterizing the energy loss. We consider Kerr solution, $\alpha = 0$, and the rotating metric with $\alpha = 0.01$. Curves represent $\frac{a}{M} = 0$, 0.25, 0.5, 0.75 and 1.0, from right to left.

References

[1] Kiselev V V 2003 *Class. Quan. Grav.* **20** 1187
[2] Toshmatov B, Stuchlik Z and Ahmedov B 2015 arXiv:1512.01498 [gr-qc]
[3] Ghosh S 2016 *European Physical Journal* C **76** 222
[4] Xu Z and Wang J 2016 *Preprint* 1609.02045 [gr-qc]
[5] Yang R J 2016 *Preprint* 1605.02320 [gr-qc]
[6] Frank J, King A and Raine D 1992 *Accretion Power.* (London: Cambridge University Press)
[7] Riffert H 1998 *Thin Accretion Disks around Black Holes.* In: *Relativistic Astrophysics* eds. H. Riffert *et al* (Germany:Vieweg+Teubner Verlag) pp 120-138
[8] Riffert H and Herold H 1995 *Astrophys. J.* **450** 508
[9] Younas A, Hussain S, Jamil M and Bahamonde S 2015 *Phys. Rev. D* **92**(8) 084042
[10] Majeed B, Jamil M and Pradhan P 2015 *Adv. High Energy Phys.* **2015** 124910
[11] Landau L D and Lifshitz E M 1991 *Hydrodynamik* (Berlin:Akademie Verlag)
[12] Kluzniak W and Kita D 2000 *Preprint* astro-ph/0006266
[13] Czapla M and Kluzniak W 2007 *Proc. of RAGtime 8/9: Workshops on black holes and neutron stars 15-19/19-21 September, 2006/2007* eds. Hledik S and Stuchlik Z (Czech Republic) pp. 27-33
[14] Hussain I and Ali S 2016 *Eur. Phys. J. Plus* **131** 275
[15] Page D N and Thorne K S 1974 *Mon. Not. R. Astron. Soc.* **191** 499
[16] Novikov I D and Thorne K S 1973 *Astrophysics of Black Holes.* In: *Black Holes* eds. C. DeWitt and B.S. DeWitt (New York: Gordon & Breach) p. 344

IOP Conf. Series: Journal of Physics: Conf. Series **869** (2017) 012068 doi:10.1088/1742-6596/869/1/012068

Monitoring survey of pulsating giant stars in the Local Group galaxies: survey description, science goals, target selection

E Saremi[1,2], A Javadi[2], J Th van Loon[3], H Khosroshahi[2], A Abedi[1], J Bamber[3], S A Hashemi[4], F Nikzat[5] and A Molaei Nezhad[2]

[1] Physics Department, University of Birjand, Birjand 97175-615, Iran
[2] School of Astronomy, Institute for Research in Fundamental Sciences (IPM), Tehran, 19395-5531, Iran
[3] Lennard-Jones Laboratories, Keele University, ST5 5BG, UK
[4] Physics Department, Sharif University of Tecnology, Tehran 1458889694, Iran
[5] Instituto de Astrofisica, Facultad de Fisica, Pontificia Universidad Catolica de Chile, Av. Vicuna Mackenna 4860, 782-0436 Macul, Santiago, Chile

Email: saremi@birjand.ac.ir

Abstract. The population of nearby dwarf galaxies in the Local Group constitutes a complete galactic environment, perfect suited for studying the connection between stellar populations and galaxy evolution. In this study, we are conducting an optical monitoring survey of the majority of dwarf galaxies in the Local Group, with the Isaac Newton Telescope (INT), to identify long period variable stars (LPVs). These stars are at the end points of their evolution and therefore their luminosity can be directly translated into their birth masses; this enables us to reconstruct the star formation history. By the end of the monitoring survey, we will have performed observations over ten epochs, spaced approximately three months apart, and identified long-period, dust-producing AGB stars; five epochs of data have been obtained already. LPVs are also the main source of dust; in combination with Spitzer Space Telescope images at mid-IR wavelengths we will quantify the mass loss, and provide a detailed map of the mass feedback into the interstellar medium. We will also use the amplitudes in different optical passbands to determine the radius variations of the stars, and relate this to their mass loss.

1. Introduction

Dwarf galaxies are the most abundant type of galaxies in the Universe and so the study of them is of great importance. In this regard, the Local Group provides an excellent opportunity for understanding dwarf galaxies because of their proximity, variety, and wide range of metallicity (0.002 $Z_\odot$ to 0.08 $Z_\odot$) [1]. The Local Group dwarfs offer the chance for a complete inventory within a galactic environment. We set out to reconstruct their formation histories, and to probe their structure and evolution.

Asymptotic Giant Branch (AGB) stars have a strong influence on the global properties of a galaxy. DUSTiNGS (DUST in Nearby Galaxies with Spitzer) finds that AGB stars are efficient dust producers even down to 0.6% $Z_\odot$ [1]. Therefore, they are drivers of galaxy chemical enrichment and evolution via the return of significant amounts of gas and dust to the interstellar medium (ISM) [2]. Several efforts have been made in the last decade to characterize the AGB populations in nearby galaxies [1,3,4]. But only a small fraction of those stars were identified in optical surveys.

Furthermore, these stars are near the end-points of their evolution, and their luminosities directly reflect their birth masses – which is what makes AGB stars powerful probes of a galaxy's star formation history. They trace stellar populations as young as ~ 30 Myr to as old as the oldest globular clusters. The most evolved AGB stars are long period variables (LPVs), and identifying them is one of the best ways to reconstruct the star formation history [5]. LPVs reach the largest amplitudes of their brightness variations at optical wavelengths, due to the changing temperature. We will identify LPVs with an optical monitoring survey of the majority of the Local Group dwarf galaxies, and then will use their luminosity distribution to reconstruct their star formation history employing a method that we have developed and successfully applied in other Local Group galaxies [4,6,7].

2. Survey description

Observations are being made with the Wide Field Camera (WFC) – an optical mosaic camera at the prime focus of the 2.5m Isaac Newton Telescope (INT) in La Palma. The field of view of WFC is about 34'×34' and it covers each galaxy in just one pointing, but dithering between repeated exposures is required to fill the gaps between the detectors. Still, even among the Andromeda system of satellites, none are near enough to one another to fit within one and the same field of view.

LPVs vary on timescales from ~100 days, for low-mass AGB stars to ~1300 days, for the dustiest massive AGB stars. Although we are not aiming to determine accurate periods, but to identify the LPVs and to determine their amplitude and mean brightness, we require monitoring over ten epochs, spaced approximately three months apart. Five epochs of data have been obtained already, and the majority of the targets have been observed more than twice.

We have selected the I-band for the monitoring survey because the spectral energy distribution (SED) of cool evolved stars peak around 1 μm, so they stand out in the I-band and the I-band is where the contrast between the LPVs and other stars is greatest. Also, the bolometric correction – needed to determine the luminosity – in this band is the smallest and the effects of attenuation by dust are minimal. However, we also observe in the V-band, as often as possible, to monitor the variations in temperature – and thus radius – and to further help constrain the SED modelling.

We have chosen exposure times that yield sufficient signal-to-noise (S/N) to detect small changes in magnitude at different epochs. The I-band amplitudes of pulsating AGB stars are > 0.1 mag. Therefore we aim for S/N = 10 for the faintest stars, equivalent to the RGB-tip.

3. Target selection

The main targets in this survey are the majority of dwarf galaxies in the Local Group that are visible in the Northern hemisphere, except some that have already been studied, such as NGC147 and NGC185 [8]. These targets contain 55 dwarf galaxies of which 22 are satellites of the Andromeda galaxy – others are satellites of the Milky Way or isolated dwarfs. Also, we have included four distant globular clusters (GC) to investigate the possibility that they are stripped nucleated dwarf galaxies.

We prioritise the targets, principally on the basis of their estimated number of AGB stars; some populous galaxies include IC10 (> 10^4 AGB stars) and Sextans A and B (> 10^3 AGB stars). Also, we are attempting to monitor the entire Andromeda system of satellites because these are all accessible to a Northern hemisphere survey. This survey benefits from the homogeneity in distances, completeness, and accuracy. The foreground populations and extinction are modest and similar between all Andromeda satellites. The main aim of this selection is to find out whether the Andromeda system could be a universal template for galaxy evolution, or just one particular case. Individual Milky Way satellites are observed as a comparison to the Andromeda system. Our targets are shown in table 1.

4. Data

4.1. Data reduction

We have used the THELI pipeline, which is an image processing pipeline with the ability to reduce multi-pointing optical images taken by a mosaic CCD camera. This pipeline works on raw images and removes several instrumental effects, implements photometric calibration and astrometric alignment, and constructs a deep co-added mosaic image complemented by a weight map of the image pixels. In this pipeline more emphasis is put on precise astrometry, than precise photometry [9]. Two examples are shown in figure 1.

Table 1. Targets [10].

Dwarf galaxy	R.A (J2000)	Dec (J2000)	$(m-M)_0$ (mag)	V (mag)	type
Andromeda I	00 45 39.8	+38 02 28	24.36±0.07	12.7±0.1	dSph
Andromeda II	01 16 29.8	+33 25 09	24.07±0.06	11.7±0.2	dSph
Andromeda III	00 35 33.8	+36 29 52	24.37±0.07	14.4±0.3	dSph
Andromeda IX	00 52 53.0	+43 11 45	23.89±0.31	16.3±1.1	dSph
Andromeda V	01 10 17.1	+47 37 41	24.44±0.08	15.3±0.2	dSph
Andromeda VI	23 51 46.3	+24 34 57	24.47±0.07	13.2±0.2	dSph
Andromeda VII	23 26 31.7	+50 40 33	24.41±0.10	11.8±0.3	dSph
Andromeda X	01 06 33.7	+44 48 16	24.23±0.21	16.6±1.0	dSph
Andromeda XI	00 46 20.0	+33 48 05	24.40±0.50	17.5±1.2	dSph
Andromeda XII	00 47 27.0	+34 22 29	24.70±0.30	18.3±1.2	dSph
Andromeda XIII	00 51 51.0	+33 00 16	24.40±0.40	18.1±1.2	dSph
Andromeda XIV	00 51 35.0	+29 41 49	24.33±0.33	15.9±0.5	dSph
Andromeda XIX	00 19 32.1	+35 02 37	24.57±0.36	15.6±0.6	dSph
Andromeda XV	01 14 18.7	+38 07 03	24.00±0.20	14.6±0.3	dSph
Andromeda XVI	00 59 29.8	+32 22 36	23.60±0.20	15.9±0.5	dSph
Andromeda XVII	00 37 07.0	+44 19 20	24.50±0.10	15.8±0.4	dSph
Andromeda XVIII	00 02 14.5	+45 05 20	25.66±0.13	16.0±9.9	dSph
Andromeda XX	00 07 30.7	+35 07 56	24.35±0.15	18.2±0.8	dSph
Andromeda XXI	23 54 47.7	+42 28 15	24.67±0.13	14.8±0.6	dSph
Andromeda XXII	01 27 40.0	+28 05 25	24.82±0.31	18.0±0.8	dSph
NGC 205 (M110)	00 40 22.1	+41 41 07	24.58±0.07	08.1±0.1	dE
NGC 221 (M32)	00 42 41.8	+40 51 55	24.53±0.21	08.1±0.1	dE
Leo IV	11 32 57.0	-00 32 00	20.94±0.09	15.1±0.4	dSph
Leo V	11 31 09.6	+02 13 12	21.25±0.12	16.0±0.4	dSph
Ursa Major I	10 34 52.8	+51 55 12	19.93±0.10	14.4±0.3	dSph
Ursa Major II	08 51 30.0	+63 07 48	17.50±0.30	13.3±0.5	dSph
WILLMAN 1	10 49 21.0	+51 03 00	17.90±0.40	15.2±0.7	dSph
Coma Berenices	12 26 59.0	+23 54 15	18.20±0.20	14.1±0.5	dSph
Canes Venatici I	13 28 03.5	+33 33 21	21.69±0.10	13.1±0.2	dSph
Canes Venatici II	12 57 10.0	+34 19 15	21.02±0.06	16.1±0.5	dSph
Bootes I	14 00 06.0	+14 30 00	19.11±0.08	12.8±0.2	dSph
Bootes II	13 58 00.0	+12 51 00	18.10±0.06	15.4±0.9	dSph
Bootes III	13 57 12.0	+26 48 00	18.35±0.10	12.6±0.5	dSph?
Draco	17 20 12.4	+57 54 55	19.40±0.17	10.6±0.2	dSph
Sextans	10 13 02.9	-01 36 53	19.67±0.10	10.4±0.5	dSph
Ursa minor	15 09 08.5	+67 13 21	19.40±0.10	10.6±0.5	dSph
Hercules	16 31 02.0	+12 47 30	20.60±0.20	14.0±0.3	dSph
Leo I	10 08 28.1	+12 18 23	22.02±0.13	10.0±0.3	dSph
Leo II	11 13 28.8	+22 09 06	21.84±0.13	12.0±0.3	dSph
Segue 1	10 07 04.0	+16 04 55	16.80±0.20	15.3±0.8	dSph
Segue 2	02 19 16.0	+20 10 31	17.70±0.10	15.2±0.3	dSph
Pisces I	01 03 55.0	+21 53 06	24.43±0.07	14.3±0.1	dIrr/dSph
Pisces II	22 58 31.0	+05 57 09	21.31±0.18	16.3±0.5	dSph
Leo T	09 34 53.4	+17 03 05	23.10±0.10	15.1±0.5	dIrr/dSph
IC 10	00 20 17.3	+59 18 14	24.27±0.18	09.5±0.2	dIrr
Sagittarius	19 29 59.6	-17 40 51	25.14±0.18	13.6±0.2	dIrr
Cetus	00 26 11.0	-11 02 40	24.39±0.07	13.2±0.2	dSph
WLM	00 01 58.2	-15 27 39	24.95±0.03	10.6±0.1	dIrr
Aquarius	20 46 51.8	-12 50 53	25.15±0.08	14.5±0.1	dIrr/dSph
Leo P	10 21 45.1	+18 05 17	25.72±0.46	---	dIrr
Leo A	09 59 26.5	+30 44 47	24.51±0.12	12.4±0.2	dIrr
UGC 4879	09 16 02.2	+52 50 24	25.67±0.04	13.2±0.2	dIrr/dSph
Pegasus	23 28 36.3	+14 44 35	24.82±0.07	12.6±0.2	dIrr/dSph
Sextans A	10 11 00.8	-04 41 34	25.60±0.03	11.5±0.1	dIrr
Sextans B	10 00 00.1	+05 19 56	25.60±0.03	11.3±0.2	dIrr
Segue 3	21 21 31.1	+19 07 02	16.10±0.10	14.9±0.5	GC
Sextans C (PAL3)	10 05 31.8	+00 04 21	---	14.3±0.7	GC
NGC 2419	07 38 07.9	+38 52 48	19.75±0.06	---	GC
PAL 4	11 29 15.8	+28 58 23	20.20±0.01	---	GC

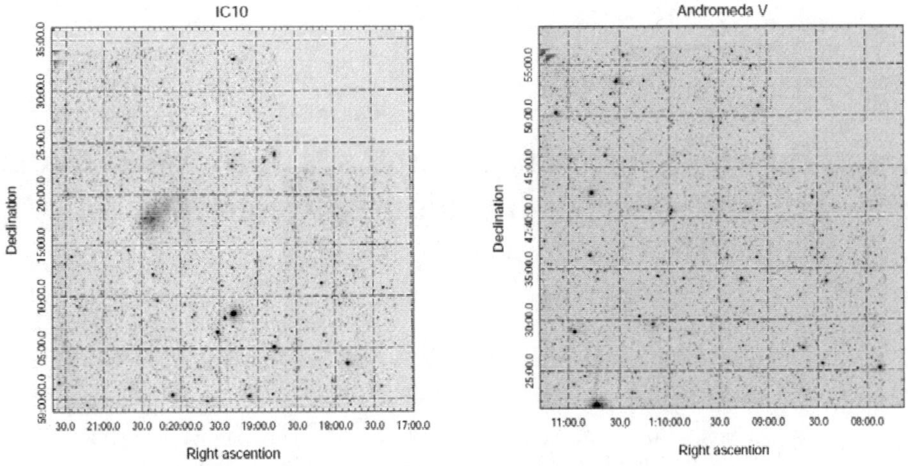

Figure 1. The IC10 dwarf galaxy and Andromeda V dwarf galaxy observed with the INT/WFC and processed with THELI.

4.2. Photometry
Photometry was obtained for all stars within each frame by automated fitting of a model of the Point Spread Function (PSF), using the DAOPHOT/ALLSTAR software suite [11]. The photometric calibration was performed using standard stars observed on some of the observing nights.

5. Science goals
The main objectives of the project are to: identify all LPVs in the dwarf galaxies of the Local Group; obtain accurate time-averaged photometry for all of them; obtain the pulsation amplitude of the LPVs; model their SEDs; study the relation between pulsation amplitude and mass loss (determined from mid-IR observations); reconstruct the star formation history.

References
[1] Boyer M L *et al* 2015 *ApJS* **216** 10B
[2] Javadi A, van Loon J Th, Khosroshahi H and Mirtorabi M T 2013 *MNRAS* **432** 2824
[3] Javadi A, van Loon J Th and Mirtorabi M T 2011a *MNRAS* **411** 263
[4] Rezaiekh S, Javadi A, Khosroshahi H and van Loon J Th 2014 *MNRAS* **445** 2214
[5] Javadi A, van Loon J Th and Mirtorabi M T 2011 *MNRAS* **414** 3394
[6] Javadi A, van Loon J Th and Mirtorabi M T 2011 *ASPC (Why Galaxies Care About AGB Stars II)* ed Kerschbaum F, Lebzelter T and Wing R. F **445** 497
[7] Javadi A, van Loon J Th, Khosroshahi H, Tabatabaei F and Golshan R H 2017 *MNRAS* **464** 2103
[8] Golshan R H, Javadi A, van Loon J Th, Khosroshahi H and Saremi E 2017 *MNRAS* **466** 1764
[9] Erben T *et al* 2005 *AN* 326 432-64
[10] McConnachie A W 2012 *AJ* **144** 4
[11] Stetson P B 1987 *PASP* **99** 191-222

IOP Conf. Series: Journal of Physics: Conf. Series **869** (2017) 012069 doi:10.1088/1742-6596/869/1/012069

Three-year of observations of Jupiter's aurora and Io plasma torus variabilities by earth orbiting extreme-ultraviolet spectroscope HISAKI

F Tsuchiya[1], K Yoshioka[2], T Kimura[3], G Murakami[4], C Tao[5], H Kita[1,6], I Yoshikawa[7], A Yamazaki[4] and Y Kasaba[6]

[1] Planetary Plasma & Atmospheric Research Center, Tohoku University, Sendai, Japan
[2] Graduate School of Science, University of Tokyo, Tokyo, Japan
[3] Nishina Center for Accelerator-Based Science, RIKEN, Wako, Saitama, Japan
[4] Institute of Space and Astronautical Science, Aerospace Exploration Agency, Sagamihara, Japan
[5] National Institute of Information and Communication Technology, Koganei, Japan
[6] Graduate School of Science, Tohoku University, Sendai, Japan
[7] Department of Complexity Science and Engineering, University of Tokyo, Kashiwa, Japan

E-mail: tsuchiya@pparc.gp.tohoku.ac.jp

Abstract. Extreme Ultraviolet spectrograph, EXCEED, on-board the HISAKI satellite is designed for observing tenuous gas and plasma around planets in the solar system. It enables us to obtain continuous and long-term data set and find time variability in the planetary magnetosphere and ionosphere with time scales of several hours to months. Here, we introduce new findings of Jupiter's UV aurora and plasma emissions from the Io plasma torus obtained from the HISAKI observation.

1. Overview of the HISAKI satellite

HISAKI was launched on 14 Sep. 2013 from the Japan Aerospace Exploration Agency's (JAXA) Uchinoura Space Center. After the initial operation of the EUV spectrograph (extreme ultraviolet spectroscope for exospheric dynamics, EXCEED), regular operation of Jupiter observation began from Dec. 2013. One of the main purposes of the mission is observation of the Io plasma torus (IPT) and Jovian aurora to investigate energy transport in the rotation-dominated magnetosphere [1-4]. The spectral range covered by EXCEED is 55 to 145 nm. The typical spectral resolution with the 140-arcsec slit was about 0.6 nm and the spatial resolution was evaluated to be 17 arcsec. The orbital period of the satellite is about 106 minutes and the inclination is 31 degrees. While observation of the planet is interrupted during Earth's eclipse of the planet and when the satellite passes over the South Atlantic Anomaly region, the data set still has enough continuity to find dynamical behaviour in the planetary magnetosphere and ionosphere. Detailed descriptions of scientific instruments are found in literatures [5-7]. Based on HISAKI's monitoring capability, coordinated observation of HISAKI with Hubble Space Telescope (HST) [8], x-ray space telescopes (Chandra, XMM/Newton, and SUZAKU) [9], and many ground-based optical/radio facilities has been conducted.

2. Hot plasma injection into the Io plasma torus

Jupiter's magnetosphere is known as a strong particle accelerator that contains ultra-relativistic electrons in its inner part. They are thought to be accelerated by whistler-mode waves excited by anisotropic hot electrons injected from the outer magnetosphere. In this context, electron transportation in the inner magnetosphere has key roles for transporting both free energy to excite the plasma wave and seed populations of the relativistic electron. However, it is still not well understood. From spectral diagnosis analysis, HISAKI show evidence for global inward transport of flux tubes containing hot plasma. High-spectral-resolution observations of the IPT enable us to obtain radial profiles of the hot electron density (figure 1). It gradually decreases with decreasing radial distance, despite the collisional time scale between hot and thermal ambient electrons is short and the hot electrons should lose their energy rapidly. This indicates a fast and continuous resupply of hot electrons responsible for keeping the hot electron population [10].

In addition to the global steady inward flow of the hot electron, simultaneous monitoring of the aurora and the IPT enables us to find definitive relationship between an explosive energy release in the magnetosphere which is manifested as an aurora transient and its connection to the inner magnetosphere. During a period from Dec. 2013 to Feb. 2014, nine pairs of brightenings of the aurora and IPT were found and showed the aurora preceded the IPT by 11.7 hours on average [11]. Physical mechanism responsible for this rapid response of the inner magnetosphere is an open question.

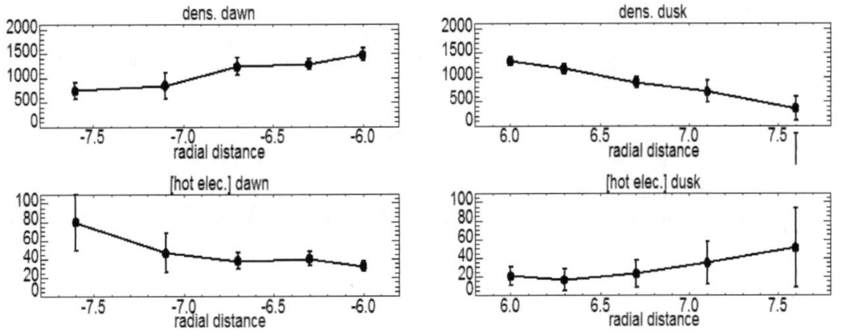

Figure 1. Plasma parameters in IPT as a function of radial distance for dawn (left) and dusk (right). (top) thermal electron density [/cc]. (bottom) hot electron density [/cc]. The hot electron density increasing with radial distance on both portions is taken as evidence of energy transport into the IPT via centrifugally driven interchange motion.

3. The solar wind impact on the inner magnetosphere of Jupiter

Because of significant rotationally-dominated plasma flow in Jupiter's magnetosphere, solar wind influence on its inner-part has been thought to be negligible. HISAKI's continuous observation enabled to find the solar wind influence on the inner magnetosphere for the first time. HISAKI found enhancements of the dusk/dawn brightness ratio of the IPT in response to rapid increase of the solar wind dynamic pressure (figure 2) [12]. The dawn-dusk asymmetry in the IPT has been reported from the Voyager observation and dawn-to-dusk electric field is one of the leading explanations. The HISAKI observation indicates that dawn-to-dusk electric field in the inner magnetosphere is modulated by the change of dynamic pressure of the solar wind and enhanced under compressed conditions.

4. Jupiter's aurora driven by internal magnetospheric process and the solar wind

Jupiter's auroral emissions reveal energy transport and dissipation through the planet's giant magnetosphere and upper atmosphere. While the main auroral emission is internally driven by planetary rotation in the steady state, transient brightening is generally thought to be triggered by compression by the external solar wind. The HISAKI and HST campaign observation in Jan. 2014 provides the first evidence that short-lived transient brightening of aurora can be internally driven (Three arrows in figure 3) [8]. These aurorae appeared under the quiet solar wind condition and HST shows the intense emissions appearing from the polar cap down to latitudes around Io's footprint, suggesting a rapid energy input into the polar region by the internal plasma circulation process. HST itself also obtained unique data set to study time variability in Jupiter's auroral structure [13, 14]. New characteristic of auroral response to the solar wind is also found. The auroral total power increases when an enhanced solar wind dynamic pressure hits the magnetosphere and the auroral total power shows a positive correlation with the duration of a quiescent interval of the solar wind that is present before a rise in the dynamic pressure, more than with the amplitude of dynamic pressure increase [15].

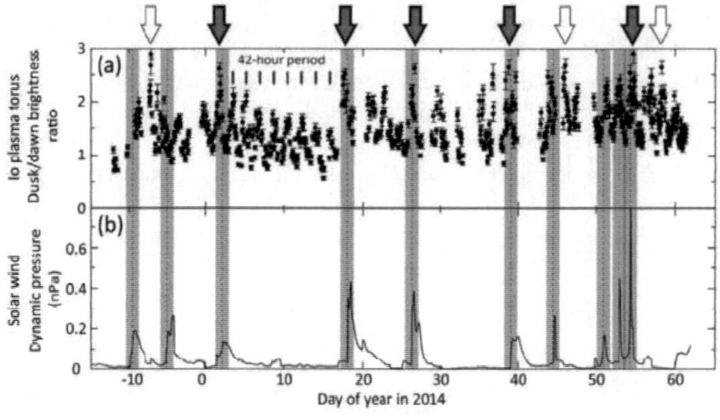

Figure 2. (a) The IPT brightness ratio in the spectral range of 64-77 nm. Arrows present the timings when the ratio exceeds 2.5. (b) Dynamic pressure of the solar wind extrapolated from Earth. Grey areas show arrival timings of abrupt increases (>0.1 nPa) with the error of the arrival times. The arrows which fall within the temporal errors of solar wind events are presented in red color.[12]

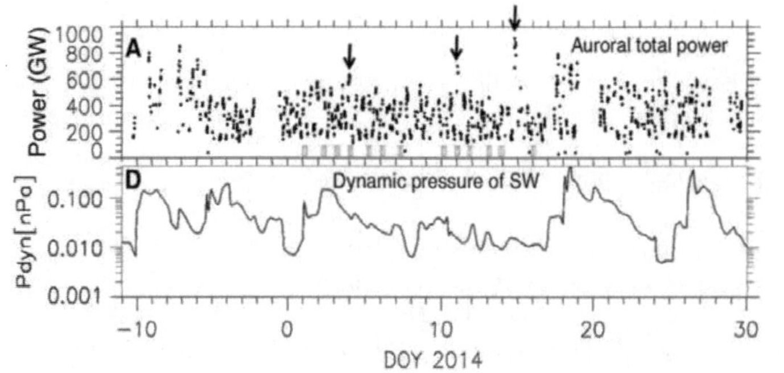

Figure 3. Total power of Jupiter's northern EUV aurora and solar wind parameters from 21 Dec. 2013 to 30 Jan. 2014. (top) The auroral total power as a function of day 2014. Vertical black arrows indicate days when the sudden brightening occurred on 4, 11, and 14. The times when northern auroral images were taken by HST/STIS are indicated by green ticks. (bottom) Solar wind variations extrapolated from those observed at Earth to Jupiter. [8]

Wide spectral coverage of the EXCEED instrument enable to include H_2 Lyman and Werner band emissions in aurora spectra and provide us aurora electron energy and total emission power [16-17]. Auroral electron energy is estimated using a hydrocarbon color ratio and emission intensity in the long wavelength range (138.5–144.8nm) provides indicator for emission power and auroral electron energy flux. HISAKI provides auroral electron parameters and their relation under different auroral activity levels. Relations between the auroral electron energy and flux estimated from the observation are distributed well along the theoretical field-aligned theory with a probable range of the magnetospheric parameters. Short- (within <1 rotation) and long-term (>1 rotation) enhancements of auroral intensity accompany increases of the electron number flux rather than the electron energy variations. The short- and long-term enhancements could be related to drivers of magnetospheric disturbances (internal and the solar wind, respectively. See also figure 3). Aurora power enhancements are mainly due to increased electron flux, not increase in electron energy, independent of drivers of aurora.

5. Plasma heating through satellite-planetary magnetic field interaction

Electromagnetic interaction between satellite and planetary magnetic field is a unique subject for Jupiter's magnetosphere. HISAKI observation found ion brightness maxima just downstream of Io. Spectral analyses show that it is caused by the increase in the hot electron population downstream of Io and 140 GW of energy is efficiently converted to hot electron production in the IPT [18].

6. Summary and future direction

The HISAKI satellite completed three-year observations of solar system planets since Dec. 2013. Advantage of EUV spectroscopy and long-term monitoring provide us many new findings on global and dynamic features of Jovian magnetosphere and reveal mass and energy transport process in the magnetosphere. In addition, HISAKI is also a unique platform to sense exosphere and ionosphere of Venus, Mars, Mercury and Earth.

In the spring of 2015, enhancement of neutral sodium cloud around Jupiter is reported [22]. The cloud enhancement directly reflects increase in Io's volcanic activity. Observation of IPT with HISAKI during this period also showed dramatic change in the sulfur and oxygen ion emissions and indicated significant plasma mass input to the inner magnetosphere. This is an unrepeatable chance to investigate response of mass and energy flows in the magnetosphere and aurora signature to the mass input. The studies are now on going. JAXA's AKATSUKI and NASA's JUNO spacecraft inserted orbits around Venus and Jupiter, respectively, and stated observation of atmosphere and magnetosphere of each planet. Collaboration between HISAKI and these orbiters must provide unique opportunities to investigate dynamic feature of upper atmosphere and magnetosphere of plants.

References

[1] Yoshikawa I *et al.* 2010 *Adv. Geosci.* **25** 29–41
[2] Yoshioka K *et al.* 2010 *Adv. Space Res.* **45** 314–321
[3] Tsuchiya F *et al.* 2010 *Adv. Geosci.* **25** 57–71
[4] Yoshioka K *et al.* 2012 *Planet. Space Sci.* **62**(1) 104–110
[5] Yoshioka K *et al.* 2013 *Planet. Space Sci.* **85** 250–260
[6] Yoshikawa I *et al.* 2014 *Space Sci. Rev.* **184**(1–4) 237–258
[7] Yamazaki A *et al.* 2014 *Space Sci. Rev.* **184**(1–4) 259–274
[8] Kimura T *et al.* 2015 *Geophys. Res. Lett.* **42**
[9] Kimura T *et al* 2016 *J. Geophys. Res.* **121**, 2308-2320
[10] Yoshioka K *et al* 2014 *Science* **345** (6204), 1581–1584
[11] Yoshikawa I *et al* 2016 *Geophys. Res. Lett.* **43** 11,552–11,557
[12] Murakami G *et al* 2016 *Geophys. Res. Lett.* **43** 12,308–12,316
[13] Badman S V *et al* 2016 *Geophys. Res. Lett.* **43** 988–997
[14] Gray R L *et al* 2016 *J. Geophys. Res.* **121** 9972–9984
[15] Kita H *et al* 2016 *Geophys. Res. Lett.* **43** 6790-6798
[16] Tao C *et al* 2016 *J. Geophys. Res.* **121** 4055-4071
[17] Tao C *et al* 2016 *J. Geophys. Res.* **121** 4041-4054
[18] Tsuchiya F *et al* 2015 *J. Geophys. Res.* **120**
[19] Yoneda M *et al* 2015 *Icarus* **261** 31-33

IOP Conf. Series: Journal of Physics: Conf. Series **869** (2017) 012070 doi:10.1088/1742-6596/869/1/012070

Search for Best Astronomical Observatory Sites in the MENA Region using Satellite Measurements

G Abdelaziz[1], R Guebsi[1], N Guessoum[2], C Flamant[1]

[1]Sorbonne Universités, Univ. Paris 06, CNRS & UVSQ UMR 8190 LATMOS, Paris, France
[2]American University of Sharjah, UAE
Email: nguessoum@aus.edu

Abstract. We perform a systematic search for astronomical observatory sites in the MENA (Middle-East and North Africa) region using space-based data for all the relevant factors, i.e. altitude (DEM), cloud fraction (CF), light pollution (NTL), precipitable water vapor (PWV), aerosol optical depth (AOD), relative humidity (RH), wind speed (WS), Richardson Number (RN), and diurnal temperature range (DTR). We look for the best locations overall even where altitudes are low (the threshold that we normally consider being 1,500 m) or where the combination of the afore-mentioned determining factors had previously excluded all locations in a given country.
In this aim, we use the rich data that Earth-observing satellites provide, e.g. the Terra and Aqua multi-national NASA research satellites, with their MODIS (Moderate Resolution Imaging Spectroradiometer) and AIRS (Atmospheric Infrared Sounder) instruments, the Defense Meteorological Satellite Program's Operational Linescan System (DMSP-OLS), and other products from climate diagnostics archives (e.g. MERRA).
We present preliminary results on the best locations for the region.

1. Introduction

The Arab world currently has only one large optical astronomy telescope larger than 1 meter in diameter (the Kottamia Observatory in Egypt). And despite the existence of a decent-size astronomical community (professionals and amateurs) in the Arab world, only a few observatories have been built in recent years, and they only host telescopes of 50-60 cm size. This state of affairs has led to calls [1] for the urgent construction of international-class observatories in the region, which is blessed with hundreds of clear nights per year. Guessoum, Alsaeed, and Abdelhafez [2] performed a preliminary search for the best sites in the Middle East and North Africa, considering locations with altitudes higher than 1,500 m and using online and published meteorological data (number of clear nights per year, humidity, temperature profile, wind speeds, light pollution from nearby towns).

That work had obvious shortcomings, however (inability to do in-situ measurements, necessary interpolations and approximations for some factors at sites of interest, etc.), issues that could be remedied by the rich data that Earth-observing satellites can provide, e.g. the Terra and Aqua multi-national NASA research satellites, with their MODIS (Moderate Resolution Imaging Spectroradiometer) and AIRS (Atmospheric Infrared Sounder) instruments. We have thus undertaken a systematic search of astronomical observatory sites using real space-based data for all factors. We look for the best locations for all countries, even those that may not have any sites higher than 1,500 m or where the combination of factors (altitude, meteorological conditions, and light pollution) had previously ended up excluding all locations in some countries.

The usage of satellite data has been proposed and applied in recent site selections for potential observatory locations in Australia, Morocco and the Canary Islands, West Africa, Turkey, and Chile [3, 4, 5, 6, 7, 8, 9] using multi-criteria decision analysis coupled with geographical information systems. Vernin et al [9] stress the importance of this approach: "satellites have the advantage of providing a long-term (longer than 5 yr) database for many sites in the world by using comparable spectrographs and techniques" and "high vertical resolution and the use of appropriate channels to detect different types and levels of clouds make the satellite data a useful tool for clear/photometric time characterization for astronomical observations." The accuracy of and the uncertainty over the data produced by satellite instruments with regard to the relevant factors has been discussed by several authors [4, 8, 9].

2. Data, Method, and Results

Table 1 lists the sources and general characteristics of the data we have used (all for 2014, except for Nighttime Lights, as 2013 was the most recent available time series, all accessed in 2015 and 2016).

Table 1. Data Sources and Characteristics (from [10, 11, 12]).

Variable Name	Product	Version	Source	Spatial Resolution	Units	Temporal Resolution
Digital Elevation Model	SRTM30	2	NASA - NGA	1' Global Relief Model	m	-
Cloud Fraction (Night time/ Descending)	AIRX3STD	006	AIRS	1°	-	Daily
Night time Lights (for 2013)	Stable Lights	4	DMSP-OLS	5 km x 5 km	-	Annually
Total Column Water Vapor (Night time/ Descending)	AIRX3STD	006	AIRS	1°	kg/m^2	Daily
Mean AOD (Combined Dark Target and Deep Blue) at 0.55 micron for land and ocean	MYD08_D3	006	MODIS-Aqua	1°	-	Daily
Relative Humidity at Surface (Night time/ Descending)	AIRX3STD	006	AIRS	1°	percent	Daily
Wind speed (time average) at 10 m above displacement height	MAT1NXSLV	5.2.0	MERRA Model	0.5 x 0.667°	m/s	Hourly
Surface Richardson Number	MAT1NXFLX	5.2.0	MERRA Model	0.5 x 0.667°	-	Hourly
Air temperature at surface (Daytime/ Ascending)	AIRX3STD	006	AIRS	1°	K	Daily
Air temperature at surface (Night time/ Descending)	AIRX3STD	006	AIRS	1°	K	Daily

IOP Conf. Series: Journal of Physics: Conf. Series **869** (2017) 012070 doi:10.1088/1742-6596/869/1/012070

Our goal is to determine, at least in first approximation, 'good' astronomical site locations in the MENA region, based on the general principles that a good observatory site has a high elevation (preferably above the inversion layer, i.e. above many of the aerosols that cause poor transparency), little atmospheric distortion (low and stable turbulence), and good 'seeing' (better suited for the implementation of adaptive optics), with low levels of CF, NTL, PWV, and AOD, which are sources of sky transparency degradation.

To evaluate the quality of each location over the entire region, we apply the following procedure:

- From the data sets presented in Table 1, we determine the value of each factor at every spot of the region, using a grid with 1° x 1° horizontal resolution (i.e. roughly 100 km x 100 km, which is the resolution provided by the satellite instruments) and vertical (altitude) levels of 50 m.
- For parameters with a different spatial grid resolution, we use a 'regridding' transformation, which is a process of linear interpolating (area-weighted average) from one inferior spatial grid resolution to 1° x 1° horizontal resolution.
- As a pre-processing stage, we use a Min-Max Normalization technique to compute the value of each parameter for every spot, which is then converted into a score of 1 to 10: 1 for best (lower levels of the physical parameter) and 10 for worst (higher levels of the physical parameter).
- For each location, we then calculate an overall score using a weighted average approach, considering the relative importance of the various factors. There is no wide agreement on the weighting or even the ordering of the physical parameters; experts ([13]) have suggested the following order: Cloud Fraction, Nighttime Lights, Total Column Water Vapor, AOD, Relative Humidity, Wind Speed, Richardson Number, Diurnal Temperature Range, and that is what we have adopted. Parameter weights vary between 0 and 1 and add up to 1. Tests are still being performed to determine the optimal weights, to give results that reflect the known quality of specific sites in the region. We will discuss this aspect of the calculation in a future publication.
- Finally, we rank the various locations according to their overall scores.

Below is a map showing the various areas by overall score as well as the locations of the best sites (shown with stars) obtained from the above procedure.

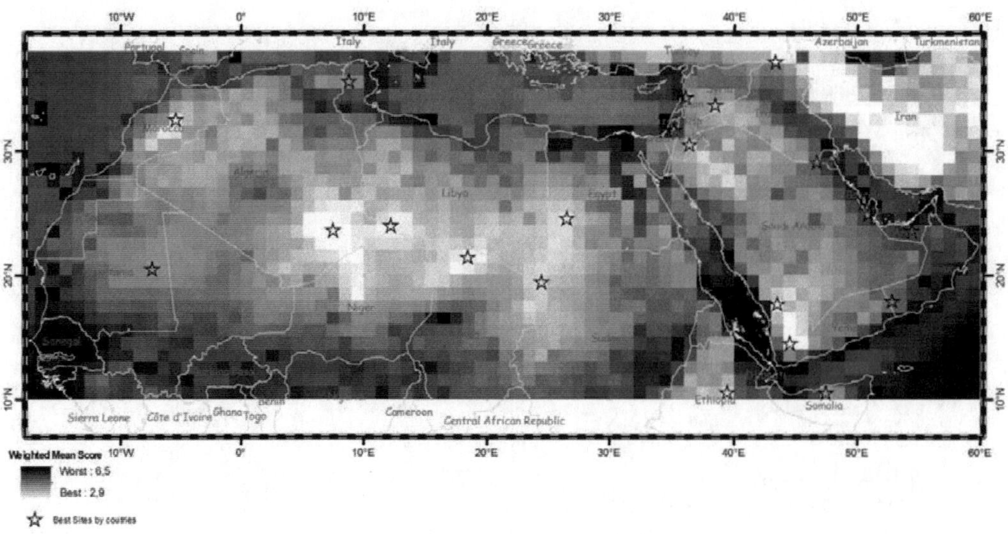

Figure 1. – Map of all locations with weighted scores: the whitest areas represent the best sites, the darkest represent the worst ones, and the stars are the very best sites.

Conclusion

We performed a systematic search for potential astronomical observatory sites in the MENA region using remotely sensed key parameters (affecting the sky transparency) as well as climate diagnostics archives and meteorological databases (parameters that affect the optical turbulence). We cover the whole region, including countries/areas where altitudes are lower than 1,500 m or where the combination of factors make locations unsuitable for observatories in principle, to help countries with limited altitudes or generally bad conditions in case educational or non-advanced observatories are to be built.

Astronomical observations depend critically on sky transparency, which we evaluate using a weighted sum of levels of cloud fraction, nighttime light pollution, precipitable water vapor, aerosol optical depth, relative humidity, wind speed, turbulence (surface Richardson number), and diurnal temperature range, all being measured from space. The score is obtained using a three-step process: a 'regridding' transformation, a normalization technique, and a weighted average.

We want to stress that this work must be regarded as a first phase in the determination of astronomical observatory locations, it must be followed up by on-site testing, with in situ measurements, using telescopes, photometers, airborne particles counters, ground meteorological stations, etc. This necessary second phase will increase the accuracy of the final list of locations and ensure that the data grid (recall the 1° x 1°, i.e. 100 km x 100 km cell size) of the space-based data did not smooth out any important local effect. Regional and international expert site-testing teams can greatly help in this endeavour.

We are conducting further investigations and refinements of this work. In particular, the effect of each factor (its weight in the overall score of a given site) requires more extensive simulations and tests. We aim to present a more detailed version of this work in the near future, with the exact locations of the best sites over the whole MENA region and for each Arab country. We hope this will facilitate and accelerate the construction of high-quality astronomical observatories in the Arab world.

References

[1] Guessoum N 2013 Time for an Arab astronomy renaissance *Nature* **498** pp 161–164

[2] Guessoum N, Alsaeed N and Abdelhafez N 2014 Preliminary search for astronomical observatory sites in the MENA region, *The Observatory* **134** pp 339-34

[3] Varela A M *et al* 2008 Astronomical site selection: on the use of satellite data for aerosol content monitoring *Mon. Not. R. Astron. Soc.* **391**:2 pp 507-520.

[4] Bounhir A 2009 *Qualification des sites Astronomiques: Nouvelle approche par l'utilisation des données satellitaires et météorologiques* Marrakech, Morocco: Université Cadi Ayyad

[5] Hotan C E, Tingay S J and Glazebrook K 2013 Testing Potential New Sites for Optical Telescopes in Australia *Publ. Astron. Soc. Aus.* **30** id.e002

[6] Erasmus, D A and van Rooyen, R 2006 A satellite survey of cloud cover and water vapor in northwest Africa and southern Spain Ground-based and Airborne Telescopes, edited by Larry M. Stepp, Proc. of SPIE Vol. 6267, 62671O

[7] Aksaker N *et al* 2015 Astronomical Site Selection for Turkey Using GIS Techniques *Exp. Astro.* **39**:3 pp 547-566

[8] Varela A M *et al.* 2012 On the Use of Remotely Sensed Data for Astronomical Site Characterization http://www.iac.es/proyecto/site-testing/images/stories/pdf/varela_et_al_2012.pdf

[9] Vernin, J *et al* 2011 European Extremely Large Telescope Site Characterization I: Overview *Publications of the Astronomical Society of the Pacific*, 123:909 pp 1334–1346

[10] GIOVANNI (from Goddard Earth Sciences Data Information Services Center) http://Giovanni.gsfc.nasa.gov/giovanni

[11] Global Modeling and Assimilation Office (GMAO)(2008), tavg1_2d_flx_Nx: MERRA 2D IAU Diagnostic, Surface Fluxes, Time Average 1-hourly V5.2.0, Greenbelt, MD, USA, Goddard Earth Sciences Data and Information Services Center (GES DISC), Accessed June 05, 2016 10.5067/4EQ54AKI405R

[12] National Geophysical Data Center https://ngdc.noaa.gov/eog/dmsp/downloadV4composites.html

[13] Ziad, A and Conan, J-M, private communications

IOP Conf. Series: Journal of Physics: Conf. Series **869** (2017) 012071 doi:10.1088/1742-6596/869/1/012071

A search for extragalactic pulsars in the local group galaxies IC 10 and Barnard's galaxy

H Al Noori[1], M S E Roberts[1,2], D Champion[3], M McLaughlin[4], Scott Ransom[5] and P S Ray [6]

[1]New York University Abu Dhabi, Abu Dhabi, UAE
[2]Eureka Scientific, Oakland, CA, USA
[3]Max-Plank-Institut fr Radioastronomie, Bonn, Germany
[4]West Virginia University, Morgantown, WV, USA
[5]National Radio Astronomy Observatory, Charlottesville, VA, USA
[6]Naval Research Laboratory, Washington, DC, USA

E-mail: haa280@nyu.edu

Abstract. As of today, more than 2500 pulsars have been found, nearly all in the Milky Way, with the exception of ~28 pulsars in the Small and Large Magellanic Clouds. However, there have been few published attempts to search for pulsars deeper in our Galactic neighborhood. Two of the more promising Local Group galaxies are IC 10 and NGC 6822 (also known as Barnard's Galaxy) due to their relatively high star formation rate and their proximity to our galaxy. IC 10 in particular, holds promise as it is the closest starburst galaxy to us and harbors an unusually high number of Wolf-Rayet stars, implying the presence of many neutron stars. We observed IC 10 and NGC 6822 at 820 MHz with the Green Bank Telescope for ~15 and 5 hours respectively, and put a strong upper limit of 0.1 mJy on pulsars in either of the two galaxies. We also performed single pulse searches of both galaxies with no firm detections.

1. Introduction and background

Almost 50 years have passed since Jocelyn Bell serendipitously discovered the first pulsar after noticing periodic fluctuations in radio telescope data [1]. Since then, over 2500 pulsars have been discovered [2]. The vast majority of these pulsars have been found in the Milky Way, with only 23 found in the Large Magellanic Cloud (LMC) and 5 in the Small Magellanic Cloud (SMC) [3], both of which are satellites of our galaxy. This work is an attempt to search for radio pulsars that are truly extragalactic, lying beyond the Magellanic Clouds.

The detection of extragalactic pulsars would open up new avenues in pulsar astronomy. What we can now learn about the ISM from Galactic pulsars, extragalactic pulsars would teach us about the intergalactic medium (IGM) and the ISM in other galaxies, which we currently know little about [4]. Additionally, the study of pulsars in other galaxies would provide us with yet another tool by which we can study stellar evolution in galaxies different from our own. Motivated by these reasons, we present searches of two local group galaxies, IC 10 and NGC 6822 (also known as Barnard's Galaxy).

1.1. IC 10
IC 10 is the closest starburst galaxy to the Milky Way, and is in fact the only starburst in the Local Group (LG). It lies fairly low in the Galactic plane ($b = -3.3°$), making distance measurements particularly difficult. However, most recent measurements have found the galaxy, using tip of the red giant branch and Cepheids to determine distance, to be in the neighborhood of 660−740 kpc away [5, 6]. The galaxy has been classified as a blue compact dwarf [7] and has quite a high rate of star formation (SF), estimated to be around 3-4 times greater than the SF rate of the SMC [8]. It is thought to be currently undergoing vigorous neutron star formation resulting from a SF burst ∼6−10 Myr ago. Recent research has shown that the younger population of IC 10, the stars resulting from the most recent SF burst, is found near the centroid of the galaxy, while the oldest population (>2 Gyr) is significantly offset from the center [9]. IC 10 shares many characteristics with the Magellanic Clouds and the SMC in particular: they both have active SF, are metal poor, and have extended HI envelopes. This makes the SMC a good template for comparison with IC 10 and gives us a good reason to choose the galaxy for this search.

The evolution of this galaxy is of particular interest as it has an extraordinary number of Wolf-Rayet (W-R) stars, with 33 stars confirmed spectroscopically while a total of ∼100 W-R stars are estimated to exist in IC 10 [10, 11]. It has an anomalously high ratio of carbon-type (WC) to nitrogen-type W-R (WN) stars, twenty times the ratio expected for a galaxy of its metallicity [12, 13], although this ratio could change as more stars are confirmed spectroscopically [14]. Although it has half the effective radius of the SMC, the galaxy has an effective W-R density 5 times that of the SMC, similar to the densities found in OB associations, and two times greater than that of any other galaxy in the LG [15].

1.2. Barnard's galaxy
Besides the Magellanic Clouds, Barnard's galaxy is the nearest dwarf irregular galaxy to us, and is a prototypical dwarf irregular galaxy. Its close distance, ∼470 kpc, makes it one of the few dwarf irregulars whose ISM and stellar population can be studied in detail [16]. This makes it an excellent candidate to search for pulsars, as probing its pulsar population could give us hints that could be extrapolated to other dwarf irregulars farther away. NGC 6822 has had a consistent, albeit modest, rate of SF for the last 400 Myr, but also experienced a recent increase in SF rate that is thought to have occured 100−200 Myr ago [17]. It has 4 W-R stars, and a W-R density similar to that of the SMC [18]. Due to the galaxy lying in the direction of the Galactic center ($b = -18.4°$), it is obscured and has high reddening and absorption, making it hard to detect things such as supernova remnants (SNRs) optically, of which only one has been detected [19].

2. Observations and results
2.1. Search methodology
We conducted three observations of IC 10 and two of Barnard's galaxy using the Green Bank Telescope in West Virginia. The observations were done using the GUPPI backend with 2048 spectral channels, observing at a central frequency of 820MHz and a bandwidth of 200MHz. The IC 10 observations were taken on the 13th, 14th and 17th of October 2009 for 6, 5, and 4.5 hours respectively, and all had a time resolution of 204.6 μs. The Barnard's galaxy observations were made on the 10th and 17th of October 2014, for approximately 2.5 hours each. We used a better time resolution with NGC 6822, for which we had a sample time of 81.92 μs. The galaxies are both well contained within the GBT beam.

Although these two galaxies are much farther than the LMC and the SMC (that have distances of 50 kpc and 60 kpc, respectively), where currently the farthest pulsars are known, we still expect to be sensitive to the most luminous pulsars at these greater distances. Assuming that

IOP Conf. Series: Journal of Physics: Conf. Series **869** (2017) 012071 doi:10.1088/1742-6596/869/1/012071

pulsars born in dwarf galaxies are similar in characteristics to the ones in the Milky Way, and that the distances we use are correct, we would optimistically hope to find ~1 in each galaxy. We base this estimate on the luminosities of Milky Way pulsars.

We used the software package PRESTO to search each observation up to a dispersion measure (DM) of 3000 pc cm^{-3} [20]. The NE2001 model for Galactic free electron density predicts a maximum Galactic contribution to the DM of 210 pc cm^{-3} in the direction of IC 10 and DM of 108 pc cm^{-3} in the direction of Barnard's Galaxy [21]. A newer model by Yao, Manchester and Wang designed to predict distances and DMs of extragalactic pulsars predicts a DM of 289 pc cm^{-3} in the direction of IC 10 and 78 pc cm^{-3} in the direction of NGC 6822. Both of these values are predominantly due to galactic effects, as the model predicts negligible DM coming from the IGM [22]. We chose to search up to a DM of 3000 pc cm^{-3}, several multiples of the predicted electron density to be safe, since very little is known about the electron densities of the IGM and the ISMs of each galaxy, and to allow for detecting fast radio bursts (FRBs), which could serendipitously occur in the direction of these galaxies. We performed an acceleration search up to a zmax of 300 to search for pulsars in tight binaries.

As we have multiple observations of each source, we are able to check the legitimacy of our pulsar candidates by making sure that they appear in all data sets. In addition to regular pulsed emission, we search for giant single pulses, which McLaughlin and Cordes predict should be detectable in each of the LG galaxies if they harbor a Crab-like pulsar [23]. Giant single pulses, such as the ones emitted by the Crab pulsar [24], can be can be over a hundred times brighter than the regular pulsed emission from pulsars [25]. We expect to be much more sensitive to giant pulses than we are to regular pulsations.

2.2. Results

Unfortunately, no strong pulsar candidates were detected, and none of the marginal candidates were seen in more than one pointing. We use the radiometer equation to calculate upper limits of 0.05 mJy and 0.07 mJy for IC 10 and NGC 6822, respectively, assuming a pulse width of ~10% of the period. A preliminary search for single giant pulses was done for each of the two galaxies, with no obvious single pulse detections. However, a more thorough single pulse search is under way.

3. Discussion

3.1. The case of IC 10

From the null result of our search, we may say that there are not many bright pulsars in IC 10. Much effort has gone into understanding the SF rate of IC 10; the literature shows that SF in the galaxy has happened in bursts rather than continuously [26], and that the most recent SF burst occured within the past 10 Myr, likely to have been as recently as 3−4 Myr ago [27, 28]. Such recent SF may explain why IC 10 is home to so many W-R stars and yet have few bright pulsars. Antoniou *et al* argue that a 10 Myr old SF burst is too young to make any pulsars, yet can produce black holes [29]. Coincidentally, IC 10 is home to the most massive known stellar mass black hole IC 10 X-1 [30, 31].

IC 10 has several large HI holes and a large synchrotron bubble that were thought to have resulted from multiple SNRs [32]. However it appears more likely that the synchrotron bubble is actually a hypernova remnant, from what could plausibly be the event that created IC 10 X-1 [33]. This further supports the hypothesis that the recent SF burst is too recent to have resulted in any pulsars. Besides the synchrotron bubble, there are no other SNRs in IC 10.

Another possibility is that IC 10's SF rate has been grossly overestimated [8]. If a top-heavy initial mass function (IMF) applies to IC 10, that could explain its high density of W-R stars, but could also mean that its SF rate could be as much as 10 times smaller than we expect.

3.2. The case of Barnard's Galaxy
Unlike IC 10, NGC 6822 has had modest SF for the past 400 Myr, which has had an almost two-fold increase over the last 100−200 Myr [34]. There is also evidence that SF rate has gone down around 10 Myr ago [17]. Since the lifetimes of most longer-period pulsars tend to be closer to 10 Myr than 100Myr, it is likely that the some of the pulsars born during the peak of Barnard's galaxy's SF peak are no longer energetic enough to emit pulses that would be detectable at this distance.

Acknowledgments
This research was carried out on the High Performance Computing resources at New York University Abu Dhabi. Many thanks to Jorge Naranjo for HPC support. Contributions to this work by PSR at NRL are supported by the Chief of Naval Research (CNR). The National Radio Astronomy Observatory is a facility of the National Science Foundation operated under cooperative agreement by Associated Universities, Inc.

References
[1] Hewish A, Bell S, Pilkington J, Scott P and Collins R 1968 *Nature* **217** 709-13.
[2] Hobbs G, Manchester R, Teoh A and Hobbs M 2004 *Conf. Proc. IAU symposium* **218** 139-40
[3] Ridley J, Crawford F, Lorimer D, Bailey S, Madden J, Anella R and Chennamangalam J 2013 *Mon. Not. R. Astron. Soc* **433** 138-46.
[4] Meiksin A 2009 *Rev.Mod.Phys.* **81** 1405-69
[5] Sakai S, Madore B, and Freedman W 1999 *Astrophys. J.* **511** 671-9
[6] Demers S, Battinelli P and Letarte B 2004 *Astron. Astrophys.* **424** 125-32
[7] Richer M *et al.* 2001 *Astron. Astrophys.* **370** 34-42
[8] Leroy A, Bolatto A, Walter F and Blitz L 2006 *Atrophys. J.* **643** 825-43
[9] Gerbrandt S, McConnachie A and Irwin M 2015 *Mon. Not. R. Astron. Soc* **454** 1000-11
[10] Crowther P, Drissen A, Royer P and Smartt S 2003 *Astron. Astrophys.* **404** 483-93
[11] Massey P and Holmes S 2002 *Astrophys. J. Letters* **580** L35-7
[12] Massey P 1999 *Proc. 193rd symposium of the IAU* **193** 429-40
[13] Massey P, Olsen K, Hodge P, Jacoby G, McNeill R, Smith R and Strong S 2007 *Astron. J.* **133** 2393-417
[14] Massey P, Neugent K and Morrell N 2015 *Proc. Potsdam Wolf-Rayet Workshop* 35-9
[15] Wilcots E, and Miller B 1998 *Astron. J.* **116** 2363-94
[16] Cioni M and Habing H 2005 *Astron. Astrophys.* **429** 837-850.
[17] Gallart C, Aparicio A, Bertelli G and Chiosi C 1996 *Astron. J.* **112** 2596-606
[18] Armandroff E and Massey P 1991 *Astron. J.* **102** 927-950.
[19] Kong A, Sjouwerman L and Williams B 2004 *Astron. J.* **128** 2783-8
[20] Ransom S 2001 *Diss. Harvard University* Cambridge, Massachusetts
[21] Cordes J, and Lazio T 2002 *arXiv* preprint astro-ph/0207156
[22] Yao J, Manchester R and Wang N 2016 "A New Electron Density Model for Estimation of Pulsar and FRB Distances" *Astrophys. J*
[23] McLaughlin M, and Cordes J 2003 *Astrophys. J.* **596** 982-96
[24] Cordes J, N Bhat, Hankins T, McLaughlin M and Kern J 2004 *Astrophys. J* **612** 375-88
[25] Cognard I, Shrauner J, Taylor J and Thorsett S 1996 *Astrophys. J. Letters* **457** L81-4
[26] Sanna N, et al. 2009 *Astrophys. J. Letters* **699** L84-7
[27] Magrini L, and Gonalves D 2009 *Mon. Not. R. Astron. Soc* **398** 280-92
[28] Vacca W, Sheehy C and Graham J 2007 *Astrophys. J* **662** 272-83
[29] Antoniou V, Zezas A, Hatzidimitriou D and Kalogera V 2010 *Astrophys. J. Letters* **716** L140-45
[30] Prestwich A, Kilgard R, Crowther P, Carpano S, Pollock A, Zezas A, Saar S, Roberts T and Ward M 2007 *Astrophys. J. Letters* **669** L21-4
[31] Silverman J, and Filippenko A 2008 *Astrophys. J. Letters* **678** L17-20
[32] Yang H and Skillman E 1993 *Astron. J.* **106** 1448-59
[33] Lozinskaya T, and Moiseev A 2007 *Mon. Not. R. Astron. Soc Letters* **381** L26-9
[34] Wyder T 2001 *Astron. J* **122** 2490-110

IOP Conf. Series: Journal of Physics: Conf. Series **869** (2017) 012072 doi:10.1088/1742-6596/869/1/012072

Spectroscopic investigation of the plasma jet interaction with water

J Q M Almarashi[1,2], A A Alhazime[1,2], M A Ellabban[1,2,3] andA-A H Mohamed[1, 2,4]

[1] Physics Department, Faculty of Science, Taibah University, Al-Madinah Al-Munawwarah, 41477, Saudi Arabia.
[2] Centre for Applied Science Research, Taibah University, Al-Madina Al-Munawwarah, 41477, Saudi Arabia.
[3] Physics Department, Faculty of Science, Tanta University, Tanta 31527, Egypt.
[4] Physics Department, Faculty of Science, Beni-Suef University, Beni-Suef 52511, Egypt.
E-mail: abdelaleamm@yahoo.com

Abstract. In this paper, the emission spectra from plasma jet-water interaction is investigated. The plasma system consists of a quartz tube which is surrounded by two copper strips separated by 14 mm, the upper stripe is connected to AC high voltage power supply and the lower is grounded. A quartz-cuvette containing the water sample is placed at 13 mm down to the tube nozzle. Emission spectra from three regions; the distance between the two electrodes (A), the distance between the grounded electrode and the sample surface (B), and through the sample (C), are investigated. The results show clear differences between the spectra emitted from the three regions. Region A emits the highest intensities for the line spectra and argon 763 nm was the maximum. As well as, O radical emission spectra were detected with the highest intensities in region B. However, new bands and lines appear in the spectra from region C, due to interaction of the jet with water, depending on the water conditions and plasma operating parameters. These results declare that plasma jet interaction with water can be used as indicator for water quality and a detector for which species play the rule in plasma sterilization too.

1. Introduction

Atmospheric pressure plasma jets have a considerable high interest form researchers due to its wide variety of biomedical and industrial applications. Plasma jet has a high potential to be used in sterilization, blood coagulation, teeth bleaching, skin treatment, water purification, and surface modifications [1,2]. The plasma jet applications cover a wide range of dimensions from micro-scales "for a precise treatments" up to large-scale objects. Moreover, plasma jet is accessible to be used for treating trenches and grooves. Recently many researchers generated plasma jets by different means and configurations such as microhollow cathode, gliding, and dielectric barriers discharge.

Emission spectroscopy is a powerful technique to diagnose the generated species in the atmospheric pressure plasma jet. Moreover, it gives the thermodynamic conditions of the generated plasma through the measurements of the plasma species temperatures (i.e. gas, rotational, vibrational, and electron temperatures) [3,4].

In gas discharge electrons are gaining their energies from external electric filed and transfer it to the surrounding plasma species in elastic and inelastic collisions. These collisions produce the ionization, excitation, to atoms and molecules. Moreover, the electron collisions could also produce reactive species

IOP Conf. Series: Journal of Physics: Conf. Series **869** (2017) 012072 doi:10.1088/1742-6596/869/1/012072

such as NOx, OH, O, and N^+ which play an important role in plasma applications practically in medicine and industry.

2. Experimental set up

The argon atmospheric pressure plasma system consists of a quartz tube which is surrounded by two identical copper strips at 31 mm and 10 mm from tube nozzle. The copper strips have 7 mm width. The upper strip is connected to a high voltage sinusoidal AC power supply while the lower is grounded, as illustrated in figure 1. A quartz cuvette containing the water sample is placed at 13 mm below the plasma jet tube nozzle. The jet is generated when 14.5 kV and 25 kHz is applied to the upper stripe at 5 SLPM argon flow. The generated plasma voltage and current waveforms are recorded using person current probe (model 6585) and high voltage probe (Tektronix P6015) with 1 Gs/s Tektronix oscilloscope (model DPO 4104B). A compact LD spectrometer system is used to record the plasma emission spectra. In this experiment, the emission spectra from three regions are investigated. The regions are: (A) the space between the two electrodes in the quartz tube, (B) the distance between the grounded electrode and the water surface, and (C) through the water in the cuvette. Optical fiber bundle is placed at 11 mm from the plasma coulomb center and used to collect the emission spectra. A CCD camera is used to capture plasma images to study plasma formation as shown in figure 1 for the jet.

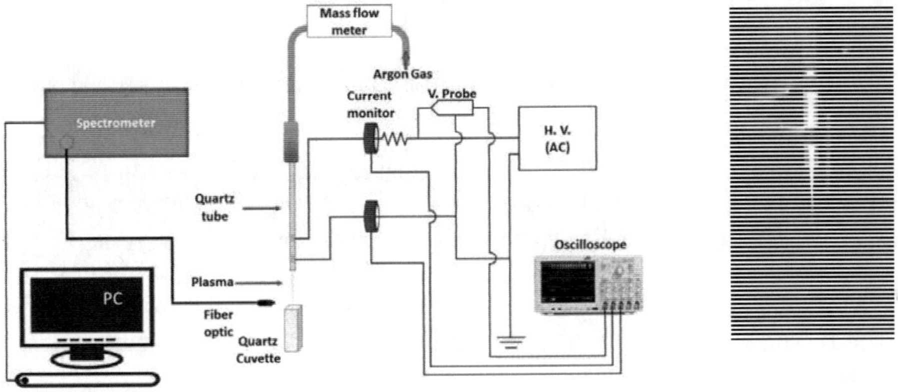

Figure 1. Experimental set up of the atmospheric pressure argon plasma jet with an image for the jet on the right side.

3. Results and discussion

3.1 Current –Voltage Waveforms

The current and voltage waveforms for the investigated plasma jet with and without discharge are shown in figure 2. An average of 16 times is used to collect the wave forms. The peak to peak voltage is 14 kV and the operating frequency is 25 kHz at 5 SLPM argon flow rate. The results show about $\pi/2$ phase difference between the current and the voltage waveforms before plasma ignition. This indicates that plasma jet system has a capacitive impedance, which means the voltage waveform lags behind the current waveform by $\pi/2$. Once the plasma is ignited the current-voltage phase difference reduces and the current waveform increased and get deformed. At breakdown, the voltage peak-to-peak value get reduced from the ignition voltage 14.5 kV to 11.7 kV at 25 kHz applied frequency. When the jet ignites an additional nonlinear load is employed to the plasma electric circuit. The plasma jet current waveforms presents a single sharp pulse per each half cycle which indicates the formation of a homogenous plasma between the two electrodes inside the quartz tube.

IOP Conf. Series: Journal of Physics: Conf. Series **869** (2017) 012072 doi:10.1088/1742-6596/869/1/012072

However, a current waveforms show repels over it which means streamers or filament discharges are formed too. The plasma jet length has been found to depend on the applied voltage and argon gas flow rate. When the plasma jet is imaged on a nanosecond time scale, it appears to be consisted of transient series of plasma pulses which is called plasma bullets [5].

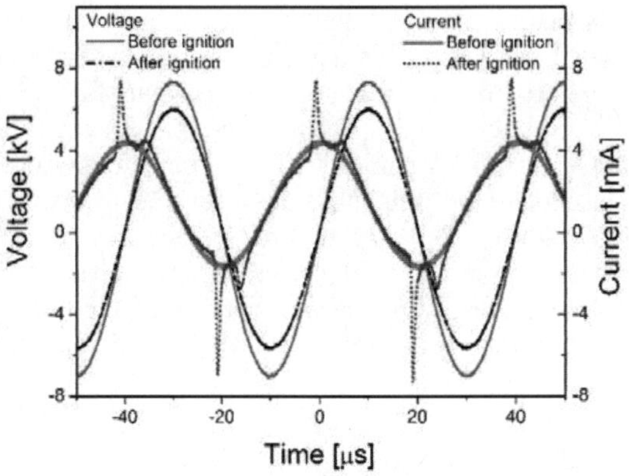

Figure 2. Current-voltage waveforms of atmospheric pressure plasma jet before and after ignition.

3.2 Plasma jet emission spectra

The emission spectra from the three investigated regions are presented in figure 3(a) and (b). The spectra from the three regions show that the neutral argon is the dominated lines from 690 to 1000 nm. The 763.51 nm line that is due to the transition $3s^2 3p^5(^2p^0_{3/2})4s \rightarrow 3s^2 3p^5(^2p^0_{3/2})4p$ has the highest intensity in three investigated regions. The O radical lines at 777.41 nm and 844.63 nm are detected in the emission spectra in the three regions with highest intensities in region B. The lines and bands (357.6, 375, 380, 405.5, 436 and 546 nm) are observed in the plasma water interaction in region C, figure 3(b).

Generally, the line spectra in region A have the highest intensities while it have the lowest in region B. The emission spectra from the water plasma jet interaction, region C, is higher than that from region B. Plasma in region A emits the highest line spectra due to the high energy transfer from the electrons, which gain their energies from the external electric field, to ambient gas in an inelastic collision process. However, plasma in region B has no external field and thus their emission spectra is minimum. Moreover, in region C the water works as a floating electrode. Therefore, a gradient potential is formed that accelerates the charged particles that increases the lines emission spectra compared with plasma in region B[6].

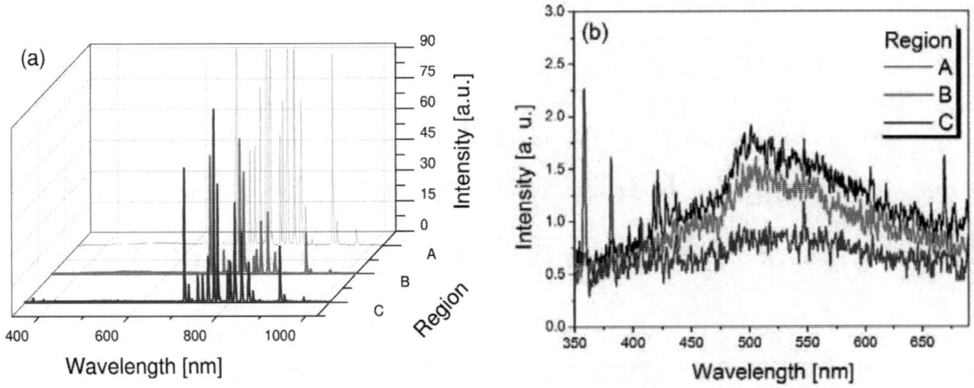

Figure 3. Plasma jet emission spectra in the range: (a) 350 to1000 nm and (b) 350 to 680 nm.

4. Conclusion

In this work, argon atmospheric pressure plasma jet interacting with water is investigated. The jet is operated at 14 kV AC sinusoidal high voltage, 25 kHz frequency and 5 SLPM argon flow rate which are optimal conditions to generate the longest jet with the lower applied voltage. The current waveform declares the generation of diffused and filamentary modes of plasma. The emission spectra show the presence of O radical which plays a very important role in plasma applications in medicine. The water sample plays as a floating electrode and new species are generated due to plasma water interaction. These new species are confirmed by the appearance of new lines and bands in the spectra from plasma interaction with water. Therefore, the plasma jet water interaction can be used as a probe for water quality.

Acknowledgement

The authors are grateful to the Deanship of Graduate Studies and Research, Taibah University, Saudi Arabia, for funding and supporting this research (project no. 4319/1435).

References

[1] Kolb J F, Mohamed A-A H, Price R O, James Swanson R, Bowman A, Chiavarini R L, Stacey M and Schoenbach K H 2008 *Appl. Phys. Lett.* **92** 241501
[2] Weltmann K D and Woedtke Th von 2017 *Plasma Phys. Control. Fusion* **59** 014031
[3] Laux C O, Spence T G, Kruger C H and Zare R N 2003 *Plasma Sources Sci. Technol.* **12** 125
[4] Machala Z, Janda M, Hensel K, Jedlovsky I, Les˘tinska L, Foltin V, Martis˘ovits V and Morvova M 2007 *J. Mol. Spectrosc.* **243** 194
[5] Kong M G, Ganguly B N and Hicks R F 2012 *Plasma Sources Sci. Technol.* **21** 030201
[6] Lee H W, Nam S H, Mohamed A-A H, Kim G C and Lee J K 2010 *Plasma Process. Polym.* **7** 274

IOP Conf. Series: Journal of Physics: Conf. Series **869** (2017) 012073 doi:10.1088/1742-6596/869/1/012073

Assessment of the potential of the new Belgo-Moroccan telescope TRAPPIST-North for high-precision exoplanet transit photometry

K Barkaoui[1], M Gillon[2], Z Benkhaldoun[1], J Emmanuel[2], T Elhalkouj[1], A Daassou[1], A Burdanov[2], L Delrez[2]
[1]Oukaimeden Observatory, LPHEA, FSSM, Cadi Ayyad University, Morocco
[2]STAR Institute, University of Liege, Belgium

Abstract. We present the observation of a transit of the exoplanet TrES-3b by the newly commissioned robotic telescope TRAPPIST-North located at Oukaimeden Observatory (Morocco). The obtained light curve reaches a photometric precison 600 ppm. Its Bayesian analysis with a Markov Chain Monte Carlo code enables us to refine the radius of the planet to $R_p = 1.346^{+0.065}_{-0.050} R_{Jup}$. These results demonstrate the high potential of TRAPPIST-North for high-photometry of exoplanet transits.

1. Introduction

When a planet crosses (transits) in front of its parent star's disk, the observed brightness of the star drops by a small amount whose amplitude depend on the relative sizes of the star and the planet. This method of detection is called transit photometry. The first extrasolar planet observed to transit its star, was HD209458b on 2000, [1]. The radial velocity method provides information about a planet's mass and the photometric method can determine the planet's radius. Combined with the radial velocity method, the transit method enables measuring the absolute mass and radius of a planet. The resulting density can than be used to constrain its nature (gas or rocky) and composition.

2. TRAPPIST project

TRAPPIST-North (TRansiting Planets and PlanetesImals Small Telescope [2] and [3]) is a 0.6 m robotic telescope that has started its operation at the Oukaimeden Observatory (Morocco) and done in collaboration with Cadi Ayyad University in June 2016. The project is led by the Origins in Cosmology and Astrophysics (OrCA) group at Liege University in Belgium. TRAPPIST-North and its twin, TRAPPIST-South, in operation in Chile since 2010, are fully dedicated to the study of planetary systems via two complementary approaches: the detection and characterization of transiting exoplanets, and the study of comets and other small bodies in the Solar System. The telescope is fully automated and is protected by a 4.2 meters diameter dome with weather station.

IOP Conf. Series: Journal of Physics: Conf. Series **869** (2017) 012073 doi:10.1088/1742-6596/869/1/012073

3. Observations, data reduction and analysis

3.1. The transit of TrES-3b

The observations, lasted for almost 3 hours on 01/06/2016, starting $\sim$ 1 hour before ingress and ending $\sim$ 1 hour after egress. TrES-3 (1SWASPJ175207.01+373246.3) has a I-magnitude $= 11.02$ and we used an exposure time of 13 seconds in the $'I+z'$ filter, in the very-near infrared (roughtly $0.9\mu m$).

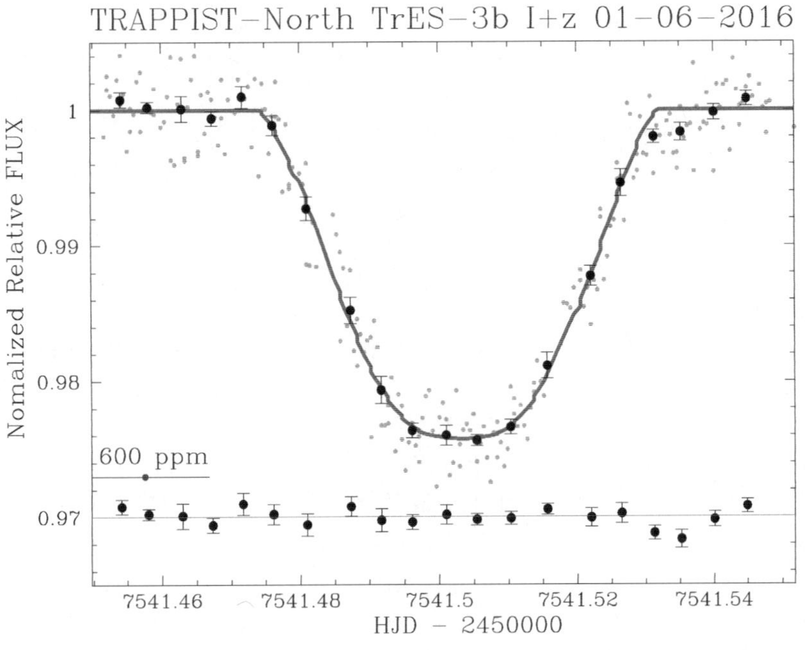

Figure 1. Top: Transit of TrES-3b observed on 01 June 2016 by TRAPPIST-North in the "$I + z$" filter, together with the best-fit transit model (red line). The normalized differential light curve (green points=unbinned, black points=binned in 7.2 min bins). Bottom : residuals. Their standard deviation is 600 ppm.

The data reduction is thus performed on-site, on dedicated computers installed in TRAPPIST-North's control room at Oukaimeden, and only some reduction products transferred to Marrakech for interactive final steps. All TRAPPIST-North images are reduced by a semiautomatic pipeline developed in Pyraf. To measure the stellar flux, the pipeline use the aperture photometry task *phot* of *IRAF/DAOPHOT* on the aligned science images. For each image, the coordinates of each star are refined by fitting 1D Gaussian functions in x and y on a region of a few pixels centered on the star's initial coordinates. The stellar fluxes are then extracted for eight different fixed apertures (multiples of the mean FWHM of the PSF over all images of the scientific sequence) centered on these refined coordinates, we used an estimate of the sky background near the star. This value is calculated from the median of pixels in the annular ring. We use the comparison stars to correct the atmospheric effect: the flux of the target star is divided by the sum of the fluxes of stable comparison stars $F_{target}/\sum F_{compstars}$ (differential photometry).

To analyze the transit light curve, we use the MCMC (Markov-chain Monte Carlo) code developed in Fortran to determine the parameters of a transiting using as input photometric

IOP Conf. Series: Journal of Physics: Conf. Series **869** (2017) 012073 doi:10.1088/1742-6596/869/1/012073

and radial velocity data. MCMC simulations are based on Bayesian inference meaning it is a method of drawing conclusions from data subject to random variations such as observational data using Bayes' rule. Bayes' rule is known in statistics to relate the odds for different events before and after conditioning another event. To implement a MCMC algorithm we need as input, a set of model parameters often called jump parameters in the framework of a code algorithm. Here the jump parameters are the transit depth dF, the transit width W, the mid-transit time T_0, the modified impact parameter $b' = a\frac{\cos i_p}{R_\star}$, the limb darkening coefficients for each filter c_1 and c_2, the modified RV semi amplitude $K_2 = KP^{1/3}$, and the orbital period P.

Table 1. System parameters for TrES-3b.

Parameters	Symbol	Value from (TRAPPIST-North)	Francis T. O' Donovan *et al*, [4]	Units
Transit epoch (BJD)	T_0	$2457541.5035^{+0.24E-03}_{-0.24E-03}$	2454185.9101 ± 0.0003	days
Orbital period	P	$1.30618608^{+0.00E00}_{-0.00E00}$	1.30619 ± 0.00001	days
Planet/star area ratio	$(R_p/R_\star)^2$	$0.02655350^{+0.23E-04}_{-0.09E-04}$	$0.027556 \pm 0.576E-04$	
Transit duration	t_T	$0.05739636^{+0.93E-03}_{-0.92E-03}$	-	days
Impact parameter	b	$0.83309007^{+0.17E-01}_{-0.13E-01}$	0.8277 ± 0.0097	$R_\star$
Orbital semimajor axis	a	$0.02266720^{+0.32E-03}_{-0.33E-03}$	0.0226 ± 0.0013	AU
Orbital inclination	i	$81.84555569^{+0.30E+00}_{-0.36E+00}$	82.15 ± 0.21	degrees
Orbital eccentricity	e	0 (fixed)	-	
Longitude of periastron	w	0 (fixed)	-	degrees
eccentricity × cos(w)	$e\cos(w)$	0 (fixed)	-	
eccentricity × sin(w)	$e\sin(w)$	0 (fixed)	-	
Stellar mass	$M_\star$	$0.91187249^{+0.39E-01}_{-0.39E-01}$	0.90 ± 0.15	$M_\odot$
Stellar radius	$R_\star$	$0.83027905^{+0.20E-01}_{-0.20E-01}$	0.802 ± 0.046	$R_\odot$
Stellar surface gravity	$\log g_\star$	$4.55987665^{+0.26E-01}_{-0.27E-01}$	4.6 ± 0.3	[cgs]
Stellar density	$\rho_\star$	$1.59462179^{+0.12E+00}_{-0.11E+00}$	-	$\rho_\odot$
Planet radius	R_p	$1.34648359^{+0.65E-01}_{-0.50E-01}$	1.295 ± 0.081	R_J

4. conclusion

Our results demonstrate that TRAPPIST-North is able to measure exoplanet transit light curves with excellent photometric precisions. The planet parameters derived from the analysis of light curve are in excellent agreement with previous works.

References

[1] Charbonneau D *et al* 2000 *ApJ* **329** L45-L48
[2] Jehin E, Gillon M, Queloz D, Magain P, Manfroid J, Chantry V, Lendl M, Hutsemkers D and Udry S *The Messenger* **145** 2-6
[3] Gillon M, Jehin E, Magain P, Chantry V, Hutsemkers D, Manfroid J, Queloz D and Udry S *European Physical Journal Web of Conferences* **11** 3
[4] T.O'Donovan F *et al* 2007 *ApJ* **663** L37-L40
[5] Gillon M *et al* 2010 *A & A* **511** A3

An optimized search for exoplanets with Kepler data

K Mohamed[1,2], A Ibrahim[2,3] and N Abd-Allah[1]

[1] Sohag University, Physics Department, Sohag, Egypt
[2] Zewail City of Science and Technology, Physics of Earth and Universe, and
[3] Cairo University, Faculty of Science, Physics Department, Giza, Egypt

Email: ai@ZewailCity.edu.eg

Abstract. The search and detection of extra-solar planets (exoplanets) have considerably improved over the past few decades. The launch of space observatories dedicated to exoplanets such as the Kepler mission and the wealth of publicly available data call for the development of efficient search methods. Here we present a search technique with a Graphical User Interface (GUI) to detect true exoplanets signals from Kepler data and filter out false ones using the optimal Box-fitting Least Squares (BLS) algorithm. The algorithm is used to detect transiting exoplanets from their light curves through searching for periodic flux variability of the host star. We present the results of the search to some Kepler catalog objects.

1. Introduction

The prospect of finding extra-solar planets (exoplanets) began to materialize in 1992 when Wolszczan and Frail discovered the first extra-solar planet around the neutron star radio pulsar PSR 1257+12 [1]. A Jupiter-mass planet orbiting around the main-sequence star 51 Pegasi was subsequently discovered by Mayor and Queloz in 1995 [2] and confirmed by Marcy et al in 1997 [3]. In recent years, the development of ground-based and space-borne observatories to search for exoplanets produced a surge in the number of known exoplanets, which is now in the thousands. Exoplanets can be detected from their gravitational influence on the host. This results in observable signatures such as a shift or wobbling in the star's position around the barycentre that is inferred by procession astrometry or by an observed Doppler shift in the star's spectrum due to its radial velocity. Exoplanets can also be detected when they transit the host star, causing a small but observable decline in the star's brightness when observed with sensitive photometers. The transit method is useful because of its ability to detect small size planets in the habitable zone of the host stars, which is the main objective of the Kepler mission.

The Kepler mission was launched to space in 2009 to observe simultaneously the brightness of more than 200,000 stars in a fixed field of view. By the completion of the mission, Kepler aims to assess the likelihood of finding terrestrial and Earth-size planets in the habitable zone of Sun-like stars, allowing to probe the diversity of the exoplanet population. The Kepler mission uses a 0.95-meter diameter telescope and a single instrument in the form of precession photometer consisting of 42 CCDs each is 50×25 mm with a resolution of 2200×1024 pixels, giving a total resolution of 94.6 megapixels. It is designed to observe a field of view of 115-square-degree of the sky.

The objective of this work is to introduce an efficient tool to search for exoplanets in the Kepler data with enhanced accuracy and reduction in the processing time. The analysis of Kepler data includes several automated stages of processing that ends with a data validation process. A manual/human stage screens thousands of computer-detected signals of candidate planets to identify true planetary signatures.

This cannot be done automatically and hence we built a search method using the Box-fitting Least Squares (BLS) that processes the data and feeds the resultant output to a Graphical User Interface (GUI) to identify the transit signals faster and more accurate. The GUI represents an efficient interactive interface that bridges the output of the search algorithm and the data validation through manual/human screening.

2. The search algorithm

The Box-fitting Least Squares (BLS) is an algorithm that is used to detect transiting exoplanets from the light curves of their host stars. The algorithm searches for transiting planet signatures (small, periodic drops in the star brightness). The principle of the BLS algorithm is based on direct Least Squares (LS) fits of step functions of the folded signal corresponding to trial periods [4]. The step function is defined as two states function High (*H*) and Low (*L*) with a period P_o. Applying that on transiting planets with fractional transit length *q*, the transit duration is qP_o, where *q* is assumed to be a small number (~ 0.01 - 0.05). For a given set of data points, the algorithm aims to find the best model while estimating five parameters: P_o, *q*, *L*, *H* and t_o, the epoch of the transit. Let us denote the data set by x_i, $i = 1, 2, ..., n$, and σ_i be the standard deviation. The noise is presented by assigning to each data point a weight w_i, which is defined by $w_i = \sigma_i^{-2} \sum_{j=1}^{n} \sigma_j^{-2}$. For a given trial period, we consider a folded time series, which is a permutation of the original time series. This series is denoted by $\overline{x_\iota}$ and the corresponding weights by $\overline{w_\iota}$. A step function is then fitted to the folded time series with the following parameters: *L,* the flux level in $[i_1, i_2]$; *H,* the flux level in $[1, i_1]$ and $[i_2, n]$. For a given window (i_1, i_2), the averaged squared deviation of the fit becomes

$$D = \sum_{i=1}^{n} \bar{x}_i^2 \overline{\omega}_i - \frac{s^2}{r(1-r)}, \quad (1)$$

where

$$r = \sum_{i=i_1}^{i=i_2} \overline{\omega}_i \text{ and } s = \sum_{i=i_1}^{i=i_2} \overline{\omega}_i \, \bar{x}_i. \quad (2)$$

Finally, the BLS frequency spectrum can be defined by the value of signal residue (*SR*) of the time series at any given trial period

$$SR = MAX \left[\frac{s^2(i_1,i_2)}{r(i_1,i_2)[1-r(i_1,i_2)\,]}\right]^{1/2}, \quad (3)$$

where *SR* represents the significance of the chosen trial period.

In practical computation, the folded time series is divided into a number of bins, and the significance ratio (*SR*) is evaluated for each of these binned values. We followed an approach to optimize the efficiency of the BLS algorithm through defining the number of bins in advance [5]. By using this approach, the efficiency of the computation is enhanced and the number of bins will define the time resolution of the evaluated LS solution. The automated BLS algorithm can show false signals. A false alarm is a signal with a planet-like signature that is not a true planet. Most of variable stars and eclipsing binaries are counted as planets because they show periodic change of the brightness in their light curves. Most of the false alarms are eclipsing binaries, thus the output of the BLS is screened manually using the high resolution light curves to filter out such sources.

For binary star systems, the drops in the brightness are not identical, which helps to distinguish binaries from planetary systems. We use the odd/even test outlined in [6] to exclude eclipsing binaries with planet-like signatures. The first and third drops in a light curve are called odds; while, second and fourth drops called evens, and so on. The principle of this test is to compare the odd and even drops in the phase folded light curves. If the light curve of a true exoplanet is folded with the exact period, the odds and the evens will be folded on each other and will be identical.

3. Results

Here we demonstrate the search method on Kepler data. The interface will first identify whether the object represents a true exoplanet candidate or not. The GUI (shown in figure 1) will then present the results for the manual/human screening. The left side of the GUI has the CONTROL PANEL which contains three interactive panels: OBJECTS CONTROL, PERIOD CONTROL, and TYPE CONTROL.

On the right side of the GUI, the INTERACTIVE PLOTS contain eight PLOT PANELS (7 to the right and 1 to the left). They are provided with an option of ZOOM IN/OUT.

The procedure for the search method includes: (a) identifying the object by searching for true planet signals which are characterized by periodic drop in the flux, (b) performing additional tests to distinguish between true signals and false alarms such as (i) using the phased light curve (PLOT #2) and the model (PLOT #3) as a first indication of a true planet signal, (ii) the (odd/even TEST) (PLOT #5), (iii) when the user suspects an overestimation or underestimation of the BLS period, it is preferable to investigate the BLS frequency spectrum (PLOT #1) with trial periods when the BLS period is not exact, (iv) compare the time series (in Julian dates) (PLOT #4) with the Centroids (PLOT #7 and #8), which shows if the drop in the flux is related to the position of the object on the CCD, (v) by examining (PLOT #4) (IN transit) one can assesses if "transits" are systematic effects of bad filtering of the data, and finally (vi) by comparing the (IN transit) points in (PLOT #4) with the model (PLOT #3) and phase folded light curve (PLOT #2).

Figure 1 shows an example with the Kepler data of a confirmed planet with Kepler Object of Interest and ID identification: KOI = 17.01 and KIC = 10874614.01. The transit planet signal can easily be identified from (PLOT #2) and (PLOT #3), and the (odd/even TEST) (PLOT #5) shows that the BLS period is exactly estimated. For further investigations, we can compare the centroids (PLOT #7 and #8) with the (IN transit) points (PLOT #4). Also, PLOT #1 and #6 show a very short period planet orbiting a star close to an F-type.

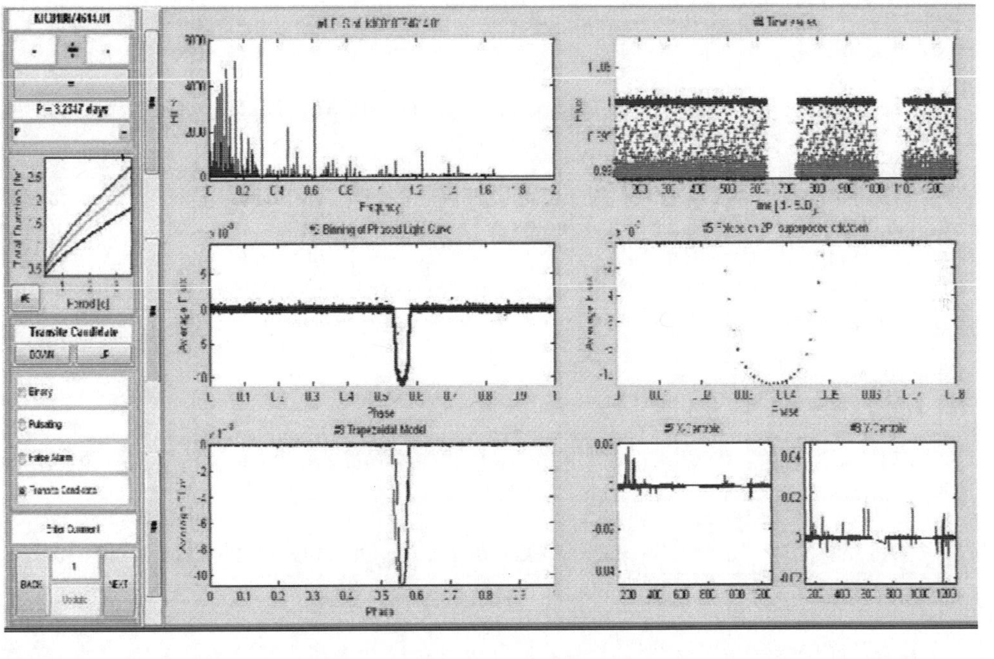

Figure 1. The output of the search method in the GUI for Kepler Catalog Object KIC= 10874614.01.

In figure 2, we show another example from the Kepler catalog of Kepler Object of Interest KOI = 509.01 and KIC = 6381846.01. The planet signal is again clearly identified from PLOT #2 and PLOT #3. The odd/even TEST (PLOT #5) shows that the BLS period is exactly estimated and the odds/evens are the same in these plots. For additional investigations, we can compare the centroids (PLOT #7 and #8) with the (IN transit) points (PLOT #4). PLOT #1 and PLOT #6 show that this is a short period planet orbiting a star close to a K-type.

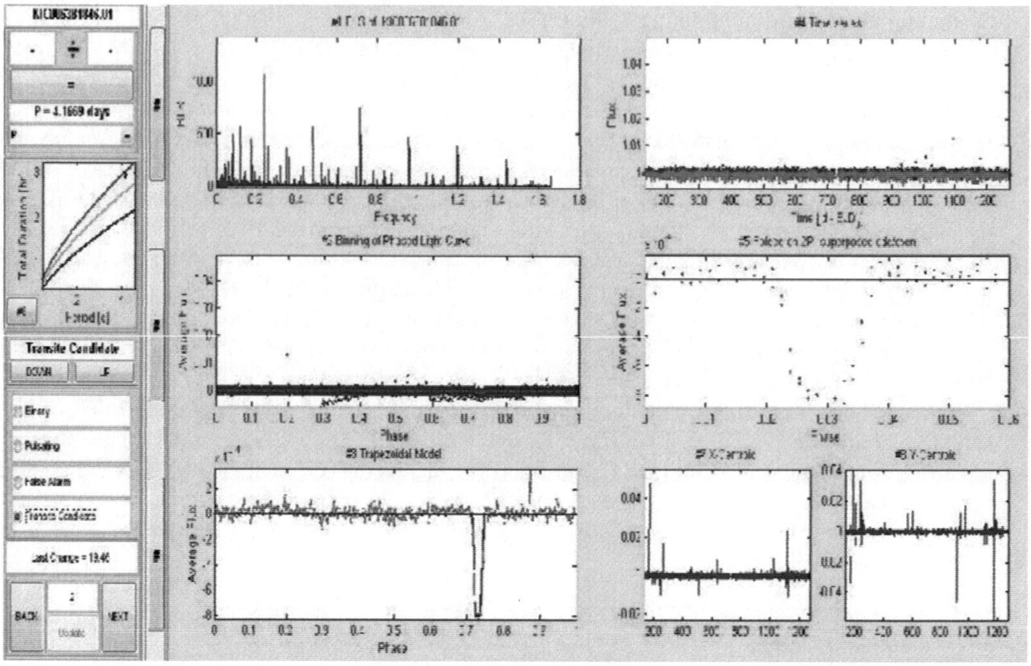

Figure 2. The output of the search method in the GUI for Kepler Catalog Object KIC= 6381846.01.

4. Conclusion

We have successfully built an exoplanet search tool for Kepler data and tested its viability in identifying true exoplanet signals through a versatile Graphical User Interface that allows for additional and customized investigations. The tool showed consistent results with ten known objects from the Kepler catalog and we showed here the results of the first two here for space limitation. Upgrading this tool to utilized large data is currently underway.

References

[1] Wolszczan A and Frail D 1992 *Nature* **355** 145
[2] Mayor M and Queloz D 1995 *Nature* **378** 355
[3] Marcy G, Butler R, Williams E, Bildsten L, Graham J, Ghez A and Jernigan J 1997 *ApJ* **481** 926
[4] Kovács G, Zucker S and Mazeh T 2002 *A&A* **391** 369
[5] Ofir A 2014 *A&A*, **561** A138
[6] Tenenbaum P, Bryson S, Chandrasekaran H, Li J, Quintana E, Twicken J and Jenkins J 2010 *Proc. SPIE* **7740** 77400J

IOP Conf. Series: Journal of Physics: Conf. Series **869** (2017) 012075 doi:10.1088/1742-6596/869/1/012075

Pattern recognition in spectra

M Gebran[1] and F Paletou[2,3]

[1]Department of Physics and Astronomy, Notre Dame University-Louaize, PO Box 72, Zouk Mikaël, Lebanon
[2]Université de Toulouse, UPS-Observatoire Midi-Pyrénées, IRAP, 31000 Toulouse, France
[3]CNRS, Institut de Recherche en Astrophysique et Planétologie, 14 av. E. Belin, 31400 Toulouse, France

E-mail: mgebran@ndu.edu.lb

Abstract. We present a new automated procedure that simultaneously derives the effective temperature $T_{\rm eff}$, surface gravity $\log g$, metallicity $[Fe/H]$, and equatorial projected rotational velocity $v_e \sin i$ for stars. The procedure is inspired by the well-known PCA-based inversion of spectropolarimetric full-Stokes solar data, which was used both for Zeeman and Hanle effects. The efficiency and accuracy of this procedure have been proven for FGK, A, and late type dwarf stars of K and M spectral types. Learning databases are generated from the Elodie stellar spectra library using observed spectra for which fundamental parameters were already evaluated or with synthetic data. The synthetic spectra are calculated using ATLAS9 model atmospheres. This technique helped us to detect many peculiar stars such as Am, Ap, HgMn, SiEuCr and binaries. This fast and efficient technique could be used every time a pattern recognition is needed. One important application is the understanding of the physical properties of planetary surfaces by comparing aboard instrument data to synthetic ones.

1. Introduction

Pattern recognition techniques are being intensively studied in Astronomy due to the large amount of data that we are receiving from space telescope/mission and on-ground instruments such as the RAdial Velocity Experiment (RAVE; [1, 2]), the Sloan Digital Sky Survey (SDSS; [3]), the Gaia ESO Survey (GES; [4]), the LAMOST Experiment for Galactic Understanding and Exploration (LEGUE; [5]), the Apache Point Observatory Galactic Evolution Experiment (APOGEE; [6]), the High Efficiency and Resolution Multi-Element Spectrograph survey (HERMES; [7]), and the Gaia Radial Velocity Spectrograph (RVS) survey (e.g. [8]). Photometric and spectroscopic surveys are becoming numerous. To give an example, the Large Synoptic Survey Telescope (LSST) project will conduct a ten-year survey of the sky that will deliver about 15 terabytes (TB) of raw data per night [9]. The total amount of data collected over the ten years of operation will be 60 PB, and processing these data will produce a 15 PB catalogue database. The need of automated techniques that can handle and analyze this amount of information becomes evident.

In this context, different semi- and fully automated techniques have been developed in order to determine fundamental parameters and/or element chemical abundances in stars. Techniques such as pixel to pixel comparison based on χ^2 algorithms are the most common [10–12]. Other

IOP Conf. Series: Journal of Physics: Conf. Series **869** (2017) 012075 doi:10.1088/1742-6596/869/1/012075

techniques, based on multivariable analysis, have shown efficiency in reducing the dimension of physical problems [13, 14].

This work compiles the results of previous studies [14–16] in which we have applied an automated techniques, based on the Principal Component Analysis (PCA), to derive simultaneously effective temperatures (T_{eff}), surface gravities ($\log g$), equatorial projected rotation velocities ($v \sin i$), Metallicities ($[\frac{Fe}{H}]$), and microturbulence velocities (ξ_t) of A, F, G, K, and M stars. This technique has proven to be fast and accurate in inverting observed high-resolution spectra. Unlike classical techniques such as χ^2 fitting, PCA is based on a dimensionality reduction for the fast search of the nearest neighbour of an observed spectrum in a learning database.

2. Observations

For this study, we have acquired spectra of stars of different spectral types. High-resolution HARPS spectra of late-type dwarf K stars and early-type dwarf M stars were taken from the ESO Science Archive Facility. High-resolution spectra of A, F, G and K stars observed with Narval and ESPaDOnS spectropolarimeters were taken from the PolarBase (http://polarbase.irap.omp.eu). Narval is a spectropolarimeter, mounted on the 2-m TBL telescope, operating in the optical domain (380–1000 nm), with a spectral resolution of R∼65 000 in its polarimetric mode. ESPaDOnS spectropolarimeter has mainly the same characteristics as Narval and is operating on the 3.6-m CFHT telescope. We have also retrieved A-type stars spectra from the ELODIE (http://atlas.obs-hp.fr/elodie/) and SOPHIE (http://atlas.obs-hp.fr/sophie/) archives. The ELODIE is a fiber-fed, cross-dispersed Echelle spectrograph that was attached to the 1.93-m telescope at Observatoire de Haute-Provence and had a resolution of R∼ 42 000. SOPHIE replaced ELODIE in September 2006. SOPHIE spectra stretch from 3870 to 6940 Å at a resolution of R∼75 000 in HR mode.

3. The learning database

Different types of learning databases were constructed for this project. One based on observed spectra and others based on calculated synthetic spectra. The first one was created after the Elodie stellar spectral library [17, 18]. The stellar parameters associated with each ELODIE spectrum were extracted from the CDS. The Synthetic databases were constructed using different sets of resolution (R∼65 000, 76 000, and 42 000 for A stars, and R∼115 000 for late-type dwarf K and early-type dwarf M stars). These different grids are constructed using a combination of different parameters such as T_{eff}, $\log g$, [Fe/H], $v \sin i$, and ξ_t. Line-blanketed ATLAS9 model atmospheres [19, 20] were calculated for the purpose of this work. ATLAS9 models are LTE plane parallel and assume radiative and hydrostatic equilibrium. ATLAS9 uses the opacity distribution function (ODF) of [21]. We included convection in the atmospheres of stars cooler than 8500 K. Convection was treated according to Smalley's prescriptions [22]. The grid of synthetic spectra was computed using SYNSPEC48 [23]. The effective temperatures of the model atmospheres vary depending of the spectral types with a typical step of 100 K. Logarithm of surface gravities vary between 2.0 and 5.0 with a step of 0.1 dex, respectively. The projected equatorial rotational velocities $v \sin i$ were varied from 0 up to 300 km s^{-1} with a non-constant step, metallicity was scaled from -2.0 dex up to +2.0 dex with respect to the solar value [24] with a step of 0.1 dex. The wavelength range that we used depend on the spectral type of the observations as explained in [14–16].

4. Principal component analysis and parameters derivation procedure

We represent all the spectra of the learning database in a matrix $\boldsymbol{S}$ of size N_{spectra} by N_λ. $\bar{S}$ will be the average of $\boldsymbol{S}$ along the N_{spectra}-axis. Then, we derive the eigenvectors $\mathbf{e}_k(\lambda)$ of the variance-covariance matrix $\boldsymbol{C}$ ($\boldsymbol{C} = (\boldsymbol{S} - \bar{S})^{\text{T}} \cdot (\boldsymbol{S} - \bar{S})$) using Singular Value Decomposition

IOP Conf. Series: Journal of Physics: Conf. Series **869** (2017) 012075 doi:10.1088/1742-6596/869/1/012075

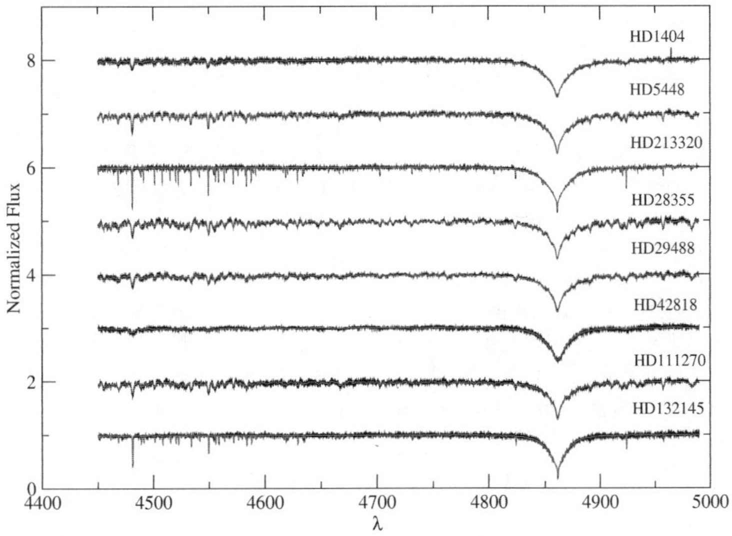

Figure 1. First neighbour fit of synthetic spectra (in red) to the observed spectra (in black). The wavelengths are in Å. Spectra were shifted up for clarity.

method or the power iteration method (see [14] for more details). The variance-covariance matrix has a dimension of $N_\lambda \times N_\lambda$. The eigenvectors of the symmetric $\boldsymbol{C}$ matrix are then sorted in decreasing eigenvalues magnitude; they are more usually referred to as "principal components". Each spectrum of the dataset $\boldsymbol{S}$ is then represented by a small number of coefficients p_{jk} defined as

$$p_{jk} = (S_j - \bar{S}) \cdot \boldsymbol{e}_k \tag{1}$$

The main point of reducing the dimensionality is the selection of $k_{max} \ll N_\lambda$. For all databases and in case of all spectral types and wavelength spectral ranges, we showed that $k_{max} = 12$ is a justified choice [14–16].

The observed spectra are denoted by $O(\lambda)$ having the same wavelength range and sampling as the learning dataset spectra. After correcting the observed spectra for radial velocities, we compute the set of projection coefficients ϱ_k defined as

$$\varrho_k = (O - \bar{S}) \cdot \boldsymbol{e}_k \tag{2}$$

The nearest neighbour is then found by minimizing a χ^2 in the low dimensional space of the coefficients

$$d_j^{(O)} = \Sigma_{k=1}^{k_{max}} (\varrho_k - p_{jk})^2, \tag{3}$$

where j covers the number of synthetic spectra. The parameters of the synthetic spectrum having the minimum d will be considered for the inversion. For each observed spectrum, an iteration is performed, according to [25], in order to correct bad normalization as explained in [14]. An example of an overall spectral fitting between the first neighbour synthetic spectrum and several observation from our sample stars is shown in figure 1.

5. Results and discussion

In order to characterize the accuracy of the PCA technique, we compared our results to previous published values using Vizier catalogue (see [26] for more details). In overall, we found very good

agreement between our inverted values and the already published ones. It is worth mentioning that depending on the effective temperature and surface gravity of the star, learning databases are constructed in different wavelength regions. As spectral lines behave differently with $T_{\rm eff}$ and $\log g$, the affected errors do also vary with spectral types. On average mean signed differences between inverted and catalog values are smaller than the adopted spectroscopic errors on $T_{\rm eff}$, $\log g$, and [Fe/H]. Typical errors of ± 150 K are found for $T_{\rm eff}$, ± 0.15 dex for $\log g$, ± 0.1 dex for [Fe/H], ± 5 km s^{-1} for $v \sin i$. Large differences between catlog and inverted values yielded to detection of peculiar and/or binary stars as shown in [14]. In some cases, large inverted $v \sin i$ is an indicator of spectroscopic binary system and large or low metallic abundances is an indicator, in case of A and late B stars, of chemically peculiar stars such as Am, Ap, HgMn, SiEuCr...
The PCA techniques is being also used in order to derive fundamental parameters of stars for the GES survey [27]. We are now investigating other multivariable techniques such as the Slice Inverse Regression (SIR) that was once used to retrieve Mars surface physical properties from spectroscopic data [28, 29].

References

[1] Steinmetz M 2003, *GAIA Spectroscopy: Science and Technology* **298** 381
[2] Munari U, Zwitter T and Siebert A 2005 *The Three-Dimensional Universe with Gaia* **576** 529
[3] York D G *et al* 2000 *AJ* **120** 1579
[4] Gilmore G *et al* 2012 *The Messenger* **147** 25
[5] Deng L C *et al* 2012 *Research in Astronomy and Astrophysics* **12** 735
[6] Majewski S R, Wilson J C, Hearty F, Schiavon R R and Skrutskie M F 2010 *Chemical Abundances in the Universe: Connecting First Stars to Planets* **265** 480
[7] Zucker D B, de Silva G, Freeman K, Bland-Hawthorn J and Hermes Team 2012 *Galactic Archaeology: Near-Field Cosmology and the Formation of the Milky Way* **458** 421
[8] Bailer-Jones C A L *et al* 2013 *A&A* **559** A74
[9] Kantor J *et al* 2007 *Astronomical Data Analysis Software and Systems XVI* **376** 3
[10] Allende P C *et al* 2006 *ApJ* **636** 804
[11] Zwitter T *et al* 2008 *AJ* **136** 421
[12] García Pérez A E *et al* 2016 *AJ* **151** 144
[13] Traven G *et al* 2016 *arXiv:1612.02242*
[14] Gebran M, Farah W, Paletou F, Monier R and Watson V 2016 *A&A* **589** A83
[15] Paletou F, Böhm T, Watson V and Trouilhet J F 2015 *A&A* **573** A67
[16] Paletou F, Gebran M, Houdebine E R and Watson V 2015 *A&A* **580** A78
[17] Prugniel P, Soubiran C, Koleva M and Le Borgne D 2007 *arXiv:astro-ph/0703658*
[18] Prugniel P and Soubiran C 2001 *A&A* **369** 1048
[19] Kurucz R L 1992 *Rev. Mex. Astron. Astrofis.* **23** 45
[20] Kurucz R L 2005 *Mem. Soc. Astron. It. Suppl.* **8** 14
[21] Castelli F and Kurucz R L 2003 *Modelling of Stellar Atmospheres* **210** 20P
[22] Smalley B 2005 *Mem. Soc. Astron. It. Suppl.* **8** 130
[23] Hubeny I and Lanz T 1992 *A&A* **262** 501
[24] Grevesse N and Sauval A J 1998 *Solar Composition and Its Evolution – From Core to Corona* 161
[25] Gazzano J C, de Laverny P, Deleuil M et al. 2010 *A&A* **523** A91
[26] Paletou F and Zolotukhin I 2014 *arXiv:1408.7026*
[27] Farah W, Gebran M, Paletou F and Blomme R 2015 *SF2A-2015: Proceedings of the Annual meeting of the French Society of Astronomy and Astrophysics* 3
[28] Bernard-Michel C, Douté S, Fauvel M, Gardes L and Girard S 2009 *Journal of Geophysical Research (Planets)* **114** E06005
[29] Watson V, Trouilhet J F, Paletou F and Girard S 2017 *IEEE (Proceeding of the International Workshop of Electronics, Control, Measurement, Signals and their application to Mechatronics)* (in preparation)

IOP Conf. Series: Journal of Physics: Conf. Series **869** (2017) 012076 doi:10.1088/1742-6596/869/1/012076

Effect of magnetic reconnection in stellar plasma

M Hammoud[1], M El Eid[1] and M Darwish[2]

[1]Physics Department, American University of Beirut, Beirut, Lebanon

[2]Mechanical Engineering Department, American University of Beirut, Beirut, Lebanon

E-mail: mmh82@aub.edu.lb

Abstract. An important phenomenon in Astrophysics is the process of magnetic reconnection (MGR), which is envisaged to understand the solar flares, coronal mass ejection, interaction of the solar wind with the Earth's magnetic field (so called geomagnetic storm) and other phenomena. In addition, it plays a role in the formation of stars. MGR involves topological change of a set of magnetic field lines leading to a new equilibrium configuration of lower magnetic energy. The MGR is basically described in the framework of the Maxwell's equations linked to Navier-Stockes equations. Nevertheless, many details are still not understood. In this paper, we investigate the MGR process in the framework of the Magnetohydrodynamic (MHD) model of a single conducting fluid using a modern powerful computational tool (OpenFOAM). We will show that the MGR process takes place only if resistivity exists. However, despite the high conductivity of the plasma, resistivity becomes effective in a very thin layer generating sharp gradients of the magnetic field, and thus accelerating the reconnection process. The net effect of MGR is that magnetic energy is converted into thermal and kinetic energies leading to heating and acceleration of charged particles. The Sun's coronal ejection is an example of the MGR process.

1. Introduction

MGR is a process of rearranging the magnetic field-lines topology leading to a restructuring of macroscopic quantities of plasma such as flow and thermal energy. It's observed in both astrophysical and laboratory plasmas. This phenomenon is important from the microphysics point of view, and for the wide range of its applications. Applying the resistive MHD approach to MGR is a challenging task, but it is worth doing, since MGR is important for the evolution of solar flares, coronal mass ejection, and is considered to occur in the formation of stars [1–5]. In addition, MGR plays a decisive role in understanding the interaction of the solar wind with the Earth's magnetosphere leading to geomagnetic storms [6], and to the formation of the auroras. NASA has launched a mission, namely "MMS", aimed at collecting data of the MGR process that takes place in the Earth's magnetosphere [7]. The theoretical approach of MHD model aims at seeking better description of the diffusion process during the MGR. In section 2, we show why resistive MHD is essential for the MGR process. In section 3, we present the simulation model. Results and conclusion are discussed in section 4.

2. Why resistive MHD?

When plasma moves, it carries the magnetic field lines along with it or vice-versa. This is what is known as the frozen-flux constraint. As shown in figure (1-a), the magnetic flux through a surface S, enclosed by a closed contour C, is given by

IOP Conf. Series: Journal of Physics: Conf. Series **869** (2017) 012076 doi:10.1088/1742-6596/869/1/012076

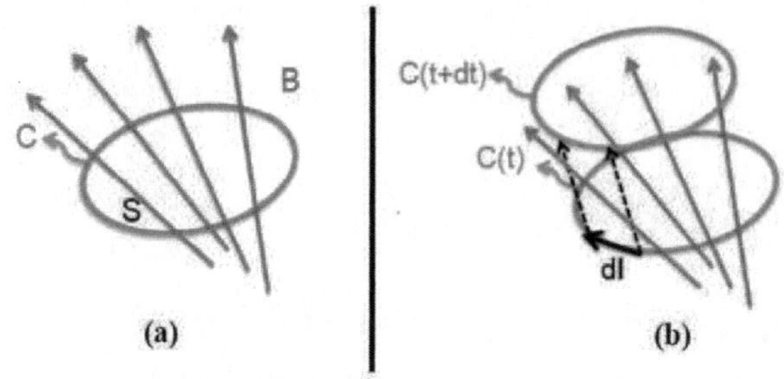

Figure 1. (a) The magnetic flux Ψ through a surface S enclosed by a contour C. (b) Varying the magnetic flux Ψ by moving the surface S with an arbitrary velocity $\vec{u} = \frac{\vec{dl}}{dt}$

$$\Psi = \int_S \vec{B}.\vec{dS} \tag{1}$$

If we want to change this magnetic flux with time, two possibilities are present:

$$\left(\frac{\partial \Psi}{\partial t}\right)_1 = \int_S \frac{\partial \vec{B}}{\partial t}.\vec{dS} = -\int_S \vec{\nabla} \times \vec{E}.\vec{dS} \tag{2}$$

or

$$\left(\frac{\partial \Psi}{\partial t}\right)_2 = \int_C \vec{B}.\vec{u} \times \vec{dl} = \int_S \vec{\nabla} \times (\vec{B} \times \vec{u}).\vec{dS} \tag{3}$$

In equation (2), the magnetic field $\vec{B}$ itself can change in time. While, in equation (3) and figure (1-b), the surface S can move with an arbitrary velocity $\vec{u}$. Combining the two contributions, we get the full time-varying magnetic flux,

$$\frac{d\Psi}{dt} = -\int_S \vec{\nabla} \times (\vec{E} + \vec{u} \times \vec{B}).\vec{dS} \tag{4}$$

Recalling Ohm's law from Maxwell's equations, $\vec{E} + \vec{u} \times \vec{B} = \eta \vec{J}$, where η and $\vec{J}$ are respectively the plasma resistivity and current density. If we neglect collisions, plasma resistivity vanishes, and magnetic flux is thus time-invariant. Therefore, the magnetic field lines must move with the plasma and the frozen-flux constraint holds. In contrast, with resistivity the plasma particles leave their own magnetic field line to get stuck on another field line, and hence the MGR mechanism occurs. So, plasma resistivity is a necessary condition to allow for MGR. However; plasma is a very good conductor, even though resistivity becomes efficient in a very fine region giving rise to sharp gradients of the magnetic field. Such a region, according to Ampère's law, holds then a large current, hence, the term current layer or current sheet.

3. Simulation model

Our simulation is carried out in a two-dimensional domain where $x \in [-16\delta, +16\delta]$ and $y \in [-32\delta, +32\delta]$. $\delta = 1.5 \times 10^6$ m is the half width of the current layer. The astrophysical plasma is treated here as a conducting fluid, so coupling between Navier-Stokes and Maxwell's equations is essential in the simulation of MGR. The obtained equations are the so-called: "Resistive MHD Model". It has four principal equations, continuity, momentum, induction and energy equations.

$$\frac{\partial \rho}{\partial t} + \vec{\nabla}.(\rho \vec{u}) = 0 \tag{5}$$

IOP Conf. Series: Journal of Physics: Conf. Series **869** (2017) 012076 doi:10.1088/1742-6596/869/1/012076

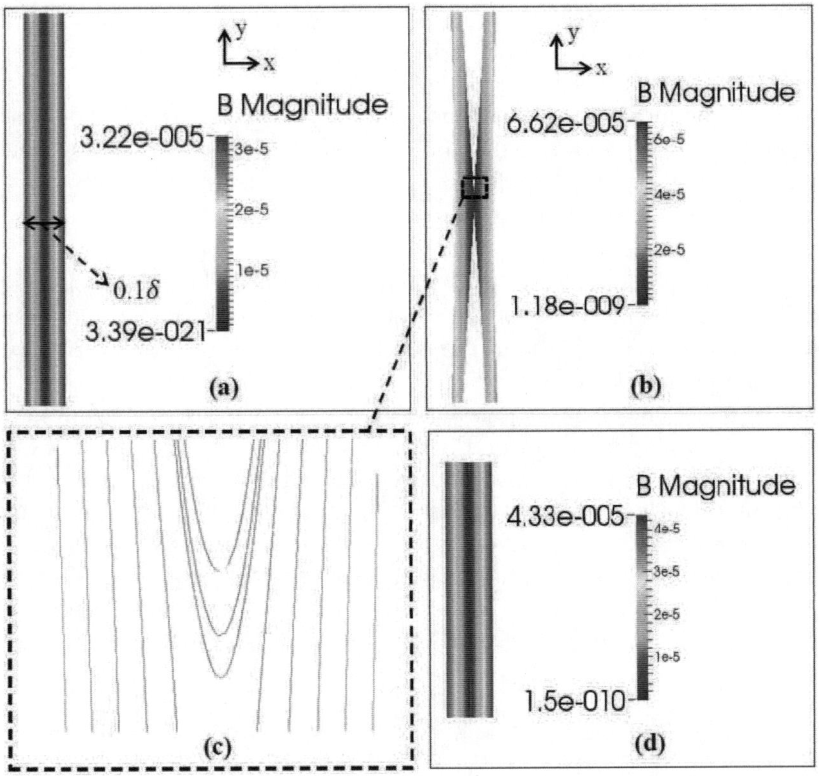

Figure 2. (a) Oppositely directed magnetic fields (downward magnetic fields in the left side and upward fields in the right side), expressed in Tesla, at time t = 0. (b) The MGR process takes place around the origin at t = 60 sec for $\tilde{\eta}$ = 200 Ω.m. (c) A zoom of the reconnection region around the origin at time t = 60 sec for $\tilde{\eta}$ = 200 Ω.m. (d) The current layer does not show any MGR process at time t = 60 sec for $\eta = 0$.

$$\frac{\partial(\rho\vec{u})}{\partial t} + \vec{\nabla}.(\rho\vec{u}\vec{u}) = -\vec{\nabla}(p + \frac{B^2}{2\mu_0}) + \vec{\nabla}.(\frac{\vec{B}\vec{B}}{\mu_0}) \tag{6}$$

$$\frac{\partial\vec{B}}{\partial t} + \vec{\nabla}.(\vec{u}\vec{B}) - \frac{\eta}{\mu_0}\vec{\nabla}^2\vec{B} = \vec{\nabla}.(\vec{B}\vec{u}) + \vec{J}\times\vec{\nabla}\eta \tag{7}$$

$$\frac{\partial\rho E}{\partial t} + \vec{\nabla}.(\rho\vec{u}E) = -\vec{\nabla}.[(p + \frac{B^2}{2\mu_0})\vec{u}] + \frac{1}{\mu_0}\vec{\nabla}.[(\vec{B}.\vec{u})\vec{B}] - \frac{1}{\mu_0}\vec{\nabla}.(\eta\vec{J}\times\vec{B}) \tag{8}$$

$E = e + \frac{1}{2}u^2 + \frac{B^2}{2\rho\mu_0}$ is the total energy per unit mass. e is the internal energy per unit mass, ρ is the plasma density (ρ_0=1.7 × 10^{-12} kg.m^{-3} for hydrogen plasma), $\vec{u}$ is the plasma velocity, $\vec{B}$ is the magnetic field (B_0 = 5 Gauss is the typical magnetic field) and μ_0 is the permeability (=$4\pi \times 10^{-7}$ SI units). The plasma here is treated as an ideal gas where plasma pressure can be written in terms of internal energy according to $p = (\gamma - 1)\rho e$ ($\gamma = 5/3$ for hydrogen plasma). The functional form of resistivity used is $\eta = \eta_0 + \tilde{\eta}e^{-(x^2+y^2)}$ [8], where η_0 = 1 Ω.m is the background resistivity and $\tilde{\eta}$ = 200 Ω.m is the enhanced resistivity around the origin to ensure the MGR process. The simulation of MGR is done using non-uniform grid spacing, from 0.3δ at boundaries down to 5$\mu\delta$ around origin. The time step is 10 ms. As initial conditions, we

IOP Conf. Series: Journal of Physics: Conf. Series **869** (2017) 012076 doi:10.1088/1742-6596/869/1/012076

use uniform density ρ_0, uniform temperature T_0 (= 2 Mk), we start with exactly zero-velocity and the initial thermal pressure is obtained by solving the equilibrium equations. The initial profile of magnetic field is chosen to be along y-direction, $B_y = B_0\tanh(x)$. As for boundary conditions, a periodic boundary condition is used in the y-direction. In the x-direction, magnetic pressure (of value 0.1 Pascal) is imposed at boundaries, while a zero-gradient condition is used for velocity, temperature and magnetic field.

4. Results and conclusion

Eqs.(5, 6, 7, 8) are four nonlinearly coupled equations. Numerical simulation of the governing equations are carried out using the open source OpenFOAM toolbox [9]. The plasma dynamics is treated here as a compressible fluid motion and the equations are solved in an Eulerian frame using the finite volume method [10].

Two runs are done to see the effect of resistivity on the MGR process. The initial profile of the magnetic field is presented by figure (2-a). Figures (2-b) and (2-c) show evidence of reconnecting field lines at time t = 60 sec if the resistivity is enhanced around the origin. However; MGR disappears in figure(2-d) where resistivity is zero. These results support what we have discussed in section 2 where the resistive MHD model is an important candidate for the MGR process. Of course, this is not the end, we still need to test the energy released during MGR [11], because it is the clue beyond some of the solar activities (solar wind, coronal mass ejection, geomagnetic storm, etc... .). But, energy analysis needs a lot of work and results, and this will be left for future work.

References

[1] Parker E N 1979 Cosmical magnetic fields: Their origin and their activity *Oxford, Clarendon Press; New York, Oxford University Press* 858.

[2] Kulsrud R M 1998 *Physics of Plasmas* **5** 1599-606.

[3] Biskamp D 2000 *Magnetic reconnection in plasmas* **3** (Cambridge monographs on plasma physics).

[4] Priest E R and Forbes T G 2000 *Magnetic Reconnection: MHD Theory and Applications.*

[5] Yamada M, Kulsrud R and J H 2010 Magnetic reconnection *Reviews of Modern Physics* **82(1)** 603.

[6] Tsurutani B T and Gonzalez W D 1997 *The interplanetary causes of magnetic storms: A review. American Geophysical Union* pp 77-89.

[7] Burch J L, Moore T E, Torbert R B and Giles B L 2016 *Space Science Reviews* **199(1-4)** 5-21.

[8] Wang L C, Li L J, Ma Z W, Zhang X and Lee L C 2015 *Phys. Lett.* A **379(36)** 2068-72.

[9] Jasak H, Jemcov A and Tukovic Z 2007 *OpenFOAM: A C++ library for complex physics simulations (In International workshop on coupled methods in numerical dynamics)* **1000** (IUC Dubrovnik, Croatia) pp 1-20.

[10] Versteeg H K and Malalasekera W 2007 *An Introduction to Computational Fluid Dynamics: The Finite Volume Method* Pearson Education.

[11] Sato T and Hayashi T 1979 *The Physics of Fluids* **22(6)** 1189-202.

IOP Conf. Series: Journal of Physics: Conf. Series **869** (2017) 012077 doi:10.1088/1742-6596/869/1/012077

A study of north-south asymmetry of interplanetary magnetic field plasma and some solar indices throughout four solar cycles

M A El-Borie[1], A A Abdel-halim[2], S Y El-Monier[2] and A A Bishara[1]

[1]Physics Dept., Faculty of Science, Alexandria University, Alexandria, Egypt
[2]Arab Academy for Science &Technology and Maritime Transport, Egypt
E-mail: elborie@alexu.edu.eg

Abstract. We provide a long epoch study of a set of solar and plasma parameters (sunspot number Rz, total solar irradiance TSI, solar radio flux SF, solar wind speed *V*, ion density n, dynamic pressure nV^2, and ion temperature T) covering a temporal range of several decades corresponding to almost four solar cycles. Such data have been organized accordingly with the interplanetary magnetic field (IMF) polarity, i.e. away (A) if the azimuthal component of the IMF points away from the Sun and T if it points towards, to examine the N-S asymmetries between the northern and southern hemispheres. Our results displayed the sign of the N-S asymmetry in solar activity depends on the solar magnetic polarity state (qA>0 or qA<0). The solar flux component of toward field vector was larger in magnitude than those of away field vector during the negative polarity epochs (1986-88 and 2001-08). In addition, the solar wind speeds (SWS) are faster by about 22.11±4.5 km/s for away polarity days than for toward polarity days during the qA<0 epoch (2001-08), where the IMF points away from the Sun. Moreover, during solar cycles 21st and 24th the solar plasma is more dense, hotter, and faster south of the HCS.

1. Introduction
The interplanetary magnetic field (IMF) is organized into alternate field sectors pointing toward or away from the Sun with relatively sharp sector boundaries, resulting from a solar field pattern in which the positive polarity is separated from the negative polarity field by a warped heliospheric current sheet (HCS). The N-S asymmetries in the interplanetary plasma, solar indices and the cosmic-ray intensities between the northern and southern sectors, with respect to the heliospheric neutral sheet, have been studied [1-9]. The N-S asymmetry distributions of major flares, sunspot areas were independent on the 11 yr solar cycle [5,6]. The N-S asymmetries of solar parameters for solar cycles (22nd-25th) have been examined and found that the southern hemisphere is dominated during cycles 22 and 23 and 24. Results of [2] showed that the solar plasma north of the current sheet was hotter, faster and less dense than south of it during the qA<0 epoch (1981-87). The N-S asymmetries of the coronal-green line brightness and the sunspot area over a period 1939-2001 have been studied [10], indicating the asymmetries should be regarded as a fundamental phenomenon of solar activity. Other study [11] showed the SWS, IMF, and the geomagnetic index Ap exhibited clear heliospheric N-S asymmetries. The N-S asymmetry for different solar activity features (solar proton events, solar active prominences, Hα flare index, soft X-ray flares, monthly mean sunspot area and sunspot number) have been examined by [12], showing the solar cycle 23rd was magnetically weak compared to solar cycle 22nd.

2. The N-S asymmetry of the solar and interplanetary parameters
We have used the hourly averages of interplanetary plasma and solar parameters (sunspot number Rz, total solar irradiance TSI, solar radio flux SF, solar wind plasma speed *V*, ion plasma density n, ion

IOP Conf. Series: Journal of Physics: Conf. Series **869** (2017) 012077 doi:10.1088/1742-6596/869/1/012077

temperature T and the dynamic pressure nV^2) collected near 1 AU which were provided by the National Space Science Data Center, over the period 1975-2013. The IMF direction is determined on an hourly basis in the geocentric solar ecliptic (GSE) coordinates. The field direction is then separated into two polarities; away (A) polarity days if the solar ecliptic azimuthal angle of the IMF daily averages lies between 45° and 225°; otherwise it is considered as toward (T) the Sun. Only days on which there were 12 or more hourly averages of field direction were considered. Also, days on which the IMF was truly of mixed polarity were removed. This may affect the accuracy of the final result and could be the reason for a sudden change in the value of N-S asymmetry. Then, we have separated the considered data into two groups according to A or T daily average IMF vector, over the period 1975-2013. The yearly average of T and A groups have been determined for each parameter.

Figure 1 displays the yearly differences between the IMF toward and away days of Rz, SWS, nV^2 and T (other parameters are not shown). The error bar is considered for each year. The times of the magnetic field reversal at the Sun's north (N) and south (S) poles are added at the top of each plot. Vertical lines display the periods of minima (dashed lines) and maxima (solid lines) solar activity.

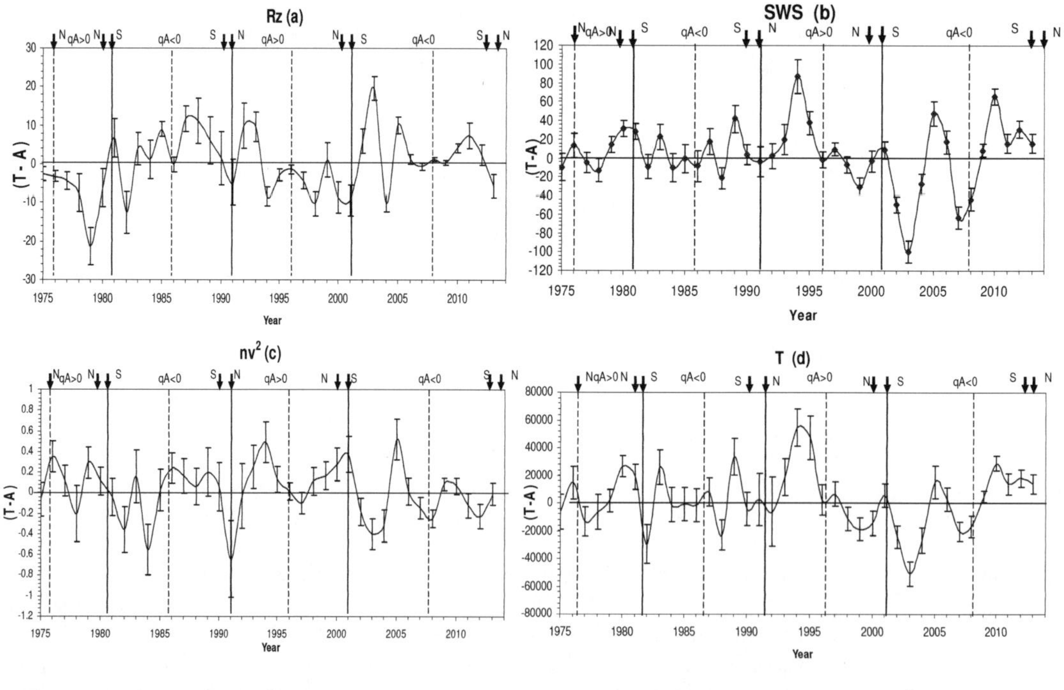

Figure 1. The yearly differences between the IMF toward and away days of Rz, TSI, nV^2 and T during the period 1975-2013. Error bars are considered for each year. The times of the magnetic field reversal at the Sun's north (N) and south (S) poles are added at the top.

From figure 1 we can notice the followings: **(1)** Plot 1a for the sunspot number Rz shows, there are 11 positive and 14 negative asymmetries. Largest positive asymmetry occurred in 2003 (19.5±3.1) during the descending phase of solar cycle 23rd. In contrast, largest negative asymmetries occurred in 1979 (-21.3±4.8) and 1982 (-12.5±5.4). Most of the asymmetries occurred when the solar northern hemisphere has toward polarity during the negative polarity epochs qA<0 (1981-88 and 2001-12). Three asymmetries occurred throughout the IMF mixed polarities (1979, 1980 and 2000). Table 1 shows the average of solar parameters for both T and A polarity days, as well as the differences T-A between the two groups over the epochs of positive and negative IMF polarity years during the solar cycles 22nd and 23rd. The sign of the asymmetry for Rz changes from one solar magnetic polarity to the other. It becomes almost positive in negative polarity epochs and negative in positive magnetic polarity epochs. Note that the sunspot number Rz has a northern dominance during the period of our study. In addition, Table 1 shows the asymmetries are great throughout the negative IMF polarity periods.

Table 1. Averages of the N-S asymmetry for the interplanetary plasma and solar parameters for positive and negative IMF polarity days during the solar cycles 22nd and 23rd

IMF & Solar Parameters	Solar Cycle 22nd (Sep 1986-May 1996)					
	Negative Polarity (1986-1988) qA<0			Positive Polarity (1991-1996) qA>0		
	Average of 256 T-days	Average of 235 A-days	T-A average	Average of 652 T-days	Average of 630 A-days	T-A average
Rz	63.28±3.2	53.09±3	10.19±4.3	54.64±2.2	59.14±2.1	-4.50±3
TSI	1365.8±0.03	1365.7±0.02	0.09±0.03	1365.8±0.02	1365.9±0.02	-0.07±0.03
SF	112.22±2.4	103.08±2.6	9.15±3.5	115.34±2.5	117.17±2.1	-1.83±3.2
SWS	439.67±6.1	433.45±5.4	6.2±8.1	472.76±4.7	445.86±4.3	26.90±6.3
n	8.28±0.3	7.79±0.28	0.49±0.4	7.72±0.2	8.59±0.23	-0.87±0.26
nV^2	2.75±0.07	2.63±0.07	0.11±0.09	2.95±0.05	2.99±0.04	-0.04±0.06
T (10^3 K)	111±4.8	112.2±4.9	-1.2±6.8	167.3±4.5	142.1±6	25.2±6
	Solar Cycle 23rd (May 1996-Jan 2008)					
	Positive Polarity (1996-1998) qA>0			Negative Polarity (2001-2008) qA<0		
IMF & Solar Parameters	Average of 387 T- days	Average of 438 A-days	T-A average	Average of 1127 T-days	Average of 1192 A-days	T-A average
Rz	32.81±1.5	35.11±1.6	-2.30±2.2	54.18±1.3	50.80±1.3	3.37±1.8
TSI	1365.7±0.01	1365.7±0.01	-0.03±0.01	1365.8±0.02	1365.7±0.02	0.07±0.03
SF	91.76±1.2	93.14±1.2	-1.38±1.7	120.8±1.4	117.56±1.3	3.30±1.9
SWS	396.66±3.3	407.48±3.6	-10.8±4.9	450.9±3.1	473.06±3.3	-22.1±4.5
n	8.03±0.2	7.75±0.2	0.28±0.28	6.05±0.1	5.65±0.1	0.40±0.1
nV^2	2.21±0.05	2.22±0.04	-0.02±0.06	2.18±0.04	2.19±0.03	-0.1±0.05
T (10^3 K)	112.2±4	126.1±4.2	-13.9±5.8	110.6±2.3	122.8±2.6	-12.2±3.5

(2) Table 1 shows the grand average differences of TSI north and south of the HCS are very low for qA< 0 and qA>0 epochs. Therefore, the averages TSI have no significant asymmetries.
(3) The SF component of toward field vector was larger magnitude than those of away field vector during the negative polarity epochs (1986-88 and 2001-08) and there is an asymmetry during the qA<0 epoch (2008-12). The SF has a northern dominance during cycles 22nd and 23rd.
(4) Plot 1b shows 15 clear positive asymmetries and seven negative orientations. The largest positive asymmetry occurred in 1994 (87.1±17.1 km/s) near the minimum of the solar cycle 23rd, in contrast a negative asymmetry occurred in 2003 (-99.6±11.7 km/s) during the descending phase of the cycle 23rd.
(5) The SWS are faster by 26.9±6.3 km/s for T days than for A days when the IMF points away from the Sun north of the current sheet, during thc qA>0 epoch (1991-96). In contrast, the SWS are faster

by 22.11±4.5 km/s for A days than those of T days when the IMF points away from the Sun south of the HCS during the qA<0 (2001-08). The asymmetry for the SWS has a northern dominance during solar cycles 21st and 24th, in contrast to southern dominance through cycles 22nd and 23rd.

(6) For the ion density (not shown), nine positive and seven negative asymmetries have been occurred. The average ion plasma density for A days north of the HCS during the 1991-96 epoch is denser by 0.87±0.26 n/cc than those of T days south of the HCS. In contrast, the ion density for A days south of the HCS during the 2008-12 epochs is 0.5±0.1 n/cc denser than that of T days north of the HCS. Furthermore, the solar plasma south of the current sheet was dense during the cycles 21 and 24 and also was more dense in north than in south of the current sheet during the cycles 22 and 23.

(7) Plot 1c shows only nineteen asymmetries. The average N-S asymmetry of nV^2 is nearly absent.

(8) The asymmetry for the ion temperature T is typically the same for the SWS with almost positive and negative asymmetry peaks. Largest positive asymmetry had occurred in 1994 with 54.6±1.3 x 10^3 °K, while negative asymmetry had occurred in 2003 with -50.7 ±8.9 x 10^3 °K.

(9) We can note that the solar plasma south of HCS was hotter and less dense than north of it during the cycles 22nd and 23rd. In contrast, the solar cycles 21 and 24 displayed the change in asymmetry orientations. Moreover, significant changes in the N-S asymmetries have been correlated with the N-S asymmetries of the solar activity in the northern and southern hemispheres.

3. Conclusion

In this paper, we report a long epoch study of a set of solar and plasma parameters to examine the asymmetries between the northern and southern HCS. Such data are organized accordingly with the interplanetary magnetic field (IMF) polarity, i.e. away (A) if the azimuthal component of the IMF points away from the Sun and T if it points towards. Our principle conclusions are the following: The sign of the north-south asymmetry in solar activity depends on the solar polarity sense. The asymmetry of Rz has a northern dominance during the considered period of our study. The grand average differences of the TSI north and south of the HCS are very low for qA< 0 and qA>0 epochs. The SF component of T field vector was larger magnitude than that of A field vector during the negative polarity epochs and no asymmetry observed in qA>0 epoch. An average SWS of toward polarity days south of HCS was faster by 26.9±6.3 km/s than those of away polarity days north of HCS through the positive polarity (1991-96). During cycles 21st and 24th the solar plasma north of HCS is less dense, hotter, and faster than south of it. In contrast during cycles 22nd and 23rd the solar plasma south of HCS is less dense, hotter, and faster than north of it. Large asymmetries in SWS, n, nV^2 and T occurred in 1994. Finally, our results indicate that the N-S asymmetries of the Sun's activity may provide a possible explanation for the observed asymmetries in interplanetary and plasma parameters.

References

[1] Oliver R and Ballester J L 1994 *Sol. Phys.* **152** 481
[2] Verma V K 1993 *J. Astrophy.* **403** 797
[3] El-Borie M A, Darwish A A and Bishara A A 1996a *Sol. Phys.* **167** 395
[4] El-Borie M A, Sabbah I S and Bishara A A 1996b *Astro. Nachr.* **317(4)** 267
[5] El-Borie M A, Abdel-halim A A, El-Monier S Y and Bishara A A 2016 *Sol. Phy.* **291** 3817
[6] Partha C and Dwivedi B N 2011 *ASI Conference Series* **2** 197
[7] De-Fang K, Zhi-Ning Qu and Qiao-Ling G 2015 *Res. A & A* **15** 77
[8] El-Borie M A and Al-Thoyaib S S 2002 *Il Nuovo Cimento* **25C** 353
[9] Wang C W, Lyons T S, Gunobunagi T, and Weygand J M 2007 *J. Geophys. Res.* **112** 12
[10] Badalyan O G, Obridko V N and Sýkora 2008 *Sol. Phys.* **247** 379
[11] Sanaikumaran V and Prabhakaran S R 2008 *Indian J. Rad. & Space Phys.* **37** 391
[12] El-Borie M A, El-Abshehy M, Talaat S and Abou-Taleb W 2012 *Astrophys.* **55** 127

IOP Conf. Series: Journal of Physics: Conf. Series **869** (2017) 012078 doi:10.1088/1742-6596/869/1/012078

On the morphological dichotomies observed in the powerful radio galaxies

H Miraghaei[1] and P N Best[2]

[1]School of Astronomy, Institute for Research in Fundamental Sciences, PO Box 19395-5531, Tehran, Iran

[2]Institute for Astronomy (IfA), University of Edinburgh, Royal Observatory, Blackford Hill, EH9 3HJ Edinburgh, U.K.

E-mail: halime@ipm.ir

Abstract. We study environment and host galaxy properties of powerful radio galaxies with different radio morphologies from compact sources to very extended double lobed radio galaxies and with different optical spectra classified as high excitation (HERG; quasar-mode) and low excitation (LERG; jet-mode) radio galaxies. We use a complete sample of morphologically classified radio sources from [1] and perform three different analyses: i) we compare compact radio sources with the extended sources from the same class of excitation. ii) we compare HERGs with the LERGs using a combined sample of compact and extended sources. iii) we investigate the origin of different morphologies observed in the very extended powerful radio galaxies, historically classified as Fanaroff-Riley (FR) radio galaxies of type I and type II by comparing a sample of FRIs with the FRIIs from the same excitation class. We discuss the results and what causes the differences in each comparison. The role of host galaxy and the central super massive black hole, and the galaxy interactions are all investigated.

1. Introduction

Radio galaxies are identified by their luminous radio emission or large-scale radio jets and lobes, which are originated from parsec-scale region near the super massive black hole (SMBH) and expand up to hundreds or thousands of kpc away. They live and evolve in a wide range of galaxies and environments and present various radio morphologies and the optical spectra. Sources with very extended radio emission are classified into type I (FRI) and type II (FRII) [2]. The peak of radio emission is located near the core in FRIs while the region of highest surface brightness are located at the edge of the radio lobes in FRIIs. In contrast, there is a class of objects with compact radio emission and no extended component classified as FR0 [3]. There is a lot of discussions about whether these three classes of radio sources have the same origin or they are fundamentally different objects [4–9].

The optical spectra of radio galaxies are also different [10]. While most of the radio galaxies have weak emission lines (LERG) in their spectrum, some of them have highly excited emission lines (HERG). Previous works claimed that the black hole feeding mechanism is responsible for these differences in the optical spectrum of radio galaxies [11].

The origin of morphological differences is not well understood up to the present time. The previous studies investigated both intrinsic and extrinsic parameters to explain the observed differences. In this study, we explore the SMBH, host galaxy and environment of radio galax-

IOP Conf. Series: Journal of Physics: Conf. Series **869** (2017) 012078 doi:10.1088/1742-6596/869/1/012078

ies to explain the radio morphological dichotomies using a complete sample of FR classified radio galaxies; crucially, we match the samples in their excitation state in order to isolate the dependencies on radio morphology only. We also consider the HERG/LERG dichotomy using morphologically matched samples. Throughout the paper we assumed a ΛCDM cosmology with the following parameters: $\Omega_m = 0.3$, $\Omega_\Lambda = 0.7$ and $H_0 = 100h$ km s^{-1} Mpc^{-1} where $h = 0.70$.

2. Sample selection

The radio source sample and the parent galaxy sample are taken from [11]. They have cross-matched SDSS DR7 with the 1.4 GHz radio surveys of NVSS and FIRST [1]. We applied the redshift cut of $0.03 < z < 0.1$ due to the spectroscopic and photometric limitations of optical data and a 40 mJy flux density cut to achieve sufficient signal-to-noise for morphological classification. Radio sources have been classified according to their radio structures into FRI, FRII and compact. The detail of FR classification is presented in [1,12] who have constructed a sample of > 1000 morphologically classified radio galaxies. FRI and FRII sources constitute an extended source radio sample to be compared with the compact objects. We keep extended objects with angular sizes above $20''$ due to the uncertainty in morphological classification at smaller sizes. The method we adopted for HERG/LERG classification is presented in [11]. Therefore, we have constructed six samples of FRI/FRII/compact HERG and FRI/FRII/compact LERG available for this study. The numbers of HERG sources are dramatically lower than the LERG sources. Figure 1 shows the redshift distribution of FRI, FRII and compact sub-samples.

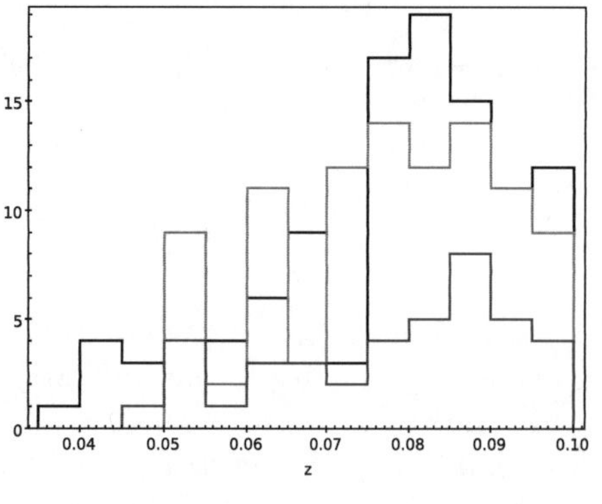

Figure 1. The redshift distribution of compact (black), FRI (red) and FRII (blue) radio galaxies used in this study.

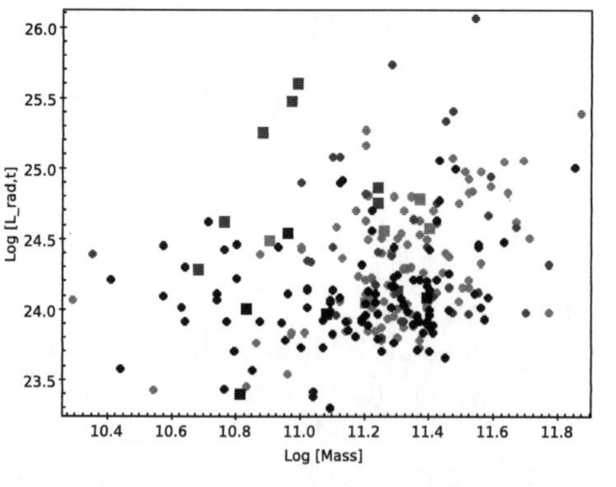

Figure 2. The total radio luminosity versus stellar mass for compact (black), FRI (red) and FRII (blue) radio galaxies separated into HERG (square) and LERG (circle).

3. The comparisons and results

In order to understand the origin of observed differences in the radio morphology and optical spectrum of the radio galaxies, we perform three comparisons. We first explore FRI/FRII dichotomy using a sample of FRI LERGs and FRII LERGs. The HERG sources have been excluded from the comparison so that the results are not biased by the HERG/LERG differences. We also remove mass and radio luminosity biases by comparing a two dimensional matched sample of FRI and FRII in stellar mass-total radio luminosity plane (figure 2). The matching strategy is randomly choosing FRI/FRII pairs in the 2D plane, which satisfy the matching

IOP Conf. Series: Journal of Physics: Conf. Series **869** (2017) 012078 doi:10.1088/1742-6596/869/1/012078

error in radio luminosity Δ log[L] = $\pm$ 0.2 and mass Δ log[M] = $\pm$ 0.1. For the final results of comparisons, we average over 1000 iterations of the random matching. The significance of differences are confirmed by the Kolmogorov-Smirnov (KS) test.

Two types of parameters have been explored in this work. The first group of parameters represents the host galaxy properties of the radio sources include *black hole mass* estimated using the velocity dispersion of the galaxy and the relation between the velocity dispersion and the black hole mass given in [13]; the host galaxy *optical size*, which is parameterized by the radius containing 50 percent of the galaxy light in the r-band (R_{50}); *surface mass density* defined by the relation: $\mu_{50} = 0.5\ M_\star/\ (\pi R_{50}{}^2)$; *concentration* of the galaxies that distinguishes the bulge-dominated systems from disk-dominated systems and is defined by the ratio of the radius containing 90 % of the galaxy light to the radius containing 50 % of the galaxy light ($C=R_{90}/R_{50}$), note that galaxies with high concentration (C>2.6) are typically bulge-dominated systems whereas galaxies with $C < 2.6$ are mostly disk-dominated systems; host galaxy *colour* (g–r); *age* of the host galaxy which is estimated by the strength of the 4000$\mathring{A}$ break of the galaxy optical spectrum, the young stellar populations have small 4000$\mathring{A}$ break while old galaxies have large 4000$\mathring{A}$ break; and the *[OIII] luminosity*.

The second group of parameters represents the environmental properties that are introduced by Sabater *et al* [14] and include the galaxy number *density* of the environment, which is parameterized by the distance to the 10th nearest neighbour of the target galaxy; *tidal interaction* estimated by the sum of tidal forces of the neighbour galaxies with respect to the internal binding forces of the target galaxy; *richness* of the galaxy group or cluster that the target galaxy belongs and is defined by the number of member galaxies in the group or the cluster. The last two environmental parameters are defined by the combination of density and tidal interaction according to the principal component analysis (PCA) in order to remove the correlation between these two parameters including *PCA1* and *PCA2*. PCA1 traces the overall interaction level and environmental density of a galaxy and PCA2 traces one-on-one interactions in which a lower value shows galaxies that are relatively isolated for their overall environment.

After constructing the matched sample, we compare the black hole, host galaxy and environment of FRIs with the FRIIs. The results show that FRIs reside in smaller galaxies with higher concentration, higher mass surface density and higher black hole to stellar mass than FRIIs. Therefore, the host galaxy in FRIs shows less disk-like structure. Comparing the environment of FRIs with FRIIs shows that FRIs lie in richer local environment having higher density, PCA1 and richness. The dense environment, particularly in galaxy groups and clusters, is the place in which highly concentrated galaxies like giant ellipticals have been formed so the result from the comparison of FRI/FRII host galaxies and environments are consistent. We can conclude that radio jets in FRI sources are probably disturbed by the dense surrounding environment and form a core dominated radio lobe with a diffused radio tail, while jets in FRIIs freely expand and form a giant radio lobe far from the center.

We apply the method that we described above to the compact LERGs and compare them with a sample of extended objects constructed by combining FRI LERGs and FRII LERGs. For this analysis, the 2D matching plane is of the stellar mass-core radio luminosity. The results show that the only significant difference is the black hole mass. Surprisingly, compact sources have lower black hole mass compared to the extended sources. So galaxies with higher black hole mass seem to be more likely to form large scale radio jets. Since there is no other significant differences, we conclude that the compact sources could be fundamentally different objects from the extended sources.

The previous comparisons were limited to the LERGs objects to remove the HERG/LERG bias. In the third analysis, we construct a 2D matched sample of HERG sources with the LERGs in stellar mass-total radio luminosity plane. We combine FRI, FRII and compact samples of each excitation class to achieve the largest possible sample of HERGs to be compared with LERGs

IOP Conf. Series: Journal of Physics: Conf. Series **869** (2017) 012078 doi:10.1088/1742-6596/869/1/012078

(of the same morphology). The results show that HERGs have younger stellar population (lower 4000Å break) with lower concentration and lower black hole mass than LERGs. They also reside in lower density, tidal interactions and PCA1 environment than those of LERGs. The results of this comparison confirms the previous studies [11] which showed HERG sources are located in star-forming disky galaxies and in low density environment compared to the LERGs which lie in bulge dominated galaxies and in dense environment. A preferred explanation is that black hole feeding mechanisms play the main role. LERGs are located in high density environments such as galaxy clusters where hot intergalactic gas is sufficiently available to feed the SMBH. This is also the place that elliptical galaxies have been formed. On the other hand HERGs are located in star forming galaxies in which cold gas is abundant and can feed the SMBH. Cold accretion in high rate results in highly excited features in the optical spectrum of radio galaxies while hot gas accretes mildly into the SMBH making LERGs.

4. Conclusion

We summarize the outcome of the above analysis as follows: The concentrated host galaxy and dense environment disrupt the radio jets creating FRI morphology compared to FRII morphology that is shaped in low density environment and low concentration host galaxies. Compared to the FRI/FRII radio galaxies that have extended emission, compact radio galaxies harbor lower mass black holes, that shows black hole mass influences launching the large scale radio jets. Black hole feeding mechanism is the main driver of HERG/LERG dichotomy.

References

[1] Best P N 2009 *Astron. Nachr.* **330** 184
[2] Fanaroff B L and Riley J M 1974 *Mon. Not. R. Astron. Soc.* **167** 31
[3] Baldi R D, Capetti A and Giovannini G 2015 *A &A* **576** 11
[4] Blandford R D and Konigl A 1979 *ApJ* **232** 34
[5] Hill G J and Lilly S J 1991 *ApJ* **367** 1
[6] Fanti R, Fanti C, Schilizzi R T, Spencer R E, Nan Rendong P P, van Breugel W J M and Venturi T 1990 *A & A* **231** 333
[7] Fanti C, Fanti R, Dallacasa D, Schilizzi R T, Spencer R E and Stanghellini C 1995 *A & A* **302** 317
[8] Zirbel E L and Baum S A 1995 *ApJ* **448** 521
[9] Alexander P 2000 *Mon. Not. R. Astron. Soc.* **319** 8
[10] Hine R G and Longair M S 1979 *Mon. Not. R. Astron. Soc.* **188** 111
[11] Best P N and Heckman T M 2012 *Mon. Not. R. Astron. Soc.* **421** 1569
[12] Miraghaei H and Best P N 2017 *Mon. Not. R. Astron. Soc.* **466** 4346
[13] Tremaine S, et al 2002 *ApJ* **574** 740
[14] Sabater J, Best P N and Argudo-Fernandez M 2013 *Mon. Not. R. Astron. Soc.* **430** 638

IOP Conf. Series: Journal of Physics: Conf. Series **869** (2017) 012079 doi:10.1088/1742-6596/869/1/012079

Monitoring of comets activity and composition with the TRAPPIST-North telescope

Y Moulane[1,2], Z Benkhaldoun[1], E Jehin[2], C Opitom[2], M Gillon[2] and A Daassou[1]

[1]Oukaimden Observatory, Laboratory of High Energy Physics and Astrophysics, Cadi Ayyad University, Marrakech, Morocco

[2]Institut d'Astrophysique et de Géophysique, Université de Liège, Liège, Belgique

E-mail: moulaneyoussef@gmail.com

Abstract. TRAPPIST-North (**TRA**nsiting **P**lanets and **P**lanetes**I**mals **S**mall **T**elescope) is a 60-cm robotic telescope that was installed in May 2016 at the Oukaimeden Observatory [1]. The project is led by the University of Liège (Belgium) and the Caddi Ayad University of Marrakech (Morocco). This telescope is a twin of the TRAPPIST-South telescope, which was installed at the ESO La Silla Observatory in 2010 [2]. The TRAPPIST telescopes are dedicated to the detection and characterization of planets orbiting stars other than our Sun (exoplanets) and the study of comets and other small bodies in our solar system. For the comets research, these telescopes have very sensitive CCD cameras with complete sets of narrow band filters to measure the production rates of several gases (OH, NH, CN, C_3 and C_2) and the dust [3]. With TRAPPIST-North we can also observe comets that would not be visible in the southern hemisphere. Therfore, with these two telescopes, we can now observe continuously the comets around their orbit. We project to study individually the evolution of the activity, chemical composition, dust properties, and coma morphology of several comets per year and of different origins (New comets and Jupiter Family comets) over a wide range of heliocentric distances, and on both sides of perihelion. We measure the production rates of each daughter molecules using a Haser model [4], in addition to the $Af\rho$ parameter to estimate the dust production in the coma. In this work, we present the first measurements of the production rates of comet C/2013 X1 (PANSTARRS) observed with TN in June 2016, and the measurements of comet C/2013 V5 (Oukaimeden) observed in 2014 with TRAPPIST-South.

1. TRAPPIST-North

1.1. TRAPPIST-North : site and instrumentation

TRAPPIST-North is a 60-cm Ritchey-Chretien robotic telescope installed in May 2016 at the Oukaimeden Observatory in Morocco and a twin of the TRAPPIST-South telescope at La Silla, Chile [1]. It is mounted on a German-equatorial direct-drive mount made by ASTELCO, company based in Munich, with a slewing speed of 30 deg/s and tracking accuracy of 1" in 4 minutes. The telescope is equipped with a 2K×2K thermo-electrically cooled Andor IKONL CCD camera with a field of view of 20'x20'. We bin the pixels 2 by 2 and obtained a resulting plate scale of 1.3"/pixel. We observed the comets with HB narrow band filters [3] isolating the cometray emission bands of OH, NH, CN, C_3 and C_2 as well as emission free dust continuum regions BC, GC and RC at three wavelengths(blue, green and red). Images are also taken with broad band B, V, Rc, and Ic Johnson-Cousin filters [3].

IOP Conf. Series: Journal of Physics: Conf. Series **869** (2017) 012079 doi:10.1088/1742-6596/869/1/012079

The Oukaimeden Observatory located at 80 km south of Marrakech (Morocco) is operated by the Cadi Ayyad University and was inaugurated in 2007. It is located in the High Atlas Mountains, at an altitude of 2750 m, with a latitude of 31° 12' 22 (N) and a longitude of -7° 51' 29 (W). The observatory offers good viewing conditions in the semi-desert area with about 250 clear nights per year [1].

1.2. TRAPPIST and comets

With TRAPPIST we study several aspects of comets. The first one is to follow the evolution of the chemical compositions of thier coma, with narrowband filters comet OH, CN, NH, C_2 and C_3 and to estimate the dust outgassing (using the Afρ proxy) with the continuum filters BC, RC and GC, of the brilliant comets (magnitude $\leq$12) week after week. In order to use the production rate calculations for these daughter molecules and gas/dust ratio to classify chemically comets. The second aspect is to study the coma morphology and determine for instance the rotation period of comets using jet that are revealed different enhancement techniques [5].

2. TRAPPIST-North monitoring of comet C/2013 X1 (PANSTARR)

Comet C/2013 X1 (PanSTARRS) was discovered on December 4th, 2013 with the Pan-STARRS 1 observatory on Mount Haleakala in Hawaii, when it was at a magnitude of 20.2. Follow-up images obtained on December 5th with the 3.6 m Canada-France-Hawaii Telescope and with a 0.51 m f/6.8 Astrograph at Siding Spring confirmed the cometary activity of this object, with a moderately condensed coma of 6" in diameter and a short tail about 2". Figure 1 shows the image of comet C/2013 X1 (PanSTARRS) in the filter CN taken by TRAPPIST-North telescope on June 14th, 2016 and the corresponding radial profile.

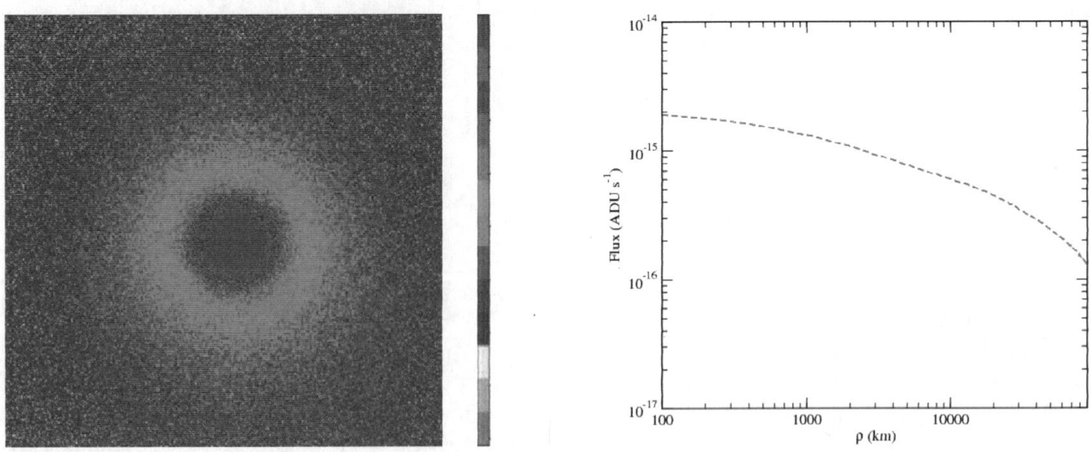

Figure 1. Original image of comet C/2013 X1 (PanSTARRS) in the filter CN taken by TRAPPIST-North telescope on June 14th, 2016 and the corresponding radial profile.

C/2013 X1 is a non-periodic comet, which describes an orbit with an inclination of 163° and an eccentricity of 1.001037. It could be a dynamically old comet; in fact, its orbit suggests that possibility as does its brightening trend since discovery [6].

Table 1. OH, NH, CN, C_3 and C_2 production rates and Afρ for comet C/2013 X1 (PANSTARRS). The blue color corresponds to the TN measurements. r and Δ are the heliocentric and geocentric distances, respectively.

UT Date	r (UA)	Δ (UA)	Production rates Q(OH) [10^{28}mol/s]	Q(CN) [10^{25}mol/s]	Q(NH) [10^{25}mol/s]	Q(C_2) [10^{25}mol/s]	Q(C_3) [10^{25}mol/s]	Afρ (BC) [10^2cm]
2016-06-07.22	1.49	0.81		12.5±0.46		16.70±0.41		8.57±0.33
2016-06-08.25	1.50	0.79	3.73±0.41					
2016-06-10.23	1.51	0.76		12.50±0.41	31.0±2.23	15.70±0.41	2.78±0.11	8.74±0.27
2016-06-14.19	1.54	0.69	3.14±0.64	12.60±0.76		15.50±0.70		09.54±0.42
2016-06-18.18	1.57	0.65	3.02±0.32			18.70±0.55	2.32±0.23	10.70±0.41
2016-06-19.11	1.58	0.65		11.80±0.53	28.4±2.00	18.10±0.76	2.60±0.14	
2016-06-19.15	1.58	0.65	3.44±0.36	12.70±0.44		22.20±0.51		11.90±0.54
2016-06-19.19	1.58	0.65	3.49±0.32	12.60±0.40		21.40±0.45		12.60±0.30
2016-06-27.13	1.65	0.67	2.75±0.24	9.09±0.37		10.70±0.60	2.24±0.18	

3. TRAPPIST-South monitoring of comet C/2013 V5 (Oukaimeden)

C/2013 V5 (Oukaimeden) is a retrograde Oort cloud comet discovered on 12 November, 2013 by the telescope MOSS [2] that is located at the Oukaimeden Observatory. C/2013 V5 is dynamically new, it came from the Oort cloud with a loosely bound chaotic orbit that was easily perturbed by galactic tides and passing stars. Before entering the planetary region, C/2013 V5 had an orbital period of several million years. After leaving the planetary region, it will have an orbital period of about 6000 years. The average production rates obtained by TRAPPIST-South observations at perihelion on September 24, 2014 (r_h=0.63 UA) are: Q(OH)=1.4±0.3 10^{28} mol/s and the rest in 10^{25} mol/s: Q(CN)=6.1±0.3, Q(C_2)=6.9±0.2, Q(C_3)=1.2±0.1, Q(NH)=16.62±1.65. The quantity of dust represented by the Afρ parameter measured in BC filter is 298±14 cm.

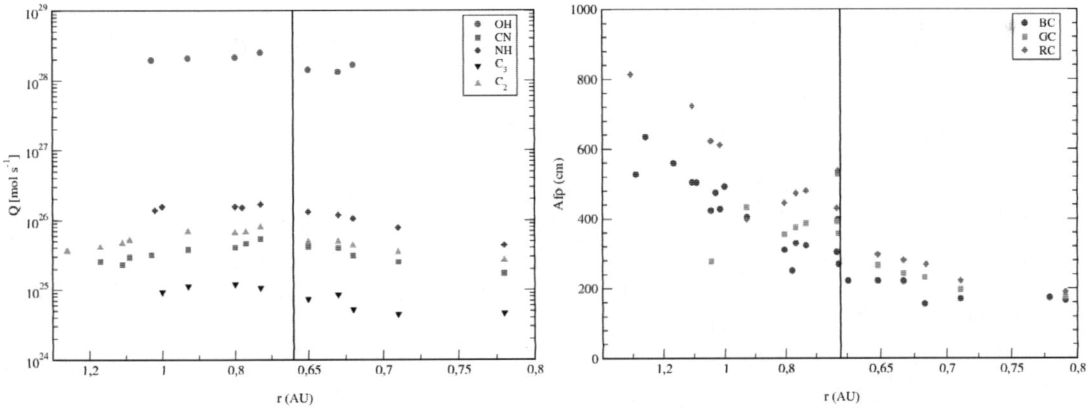

Figure 2. OH, CN, NH, C_2 and C_3 production rates and Afρ parameter measured from narrow band BC, GC, and RC continuum filters as a function of the heliocentric distance. The vertical line in the middle shows the perihelion (r=0.62 AU) of comet C/2013 V5. Note that the scale of the graphic for the heliocentric distance is not the same before and after perihelion.

The ratio of the production rates compared to OH and CN help to classify the comets. A'Hearn *et al* [7] showed that there are two main categories of comets: one category is characterized by a depletion of carbon chain species, the other is typical having no depletion of C_2 and C_3. The mean value of the ratio $Q(C_2)/Q(CN)$ for a typical comet is 1.5, the limits of this ratio for a comet showing no depletion is between 0.7 and 2. The average obtained for Oukaimeden comet during its 2014 passage, is $Q(C_2)/Q(CN)$=1.36, so this comet has a typical composition [8].

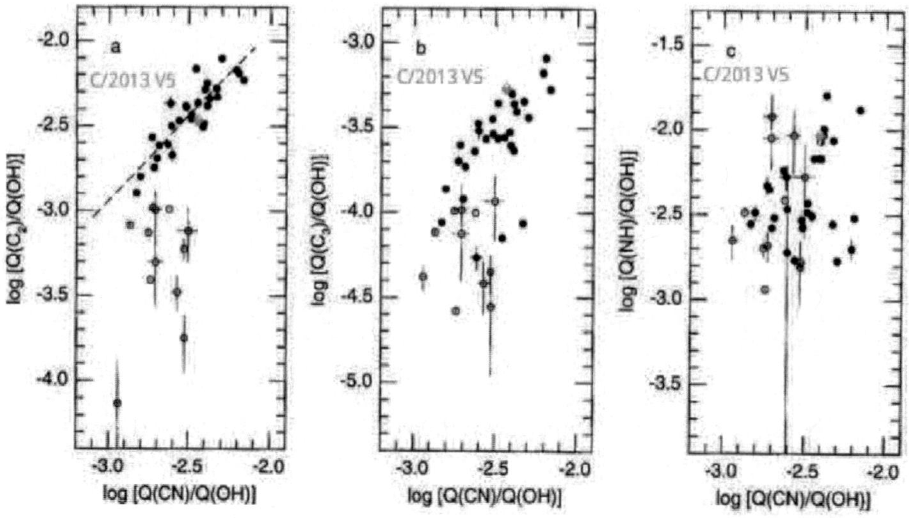

Figure 3. Logarithm of the ratio of C_2 production rate, C_3 and NH relative to OH as a function of the logarithm of relative CN production rate relative to OH, for a lot of comets.These data for 41 comets and C/20113 V5 marked in blue. Error bars are shown on all points but in most cases are smaller than the points. Filled symbols indicate comets in the typical taxonomic class.

References

[1] Benkhaldoun Z, Abahamid A, El Azhari Y and Lazrek M 2005 *Astronomy and Astrophysics* **441** 839–843
[2] Jehin E, Gillon M, Queloz D, Magain P, Manfroid J, Chantry V, Lendl M, Hutsemékers D and Udry S 2011 *The Messenger* **145** 2–6
[3] Farnham T L, Schleicher D G and A'Hearn M F 2000 *Icarus* **147** 180–204
[4] Haser L 1957 *Bulletin de la Societe Royale des Sciences de Liege* **43** 740–750
[5] Martin M P, Samarasinha N and Larson S 2014 *AAS/Division for Planetary Sciences Meeting Abstracts* (*AAS/Division for Planetary Sciences Meeting Abstracts* vol 46) p 209.31
[6] Manzini F, Oldani V, Behrend R, Ochner P, Baransky A and Starkey D 2016 *Planetary and Space Science* **129** 108–117
[7] A'Hearn M F, Millis R C, Schleicher D O, Osip D J and Birch P V 1995 *Icarus* **118** 223–270
[8] Ahearn M F, Schleicher D G, Millis R L, Feldman P D and Thompson D T 1984 *Astrophysical Journal* **89** 579–591

Frontiers in Theoretical and Applied Physics/UAE 2017 (FTAPS 2017) IOP Publishing
IOP Conf. Series: Journal of Physics: Conf. Series **869** (2017) 012080 doi:10.1088/1742-6596/869/1/012080

Investigating the Gamma-Ray Burst – Supernova Connection

N Guessoum[1], O Alarayani[1], K Al-Qassimi[1], M G AlShamsi[1], N Sherif[1], H Hamidani[2], H Zitouni[3], W J Azzam[4]

[1]American University of Sharjah, UAE
[2]University of Tokyo, Japan
[3]Faculté des sciences, Université Dr Yahia Fares, Médéa, Algeria
[4]Department of Physics, College of Science, University of Bahrain, Bahrain

Email: nguessoum@aus.edu

Abstract. We have undertaken a literature search for associations of gamma-ray bursts (GRBs) with supernovae (SNe), and we have constructed a much larger table of cases than has been published until now. The table contains a suite of physical properties for the GRBs (and the SNe when available), which will allow us and others to infer valuable knowledge about the GRBs and their physical mechanisms. From this basic table, we have undertaken a very preliminary examination, looking at the intrinsic GRB properties at hand, i.e. duration, isotropic energy output, peak spectrum energy, fluence, spectral index, redshift etc., and we present initial results. Future analyses will be performed to try to determine whether GRBs with no associated SNe constitute a subclass or category of bursts with particular characterizing properties.

1. Introduction

With the consensus that "long" gamma-ray bursts (of duration more than ~ 2 seconds) are due to hypernovae (very big stars exploding as described by the collapsar model), the search for correlations between GRBs and SNe has become a lively topic in this very active field. Very few SNe have been found for GRBs, however, whether by direct detections or by catalog searches, which may or may not be surprising: perhaps this is due to the large distances at which GRBs occur, or perhaps to special properties for some GRBs, making them more or less amenable to an SN association. This has led researchers to ask what characteristics make some GRBs manifest themselves as supernovae and others not. To investigate this, lists of GRBs with/without SNe (when searches were performed) needed to be constructed.

Bosnjak et al. [1] constructed one such list using only the BATSE catalogs of GRBs and based on the following approach: a) spectra were searched for a late re-brightening ("bump") in the optical afterglow of the GRB, which could be the signature of an underlying SN; b) SNe catalogs were searched for coincidence (spatial and temporal) with detected GRBs (although positions are often tainted by large uncertainties – more than 8 degrees in many cases). That list contained 36 GRBs, all but three with suggested SN associations. However, contrary to the consensus in the community, about half of the SNe listed in the table are of type Ia, making them highly doubtful.

Hjorth [2] produced another such list, updated and expanded from [3] and [4]; it too contained 36 GRB, 12 of them with SN associations. Most recently, Cano et al. [5] published a new list, similar to [2] but with more physical parameters (e.g. V-magnitude, bolometric luminosity), but often available for only a few GRBs.

IOP Conf. Series: Journal of Physics: Conf. Series **869** (2017) 012080 doi:10.1088/1742-6596/869/1/012080

2. Research, New Data, and Results

To produce a new list, we used the following approaches:

- Search publication databases (NASA/ADS, Astro-ph ArXiv, Google Scholar) to collect information on GRBs, related SNe, and their physical properties;
- Search for corresponding objects (SNe for GRBs or vice-versa) by close dates and celestial locations within databases of astronomical objects (e.g. the Swift GRBs database, the Central Bureau for Astronomical Telegrams, the Gamma-ray Coordinates or Coordination Network, the Open Supernova Catalog for SNe).

Our table of new GRB-SN cases contains 95 GRBs, about 25 of them new and not previously reported, most of them with SN associations. 'No SN found' is a rather difficult criterion to implement for the simple reason that it is often impossible to know whether a GRB's afterglow was observed for long enough (tens of days) for any SN "bump" to be seen or not. The double list gives physical parameters for the GRBs (duration, redshift, isotropic gamma-ray energy, spectral index, fluence, etc.). We must stress that this data is preliminary; most but not all the association cases are robust; checks are still being made, as well as searches for physical parameter values in order to ascertain and complete the data as strongly and fully as possible. Likewise, finding GRBs where SNe were really looked for and were not found is not straightforward.

A quick analysis of all the GRBs with/without SNe reveals the following:

- The GRBs for which an SN has been ascertained are closer than those without: average redshift of 0.34 ($\pm$ 0.33), compared to 0.57 ($\pm$ 0.52). The (statistically not significant) difference is not surprising: it is easier to detect supernovae for closer GRBs, considering the low fluxes.
- GRB duration sets are very similar for GRBs with/without SNe, both in the observer's frame: 40.1 $\pm$ 47.8 vs. 47.6 $\pm$ 46.6 s (leaving out 3 bursts longer than 1,000 s and half a dozen bursts with T90 $<$ 1 s, as short GRBs are not supposed to result from explosions and hence should not be associated with SNe), and in the rest frame: 29.9 $\pm$ 32.6 vs. 29.8 $\pm$ 33.9 s.
- No difference was found in the peak V-magnitude, in the photon index, or the spectral index.

There are more analyses to be performed on the GRBs with/without SNe: a) Bosnjak et al. [1] reported that most of the SN-associated bursts that they analyzed have single-peak spectra; the 29 new ones we have dug up should be checked for this characteristic; b) how do the two groups differ in fitting (or not) the GRB correlations (e.g. Amati, Yonetoku, etc.)? c) taking into account the distances/redshifts, how do these groups differ in terms of luminosity/magnitude in the optical and the gamma domains? These are questions that we plan to explore in the near future.

3. Conclusions

In our (preliminary) investigation of the GRB-SN connection, we have collected some 30 new GRBs, most of them with a supernova counterpart. This is a substantial increase in the number of previously listed/published GRBs with/without associated SNe [1, 2, 5]. We have performed only preliminary checks on these objects to ascertain the SN association, and we shall strengthen those claims (positive or negative) in the near future.

We have performed very basic and quick comparisons between the two groups and did not find any surprising or significant differences (in durations and spectral indices). GRBs with SNe were found to be closer, which is normal observational bias. More substantial analyses, for instance of the spectra (single peak or not) and of the physical parameters that should correlate à la Amati, Yonetoku, and other proposed relations, will be performed in the near future.

If GRBs with SNe turn out to have different properties than those without SNe, the implications would be very significant. At minimum, it would give observers an indicator of which GRBs one should target when instrumentally searching for SNe. More importantly, it might imply a somewhat different physical process at work in the explosion and perhaps even a different class of objects that undergo such bursts and supernova explosions. For instance, Bosnjak et al. [1] do not rule out Type Ia supernovae in the associations they make with GRBs even though the general consensus is that (long) GRBs should be associated with Type Ic, or at least Type Ibc, supernovae.

Table 1a – Sample of GRBs with SNe associations

GRB	SN	Identification	z	T_{90} (sec)	$T_{90\text{-rest-frame}}$ (sec)	M_{Vpeak}	E_{giso} (erg)	Spectral Index (Gamma)	Photon Index	Fluence (erg/cm^2)
920613	1992ae		0.075	129 ± 8.4	120 ± 7.8					1.86×10^{-6}
951107	1995bc		0.048	43.52	41.53			1.62		1.45×10^{-6}
970514	1997cy		0.063	1.3	1.2		4×10^{48}			4.1×10^{-5}
980425	1998bw		0.0085	18	17.8	−19.15	8.57×10^{48}	2.56 ± 0.7		$(3.04 \pm 0.21) \times 10^{-6}$
991002	1999eb	S&T M [6]		1.9	1.9			1.12		6.2×10^{-5}
011121	2001ke	LCB [7]	0.362	47	34.5	−19.05	7.80×10^{52}			
030329	2003dh	LCB [7]	0.1685	22.76	19.48	−19.2	1.50×10^{52}			
050525A	2005nc	LCB [8]	0.606	8.84	5.5	−18.8	2.50×10^{52}	2.09	1	153.0×10^{-7}
060218	2006aj	LCB [9]	0.0334	2100	2000	−18.7	5.30×10^{49}	3.52	2.26	15.70×10^{-7}
080319B			0.938	124.9	64.4	−18.5	1.14×10^{54}	1.81	1.04	810×10^{-7}
081007	2008hw	LCB [10]	0.53	9.01	5.89	−19.2	1.60×10^{51}	2.1	2.51	7.1×10^{-7}
090618		LCB [11]	0.54	113.34	73.6	−18.4	2.57×10^{53}	1.83	1.42	1050×10^{-7}
091127	2009nz	LCB [12]	0.49	7.42	4.98	−19.0	1.10×10^{53}	1.8	2.05	90×10^{-7}
100316D	2010bh	SN Obs [2]	0.0591	1300	1200	−18.25	5.90×10^{49}	2.49	2.29	3.0×10^{-7}
101219B	2010ma	Spectro [13]	0.552	51	32.86	−19.37	4.20×10^{51}	1.86	1.56	21×10^{-7}
130427A	2013cq	LCB [14]	0.3399 ± 0.0002	162.8	121.52	-12.1	9.6×10^{53}	1.79	1.21	4.2×10^{-3}
130831A	2013fu	LCB [11]	0.4791	32.5 ± 2.5	22.0 ± 1.7	-14.4	$(4.6 \pm 0.2) \times 10^{51}$	1.79	1.93	$(6.5 \pm 0.2) \times 10^{-6}$
150818A		Spectro [15]	0.282	123.3	96.18	> -19.36		1.98	1.96	42×10^{-7}

Identification: S&T M = Spatial & Temporal Match; LCB = Light Curve Bump; SN Obs = SN Observed; Spectro = Spectroscopy.

IOP Conf. Series: Journal of Physics: Conf. Series **869** (2017) 012080 doi:10.1088/1742-6596/869/1/012080

Table 1b – GRBs with no SNe found

GRB	z	T_{90}	T_{90} in rest frame	M_{Vpeak}	Spectral Index (Gamma)	Photon Index	Fluence (erg/cm^2)
000630		20					(15.1±1.1)x10^{-6}
050824	0.828	22.6	12.36	20.02	1.94709	2.76	2.66 x10^{-7}
060505	0.089	4	3.67				
060614	0.125	108.7	96.62	19.54	1.89544	2.02	2.04 x10^{-5}
070419A	0.971	115.6	58.65	20.0	2.18444	2.35	5.58 x10^{-7}
090902B	1.822	21	7.44				
100418A	0.624	7	4.31	19.90	2.26662	2.16	3.4 x10^{-7}
111005A	0.01326	26	25.66			2.03	6.2 x10^{-7}
111228A	0.716	101.2	58.97	17.73	2.03824	2.27	85 x10^{-7}

References

[1] Bosnjak Z et al. 2006 Gamma-ray bursts associated with supernovae: a systematic analysis of BATSE GRB candidates *A & A* **447**, 121-132

[2] Hjorth J 2013, Table of GRB–SN connections http://www.dark-cosmology.dk/GRBSN/GRB-SN_Table.html

[3] Hjorth J and Bloom J S 2012 The Gamma-Ray Burst - Supernova Connection *Gamma-Ray Bursts*, Cambridge Astrophysics Series 51, eds. Kouveliotou C Wijers R A M J and Woosley S Cambridge: Cambridge University Press 169-190

[4] Hjorth J 2013 The supernova–gamma-ray burst–jet connection *Phil Trans R Soc A* **371**: 20120275

[5] Cano Z et al. 2016 The Observer's Guide to the Gamma-Ray Burst-Supernova Connection. Eighth Huntsville Gamma-Ray Burst Symposium held 24-28 October 2016 LPI Contribution No. 1962 id.4116

[6] IAU Circular 7269 http://www.cbat.eps.harvard.edu/iauc/07200/07269.html

[7] Zeh A *et al.* 2004 A Systematic Analysis of Supernova Light in Gamma-Ray Burst Afterglows *Astrophys J* **609**, 952–961

[8] Della Valle M. *et al.* Hypernova Signatures in the Late Rebrightening of GRB 050525A. *Astrophys. J.* **642**, L103–L106

[9] Campana S et al. (2006) The association of GRB 060218 with a supernova and the evolution of the shock wave *Nature* 442, 31 August 2006 pp. 1008-1010

[10] Jin Z-P et al. (2013) GRB 081007 and GRB 090424: The Surrounding Medium, Outflows, and Supernovae *Astrophys. J.* **774**, 114

[11] Cano Z et al. (2011) A tale of two GRB-SNe at a common redshift of $z = 0.54$ *Mon. Not. R. Astron. Soc.* **413,** 669–685

[12] Cobb B E et al. (2010) Discovery of SN 2009nz Associated with GRB 091127 *Astrophys. J. Lett.*, **718**, L150–L155

[13] Sparre M et al. (2011) Spectroscopic Evidence for SN 2010ma Associated with GRB 101219B *Astrophys. J. Lett.*, **735**, L24-L28

[14] Ruffini R et al. (2015) GRB 130427A and SN 2013cq: A Multi-wavelength Analysis of an Induced Gravitational Collapse Event *Astrophys. J.*, **798**, 10-20

[15] GCN 18213 https://gcn.gsfc.nasa.gov/gcn3/18213.gcn3

IOP Conf. Series: Journal of Physics: Conf. Series **869** (2017) 012081 doi:10.1088/1742-6596/869/1/012081

Microphysical characteristics of atmospheric particulate matter from NASA's MODIS, MISR, and AERONET observations

Nashaat Gad[1], Mohammed Shokr[2] and Alaa I Ibrahim[1,3]

[1] Physics of Earth and Universe and Centre for Fundamental Physics, Zewail City of Science and Technology, Giza, Egypt
[2] Environment and Climate Change Canada, Toronto, Ontario, Canada
[3] Cairo University, Faculty of Science, Physics Department, Giza, Egypt

E-mail: ai@ZewailCity.edu.eg

Abstract. We present a comparative study of atmospheric particulate matter (also known as aerosols) observed by satellite remote sensing and ground-based observations. We compare satellite measurements obtained by NASA's Moderate Resolution Imaging Spectro-Radiometer (MODIS) and Multi-angle Imaging Spectro-Radiometer (MISR) instruments against the ground-based aerosol sun-photometer data from the Aerosol Robotic Network (AERONET) station in Cairo, Egypt from 2003 to 2014 to build a long-term database for climatological studies and to improve upon the accuracy and coverage achievable from the satellite data. We deduce microphysical and geometrical properties about the dominant aerosols based on key optical properties including aerosol optical depth (AOD), single scattering albedo (SSA), and Ångström exponent (AE). This has allowed us to place important constraints on the type of aerosols (natural, anthropogenic, and biogenic).

1. Introduction

Atmospheric particulate matter or aerosols are solid or liquid particles suspended in the atmosphere. Often observable as dust, smoke, and haze, they are ubiquitous in the air. Aerosols come from natural or biogenic sources, arising from plant debris, mineral and humic matter, and anthropogenic sources, arising from a variety of combustion sources. They affect the Earth's energy budget and climate by scattering and absorbing radiation; modifying radiative properties of clouds; and altering the intensity of sunlight scattered back to space, absorbed in the atmosphere, and arriving at Earth's surface.

The immense diversity of aerosol types, composition and size as well as their highly heterogeneous distribution in time and space represent a great challenge in identifying the dominant aerosol type. The eastern Mediterranean is considered a crossroad of air masses carrying mineral dusts from north Africa, anthropogenic aerosols from many populated urban and industrial centers, maritime aerosols from sea spray and biomass burning from agriculture areas [1, 2].

Remote sensing from satellite and ground-based sensors is used to infer microphysical and geometrical properties of aerosols based on measurements of their optical properties. These instruments measure key optical properties such as aerosol optical depth (AOD), single scattering albedo (SSA), and Ångström exponent (AE) at different wavelengths [3]. AOD is a quantitative estimate of how much aerosol is encountered throughout the atmospheric column. SSA describes the fraction of light scattered by the aerosols compared to the total scattered and absorbed. AE defines the spectral dependence of the

IOP Conf. Series: Journal of Physics: Conf. Series **869** (2017) 012081 doi:10.1088/1742-6596/869/1/012081

AOD. It is inversely related to the average size of the particles in the aerosol; the smaller the particles, the larger the exponent.

Here we examine satellite aerosol measurements obtained by NASA's MODIS and MISR instruments against measurements from the ground-based aerosol sun-photometer data available from AERONET station in Cairo, Egypt. Our results reveal the microphysical properties of the dominant aerosols and hence constrain the class of the aerosols.

2. Observations and data

Remote sensing data at level 2 (10 km resolution) were obtained from MODIS and AOD was collected at 550 nm wavelength using Collection 6 algorithm, which is suitable over land and water surfaces [4]. MISR data, also acquired from level 2 (17.6 km), provides observations with nine angular views [5]. All data were taken over Cairo, Egypt covering from January 1st 2003 to December 1st 2014.

Retrieved data include a number of physical parameters as well as microphysical properties (size and shape) [6]. For example, MISR retrieves the fraction of AOD due to "fine" (particle radii < 0.35 μm), "medium" (particle radii between 0.35 and 0.7 μm) and "large" (particle radii > 0.7 μm) particles as well as the fraction of AOD due to "spherical" and "non-spherical" particles from measurements in the four spectral bands [7]. The daily time series of the satellite data were obtained over five domains, all centered at the center of Cairo (30.05° N, 31.24° E) with radii 10, 25, 50 and 100 km. This is used to estimate the spatial gradients of AOD in each of the four cardinal directions away from the city center. The AERONET data were obtained from the station at the site of the Egyptian Meteorological Authority near the center of Cairo. Details about the data standards and the uncertainty of the products can be found in [8]. The three parameters (AOD, SSA, and AE) were derived from each sensor. In addition, the daily and monthly average of the volume particle size distribution were obtained from the AERONET data.

3. Results and discussion

Comparison of AOD from AERONET versus AOD from MODIS and MISR, both averaged over 50×50 km^2 around the center of Cairo, is shown in figure 1. The trend lines are shown (generated with zero intercept). Both MODIS and MISR underestimate the AOD but MISR underestimates it more. Bias and standard deviation of the difference between AOD from AERONET and MODIS are 0.05 and 0.143, respectively; and the corresponding values for MISR are 0.091 and 0.197. The scattering of the MODIS data is slightly higher (as represented by the R^2 value). To obtain more information about these biases, the monthly average of the bias is plotted in figure 2 (left panel). It is interesting to note the agreement between the satellite estimates and the AERONET estimate in the months of March-June. The cause of this observation is apparent from the data in the right panel, where the monthly average of AOD from AERONET and the fraction of the fine aerosols (also estimated from AERONET measurements) are plotted. Obviously, the above-mentioned agreement cannot be explained by the value of the AOD but it is certainly related to the dominant size of the aerosols. During the aforementioned months, the fine aerosol fraction is relatively small, meaning that the dominant size was large. These months feature the frequent sand storms that originate from the western desert. It can be concluded that estimates of AOD from satellite observations are more accurate if the dominant aerosols are coarse.

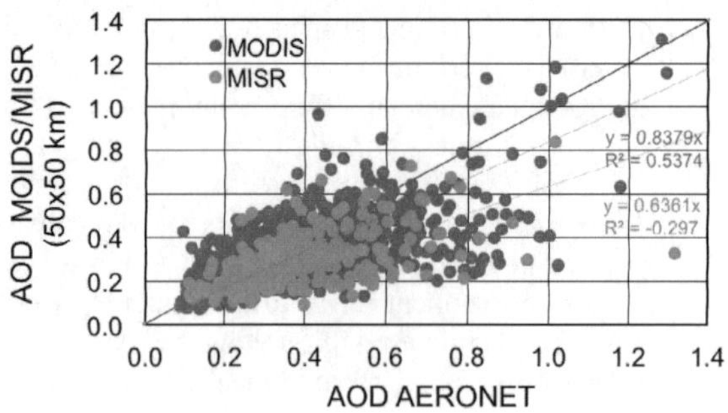

Figure 1. Scatter plot of AOD from AERONET station in Cairo versus the same parameter estimated from MODIS and MISR. Equations of trend lines are shown next to each trend line, along with the residue (R^2 value).

IOP Conf. Series: Journal of Physics: Conf. Series **869** (2017) 012081 doi:10.1088/1742-6596/869/1/012081

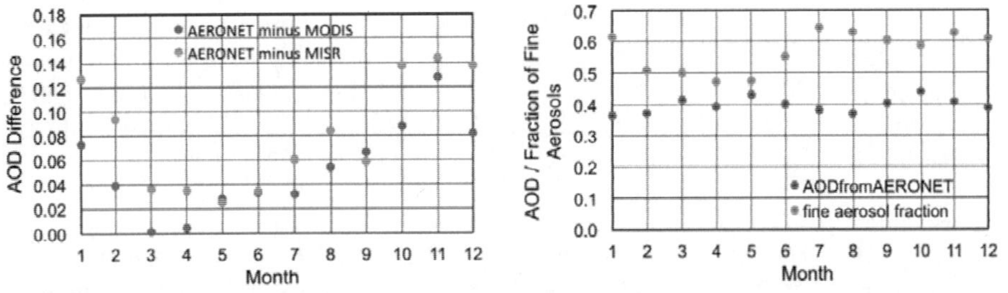

Figure 2. Difference between AOD from AERONET (500 nm) minus AOD from MODIS (550 nm) and MISR (558 nm), broken down by monthly average (left) and AOD from AERONET along with the fraction of the fine aerosols (also estimated from AERONET) (right). Data obtained from 11 years (2004-2014).

In a mega city such as Cairo, human activities, including traffic and industrial parks, are usually reduced farther away from the city's core. This hypothesis is tested using the data in table 1, which show the temporal average of AOD from MODIS over Cairo from 29 October 2004 until December 31st, 2014. The Data were averaged over pixels included the given five domains. It can be seen that the average AOD decreases as the domain increases. The rate is higher near the center. The difference of AOD estimates between the AERONET station and MODIS is negligible at the city's core (where the station is located) and increases farther away. Once again, the rate is higher near the core.

Table 1. Average AOD from 29 October 2004 to 31 December 2014.

AOD	Domain size in km^2, centered at the center of Cairo				
	20x20	50x50	100x100	150x150	200x200
MODIS	0.383	0.327	0.319	0.314	0.314
AERONET minus MODIS	0.001	0.051	0.069	0.079	0.079

The single scattering albedo, a measure of how much scattering loss relative to the total loss (due to scattering and absorption) was obtained from AERONET (at 500 nm wavelength), MODIS (at 500 nm) and MISR (at 558 nm). The difference between the estimate from AERONET and MODIS or MISR is shown in figure 3. Ideally, that difference should be zero but the data show that both MODIS and MISR overestimate this parameter with respect to AERONET. The overestimation is higher at low values of SSA, when the dominant aerosols are absorptive. AERONET data agree better with MODIS estimates. The average difference between AERONET and MODIS estimates is -0.042, and between AERONET and MISR is -0.092.

Monthly average volume particle size distribution over Cairo were obtained from the AERONET station (figure 4). Aerosol particles > 1 μm are produced by windblown dust and sea salt while particles < 1 μm are mostly formed by condensation processes such as transformation of sulfur dioxide gas to sulfate particles or by formation of soot and smoke during burning processes. Coarse aerosols exist with higher concentrations throughout the year but it reaches a minimum value in the months of November, December and January. In February it starts to increase before it peaks in April and May; the time of the intensive sand storms blowing from the Sahara through the western desert of Egypt. Fine aerosols, on the

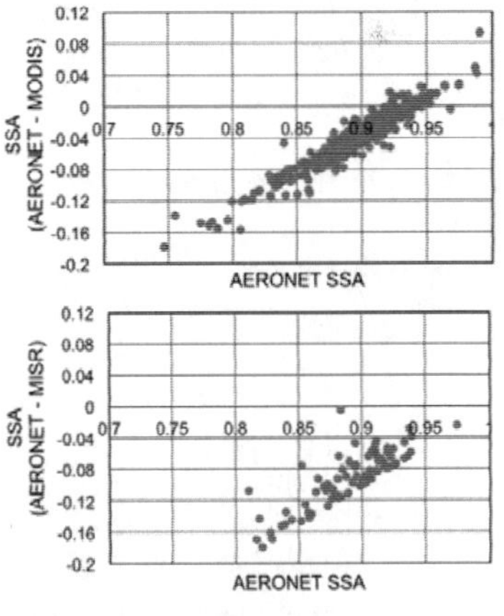

Figure 3. Difference of SSA from AERONET and satellites (MODIS or MISR) versus the value from AERONET. Higher difference is observed at low SSA. MISR data correlates better (0.72) than MODIS (0.38).

IOP Conf. Series: Journal of Physics: Conf. Series **869** (2017) 012081 doi:10.1088/1742-6596/869/1/012081

other hand, exist with relatively low concentration during winter and spring months then increase in the summer and fall seasons perhaps due to the increased traffic, combined with the stagnant. Depending on its type, fine aerosols can be health-hazard as they are mostly produced from combustion processes, which may contain toxic metals and organic pollutants [9]. Ultra-fine particles of less than 0.1 μm (also called nano-particles), are especially targeted for their effects on human health [10]. Figure 4 confirms the presence of these particulate matter in sensible quantities throughout the year over Cairo.

4. Conclusion

Compared to AERONET measurements, both MODIS and MISR underestimate the AOD, with more underestimation by MISR. Comparison gives best results at the center of the city (where the AERONET station is located). The average AOD over the 12-year period at the center is 0.38. Satellite measurements show that this average decreases by 15% at 25 km from the core of the city and continues to decrease albeit at a much smaller rate beyond this distance. Monthly-average of AOD shows that values of this parameter from AERONET agree well with the corresponding values from either MODIS and MISR during March to June. It has been shown that these months feature dominant aerosols of coarse size. Therefore, it is concluded that the disagreement is more pronounced due to presence of fine aerosols.

Comparison of satellite versus AERONET estimates of the Ångström exponent reveals agreement of the trend between AERONET and MODIS/MISR. Both MODIS and MISR overestimate this parameter with respect to AERONET. The overestimation is higher at low values of SSA, i.e. when the dominant aerosols are absorptive.

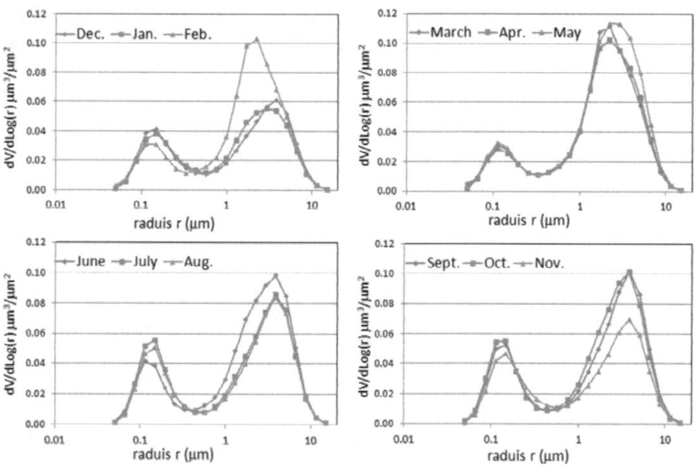

Figure 4. Monthly average of volume particle size distribution over Cairo in the range of sizes between 0.05 μm and 15 μm obtained from the AERONET data (2003-2014). Data from each season are included in one panel. Bimodal distribution is apparent in every season.

References

[1] Hatzianastassiou N, Gkikas A, Mihalopoulos N, Torres O and Katsoulis B 2009 *J. Geophys. Res.* **114** D24202

[2] Lelieveld J *et al* 2002 *Science* **298** 794

[3] Osunsanya T, Prescott G and Seaton A 2001 *Occup. & Env. Med.* **58 (3)** 154

[4] Remer L, Kaufman Y, Tanré C, Martins J, Li R, Ichoku C, Levy R, Kleidman R, Eck T, Vermote E and Holben B 2005 *J. of Atmosph. Sci.* **62** 947

[5] Diner D, Abdou W, Bruegge C, Conel J, Crean K and Gaitley B 2001*Geophys. Res. Lett.* **28** 3127

[6] Martonchik J, Diner D, Kahn R, Gaitley B and Holben B 2004 *Geophys. Res. Lett.* **31** L16102

[7] Bull M, Matthews J, McDonald D, Menzies C, Moroney C, Mueller K, Paradise S and Smyth M 2011 MISR data product specification, Jet Propulsion Laboratory JPL D-13963, Rev. S 324

[8] Holben B, Eck T, Slutsker I, Tanré D, Buis J, Setzer A, Vermote E, Reagan J, Kaufman Y, Nakajima T, Lavenu F, Jankowiak I and Smirnov A 1998 *Remote Sens. Environ.* **66** 1

[9] Davidson C, Phalen R and Solomon P 2005 *Aerosol Sci. & Tech.* **737** 749

[10] Lenoble J, Remer L and Tanré D 2013 *Aerosol remote sensing Springer-verlag, Heidelberg New York, London*, 328p

IOP Conf. Series: Journal of Physics: Conf. Series **869** (2017) 012082 doi:10.1088/1742-6596/869/1/012082

Investigating the origin of high-energy cosmic-ray electrons with Monte Carlo simulation

R Attallah

Badji Mokhtar University, Physics Department, P. O. Box 12, Annaba 23000, Algeria

E-mail: reda.attallah@univ-annaba.dz

Abstract. Due to severe radiative energy losses during propagation, high-energy cosmic-ray electrons can reach Earth only from nearby sources. Although these sources clearly manifest themselves in the special features of the energy spectrum observed by recent space-borne experiments, especially the increase in the positron fraction, their exact nature is still a matter of debate. The standard method for interpreting cosmic-ray electron data consists in solving appropriate transport equations. It can be supplemented with a Monte Carlo approach taking advantage of the intrinsic random nature of cosmic-ray diffusive propagation. This analysis gives valuable information on the electron-by-electron fluctuations and hence allows to address the issue from a different angle. Here we show how to implement a fully three-dimensional time-dependent Monte Carlo simulation of the propagation of high-energy cosmic-ray electrons from nearby sources and discuss the "single-source" astrophysical scenario.

1. Introduction

High-energy cosmic-ray electrons are subject to strong energy losses, mainly by synchrotron radiation in magnetic fields and inverse Compton scattering in radiation fields [1]. As a result, they can only originate from nearby sources, which are yet to be identified. It is very likely that these sources are behind the special spectral features of cosmic-ray electrons observed very recently by several space-borne experiments (PAMELA, Fermi-LAT and AMS-02). The most noteworthy of these features is certainly the increase in the positron fraction above ~ 10 GeV [2–5]. This remarkable result is not consistent with conventional models, which rather predict a positron fraction falling smoothly with energy [6]. Several scenarios have been put forward to account for these spectral anomalies. To some, they may be attributed to the contribution from some local sources such as supernova remnants (SNRs) [7,8] or pulsars [9,10]. For others, they can result from the annihilation or decay of dark-matter particles [11–13]. However, none of these explanations is yet conclusive [14].

The energy spectrum of cosmic-ray electrons is usually interpreted in the context of propagation models and the standard method consists of solving transport equations [15,16]. However, the intrinsic random nature of the diffusive propagation of cosmic-ray electrons makes it possible to use Monte Carlo simulation [17]. This approach is particularly suitable at very high energy where only a few sources are expected to dominate the energy spectrum. It complements the traditional method by providing valuable information about the electron-by-electron fluctuations. Here we show how to implement a fully 3-dimensional time-dependent Monte Carlo simulation of the propagation of high-energy cosmic-ray electrons. We have considered only the energy region above 10 GeV where the solar modulation is negligible and focused on the astrophysical scenario, assuming a pulsar/SNR origin. We have used a two-component approach and separated the local source contribution from the contribution of distant sources [18]. The latter is estimated with the public code GALPROP [19]. To speed up calculations we have benefited from C++/MPI parallel programming on an HPC cluster system.

2. Simulation setup

Assuming homogeneous and isotropic diffusion, a single electron is injected into space from a selected source with a random three-dimensional direction. Its energy is sampled according to a power law with an exponential cutoff ($E_{\rm cut}$),

$$Q(E) = Q_0 E^{-\gamma} \exp\left(-E/E_{\rm cut}\right), \tag{1}$$

where γ is the spectral index and Q_0 a normalization factor. Here $\gamma = 2$ and $E_{\rm cut} = 5$ TeV. The path length l is generated according to an exponential probability distribution ($\propto \exp -l/\lambda$), where λ ($\propto E^{0.33}$) is the mean free path. After traveling the distance l, the electron energy is adjusted taking into account the energy loss rate (synchrotron+IC+bremsstrahlung) given by

$$-\frac{{\rm d}E}{{\rm d}t} = a_0 E + b_0 E^2, \tag{2}$$

with $a_0 \simeq 3.7 \times 10^{-16}$ s^{-1} and $b_0 \simeq 1.3 \times 10^{-16}$ GeV^{-1} s^{-1} [20]. The elapsed time (t) is calculated by considering a delay of 20 kyr with respect to the source birth. The electron continues to propagate until it crosses a shell of 1 pc radius centered on the Sun, in which case it is recorded (observed). If the electron reaches the boundaries of the confinement region, it is discarded (free escape) and a new one is simulated starting from the source. The geometry of the confinement region has the shape of a cylindrical slab with a 20 kpc radius and a 4 kpc height. If the electron does not reach the solar system or the boundaries of the confinement volume within a maximum time of 2×10^7 yr, corresponding to a minimum energy of 10 GeV, it is also discarded and the next electron is considered.

3. Results

As a first step of our analysis, we calculated the energy spectra of the observed cosmic-ray electrons from the closest objects. As illustrated in figure 1, Monogem is predominant. B1742-30, Geminga and B1822-09 also contribute but are all weaker. By the way, the observation by HESS of a steepening in the cosmic-ray electron energy spectrum at ~ 1 TeV [21, 22] supports the scenario of middle-aged objects, just like Monogem, B1742-30, Geminga and B1822-09.

In the next step, we calculated the flux of cosmic-ray electrons from only Monogem. Assuming a two-component model, we superimposed the contribution from this source on the distant-source spectrum (background), given here by GALPROP [23, 24]. As can seen in figure 2, the contribution from Monogem reproduces well the increase in the positron fraction. The other candidate sources also reproduce the data, but after tuning the injection parameters.

To assess the degree of anisotropy induced by the candidate sources, we analyzed the distribution of arrival directions by using the spherical harmonic method in galactic coordinates within the framework of HEALPix [25]. Mollweide view for Monogem is illustrated in figure 3. In case of the absence of any background the obtained amplitudes of dipole anisotropy, given in table 1, are not consistent with Fermi data, which set the upper limit of δ above 60 GeV at 0.5% [26]. Obviously, the observations do not support the scenario of one single dominating source at high energy. In case of the existence of an overwhelming background (one order of magnitude higher than the source signal), the obtained values are more in line with the experimental bound. Moreover, the calculations indicate clearly that there is not just one but several key objects, namely Monogem, B1742-30, Geminga and B1822-09 (see figure 1).

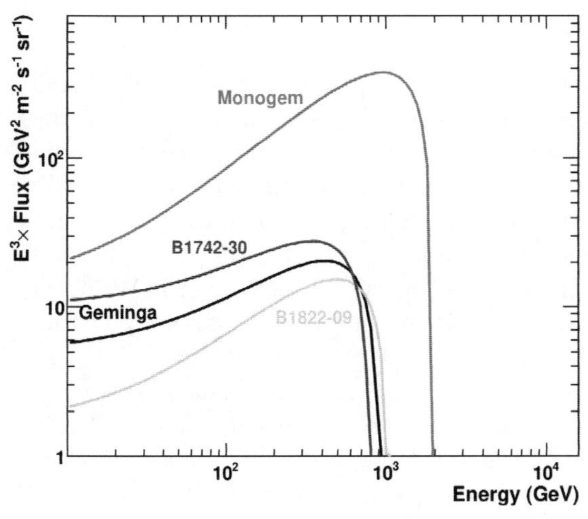

Figure 1. Electron energy spectra from the candidate sources.

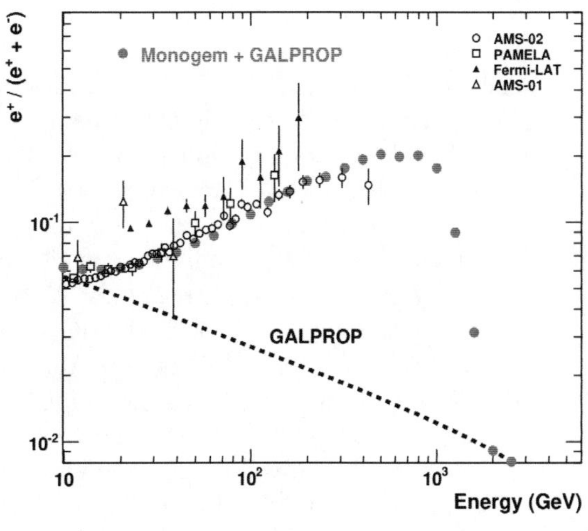

Figure 2. Positron fraction induced by Monogem.

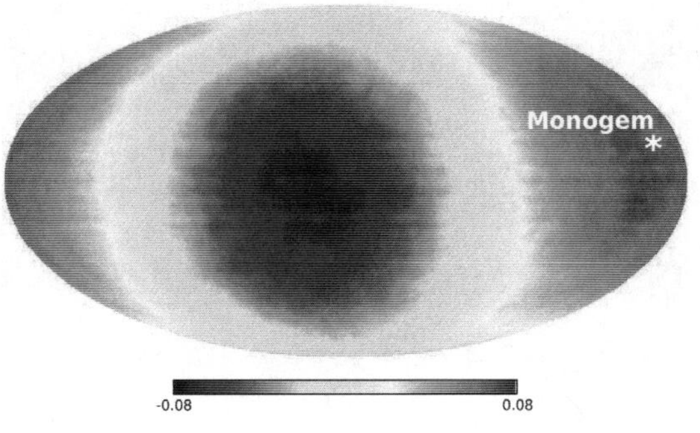

Figure 3. Mollweide view of Monogem dipole anisotropy.

Monogem and Geminga are located too close on the sky dome, and so are B1742-30 and B1822-09. But the first duo is roughly diametrically opposite to the second one, which would offset the effect of anisotropy. Indeed, if we put aside Geminga and B1822-09, and assume that only Monogem and B1722-30 contribute to the electron spectrum at high energy, 50% each, the dipole anisotropy is reduced by a factor 3. In sum, the non-observation of anisotropy, which challenges to some extent the astrophysical scenario, can be explained simply if we assume the existence at high energy of at least two dominant sources in such a configuration that one cancel out the effect of the other, and/or the existence of an overwhelming background.

Table 1. Amplitude of the dipole anisotropy (δ) for the candidate sources in case of the absence of any background and, in brackets, in case of the existence of an overwhelming background.

	Monogem	B1742-30	Geminga	B1822-09
δ (%)	7.1 (0.8)	1.8 (0.3)	2.2 (0.3)	3.6 (0.3)

4. Conclusion

The problem of the spectral peculiarities observed at high energy for cosmic-ray electrons is generally tackled by solving the transport equation describing the galactic propagation of these particles. This work demonstrates the feasibility and the relevance of fully three-dimensional time-dependent Monte Carlo approach, which can supplement the standard method by providing additional information on the electron-by-electron fluctuations. It is particularly efficient at investigating deeply the "single-source" astrophysical scenario.

The calculations show in the first place that middle-aged nearby pulsars, such as Monogem, B1722-30, Geminga and B1822-09, may well be the best potential sources of high-energy cosmic-ray electrons. These objects reproduce correctly the high-energy spectral data. The non-observation of anisotropy, which conflicts with the thesis of one single dominant object, can be explained simply if we consider the geometrical configuration of the candidate sources. Indeed, Monogem and Geminga are located close together on the sky dome, but B1742-30 and B1822-09 are situated just in the opposite side. In this way, anisotropy can be considerably diluted. In addition, an overwhelming isotropic background, likely to arise from other distant sources, can also reduce anisotropy significantly.

Acknowledgments

We gratefully acknowledge computing support provided by the Research Center on Scientific and Technical Information (CERIST) in Algiers (Algeria) through the HPC platform Ibnbadis.

References

[1] Blumenthal G R and Gould R J 1970 *Rev. Mod. Phys.* **42** 237
[2] Adriani O *et al.* 2009 *Nature* **458** 607
[3] Ackermann M *et al.* 2012 *Phys. Rev. Lett.* **108** 011103
[4] Aguilar M *et al.* 2013 *Phys. Rev. Lett.* **110** 141102
[5] Accardo L *et al.* 2014 *Phys. Rev. Lett.* **113** 121101
[6] Moskalenko I V and Strong A W 1998 *Astrophys. J.* **493** 694
[7] Blasi P 2009 *Phys. Rev. Lett.* **103** 051104
[8] Ahlers M *et al.* 2009 *Phys. Rev.* D **80** 123017
[9] Hooper D *et al.* 2009 *J. Cosmology Astropart. Phys.* **01** 025
[10] Malyshev D *et al.* 2009 *Phys. Rev.* D **80** 063005
[11] Kane G *et al.* 2009 *Phys. Lett.* B **681** 151
[12] Arkani-Hamed N *et al.* 2009 *Phys. Rev.* D **79** 015014
[13] Cholis I *et al.* 2009 *Phys. Rev.* **80** D 123511
[14] Moskalenko I V 2013 *Nucl. Phys. B(Proc. Suppl.)* **243-244** 85
[15] Ginzburg V L and Ptuskin V S 1976 *Rev. Mod. Phys.* **48** 161
[16] Strong A W *et al.* 2007 *Ann. Rev. Nucl. and Part. Sci.* **57** 285
[17] Attallah R 2016 *J. Cosmol. Astropart. Phys.* **JCAP12** 025
[18] Atoyan A M *et al.* 1995 *Phys. Rev.* D **52** 3265
[19] Vladimirov A E *et al.* 2011 *Comp. Phys. Comm.* **182** 1156
[20] Longair M S 2011 *High Energy Astrophysics* (Cambridge University Press)
[21] Aharonian F *et al.* 2008 *Phys. Rev. Lett.* **101** 261104
[22] Aharonian F *et al.* 2009 *Astron. Astrophys.* **508** 561
[23] Strong A W and Moskalenko I V 1998 *Astrophys. J.* **509** 212
[24] Ptuskin V S *et al.* 2006 *Astrophys. J.* **642** 902
[25] Górski K M *et al.* 2005 *Astrophys. J.* **622** 759
[26] Ackermann M *et al.* 2010 *Phys. Rev.* D **82** 092003

IOP Conf. Series: Journal of Physics: Conf. Series **869** (2017) 012083 doi:10.1088/1742-6596/869/1/012083

BeePol: an imaging polarimeter for the Farid & Moussa Raphael Observatory

R Hajjar[1] **and B Wehbé**[12]

[1] Department of Physics & Astronomy, Notre Dame University–Louaize, PoBox 72 Zouk Mikael, Lebanon

[2] Centro de Astrofisics, University of Porto, Portugal

E-mail: rhajjar@ndu.edu.lb

Abstract. BeePol is an imaging polarimeter designed for the 60-cm telescope of the Farid & Moussa Raphael Observatory (FMRO) at Notre Dame University Louaize (NDU). It is based on the use of a Wedged Double Wollaston prism (WeDoWo), allowing for the simultaneous measurement of the I, Q, and U stokes parameters. When coupled to an Apogee Alta F42, 2k× 2k back illuminated, thinned, CCD camera, the polarimetric module allows for polarimetric imaging of a $1.25' \times 10'$ field of view. Calculations done for the observing conditions at the FMRO show that the instrument may detect 1% polarization at 3σ for objects of magnitude 9, with an integration time of 1 hour. BeePol aims to serve as the prototype for an inexpensive imaging polarimeter of interest to small observatories.This paper reports on the basic design of the instrument, and ongoing work to build, test it, and start science observations.

1. Introduction

The FMRO, described in [1], is a small campus observatory with photometric and spectroscopic capabilities. The observatory's most important use is for education, and for research in observational astrophysics. We wished to explore the possibility to measure all properties of light by adding polarimetry to the range of possible observational projects that may be carried at the FMRO. One of the primary use of polarization in astronomy is related to the determination of magnetic fields in various astrophysical media. Polarization of background stars by aligned dust grains serves to determine field properties in large interstellar clouds, an important parameter in understanding star formation. Furthermore, polarization as a function of wavelength affords information about dust properties in various media, such as cometary tails, interstellar and circumstellar environments. Polarization is also used to reveal information about the geometries of scattering media in unresolved objects. With the advent of CCDs, 2-D polarization measurements where introduced, mainly for extended objects or the simultaneous measurements of multiple stars in fields. Imaging polarimetry allowed for detailed studies of comets, galaxies, and reflection nebulae, mostly around YSOs, among other. In order to introduce some of the different possible studies to the capabilities of the FMRO, we decided to design and build a small polarimetric module we called BeePol. It is meant to be based on the simplest possible design with, as much as possible no moving parts, and using off the shelf parts to the maximum extent possible. In what follows we describe BeePol, and the science that may be possible with the instrument installed at the FMRO telescope. Section 2 described the basic

IOP Conf. Series: Journal of Physics: Conf. Series **869** (2017) 012083 doi:10.1088/1742-6596/869/1/012083

layout of BeePol, section 3 discusses the limitations set by the proposed set-up, and identifies possible targets, and we conclude with a brief description of ongoing work to build BeePol.

2. Design of BeePol

BeePol is based on the IMPOL polarimeter [2]. We have replaced the rotating half-wave plate and the Wollaston prism by a wedged double Wollaston (WeDoWo) as described by Oliva [3]. With a WeDoWo, one obtains simultaneous non-overlapping polarized images of the field of view (FOV) at each of 0°, 45°, 90°, and 135° position angles. This allows for the simultaneous calculation of the first three stokes parameters, I, Q, and U. Linear polarization is thus instantaneously determined in the FOV. Figure 1, from [3] shows the principle of operation of the WeDoWo.

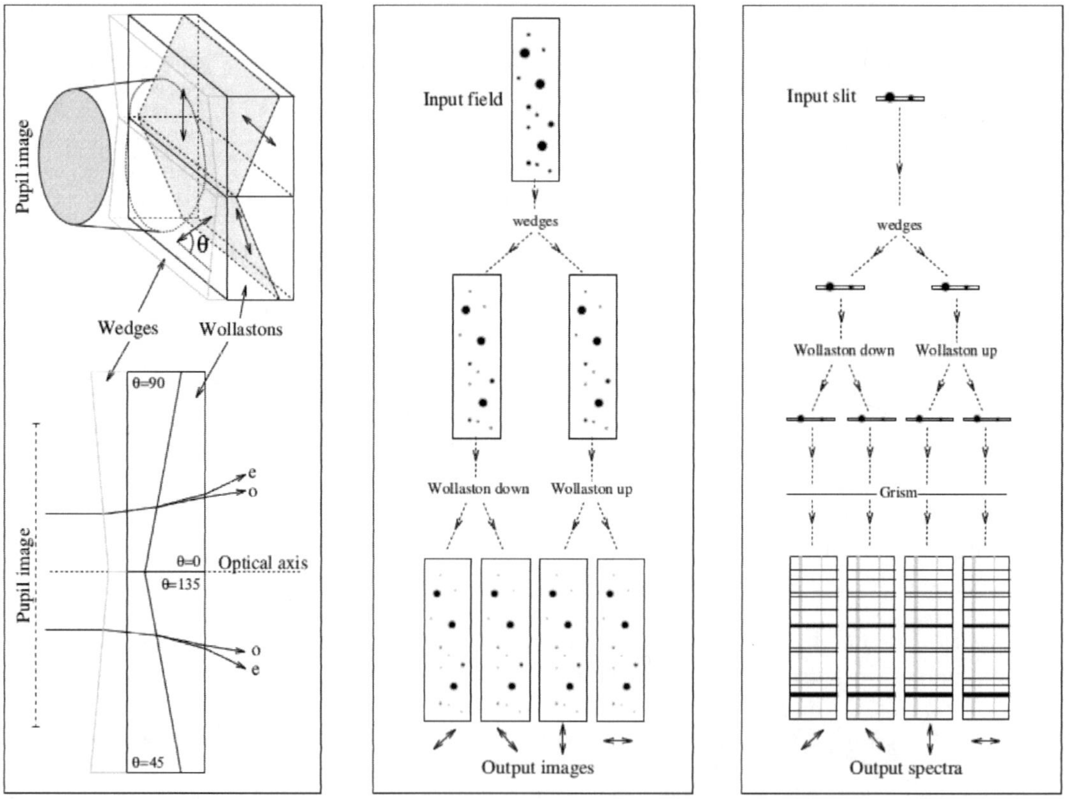

Figure 1. The basic principle of operation of the the WeDoWo in imaging and spectropolarimetry, figure from [3]

The basic optical layout of BeePol is shown in figure 2. It uses two commercially available achromats as collimating and imaging lenses. The filter wheel was placed before the CCD, rather than before the module to account for the mechanical properties of its box and reduce flexure within the device. Based on the appropriate equations found in [3], we have calculated the field of view and prism angles of the We-Do-Wo. In view on available lenses, and the need to have a reasonable field of view, we have adopted a pupil size of 2.5 cm.

The polarimetric module will be used with the FMRO main imager, an Apogee Alta F-42 [5]. When installed at the 24-inch telescope, the camera provides for a scale of 0.68" per pixel. We have striven to preserve a similar plate scale when used with BeePol. The pixel size on

IOP Conf. Series: Journal of Physics: Conf. Series **869** (2017) 012083 doi:10.1088/1742-6596/869/1/012083

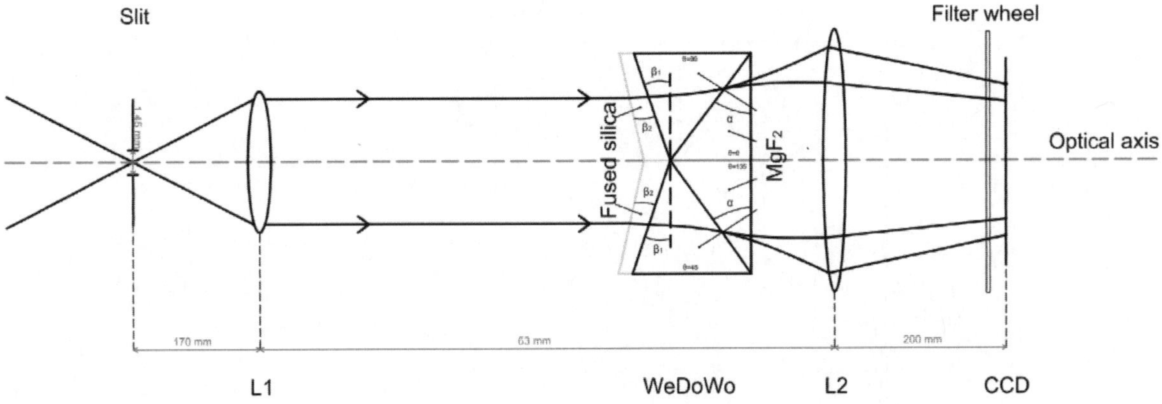

Figure 2. Optical layout of BeePol (not to scale), from [4]

the sky was used as the main constraints to determine the field of view that will create four non-overlapping images on the CCD. BeePol will thus be able to measure polarization on a field of $1'15" \times 10'$ in a single exposure.

3. Science with BeePol

BeePol is meant to be used on a small telescope. In order to assess its capability, we have computed the expected detection limits of the instrument when used at the FMRO. The uncertainty on a dual-beam polarimetric measurement is given by [6]

$$\sigma_P = \frac{1}{\sqrt{\frac{N}{2}}SNR}$$

where N is the number of half-wave plate positions used. In our case, with a WeDoWo, $N = 1$. SNR is the signal to noise ratio on the total intensity I. To determine the capabilities of BeePol, we have assumed that the detection limit is $3\sigma_P$. Furthermore, when calculating the exposure time from the SNR, we have assumed a seeing of 2", and a sky brightness of 18 mag/arcsec^2 in all filters (UBVRI). The equation used is:

$$SNR = \frac{n_* \times T \times QE \times t}{\sqrt{(n_* + n_{sky} \times N_{pix}) \times T \times QE \times t + (RN^2 + DC \times t) * N_{pix}}}$$

where n_* and n_{sky} are the photon rates arriving at the telescope from the star and the sky background, respectively, T the transmission of the optics, QE the quantum efficiency of the detector, RN and DC the read noise and dark currents, N_{pix} the number of pixels covered by the seeing disk, and t the exposure time. Results are summarized in figure 3. Based on these expected performance indicators, the best targets should be bright and highly polarized. In that regard, the list of potential targets that may lend themselves to work with BeePol are:

- Comets at high enough phase angles;
- Bright YSOs and their associated reflection nebulae;
- Objects were mass loss and/or rotation produces highly non-spherical geometries such as post-AGB stars or WR+O binary systems.

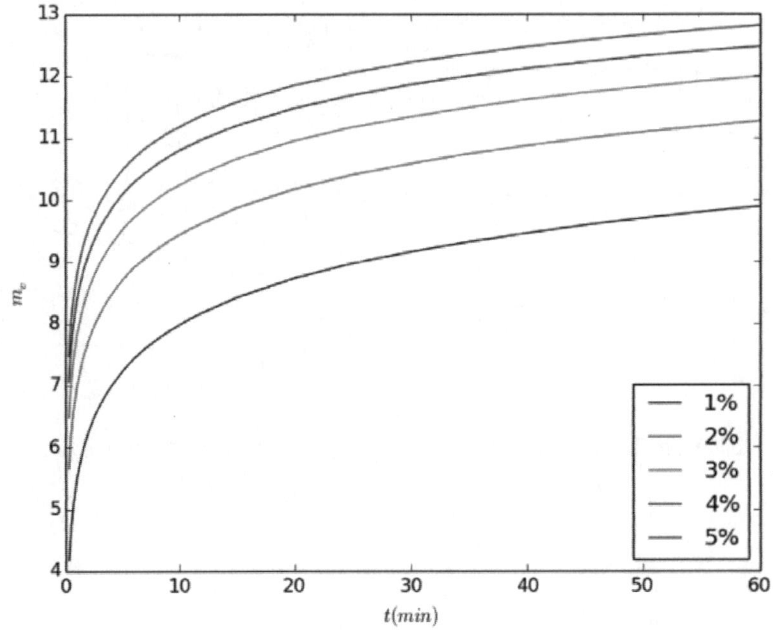

Figure 3. Polarization detection limits for BeePol. The y axis represents stellar magnitudes in V, exposure time is on the horizontal axis, from [4]

One should also keep in mind that the interest of small telescopes resides in the availability of time, and thus in the ability to provide long-term monitoring of such objects.

4. Conclusion

We have presented the design of BeePol, a polarimetric imaging module designed for used on the 24-inch telescope of the FMRO. The design is based on the principle of a low-cost instrument that will allow for interesting polarization measurements. We are currently in the process of building and testing the instrument. Tests will first be performed on an optical bench. The module's housing will be built at the workshops of NDU. It will then be tested on the telescope. We are aiming for first light in the summer of 2017.

Acknowledgments

BeePol is built with support from the Lebanese CNRS through grant 01-08-15. We thank Ghazi Asmar, Najib Metni, and Manuel Abreu for useful comments and continued help in this ongoing project.

References

[1] Hajjar R 2017 *Frontiers in Theoretical and Applied Physics 2017*
[2] AN Ramprakash R Gupta A S and Tandon S 1998 *A& AS* **128** 69
[3] Oliva E 1997 *A& AS* **123** 589
[4] Wehbe B 2014 *BeePol: An Imaging Polarimeter for the Farid and Moussa Raphael Observatory* Master's thesis Université Saint Joseph
[5] URL `http://www.andor.com/pdfs/specifications/Apogee_Alta_F42_Specifications.pdf`
[6] Patat F and Romaniello M 2006 *PASP* **118** 146

The Farid & Moussa Raphael Observatory

R Hajjar

Department of Physics & Astronomy, Notre Dame University–Louaize, PoBox 72 Zouk Mikael, Lebanon

E-mail: `rhajjar@ndu.edu.lb`

Abstract. The Farid & Moussa Raphael Observatory (FMRO) at Notre Dame University Louaize (NDU) is a teaching, research, and outreach facility located at the main campus of the university. It located very close to the Lebanese coast, in an urbanized area. It features a 60-cm Planewave CDK telescope, and instruments that allow for photometric and spetroscopic studies. The observatory currently has one thinned, back-illuminated CCD camera, used as the main imager along with Johnson-Cousin and Sloan photometric filters. It also features two spectrographs, one of which is a fiber fed echelle spectrograph. These are used with a dedicated CCD. The observatory has served for student projects, and summer schools for advanced undergraduate and graduate students. It is also made available for use by the regional and international community. The control system is currently being configured for remote observations. A number of long-term research projects are also being launched at the observatory.

1. Introduction

The Farid & Moussa Raphaël Observatory was inaugurated in July 2013. The observatory was initially conceived with the start of a joint masters degree in astrophysics between NDU and Université St-Joseph, two private Lebanese universities. The initial proposal meant to develop a versatile facility capable of assuming many roles. It was designed to serve outreach activities like school and general public visits during observing nights, education in undergraduate, and graduate classes, as well masters theses work. The observatory was meant to be suitable for research projects, since small observatories are considered important in long-term monitoring of variable and transient phenomena. We also aimed at building a cost effective facility. Most of the selected options where based on these essential constraints, knowing that the FMRO was to be built on campus. A gift by Mr. Gilbert Chaghoury helped finance the purchase of equipment and construction costs. The location of the observatory is provided in table 1. It is in continual development as we are constantly improving on its operations, as well as adding to our set of instruments and/or software suites. In what follows we will describe the equipment of the observatory, showcase some of the work done by students, and conclude with future prospects.

2. Equipment

2.1. Telescope and mount

The Observatory is centered around a Planewave CDK24, a corrected Dale-Kirkham Astrograph telescope, with a diameter of 61 cm and a focal ratio of 6.8. The telescope provides for a 70 mm corrected field of view, equivalent to 58 arcminutes. Instruments are attached to a 3.5 inch,

Table 1. Basic geographical data of the FMRO

Latitude	33° 56′ 58.54″ N
Longitude	35° 36′ 31.61″ E
Altitude	149 m

temperature compensated, Hedricks focuser at the back-end. The telescope is mounted on a Astro-Physics 3600GTOPE german equatorial mount, equipped with high-precision encoders. Initial work have shown that the setup allows for unguided 5-minutes exposure for objects at relatively large altitudes. The system is housed inside a Technical Innovations 15 feet ProDome. The dome may be slaved to the telescope. Figure 1 shows the telescope on the mount, and the building and dome of the FMRO.

Figure 1. The Planewave CDK 24 on the Astro-Phyiscs 3600GTOPE mount at the FMRO, and a view of the building and dome

2.2. Instruments

The FMRO is currently equipped for photometry and imaging, and spectroscopy. The main imager is an Apogee Alta F-42 camera equipped with a 2k×2k thinned, back-illuminated, e2V CCD chip. It has $13.5 \times 13.5\mu m^2$ pixels, for a total imaging area of $27.6 \times 27.6 mm^2$. The field of view of the camera is $23' \times 23'$, with a scale of 0.68"/pixel. The camera is housed in Apogee's deep cooling housing, D9. It is thermo-electronically cooled up to a temperature of 60°C below ambient [1]. The observatory possesses two sets of photometric filters: the Johnson-Cousins UBVRI [2] [3], and the Sloan u'g'r'i'z'[4]. These are 50mm Astrodon square filters [5]. They are mounted on two carousels to be used with the Optec Inc IFW-3 filter wheel [6].

Spectroscopy at the FMRO is served by two spectrographs, both designed and built by Shelyak. They were designed as part of a pro-am collaboration. We have the Littrow high resolution spectrograph, LHIRES III, with 5 gratings: 2400, 1800, 1200, 600, and 300 l/mm. With the 2400 l/mm gratings, the instrument provides for a resolution $R = 17000$ at Hα. The second spectrograph, eShel , is a fiber-fed visible echelle spectrograph covering the wavelength range $450\mu m \leq \lambda \leq 700\mu m$. The resolution averages 12000 over the whole wavelength range [7].

IOP Conf. Series: Journal of Physics: Conf. Series **869** (2017) 012084 doi:10.1088/1742-6596/869/1/012084

A Santa Barbara Instruments Group STT 1603ME CCD camera is dedicated for spectroscopy. The CCD is a Kodak KAF-1603 sensor with 1536 × 1024, $9 \times 9\mu\text{m}^2$ pixels.

3. Students Projects

The FMRO have been used extensively for graduate student work, mostly within a course on observational astrophysics. Projects from the detection of the transit of an exoplanet to the determination of the type of a suspected variable, or CMDs of open clusters have been attempted or completed. We present three examples of students projects done with the FMRO.

Figure 2 shows the first visible light CMD of the open cluster, Dolitze 9. Based on data obtained at the observatory, and other from the literature, one graduate student is completing his master thesis on the cluster.

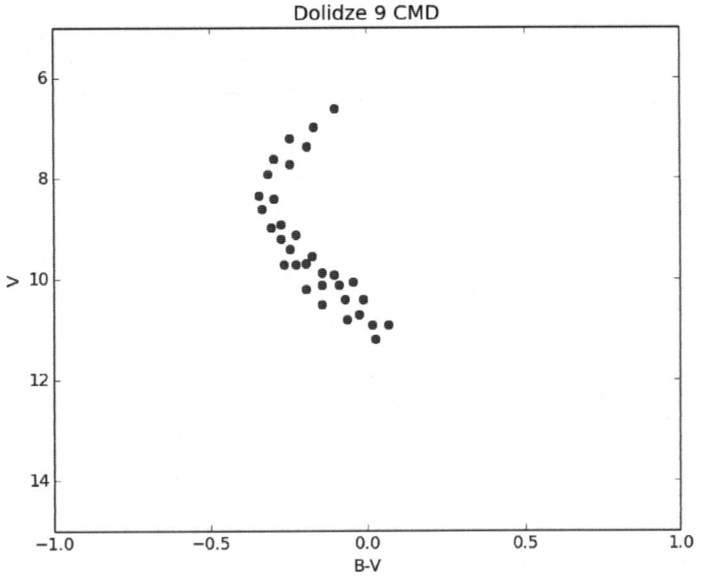

Figure 2. The first CMD of Dolitze 9, a sparsely populated open cluster

HD 133159 is listed as a suspected variable. A graduate student plans to try to identify its type. Figure 3 shows a preliminary light curve of the object. The night during which data was obtained was not photometric. Magnitudes are based on differential photometry using a single comparison star.

KOI 1150 is from the catalog of Kepler objects of interest. Its transit was observed at the FMRO for the course project of another graduate student. The star, K1150.01 has a Kepler magnitude of 13.326. The exoplanet orbits the star in 0.677375164 days and the transit duration is 1.8243 hours. Figure 4 shows the light curve obtained at the FMRO. Error bars are inaccurately reported by the student.

Most recently, the observatory has registered with the Gamma-ray Coordinates Network to begin contribution to GRB follow-up observations.

4. The Future of the FMRO

Although located in a light polluted area, about 15 kilometers to the north of the Lebanese capital, Beirut, and in an urbanized area, the FMRO have shown potential to produce interesting

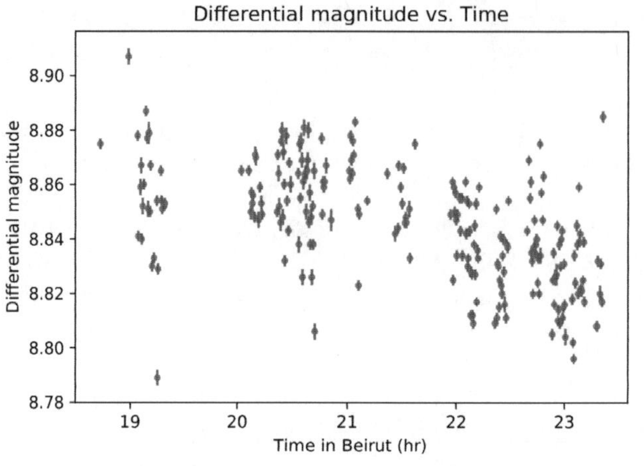

Figure 3. Light curve of HD 133159 in the V band from differential photometry, based on one night of observations

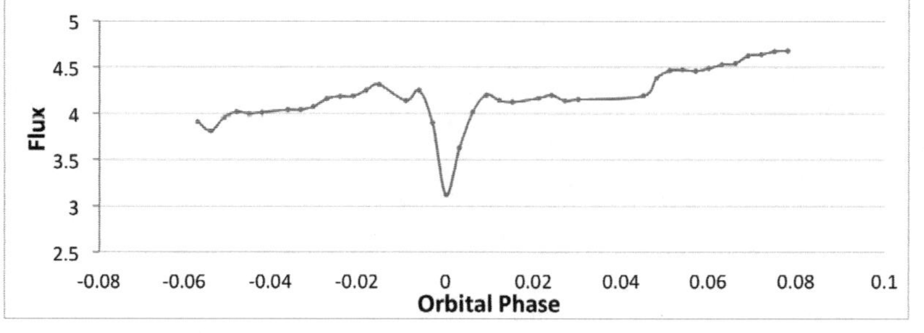

Figure 4. Transit of the exoplanet orbiting K1150.01 from the Kepler Objects of Interest

work in the areas for which it was developed. Although all of the examples shown are photometric projects, the two spectrograph have been successfully tested on the telescope. Future work will develop also around spectroscopy. We are also developing polarimetric capabilities with the development of BeePol [8].

We are also developing appropriate controls and selecting software interfaces to be able to make the observatory available to remote observations. We also plan to explore automated observations with the aim of achieving robotic operation for parts of the year. The observatory, in that regard, is considered a test bed for the future operation of a planned observatory to be installed on a mountain site.

References

[1] URL `http://www.andor.com/pdfs/specifications/Apogee_Alta_F42_Specifications.pdf`
[2] Johnson H and Morgan W 1953 *ApJ* **117** 313
[3] Cousins A 1976 *Monthly Notices of the Astron. Soc. Southern Africa* **35** 70
[4] M Fukugita T Ichikawa J G M D K S and Schneider D 1996 *AJ* **111** 1748
[5] URL `http://www.astrodon.com/index.html`
[6] URL `http://www.optecinc.com/astronomy/catalog/ifw/ifw-3.htm`
[7] Eversberg T 2016 *PASP* **128** 1
[8] Hajjar R and Wehbe B 2017 *Frontiers in Theoretical and Applied Physics 2017* (J. of Phys. Conf. Series)

A Geostationary air quality monitor for the Middle East

R M Suleiman[1], K Chance[1] and X Liu[1]

[1]Harvard-Smithsonian Center for Astrophysics, 60 Garden Street, Cambridge, MA 02138, USA
Email: rsuleiman@cfa.harvard.edu

Abstract. Atmospheric pollution measurements from space have been evolving from low-earth-orbits (LEO) to geostationary orbits (GEO), to track the diurnal variation of atmospheric emissions. There are three GEO instruments in development. TEMPO is NASA's first Earth Venture Instrument, to be launched during 2018-2021. It will measure atmospheric pollution for greater North America using ultraviolet and visible spectroscopy. The European Sentinel-4 and the South Korean GEMS will measure atmospheric pollution for Europe and Southeast Asia, respectively. In this paper, we describe NASA's TEMPO instrument and we propose a similar instrument in GEO to provide hourly atmospheric pollution measurements for the Middle East at very high spatial resolution. The proposed Middle-East TEMPO instrument will cover Arab Countries, and parts of Asia and Europe. The measurement will include the spectra required to retrieve O_3, NO_2, SO_2, H_2CO, $C_2H_2O_2$, H_2O, aerosols, cloud parameters, and UVB radiation.

1. Introduction

NASA selected TEMPO as the first Earth Venture Instrument in 2012 for launch during 2018-2021. It will measure atmospheric pollution for greater North America from space using ultraviolet and visible spectroscopy. TEMPO [1], [2] measures from Mexico City and Cuba to the Canadian tar sands, and from the Atlantic to the Pacific, hourly and at high spatial resolution (2 km N/S × 4.5 km E/W at 36.5°*N*, 100°*W*).

TEMPO provides tropospheric measurements that include the key elements of tropospheric air pollution. Since LEO instruments can only measure each point on Earth once per day and lack time resolution [3], TEMPO measurements will be from a GEO orbit, in order to capture the high variability in the diurnal cycle of emissions. GEO measurements will also provide much smaller spatial footprints, to resolve pollution sources at small spatial scales. TEMPO will be hosted on a commercial GEO host spacecraft to provide a modest cost mission. TEMPO will be launched as the North American component of the global geostationary constellation of pollution monitoring together with European Sentinel-4 and Korean GEMS, which will all be launched during 2018-2022. TEMPO and its European and Asian constellation partners will make the first tropospheric trace gas measurements from GEO. TEMPO's measurements from GEO are built on the heritage of six [4], [5] spectrometers flown in LEO [6], [7]. The TEMPO retrieval algorithms are being developed by TEMPO Science Team members and based on the operational algorithms currently running in operational environments [8], [9].

2. Middle-East Tropospheric emissions: Monitoring of pollution (ME-TEMPO)

In this section, we describe NASA's TEMPO instrument and we highlight how a TEMPO instrument can be applied to the Middle East.

2.1. NASA's TEMPO Instrument Description and Data Products

NASA's TEMPO will measure atmospheric pollution for greater North America from space using ultraviolet and visible spectroscopy.

TEMPO measurement will include the spectra required to retrieve O_3, NO_2, SO_2, H_2CO, $C_2H_2O_2$, H_2O, aerosols, cloud parameters, and UVB radiation. TEMPO will provide near- real-time air quality products that will be made widely, publicly available.

Table 1. TEMPO Baseline Products.

Species/Products		Typical Value[a]	Required Precision[a]	Expected Precision[b]	
				Worst[a]	Nominal[a]
O_3 Profile	0-2 km (ppbv)	40	10	9.15	9.00
	FT (ppbv)[c]	50	10	5.03	4.95
	SOC[c]	8x10^3	5%	0.81%	0.76%
Total O_3		8x10^3	5%	0.81%	0.76%
NO_2[d]		6	1.00	0.65	0.45
H_2CO (3/day)		10	10.0	2.30	1.95
SO_2 (3/day)		10	10	8.54	5.70
$C_2H_2O_2$ (3/day)		0.2	0.4	0.23	0.17
AOD		0.1 1	0.05	0.041	0.034
AAOD		0 – 0.05	0.03	0.025	0.020
Aerosol Index (AI)		-1 +5	0.2	0.16	0.13
Cloud Fraction		0 – 1	0.05	0.015	0.011
Cloud Top Pressure (hPa)		200 – 900	100	85.0	60.0

Spatial resolution: 8×4.5 km^2 at the center of the FOR. Time resolution: Hourly unless noted. Threshold products are at 8×9 km^2, at 80-minute time resolution.

[a] [1]Units are 10^{15} molecules cm^{-2} for gases and unitless for aerosols and clouds unless specified.

[b] Expected precision is viewing condition dependent. Results are for worst and nominal cases.

[c] FT = free troposphere, 2km – tropopause; SOC = stratospheric O_3 column.

[d] = background value. Pollution is higher, and in starred constituents, the precision is applied to polluted cases.

TEMPO is a dispersive grating spectrometer that measures solar back-scattered light in the UV-Vis spectral range to measure trace gases, aerosols, and clouds[1,2]. The TEMPO instrument is based on LEO instrument subassembly heritage with modification to geostationary GEO operations. A scan mirror steps the spectrometer slit across the Field Of Regard (FOR) from East to West. A three-mirror telescope images the scene onto the slit of an Offner-type spectrometer. Spectra are imaged onto two $2K \times 1K$ CCD focal plane arrays. One array measures 2K ground pixels from 290-490 nm and the other from 540-740 nm [1], [2]. Figure 1 shows the range of expected Earth radiance spectra to be measured by TEMPO, expressed as reflectances. The spectra [1], [2] are derived from ESA GOME-1 measurements [6], [7] and cover varied scenes measured over the Earth.

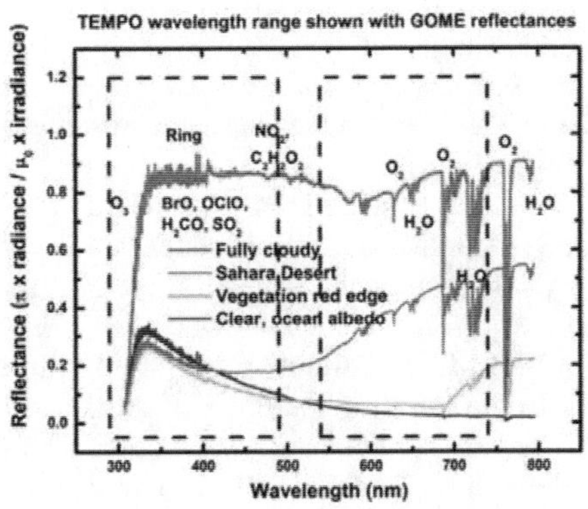

Figure 1. Earth reflectance spectra derived from the ESA GOME-1 instrument for the range of conditions to be monitored by TEMPO.

TEMPO will measure as standard baseline data products O_3 profiles, total O_3, NO_2, H_2CO (3/day), SO_2 (3/day), $C_2H_2O_2$ (3/day), the aerosol optical depth (AOD), the aerosol index (AI), cloud fraction, and cloud top pressure (see table 1).

The improved spatio-temporal resolution of the TEMPO measurements is ideally suited for constraining regional air quality (AQ) prediction systems employing global chemical data assimilation systems developed to utilize LEO trace gas data [10]. These prediction systems will benefit significantly from TEMPO's multiple observations of a given region each day at a horizontal resolution that is commensurate with regional AQ prediction. Additionally, TEMPO UV-Vis measurements of tropospheric O_3 [11] can substantially improve the analysis and assimilation of surface O_3 concentrations, reducing errors by 50% [12]. TEMPO's high spatio-temporal resolution allows a more detailed assessment of emission inventories, than is possible with LEO observations [13].

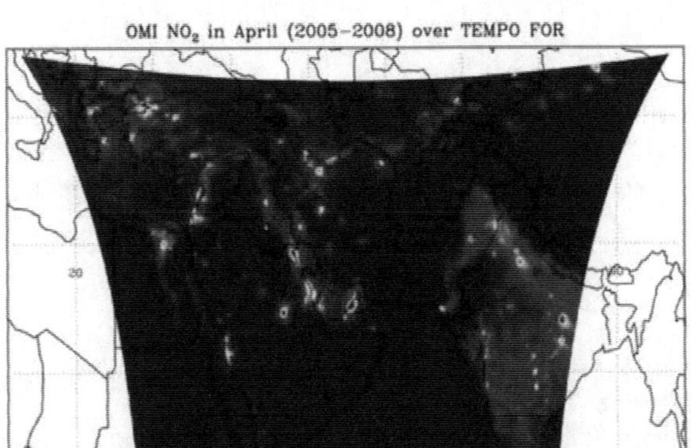

Figure 2. NO_2 concentrations from OMI instrument on AURA over the FOR for a Middle Eastern TEMPO.

2.2. *Application to the Middle East*

The TEMPO instrument concept and design can be applied to the Middle-East region without any significant changes to optics or the overall telescope design. The FOR of a future Middle-East

TEMPO-like instrument will extend from (*5°N, 20°E*) to (*40°N, 90°E*) with the center of the FOR being at (*24.4°N, 24.4°E*) (UAE/Sharjah as the center of FOR). The spatial resolution for a Middle Eastern instrument is 1.74 km N/S × 4.7 km E/W native pixel resolution (7.87 km^2). This spatial resolution will be better than NASA's TEMPO for the same instantaneous field of view since it will be closer to the equator.

As shown in figure 2, a Middle-East TEMPO-like instrument FOR will cover a region from Djibouti (south) to the Black Sea (north) and from Calcutta, India (East) to Croatia (west). This FOR will have an overlap with ESA's Sentinel-4 and Korea's GEMS instruments. The FOR shown in figure 2 will be scanned every hour, where each 1.74 km × 4.52 km pixel is a 2K element spectrum from 290-740 nm. The Middle-East TEMPO instrument will scan 1250 steps from the east to the west where each step is 110 μrad. In the north to south it will measure 2000 pixels, each 40.6 μrad N/S. This will produce 2.5×10^6 spectra per hour of observations. The ground s ample area (GSA) for various locations within the FOR are shown in table-2. The data products will be measured hourly with the spatial resolution demonstrated in table 1.

Table 2. The ground sample area (GSA), assuming 2000 pixels N/S, and the spatial resolution for various locations within the FOR.

Location	N/S (km)	E/W (km)	GSA (km^2)
25°N, 55°E	4.52	1.74	7.87
Mecca	4.52	1.75	7.91
Sharjah	4.77	1.69	7.87
Riyadh	4.60	1.74	10.2
Jerusalem	5.05	1.99	9.73
Cairo	5.26	1.97	9.90
Istanbul	5.57	2.52	12.77
New Delhi	5.15	1.91	9.48

3. Discussion

A future instrument like TEMPO over the Middle East will provide high spatial resolution measurements of Earth's atmosphere. Additionally, it will provide a significant Arab contribution (along with TEMPO (US), Sentinel- 4 (ESA) and GEMS (Korea)) to the understanding of the diurnal cycle of emissions, photochemistry, and dynamical transport coupling air quality and climate change.

4. References

[1] Chance K V *et al* 2013 *Proc. SPIE* **8866** 88660D
[2] Zoogman P *et al* 2017 *J. Quant Spectrosc. Radiat. Transfer* **186** 17
[3] Fishman J *et al* 2008 *Bull. Am. Met. Soc.* **89** 805
[4] Dittman M G *et al* 2002 *Proc. SPIE* **4814** 111
[5] Seftor C J *et al* 2014 *J. Geophys. Res.* **119** 4413
[6] Chance K V *et al* 1991 *Proc. SPIE Remote Sensing of Atmospheric Chemistry* **1491** 151
[7] Bednarz F 1995 *European Space Agency Publication SP-1182, ESA ESTEC Noordwijk The Netherlands*
[8] Bhartia P K and C G Wellemeyer 2002 *OMI Algorithm Theoretical Basis Document* **2**
[9] Chance K V *et al* 2002 *OMI Algorithm Theoretical Basis Document*, **4**
[10] Carmichael G R. *et al* 2008 *J. Comp. Phys.* **227** 3540
[11] Kazadzis S *et al* 2009 *Atmos. Chem. Phys.* **9** 585
[12] Zoogman P *et al* 2001 *Atmos. Env.* **45** 7143
[13] Kim S -W *et al* 2006 *Geophys. Res. Lett.* **33** L22812

Searching for Compton-thick AGNs with XMM-Newton observations

R Mostafa[1], M Guainazzi[2] and A Ibrahim[3,4]

[1] Fayoum University, Physics Department, Science Faculty, Fayoum, Egypt
[2] European Space Research and Technology Centre, Noordwijk, The Netherlands
[3] Zewail City of Science and Technology, Physics of Earth and Universe, and
[4]Cairo University, Faculty of Science, Physics Department, Giza, Egypt

E-mail: rma04@fayoum.edu.eg

Abstract. Most Active galactic nuclei (AGNs) are obscured by large column densities of cold and neutral gas. If the X-ray obscuring matter has a column density equal to or larger than the inverse of the Thomson cross-section ($N_H \geq \sigma_T \approx 1.5\times10^{24}$ cm^{-2}), then the source is identified as a Compton-thick AGN. One of the characteristics of Compton-thick AGN is the presence of Fe Kα emission line in their spectra with a large equivalent width. Using this criterion with XMM-Newton observations we identified Compton-thick AGNs by following a selection method, the FLEX algorithm, developed by Maccacaro *et al* [1] to search for X-ray line emitting objects (XLEOs). This technique detects the sources having significant excess of photons resulting from the iron emission line. Here we present the results from applying this method on the 28 highly absorbed AGNs recently detected by Corral *et al* [2]. Of these 28 AGN, 15 are candidate Compton-thick AGN. We applied the detection algorithm on a pilot sample of 40 XMM-Newton observations. Our results confirm the Compton-thick nature of 14 of Compton-thick AGN, based on the observed properties of the Fe Kα emission line. We use the characteristics of the observed lines to diagnose the AGNs and their environments.

1. Introduction

Compton-thick active galactic nuclei (AGNs) are obscured by large absorbed column density ($N_H \geq 1.5\times10^{24}$ cm^{-2}). According to the standard unification model, in Seyfert 2 galaxies the powerful engine is obscured by a molecular torus surrounding the accretion disc [3]. Detection of Compton-thick sources can be probed by the presence of a strong iron Kα line complex at 6.4-7 keV and Compton reflection component. At these high column densities, the iron line equivalent width (EW) reaches values of order 1 keV and in case of high inclination angles and small torus opening angles it increases for several keV. A smaller opening angle means the torus obscures a larger solid angle around the nucleus, and hence more Fe K-edge photons are captured and generate Kα photons. High inclination angle closer to the equatorial plane see larger EWs in the Kα line because the continuum is most severely attenuated along that direction. Large optical depth promotes Kα production by absorbing more hard X-rays: the continuum at Kα is suppressed, while photons well above the Fe K-edge are absorbed and trigger

IOP Conf. Series: Journal of Physics: Conf. Series **869** (2017) 012086 doi:10.1088/1742-6596/869/1/012086

fluorescence but for optical depth more than ≈ 4 the EW decreases because the optical depth from the points of Kα production to the surface is too large for the fluorescence photons to escape [4].

A recent selection technique to search for highly obscured AGNs was discussed by Corral *et al* [2]. They applied automated X-ray spectral fits implemented for the XMM-Newton spectral fit database and the selection of highly obscured candidates was based on the presence of either (a) flat rest-frame spectra from a simple power-law fit; (b) flat observed spectra from an absorbed power-law fit; (c) an absorption turnover, indicative of a high rest-frame column density; or (d) the presence of an Fe Kα line with a large equivalent width (EW >500 eV). More detailed manual spectral fits result in a detection of 28 heavily obscured with column density higher than 10^{23} cm^{-2}. A novel search method FLEX (Finder for Line Emitting X-ray sources) to search for X-ray line emitting objects (XLEOs) in XMM-Newton observations has been developed by Maccacaro *et al* [1]. They detected three sources, most likely highly absorbed AGNs ($N_H > 10^{24}$ cm^{-2}) in a test-run analysis of 13 XMM-Newton observations. The idea is based on the extension to the energy axis of the usual source detection techniques and a search for a significant excess is performed along the energy axis. Here, we apply this selection method technique on the 28 highly absorbed AGNs detected by [2] to test the reliability of this selection technique.

2. Data reduction and detection method

The sample is composed of 40 observations from XMM-Newton serendipitous source catalog (3XMM-DR4) corresponding to 28 sources, which were detected as heavily absorbed sources by Corral *et al* [2]. We reduced the EPIC-PN data using SAS v15.0.0. Data were first processed through the pipeline chains, and then filtered. Events with flag=0 and pattern between 0 and 4 were used. The selected energy range (2.3-6.4 keV) includes all sources with redshift interval 0-1.8. For each of the 41 observations, PN images are created in 23 energy bands in the energy range (2.1-6.7 keV), we increase one bin on each side, each band in 200 eV. For one source, 3XMM141546.2+112943, we extended the lower energy limit to be (1.7-6.4 keV) due to its high redshift (z = 2.56). Then we applied the detection meta-task edetect chain for all 23 energy bands. The three-dimensional detection cell has a box size 5×5 pixels and energy size 200 eV. We rejected the energy bands of sources with detection likelihood smaller than 11.5 (corresponding to probability greater than ~10^{-5}). We find 38 observations out of 41, corresponding to 27 sources, having excess source counts with a probability of being fluctuation of the estimated background level smaller than ~ 10^{-5}.

3. Spectral analysis and results

In order to distinguish Compton-thick from other AGN types that exhibit the iron emission line we have to differentiate the excess counts of iron line from the continuum counts in the energy bands where excess is detected with a probability of being a random noise fluctuation less than ~ 10^{-5}. To do this, we applied the following method. Only EPIC-PN data is used. For each observation, we extracted the source and background spectrum from a circular region with the same area. Source spectra were checked for pile-up. All spectra were binned to 1 counts bin^{-1} and Cash-statistics [5] was used for modeling the spectra in XSPEC v12.9.0. A simple model composed of absorbed power-law plus a Gaussian emission line whose energy is fixed to 6.4 keV rest-frame and width to 0.1 keV [zwabs*(pow+zgau) in XSPEC] were applied to the spectra in the energy range 2.0 - 10.0 keV. We determined the flux from both the best-fit model and the model with line normalization set to zero in each detected energy band for the 38 observations. We then estimated the continuum counts from the ratio of both fluxes and the source counts from edetect_chain.

The results are listed in table 1. For each source, the energy range with the most excess from the Kα iron line and the fraction of this excess with respect to the source counts are shown in table 1. Three sources are discarded since we did not detect any excess from the line. In type2 AGN the torus

IOP Conf. Series: Journal of Physics: Conf. Series **869** (2017) 012086 doi:10.1088/1742-6596/869/1/012086

Table 1. Detected sources

Source Name	Energy band (keV)	Source counts	Line (%)	EW (keV)	Classification	A. Corral Classification
3XMMJ080741.0+390015[(a)]	6.1-6.3	10.5	77.7	$0.927^{+0.76}_{-0.63}$	CT	CT
3XMMJ215649.5-074531[(b)]	5.9-6.1	11.2	61.6	$0.316^{+0.84}_{-0.32}$	CT candidate	CT
3XMMJ093551.5+612111[(b)]	6.1-6.3	33.5	67.4	$0.907^{+0.63}_{-0.47}$	CT	CT
3XMMJ093952.7+355358[(b)]	5.5-5.7	44.2	62.8	$0.862^{+0.73}_{-0.86}$	CT	CT
3XMMJ103408.5+600152[(b)]	5.9-6.1	17.7	90.8	$1.647^{0.56}_{-1.65}$	CT	CT
3XMMJ113549.0+565708[(a)]	6.1-6.3	8.15	55.8	$0.119^{+1.06}_{-0.12}$	CT candidate	CT
3XMMJ115704.8+524903[(a)]	6.1-6.3	29	85.6	$1.445^{+2.19}_{-0.83}$	CT	CT
3XMMJ121839.4+470626[(b)]	5.7-5.9	11	74.3	$0.349^{+0.92}_{-0.35}$	CT candidate	CT
3XMMJ122546.7+123943						
OBID 0110930301[(a)]	6.3-6.5	193.5	61.3	$0.538^{+0.15}_{-0.12}$	-	CT
OBID 0110930701[(b)]	6.3-6.5	748	51.4	$0.229^{+0.08}_{-0.06}$	-	-
OBID 0675140101[(a)]	6.3-6.5	3234	25.3	0.083 ± 0.01	-	-
3XMMJ131104.6+272806[(b)]	5.1-5.3	9.26	90.5	$1.937^{+1.95}_{-1.35}$	CT	CT
3XMMJ140700.3+282714[(b)]	5.9-6.1	24.7	75.8	$1.100^{+0.69}_{-0.58}$	CT	CT
3XMMJ141546.2+112943						
OBID 0112250301[(b)]	1.7-1.9	15	21	$0.018^{+1.76}_{-0.02}$	CT candidate	CT
OBID 0112251301[(b)]	1.7-1.9	21.5	54.4	0.239 ± 0.25	-	CT
3XMMJ150754.3+010817[(a)]	5.9-6.1	17.4	81.4	$1.095^{+1.94}_{-0.81}$	CT	CT
3XMMJ153457.2+233011						
OBID 0101640901[(b)]	6.3-6.5	6	54.8	$0.135^{+5.99}_{-0.13}$	CT candidate	CT
OBID 0205510401[(b)]	6.3-6.5	13.1	77.1	$3.662^{+7.60}_{-3.66}$	CT	CT
3XMMJ082443.2+295923[(a)]	6.1-6.3	109.6	61.8	$0.272^{+0.14}_{-0.11}$	-	-
3XMMJ083139.0+524205						
OBID 0092800201[(a)]	5.9-6.1	42	51	$0.436^{+0.49}_{-0.26}$	-	-
OBID 0502220201[(a)]	5.9-6.1	21.1	52	$0.470^{+0.52}_{-0.41}$	-	-
OBID 0502220301[(a)]	5.9-6.1	12.9	38	$0.118^{+0.28}_{-0.12}$	-	-
3XMMJ084002.3+294902[(b)]	5.9-6.1	89.5	65.5	$0.405^{+0.31}_{-0.27}$	-	-
3XMMJ095906.6+130134[(b)]	6.1-6.3	61.3	36.6	$0.144^{+0.29}_{-0.14}$	-	-
3XMMJ104930.9+225752[(b)]	6.1-6.3	65.4	40	$0.127^{+0.17}_{-0.13}$	-	-
3XMMJ113240.2+525701						
OBID 0200431301[(b)]	6.1-6.3	28.8	50.9	$0.425^{+0.57}_{-0.42}$	-	-
OBID 0200430501[(b)]	6.1-6.3	24.4	59	$0.203^{+0.22}_{-0.20}$	-	-
3XMMJ120429.6+201858[(b)]	6.1-6.3	15.5	62.2	$0.463^{+0.64}_{-0.46}$	CT candidate	-
3XMMJ080535.0+240950[(a)]	5.9-6.1	13	63.3	$0.381^{+0.57}_{-0.38}$	-	-
3XMMJ123843.4+092736[(b)]	5.9-6.1	81.3	39.2	$0.049^{+0.10}_{-0.05}$	-	-
3XMMJ132348.4+431804[(a)]	6.1-6.3	89.1	62.2	$0.382^{+0.26}_{-0.17}$	-	-

[a] zwabs*(pow+zgau)
[b] zwabs*(pow+zgau)+pow

intercepted along the line of sight and the primary radiation is absorbed up to energies that depend on the column density of the obscuring material. A fraction of this direct radiation is reprocessed producing Fe Kα emission lines. For high column densities (Compton thick) almost the primary radiation is absorbed, enhancing the production of the iron line. So we anticipate that sources with a continuum spectrum dominated by high excess counts from the emission line to be a Compton-thick source. Therefore, sources which are highly dominated by the iron line by a fraction larger than 75% are most likely to be Compton-thick AGNs. The critical parameter here is the equivalent width of the kα iron line with respect to the underlying continuum. Therefore, we fitted the spectra in the energy range 4.0-8.0

keV (1.5-8.0 keV for high redshift source) with the two-component model: absorbed power-law and Gaussian line using C-stat in XSPEC. Another component: the unabsorbed power-law, which represents the scattered contribution, is added if it is significantly required (indicated by the goodness of fit in XSPEC). The two power-law indices are tied to have the same value. A prominent Fe Kα line with large EW > 1 keV is a common feature of Compton-thick AGN [4]. So we classified the 6 sources with EW > 1 keV as Compton-thick sources. If we take into consideration, the 90% confidence level values of EW another 5 sources are classified as Compton-thick candidate. We finally considered an additional three sources as CT sources with an EW > 550 eV according to the prediction that EW reaches a maximum value of 550 eV [6].

4. Discussion

We have applied a modified method presented by Maccacaro *et al* [1] to search for Compton-thick AGN based on the detection of the excess from Fe k-α emission line. This is followed by spectral fitting to estimate the equivalent width of Fe Kα emission line with respect to the continuum. From 40 observations, we detected the sources in 32 observations, corresponding to 24 individual sources. We classified the 24 sources as 9 CT if EW is larger than 550 eV and 5 CT candidates if EW is larger than 1 keV within the 90% confidence level. Out of 14 CT and CT candidates, 13 are consistent with being CT AGN as detected by Corral [2]. Only one CT candidate in our sample (3XMMJ120429.6+201858) was not classified as CT AGN by Corral [2] whereas we did not detect two of their CT AGNs (3XMMJ091804.2+514113 and 3XMMJ122546.7+123943). For 3XMMJ091804.2+514113 we did not find any line excess, probably because of its very low counts. Although we detected a line excess in 3XMMJ122546.7+123943 the EW do not obey our CT classification. In addition, those two sources disagree with a previous classification as reported by Corral *et al* [2].

Moreover, we find that all 7 detected sources with line fraction > 75% are classified as CT sources. This is consistent with in the Compton-thick AGN where the intrinsic continuum is strongly reprocessed. For the other 17 remaining sources, we could not classify them through this prediction and the estimation of their equivalent width is needed. Finally, our results from this detection technique agree with that reported by Corral *et al* [2].

References

[1] Maccacaro T, Braito V, Della Ceca R, Severgnini P and Caccianiga A 2004 *ApJ* **617** L33
[2] Corral A *et al* 2014 *A&A* **569** A71
[3] Comastri A 2004 *in Astrophysics and Space Science Library* vol 308, ed Barger A J chapter 8 pp 245
[4] Levenson N A, Krolik J H, Zycki P T, Heckman T M, Weaver K A, Awaki H and Terashima Y 2002 *ApJ* **573** L81
[5] Cash W 1979 *ApJ* **228** 939
[6] Zycki P T and Czerny B 1994 *MNRAS* **266** 653

Frontiers in Theoretical and Applied Physics/UAE 2017 (FTAPS 2017) IOP Publishing
IOP Conf. Series: Journal of Physics: Conf. Series **869** (2017) 012087 doi:10.1088/1742-6596/869/1/012087

Star formation histories in NGC 147 and NGC 185

R Hamedani Golshan[1,2], A Javadi[1] and J Th van Loon[3]

[1]School of Astronomy, Institute for Research in Fundamental Sciences (IPM), Tehran, 19395-5531, Iran

[2]Department of Physics, Isfahan University of Technology, Isfahan, 84156-83111, Iran

[3]Lennard-Jones Laboratories, Keele University, Staffordshire, ST5 5BG, UK

E-mail: rhgolshan@ipm.ir

Abstract. NGC 147 and NGC 185 are two of the most massive satellites of the Andromeda galaxy (M 31). With similar mass and morphological type dE, they possess different amounts of interstellar gas and tidal distortion. The question therefore is, how do their histories compare? We present the first reconstruction of the star formation histories of NGC 147 and NGC 185 using long-period variable stars (LPVs). LPVs are low- to intermediate-mass stars at the asymptotic giant branch, which their luminosity is related to their birth mass. Combining near-infrared photometry with stellar evolution models, we construct the mass function and hence the star formation history. For NGC 185 we found that the main epoch of star formation occurred 8.3 Gyr ago, followed by a much lower, but relatively constant star formation rate. In the case of NGC 147, the star formation rate peaked only 7 Gyr ago, staying intense until $\sim$ 3 Gyr ago, but no star formation has occurred for at least 300 Myr. Despite their similar masses, NGC 147 has evolved more slowly than NGC 185 initially, but more dramatically in more recent times.

1. Introduction

Dwarf galaxies are the most abundant type of galaxies in the universe; since they come in two main flavors: dwarf Irregulars (dIrrs) and dwarf Spheroidal/Ellipticals (dSph/dEs). Principally they are distinguished by the significant interstellar medium in dIrrs and almost complete lack thereof in dSphs/dEs, but their star formation histories also differ, with dIrrs being "younger" [1].

NGC 147 and NGC 185, dEs, are satellites of Andromeda (M 31). They share fundamental properties such as luminosity ($M_V \sim 16$) and velocity dispersion (25 km s^{-1}), but differ in many other aspects. NGC 185 contains some gas, dust and shows evidence for recent star formation in its central regions, while NGC 147 is destitute of gas and dust and shows no sign of recent star formation activity [2, 3].

The Star Formation History (SFH) is of the most important tracers of the galaxies' evolution. We have developed a novel method to use Long-Period Variable stars (LPVs) to reconstruct the SFH [4–7]. LPVs are the most evolved stars with low to intermediate mass, at the tip of the Asymptotic Giant Branch (AGB). They represent the most luminous phase in their evolution, $\sim$ 3000–60,000 $L_\odot$, and reach their maximum brightness at near-infrared wavelengths. Since the maximum luminosity on the AGB relates to the star's birth mass, we are able to use the brightness distribution function of LPVs to construct the birth mass function and hence the Star Formation Rate (SFR) as a function of time. We present the results from application of this method to the Local Group dEs NGC 147 and NGC 185.

IOP Conf. Series: Journal of Physics: Conf. Series **869** (2017) 012087 doi:10.1088/1742-6596/869/1/012087

2. Data and Method

2.1. Data

We take advantage of a number of published photometric catalogues; the optical V and i band magnitude of AGBs in [8], infrared K- band magnitude of LPVs in [9] and the J, H and K-band magnitude of carbon stars in NGC 147 and NGC 185 [10, 11].

The K-band photometry is the best suited for calculating the SFR; since not only spectral energy distributions of stars peak at near-infrared wavelengths but also interstellar and circumstellar extinction are much less important at infrared wavelengths than optical. Carbon stars also can be used to trace (part of) the AGB, therefore we have cross matched the LPV and carbon stars catalogues, using the average of K-band magnitudes for stars in common. We can thus examine the location of the AGB stars in a colour–magnitude diagram (CMD), in comparison with isochrones from the stellar evolution models from the Padova group [12], which we use later in our technique to derive the SFH. The CMDs of NGC 47 and NGC 185 are presented in figure 1.

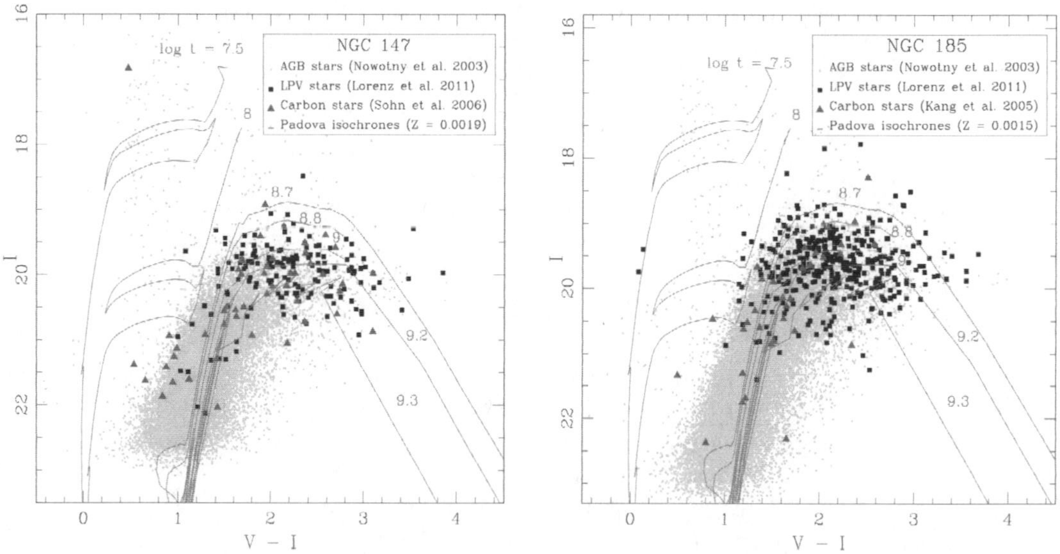

Figure 1. CMD of NGC 147 (*left panel*) and NGC 185 (*right panel*). Isochrones from [12] are labelled with logarithmic ages.

2.2. Method: star formation history

The SFH of a galaxy is a measure of the rate at which gas mass was converted into stars over a time interval in the past. Or in other words, it is the SFR, ξ (in $M_\odot$ yr^{-1}), as a function of time. We used the following relation to calculate the SFR based on LPV counts

$$\xi(t) = \frac{\int_{\rm min}^{\rm max} f_{\rm IMF}(m) m\, dm}{\int_{m(t)}^{m(t+dt)} f_{\rm IMF}(m)\, dm}\,\frac{dn'(t)}{\delta t}, \tag{1}$$

where f_{IMF} is the initial mass function, $dn'(t)$ and δt are a statement of numbers of LPVs and pulsation duration respectively. Detailed procedure to obtain above relation can be found in [4–7].

LPVs reach their maximum luminosity that relate their birth mass. Theoretical models of Padova [12] are appropriate tools to transform this maximum luminosity to birth mass. We have applied Padova models to construct mass K-band magnitude relation. The left panel in figure 2 represents this relation for the mean metallicity of NGC 147; $Z = 0.0019$ (more information for NGC 185 can be found in [7]).

Mass-age relation, like the ones in the second panel of figure 2 are obtained from isochron tables of Padova model too. These graphs relate the birth mass of variable stars that are observable at present time to the time spent from their birth.

A correction factor is appeared in equation 1, this factor is a statement of the duration in which star pulsates with large amplitude. For an LPV to be detected in the observations, pulsation duration is an important factor. Massive stars spend less time in this phase than lower mass stars. By using isochrons table and fitting a multiple Gaussian function, we are able to achieve pulsation duration as a function of Mass as shown in the third panel of figure 2.

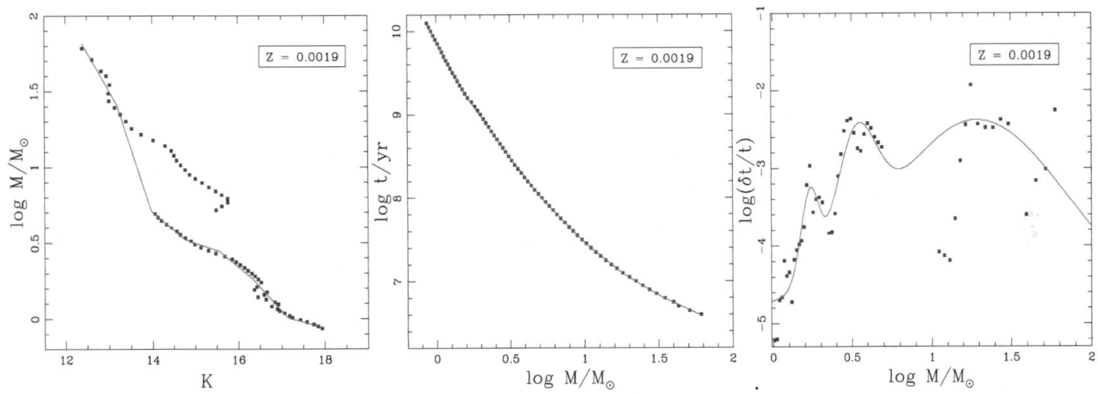

Figure 2. Mass–luminosity relation for $Z = 0.0019$ (*left panel*), mass–age relation (*middle panel*)m and mass–pulsation duration relation (*right panel*). The squares refer to models from [12].

Being in hand the coefficients of fitted lines and curves on dots - resulting from Padova models - and using equation 1, we are able to calculate SFR of different time bins and achieve SFH of galaxies based on LPVs count (The tables of coefficients are presented in [7]) for different metallicities).

3. Results and Conclusions

We have applied a novel method of [4] using long-period variable AGB stars to derive the SFH in NGC 147 and NGC 185 (see figure 3). Our main findings are:

- Star formation started earlier in NGC 185 than in NGC 147, peaking around 8.3 Gyr ago in NGC 185 and 7 Gyr ago in NGC 147.
- Star formation continued in NGC 185 over the past 6 Gyr, albeit at a much lower rate, until as recent as 200 Myr ago in the centre; in NGC 147, on the other hand, while star formation was significant between 3–6 Gyr ago, no star formation is seen for the past 300 Myr.
- The total stellar mass is $> 2.42 \times 10^8$ M$_\odot$ for NGC 185 and $> 1.16 \times 10^8$ M$_\odot$ for NGC 147; the true values are unlikely to be higher by more than a factor of a few.
- Of the total stellar mass, 70% (90%) was in place by $\log t = 9.90$ (9.56) in NGC 185 and by $\log t = 9.71$ (9.55) in NGC 147 (see figure 4).
- Our conclusions were obtained completely independently, using different data and a different method, yet they are corroborated by previous work.

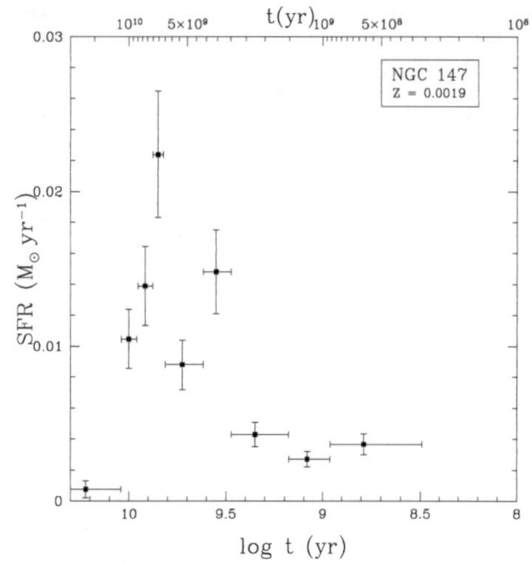

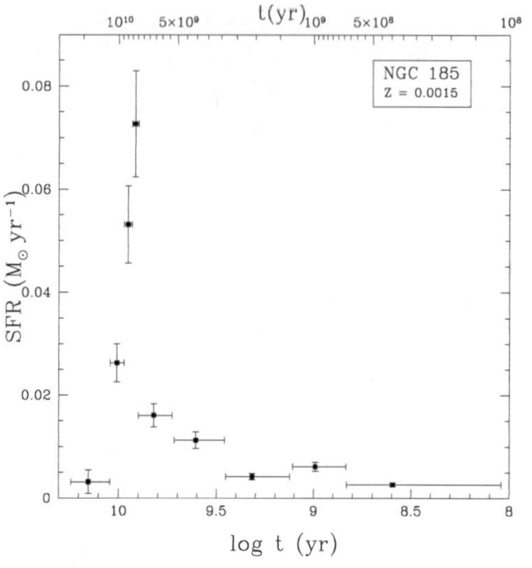

Figure 3. SFHs in the central 6′.5 × 6′.5 regions of NGC 147 (*left*) and NGC 185 (*right*), from LPV counts assuming a constant metallicity.

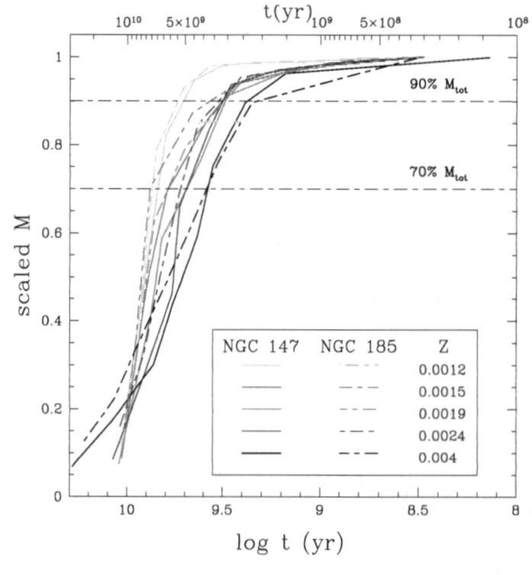

Figure 4. Cumulative SFH in NGC 147 and NGC 185 assuming different mean metallicity.

References

[1] Battinelli P and Demers S 2006 *A&A* **447** 473
[2] Young L M and Lo K Y 1997 *ApJ* **476** 127
[3] De Looze I *et al.* 2016 *MNRAS in press*
[4] Javadi A, van Loon J Th and Mirtorabi M T 2011 *MNRAS* **414** 3394
[5] Rezaeikh S, Javadi A, Khosroshahi H and van Loon J Th 2014 *MNRAS* **445** 2214
[6] Javadi A, van Loon J Th, Khosroshahi H G, Tabatabaei F and Hamedani Golshan R 2017 *MNRAS* **464** 2103
[7] Hamedani Golshan R, Javadi A, van Loon J Th, Khosroshahi H and Saremi E 2017 *MNRAS* **466** 1764
[8] Nowotny W, Kerschbaum F, Olofsson H and Schwarz H E 2003 *A&A* **403** 93
[9] Lorenz D, Lebzelter T, Nowotny W, Telting J, Kerschbaum F, Olofsson H and Schwarz H E 2011 *A&A* **532** 78
[10] Sohn Y J, Kang A, Rhee J, Shin M, Chun M S and Kim H I 2006 *A&A* **445** 69
[11] Kang A, Sohn Y J, Rhee J, Shin M, Chun M S and Kim H I 2005 *A&A* **437** 61
[12] Marigo P, Girardi L, Bressan A, Groenewegen M A T, Silva L and Granato G L 2008 *A&A* **482** 883

IOP Conf. Series: Journal of Physics: Conf. Series **869** (2017) 012088 doi:10.1088/1742-6596/869/1/012088

Metallic line doubling in the spectra of the variable RR Lyrae star

F L Sefyani[1], A Benhida[1], Z Benkhaldoun[2], T De France[3], D Gillet[4], P Mathias[5], M Lazrek[2], Y Eljariri[2], K Chafouai[2], A Daassou[2]

[1]Oukaimeden Observatory, LPHEA FSTG, Cadi Ayyad University, BP 549 Marrakech, Morocco

[2]Oukaimeden Observatory, LPHEA FSSM, Cadi Ayyad University, BP 2390 Marrakech, Morocco

[3]AAVSO (American Association of Variable Star Observers), 49 Bay State Rd, Cambridge, MA

[4]Observatoire de Haute-Provence - CNRS/PYTHEAS/Université d'Aix-Marseille, 04870 Saint Michel l'Observatoire

[5]CNRS UMR5277 Institut de Recherche en Astrophysique et Planétologie, 14 Avenue Edouard Belin, 31400 Toulouse, France

E-mail: `sefyani@uca.ma`

Abstract. We present the observation of line doubling absorption in the spectra of the RR Lyrae variable star on the metal lines of FeII (λ = 4923.921Å) as well as of D1 and D2 lines of sodium. For the emission, we observed the lines of hydrogen H_α and H_β with a very high intensity and the two lines of HeI (λ = 5875.66 Å and λ = 6678.15Å). This transmission/helium I remission [1, 2] is directly related to the intensity of the shock wave through the atmosphere of the star during the phase of maximum Blazhko cycle [3]. During the expansion phase of the photosphere of the star and during the passage of the shock wave we witnessed a disappearance of the absorption lines of neutral FeI (λ = 4934.006Å and λ = 4920.503Å) and their reappearance in phase 1.00. These observations were made with a spectrograph resolution scale of about 12,000 installed on the C14 telescope at the Oukaimeden observatory (J43) during the different star pulsation cycle. We noticed that some of them almost coincide with the maximum Blazhko cycle. The doubling line were interpreted by Schwarzschild [4] on the basis of a two-layer atmosphere. This interpretation could measure the speed of the shock wave derived from the difference between the two red and blue spectral components on H_α hydrogen lines, D3 helium, D1 and D2 lines of sodium and FeII during the observed doubling of lines. At phase 1.00, the shock wave reached the maximum speed of 160 km.s^{-1} confirming the hypersonic regime occur during this maximum Blazhko cycle.

1. Introduction

The Astronomical Observatory of Oukaimeden (J43) is a research entity belonging to the Cadi Ayyad University. It is located at an altitude of 2700 m on the High Atlas range 78 km south of Marrakech (7°52′52″ west and 31°12′32″ north).

RR Lyrae (HD 182989) is the brightest star of the sky in its class. It is also the most observed star in spectroscopy and photometry. Its light curve shows a period of 0.5668 days and is subject to amplitude and phase modulation discovered by Bazhko in 1907 [3]. At present, it is well established that the shock waves pass through the atmosphere of the pulsating variable star RR Lyrae during the periodic pulsation movements. Among the observational facts in favor of these waves, there is the discontinuity in the radial velocity curve, the doubling of absorption lines, the appearance of emission lines and the Van Hoof phenomenon [5]. The doubling of lines was interpreted by Schwarzschild in 1952 [4] on the basis of a two-layer atmosphere: during compression (decreasing curve of light), all the layers of the atmosphere move inwards, the front of shock wave penetrates the lower layer of the atmosphere and a reversal of direction of motion occurs. This results in a rapid movement of the atmospheric layers from the inside to the outside in the phase interval [0.9 − 0.1]. Historically, the doubling of metal lines was observed for the first time by Sanford in 1952 [6] on W. Virginis. In 1992, Lbre and Gillet [7] observed the doubling of the sodium D lines.

In 1996, Chadid and Gillet [8] detected the doubling on metallic lines of FeI ($\lambda = 4934.006$Å) and TiII ($\lambda = 5188.700$Å), the physical interpretation of which was presented by Fokin and Gillet in 1997 [9]. Phenomenon results from the passage of the shock wave in the atmosphere of the star during phase 0.9 at the beginning of the expansion of its atmosphere. In 1997, Chadid and Gillet [10] also detected the doubling in the metallic lines of FeII ($\lambda = 4923.921$Å) and MgI ($\lambda = 5183.604$Å). In 2008, Chadid *et al* [11] detected for the first time the disappearance of the metallic FeI lines and their reappearances during the pulsation cycle of the star S Area. As for us, we have observed this doubling on the FeII line ($\lambda = 4923.921$Å) at phase 0.13 of Blazhko Cycle.

2. Observations and data Reduction

Measurements of the spectra of the star RR Lyrae were performed on 12 October 2013 and 28-29-30 October 2015 as well as during the night of 6 November 2015. The period of each pulse is 0.566835616 day ie 13 h 36 min. The spectra were acquired using the C14 telescope. The Scale Spectrograph, connected to the 50 μm diameter optical fiber telescope, has a resolution of 12000 with a dispersion of 16Å / pixel and is equipped with a QSI camera using a 3.2 meg pixel CCD at 6.8 μm per pixel. Calibration is performed with a Thorium-Argon lamp. The camera is cooled to $-20C$ throughout the measurements. The exposure time is 300s for each recording. The spectra were filtered with a Gaussian filter at $\sigma = 0.6$ pixels. The spectral dispersion is about 0.1 Å/pixel. The measured signal-to-noise ratio is about 30 dB. The time resolution is: $\Delta t/P \approx 0.6\%$. The spectrograph is equipped with a diffraction grating for dispersing light between 4100 Å and 7200 Å between orders 32 and 52. The curves of the line profiles established according to Stellar Rest Frame. The velocities presented are taken with respect to the barycenter of the star established in the heliocentric reference frame. The value of the barycentric velocity of the star in the heliocentric reference frame is −73.5 km/s from the results of Chadid & Gillet [8]

3. Metallic Line Doubling

It is well established that the shock waves pass through the strip formation zone during each pulsation cycle of the star atmosphere. These shock waves generate a phenomenon of overall motion of the atmospheric layers. Once these layers arrive at the level of the photosphere, they rebound and give rise to a migration of the metal lines of absorption of red to blue. The measurements carried out by Chadid and Gillet in 1996 [8] confirm, for the first time, the presence of doubling of metal lines on the star rr-lyrae at phase 0.926. These are the neutral FeI ($\lambda = 4394.006$Å) and ionic TiII ($\lambda = 5188.700$Å) lines. They also observed the doubling in the form of broadening of lines on the FeII ($\lambda = 4923.921$Å) and on the MgI ($\lambda = 5183.604$Å). The intensity of these duplications depends on the Blazhko phase according to the study made by Chadid and Gillet in 1997 [10]. This suggests that the doubling on the FeII is synonymous with the presence of a very strong shock wave observable at the maximum of the Blazhko cycle. During the night of 12 October 2013 and at phase 0.853, at the beginning of the observations,

IOP Conf. Series: Journal of Physics: Conf. Series **869** (2017) 012088 doi:10.1088/1742-6596/869/1/012088

the doubling of the FeII metallic lines had indeed begun (figure 1). The ion FeII absorption line (λ = 4923.921Å) reached the middle of the doubling at phase 0.941. On the other hand, no duplication of metal lines is observed on the spectra corresponding to phases 0.38 and 0.58 of the Blazhko cycle, just a phenomenon on broadening occur at this phase. Moreover, we observed the phenomenon of disappearance of the neutral FeI lines (λ = 4934.006Å and λ = 4920.503Å) and their reappearances in phase 1.00.

The doubling of metal lines detected by Preston in 2011 [12] on the RR-lyrae type stars shows that the emission of HeII (λ = 4685.68Å) appears during the pulsation cycles in which the doubling of the metal lines occurs. This suggests that this phenomenon of doubling is physically synonymous with intense dynamic activity of the pulsating movements of the star's atmosphere. On the other hand, Chadid *et al* observed for the first time on the star S Arae [11] the disappearance of absorption lines of FeI (λ = 4920.503Å), FeI (λ = 4934.006Å) and neutral TiI (λ = 4918.6Å) during the passage of the shock wave. This same phenomenon was observed also on the star RR-lyrae but concerning just the lines of the FeI neutral (λ = 4934.006Å and λ = 4920.503Å). Thus, it appears that the ionization of the FeI (λ = 4934.006Å and λ = 4920.503Å) during the expansion phase of the star photosphere reflects the intensity of the shock wave that passes through this atmosphere (figure 2).

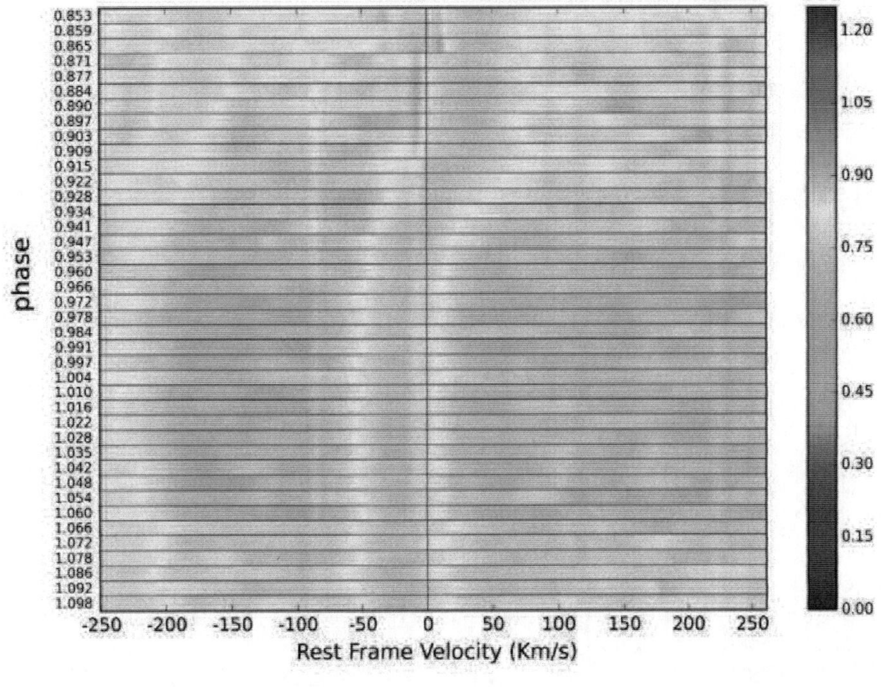

Figure 1. Time series of FeII (λ = 4923.921Å) line of RR Lyr, ordered by increasing phase from top to bottom. On the left, time series of FeI (λ = 4920.503Å). Spectra are interpolated to provide a 2-dimensional map of the (λ = 4923.921Å) profile in the [-250; 250] km /s velocity range. Velocities are given in the stellar rest frame with positive velocities corresponding to inward motion (toward the photosphere). The pulsation phase is given on the left side. On the right side, the color bar representing the flux is from 0.0 to 1.2.

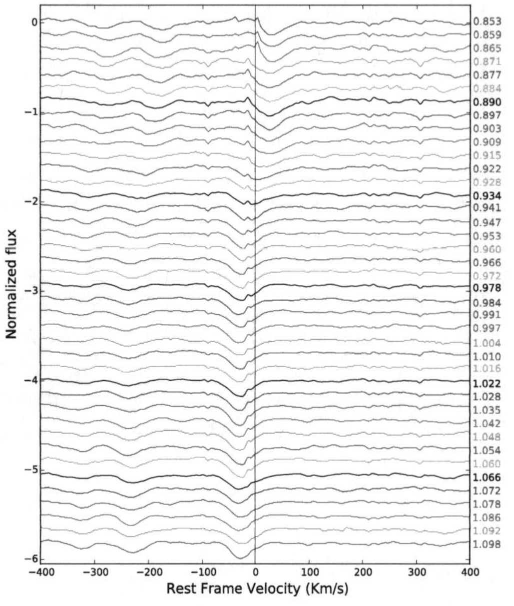

Figure 2. Evolution of the (λ = 4923.921Å) line profile of RR Lyrae at phase 0.13 of Blazhko Cycle. The pulsation phases are indicated on the right side. The vertical line represents the zero velocity in the stellar rest frame. The profiles are arbitrarily shifted in flux.

4. Conclusion

We have observed that the doubling of the ionic FeII line (λ = 4923.921 Å) occurs in phase 0.941 during the night of 12 October 2013 corresponding to the Blazhko phase 0.13. On the other hand, we did not observe any doubling of metal lines in the spectra corresponding to phases 0.38 and 0.58 of the Blazhko cycle. Neutral FeI lines at (λ = 4934.006 et) and (λ = 4920.503Å) showed broadening at the time of transition from the red component of the line to the blue component due to the low amplitude intensity compared to the intensity of the ionic FeII lines. The shock wave is sufficiently intense to ionize the majority of iron atoms involved in the process of line doubling. During the passage of the shock wave we witnessed a disappearance of the absorption line of the neutral FeI (λ = 4934.006Å and λ = 4920.503Å) in favor of an increase in the intensity of the ionic FeII line (λ = 4923.921Å). From phase 1.00, the neutral FeI line reappears (λ = 4934.006Å and λ = 4920.503Å).

References

[1] Gillet D, Sefyani F L, Benhida A, Fabas N, Mathias P, Benkhaldoun Z and Daassou A 2016 *A&A* **587** A134
[2] Gillet D, Fabas N and Lbre A 2013 *A&A* **553** A59
[3] Blazhko S 1907 *Astron. Nachr.* **175** 325
[4] Schwarzchild M 1952 152 Transaction of the IAU VIII, Ed. P. Th. Oosterhoff, Cambridge University Press., 811
[5] Van Hoof A and Struve O 1953 *PASP* **65** 158
[6] Sanford R F 1952 *Transaction of the IAU* **8** 809
[7] Lbre A and Gillet D 1992 *A&A* **255** 221
[8] Chadid M and Gillet D 1996 *A&A* **308** 481
[9] Fokin A B and Gillet D 1997 *A&A* **325** 1013
[10] Chadid M and Gillet D 1997 *A&A* **319** 154
[11] Chadid M, Vernin J and Gillet D 2008 *A&A* **491** 537
[12] Preston G 2011 *Astron. J.* **241** 1

Earth's bow shock dynamics and structure scales based on MMS multi-spacecraft data

T Y Alrefay

Space and Aeronautics Research Institute, KACST, Riyadh, Saudi Arabia

E-mail: talrefay@kacst.edu.sa

Abstract. Space plasmas studies on bow shock dynamics and structure scales continue to attract intense theoretical and experimental investigations. The Earth's bow shock is the closest shock accessible to scientists through various satellite missions. These missions have enabled investigations on different physical phenomena associated with solar-terrestrial interaction. Access to the interplanetary medium through satellites has provided access to valuable spatial and temporal data on the Earth bow shock, and has furthered understanding on certain aspects of shock physics that were inaccessible until now. The main objective of this study is to quantify the dynamics and structure scales of the Earth's bow shock using data obtained by the MMS multi-spacecraft during shock crossing.

1. Introduction

The initial investigations on shock motion date back to the 1960's [2, 3]. Single satellite observations have estimated shock speeds of $\sim$ 10 km/s on an average. Shock motion may result from several disturbances in the solar wind ram pressure. Völk and Auer's [7] work forms the sole exhaustive study, although restricted to the nose region, of the impact of the solar wind conditions on variations in the shock motion. Before multi-spacecraft missions such as Cluster and MMS, all shock motion measurements were performed on data collected by a single or dual spacecraft. Meziane et al. [1] analyzed the data from 133 shock crossings collected by Cluster. The study found that the shock velocity distribution could be satisfactorily described by a zero-mean Gaussian for speeds less than $\sim$ 80 km/s as shown in Figure (2). This implies that the shock move under solar wind ram pressure, earthward and sunward, fluctuates randomly. In their study, Meziane et al. [1] showed that the distribution of the shock normal for 133 crossings derived from timing analysis [4] indicated that the shock motion is along the $x - GPE$ [6]. Consequently, the shock move inward (outward) along the Sun-Earth direction is shown in Figure (1). Alrefay [6] showed that the propagation of Alvén waves interacting with the shock induced shock speed of the order of 10 km/s or less on an average. Based on the systematic inward-outward shock motion, Meziane et al. and Alrefay [5, 6] constructed the probability function (PDF) for the shock velocity near the nose region and compared it with the distribution of the ram pressure fluctuations and derivative of the ram with respect to time fluctuations as shown in Figure (3, 4). The ram pressure temporal changes provided shock velocity values that agreed with the observations [6]. Because the technique of measurement comprises determination of time delays as explained in the next section, a fundamental assumption stipulates that the shock front is quasi-stable during successive measurements. Depending on the involved time scales, the validity of this assumption may be violated given the shock reformation process. Quasi-perpendicular supercritical shocks usually appear as a sharp transition. However, high-precision time magnetic field measurements indicate that the shock transition layer presents short time variations of the different structures making up the shock. This raises serious difficulties in the analysis of shock motion as derived from successive shock crossings. In this study, we focus on the shock dynamics and shock structure scales by using multipoint measurements provided by the MMS quartet, which comprise high-precision time magnetic field measurements. This allows for an in-depth investigation on the effects of a shock's ramp scales and obtained shock motions. We provide two examples of shock crossings observed by the MMS spacecraft and its dynamics.

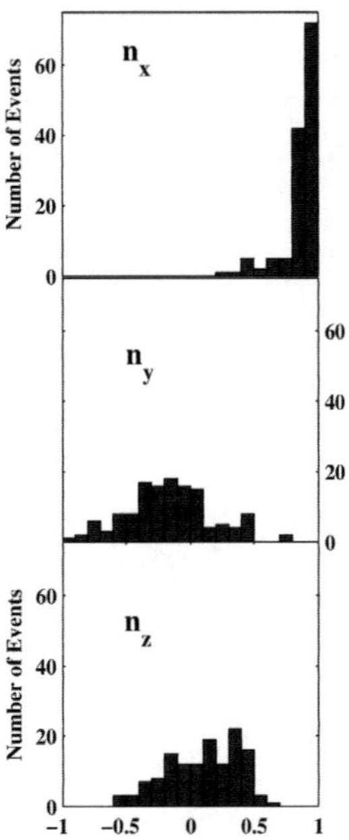

Figure 1. Histogram for the direction of shock motion along the three Cartesian coordinates

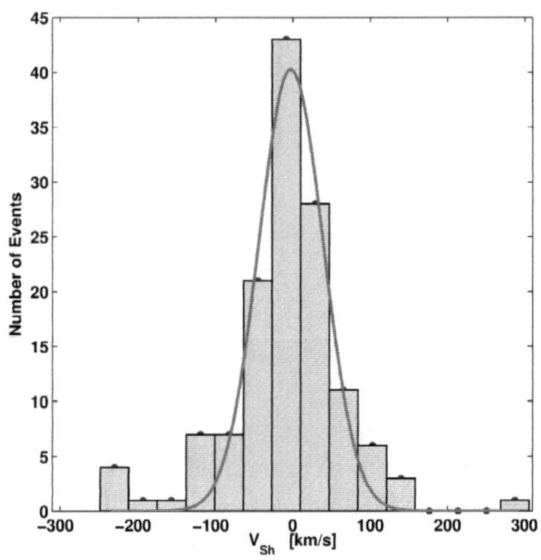

Figure 2. Shock velocity distribution

2. MMS Data analysis (Timing Analysis Method)

The MMS Multi-Spacecraft provides full three-dimensional shock velocities, shock normals, and shock structure scales. The time delay between successive shock crossings is determined by using the timing analysis and cross-correlation between the MMS spacecraft signals assuming that the bow shock is planar and moving with constant velocity at the scale of the MMS quartet during the time of interest [8]. If the same bow shock passes the MMS quartet, the relative position and successive time delay can be used to construct the bow shock normal $\hat{n}$ and velocity $\vec{V_{sh}}$ from the solution of the following system [4].

$$\begin{pmatrix} \vec{r}_{13} \\ \vec{r}_{23} \\ \vec{r}_{43} \end{pmatrix} \cdot \frac{1}{V} \begin{pmatrix} n_x \\ n_y \\ n_z \end{pmatrix} = \begin{pmatrix} t_{13} \\ t_{23} \\ t_{43} \end{pmatrix} \tag{1}$$

In equation (1), the (n_x, n_y, n_z) are the components of the normal vector $\hat{n}$, and V_{sh} is the shock speed. During the time delay t_{13}, the bow shock plane moves by a distance of $V(t_1 - t_3)$ along

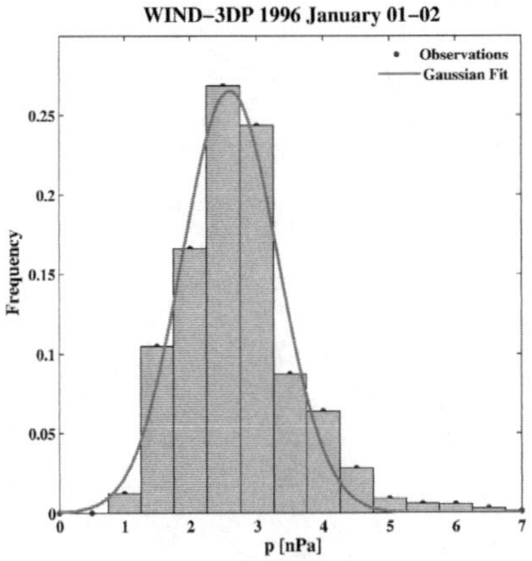

Figure 3. Solar wind ram pressure distribution

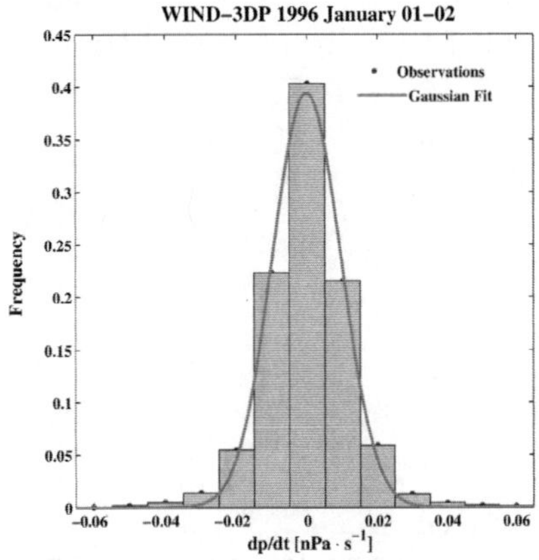

Figure 4. Solar wind dp/dt distribution

the normal between MMS3 and MMS1, which is equal to the projection of the inter-spacecraft separation $\vec{r}_1 - \vec{r}_3$ onto the normal [10]. This method is used here for determining the normal and velocity along the normal of the Earth's bow shock. The timing method assumes that the bow shock structure is planer over the MMS spacecraft configuration scale [8]. To improve the determination of the time delay between two successive shock crossings, high frequency-IMF fluctuations, when present, are removed or suppressed using appropriate techniques such as low band-pass filtering or moving averages as shown in Figure (5). For this crossing, a time resolution of 16 Hz is used for the interplanetary magnetic field data, which is good enough when the spatial ramp thickness is not too small. However, some studies show that the thickness of the shock ramp could be very small and highly dynamic [9]. Figure (6) shows the scale of the shock ramp structure; the time domain is translated into distance assuming that the shock structure moves with a constant velocity obtained by timing analysis of the interplanetary magnetic field. For this shock crossing of the MMS spacecraft, the results of timing analysis as well as solar wind condition and shock geometry are given in Table (1)

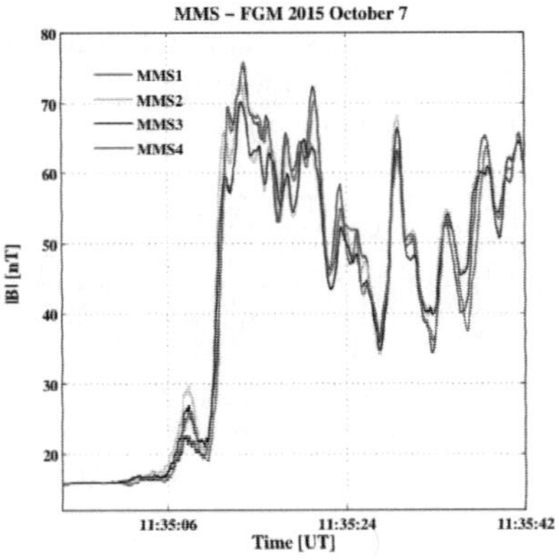

Figure 5. An Example of a bow shock crossing where moving averages are used on the time series of IMF measurements.

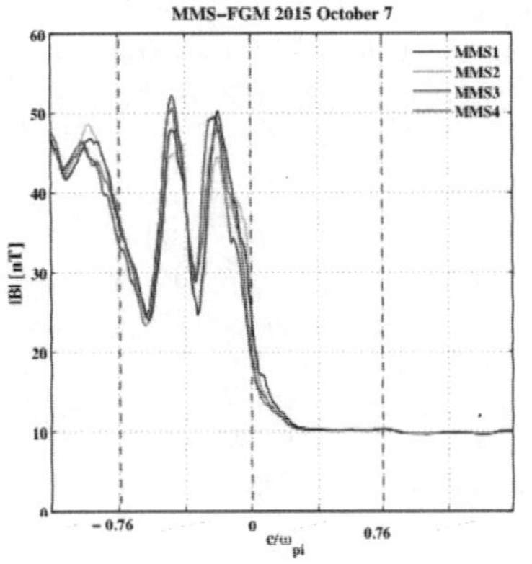

Figure 6. An Example of a bow shock crossing where time is converted to distance and normalized to upstream proton inertia length.

Table 1. Two Shocks Parameters (2015 Oct 07)

Parameters	Unit		
Crossing Time	hh:mm:ss	11:35:10	11:37:26
$\hat{n}$	-	(0.895 0.447 0.0)	(0.873 -0.486 -0.004)
V_{sh}	km/s	+3	-4
θ_{Bn}	deg	79°	79°
M_A	-	6.30	10.60
M_s	-	6.70	5.60
β	-	1.20	1.70
c/ω_{pi}	km	41	37
Standoff distance	R_E	10.40	11.50

3. Conclusions

This paper described preliminary investigations on the Earth's bow shock crossing using data from MMS multi-spacecraft. The dynamics and shock structure scales were obtained by applied timing analysis. Several shock crossings were identified and carefully characterized. In the future, detailed investigation on non-stationary shock will be carried out using high-resolution data on interplanetary magnetic field and solar wind conditions provided by the MMS spacecraft quartet. This will lead to significant progress in understanding the shock transition layer.

References

[1] Meziane K, Alrefay T and Hamza A 2014 On the shape and motion of the Earth's bow shock, *Planetary and Space Science*,**vol. 93**, pp. 1-9
[2] Holzer R E, McLeod M G and Smith E J 1966 Preliminary results from the OGO 1 search coil magnetometer: Boundary positions and magnetic noise spectra *Journal of Geophysical Research* **vol. 71, no. 5** pp. 1481-1486
[3] Greenstadt E W, Green I M, Inouye G, Hundhausen A, Bame S and Strong I 1968 Correlated magnetic field and plasma observations of the earth's bow shock *Journal of Geophysical Research* **vol. 73 no. 1** pp. 51-60
[4] Schwartz S J 1998 Shock and discontinuity normals, Mach numbers, and related parameters *ISSI Scientific Reports Series* **vol. 1** pp. 249-270
[5] Meziane K, Hamza A, Maksimovic M and Alrefay T 2015 The Earth's bow shock velocity distribution function,*Journal of Geophysical Research: Space Physics* **vol. 120, no. 2** pp. 1229-1237
[6] Alrefay T Y S 2014 *A cluster Multi-spacecraft Study of Earth's bow shock* (Fredericton University of New Brunswick Scholar Repository)
[7] Völk H J and Auer R 1974 Motions of the bow shock induced by interplanetary disturbances *Journal of Geophysical Research* **vol. 79, no. 1** pp. 40-48
[8] Horbury T, Cargill P, Lucek E, Eastwood J, Balogh A, Dunlop M, Fornacon K and Georgescu E 2002 Four spacecraft measurements of the quasiperpendicular terrestrial bow shock: Orientation and motion *Journal of Geophysical Research: Space Physics* **vol. 107** no. A8
[9] Hobara Y, Balikhin M, Krasnoselskikh V, Gedalin M and Yamagishi H 2010 Statistical study of the quasi?perpendicular shock ramp widths*Journal of Geophysical Research: Space Physics* **vol. 115** no. A11
[10] Harvey C 1998 Spatial gradients and the volumetric tensor *ISSI Scientific Reports Series* **vol. 1** pp. 307-322

IOP Conf. Series: Journal of Physics: Conf. Series **869** (2017) 012090 doi:10.1088/1742-6596/869/1/012090

Probing the accretion induced collapse of white dwarfs in millisecond pulsars

A Taani[1] and A Khasawneh[2]

[1]Applied Science Department, Aqaba University College, Al-Balqa Applied University, Aqaba, Jordan

[2]Royal Jordanian Geographic Center, Amman, 11941 Jordan

E-mail: ali82taani@gmail.com

Abstract. This paper investigates the progenitors of Millisecond Pulsars (MSPs) with a distribution of long orbital periods ($P_{orb} > 2$ d), to show the link between white dwarf (WD) binaries and long orbits for some binary MSPs through the Accretion Induced Collapse (AIC) of a WD. For this purpose, a model is presented to turn binary MSPs into wide binaries and highly circular orbits ($e < 0.1$) through the asymmetric kick imparted to the pulsar during the AIC process, which may indicate a sizeable kick velocity along the rotation of the proto-neutron star. The results show the effects of shock wave, binding energy, and mass loss ($0.2M_{\odot}$). The model shows the pulsar systems are relevant to AIC-candidates.

1. Introduction

A number of channels of Millisecond Pulsars (MSPs) formation have been discussed [1–3]. The Recycling process is usually considered as a standard model to produce MSPs. An old and quiescent Neutron Stars (NSs) is spun up to millisecond periods, through the accretion of matter from a companion when the system is in the low-mass X-ray binary (LMXB) phase of its evolution [4, 5]. Another often discussed theory involves the Accretion Induced Collapse (AIC) of an O-Ne-Mg white dwarf (WD) in binaries [6–10], which can indicate that (1) the WD is massive enough to be an O-Ne-Mg one, and thus in principle may be able to collapse to an NS (e.g. Cal 83 $\sim 1.3M_{\odot}$ and RX J0648.04418 $\sim 1.28M_{\odot}$), and (2) the mass transfer rate is large enough to make the WD grow -by considering the critical value ($\Delta M_{crit} \sim 0.1M_{\odot} - 0.2M_{\odot}$) [11], because at low accretion rates Nova explosions and a WD will not grow. But to make a Type Ia SNe (a thermonuclear explosion of a CO WD that has grown to the Chandrasekhar limit), the WD needs to grow, so it needs a high accretion rate, such that steady nuclear burning on the surface ensues and the WD grows [12].The time lines for the two routes of MSP evolutions, recycling and AIC, are depicted in figure 1.

2. Estimation of the new orbital period

The relation between the orbital eccentricity and the amount of mass ΔM_{SNe} ejected in the SNe is

$$e = \frac{\Delta M_{SNe}}{M_1 + M_2}, \tag{1}$$

where M_1 and M_1 are the masses of the first- and the second-born NSs.

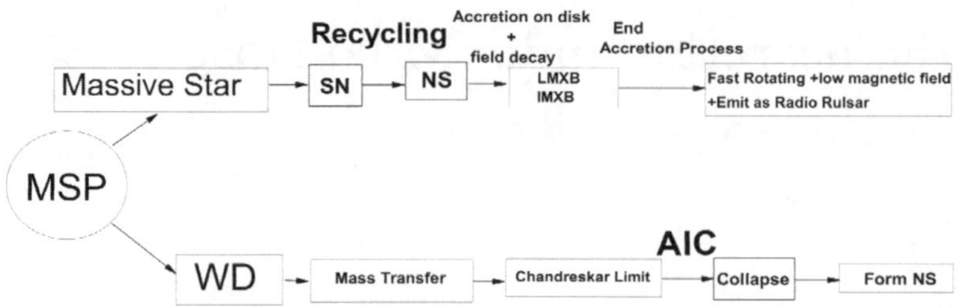

Figure 1. The flow-chart illustrates various evolutionary scenarios for MSPs involving recycling and AIC.

According to the energy conservation law [13]

$$\frac{GM}{R_{WD}} - \frac{GM}{L} = \frac{1}{2}[v_{final}^2 - v_{initial}^2], \tag{2}$$

where G is the gravitational constant, M the mass companion, R_{WD} the WD radius, L the distance between the companions, $v_{initial}$ the initial explosion velocity for the WD (it has been supposed that the $v_{initial} \sim 0.1c$ [14], and v_{final} the velocity when the companion received the material on the surface; because the SNe ejecta may either directly strip material from the companion by direct transfer of momentum or evaporate the envelope through the conversion of the blast kinetic energy into internal heat [15]. It has been assumed that the amount of ejected mass during the explosion process by a WD is $0.1\mathrm{M}_\odot$.

$$\Delta M_{SNe} = \frac{R^2}{L^2} \times 0.1 M_\odot. \tag{3}$$

During the SNe, the WD will eject matter in every direction, so ΔM_{SNe} is the amount of mass received by the companion (accreted material) and R is the radius of the companion. When $L \gg R_{WD}$, after some calculations, equation (2) will be

$$\frac{R_S\, c^2}{2R_{WD}} = 10^{-4} c^2, \tag{4}$$

where R_S is Schwartzchild radius.

It has been assumed that the explosion to be instantaneous (short duration) like in very close binaries [13], in the range of LMXBs, Cataclysmic Variable-type binaries (CVs) and close double pulsars, such as the PSR 1913+16, J0737- 3039AB and the close WD-pulsar system PSR 0655+64 [12]. Hence $v_{final} \approx v_{initial} \approx 0.1c$

$$v_k = \frac{\Delta M v_{final}}{M} = \frac{R^2}{L^2}(M/0.1M_\odot)^{-1} \cdot v_{final}, \tag{5}$$

where v_k is the kick velocity of the companion. It should be noticed that if the angular momentum of the accreted material equals the orbital angular momentum, due to the instability in the accretion shock around a proto-NS. This would enhance the possibility for the companion to be kicked away [16]. An assumption here is that a more energetic explosion would be able to impart a greater initial angular velocity onto the proto-NS than a lesser explosion, and the result is typically an eccentric orbit with an orbital period of a days. The escape velocity for the companion in the gravitational field of the WD is

$$v = \sqrt{\frac{2GM_{WD}}{L}} \simeq \sqrt{\frac{R_S}{L}}\; c. \tag{6}$$

Hence, this is the largest kick velocity that a system can attain after a symmetric SNe explosion in case of $v_k \gg v$. Now let us define the η parameter as the ratio between two velocities

$$\eta = \frac{v}{v_k} \sim 10^{-2}(\frac{R}{L})^2(\frac{L}{R_S})^{1/2}(\frac{M}{M_\odot})^{-1}. \tag{7}$$

Assuming canonical values for $L \sim 3 \times 10^{10}$ cm, $R \sim 10^9$ cm and $R_S = 3 \times 10^5$ cm, $\eta \sim (\frac{M}{0.1M_\odot})^{-1} = 1$. The energy difference between the initial and final positions of the companion in the explosion is

$$\frac{GM_{WD}}{2L} - \frac{GM_{WD}}{2L_1} = \frac{1}{2}v^2. \tag{8}$$

Then by substituting R_S, equation (6) and (7) into equation (8),

$$\begin{aligned}\frac{R_S}{2L} - \frac{R_S}{2L_1} = \frac{v^2}{c^2} &\rightarrow \frac{v_k^2}{2c^2} - \frac{v^2}{c^2} \\ &= \frac{R_S}{L}[1/2 - \eta^2],\end{aligned} \tag{9}$$

where $P_{orb} \propto L^{3/2}$, then

$$L_1 = \frac{L}{1-2\eta^2} \Rightarrow \frac{P_{orb1}}{P_{orb}} = [1-2\eta^2]^{-3/2}, \tag{10}$$

where L_1 is the new distance after the explosion and P_{orb1} is the new orbital period corresponding to L_1. Figure 2 demonstrates that $\mathrm{P_{orb1}}$ will be evolved and kicked up to relatively long orbital period. (i.e. PSR J0900-31, $\mathrm{P_{orb}}$ = 18.7 d, e = 1.03×10^{-5} and PSR J1600-30, $\mathrm{P_{orb}}$ = 14.35 d, and e = 1.74×10^{-4}). The above described model could actually be used for wide and circular binary MSPs, since the AIC process in a close binary will not induce a sizeable kick velocity to the thus formed NS [17]. In our binary MSPs, we have taken the latest observational results from the ATNF pulsar catalog [18], and it has been found that $\sim$ 67% (73 out of 323) of all MSPs could be formed through the AIC process.

3. Summary and Conclusions

The dynamical instability occurs in binary systems through the AIC, is driven by dissipative processes such as kicks, mass loose and shock wave. These effects may have acquired enough energy to the companion star and then can account for the differences in their distributions. Consequently, the AIC decreases the mass of the NS and increases the orbital period leading to orbit circularization. It turns out that the contribution of AIC process which potentially presents in the binary pulsars, is about $\sim$ 67% after the final substantial orbital eccentricities. In addition, in some cases the kick is sufficient to disrupt the system, producing a population of isolated MSPs.

Acknowledgements

A. Taani would like to thank the Deanship of Scientific Research at Al-Balqa Applied university for funding his participation to this conference.

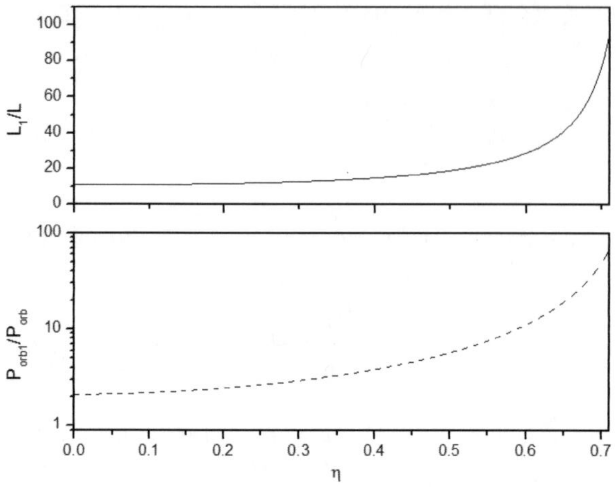

Figure 2. The P_{orb1}/P_{orb} and L_1/L as a function of η.

References

[1] Bhattacharya D and van den Heuvel E P J 1991 *Phys. Rep.* **203** 1-123
[2] Lyne A G and Graham-Smith F 2007 Pulsar Astronomy Cambridge Astrophysics Series, Cambridge University Press 3ed Edit
[3] Lorimer D R 2008 *Living Rev. Relativity* **11** 8-98
[4] Alpar A, Cheng F, Ruderman A and Shaham J 1982 *Nature* **300** 728-730
[5] Taani A, Zhang C M, Al-Wardat M and Y H Zhao 2012a *Astronom. Nachri.* **333** 53-58
[6] Nomoto K 1987 *Astrophys. Journ.* **322** 206-214
[7] Taani A, Zhang C M, Al-Wardat M and Y H Zhao 2012b *Astrophy&Spac Sci* **338** 295-301
[8] Wickramasinghe D, Hurley J, Ferrario L and et al. 2009 *J. Phys. Conf. Ser.* **172** 12037-12040
[9] Tauris T, Sanyal D, Yoon S-C and Langer N 2013 *Astron. Astrophys* **558** 39-50
[10] Freire P and Tauris T 2014 *Month. Not. Roy. Ast. Soc.* **438** 86-90
[11] Taani A 2015 *JJP* **8** 3-10
[12] van den Heuvel E P J 2009 Astrophysics and Space Science Library **359** 125-170
[13] Camenzind M 2007 Compact objects in astrophysics white dwarfs, neutron stars, and black holes 1st ed. (Springer, Verlag Berlin Heidelberg
[14] Metzger B, Piro A and Quataert E 2009 *Month. Not. Roy. Ast. Soc.* **396** 304-314
[15] Kiel P and Hurley J 2009 *Month. Not. Roy. Ast. Soc.* **395** 2326-2346
[16] Podsiadlowski P 2001 The evolution of Binary Systems, in Accretion Processes in Astrophysics, (ed.) Matrtinez-Pais, I G Shahbaz T and Velazquez, J C Cambridge University Press
[17] Kalogera V 1996 *Astrophys. Journ.* **471** 352-365
[18] Manchester R N *et al* 2005 *Astron. J.* **129** 1993-2008

IOP Conf. Series: Journal of Physics: Conf. Series **869** (2017) 012091 doi:10.1088/1742-6596/869/1/012091

The lithium problem: new insight in the big bang nucleosynthesis (BBN) beyond the standard model

T R Makki and M F El Eid
Department of Physics, American University of Beirut, Lebanon.
E-mail: trm03@mail.aub.edu

Abstract. The production of the light elements in the framework of the standard Big Bang nucleosynthesis model (SBBN) matches the observed abundances except in case of ^{7}Li ,where observations lie a factor 2.4-4.3 below SBBN+WMAP(*Wilkinson Microwave Anisotropy Probe*) predictions. This so-called "Lithium problem" needs to be resolved beyond the SBBN. In this contribution we focus on the effect of degenerate neutrinos and the addition of dark component, including dark energy density and dark entropy. We find that the effect of the degeneracy parameter is significant if chemical potentials of neutrino families are different. Concerning the dark component, the major effect comes from adding dark entropy.

1. Introduction

The standard Big Bang nucleosynthesis model (SBBN) represents one cornerstone of the Big Bang theory. It leads to the synthesis of light elements (^{2}D, ^{3}He , ^{4}He , ^{6}Li , ^{7}Li and ^{7}Be) during the first three minutes after Big Bang. The SBBN is a parameter-free theory, since it depends only on the baryon to photon ratio η which is meanwhile well determined by the WMAP. It is found that $\eta = (6.23 \pm 0.17) \times 10^{-10}$ [1, 2]. The observed abundances of ^{4}He , ^{3}He and ^{2}D are in good agreement with SBBN+WMAP predictions but not in case of ^{7}Li. The observation of ^{7}Li in old Halo stars [3], yields $^7\text{Li/H} = (1.23^{+0.68}_{-0.32} \times 10^{-10})$ where the SBBN prediction [4] $^7\text{Li}/H = (4.681 \pm 0.335) \times 10^{-10}$.

Some dispersion is found for deuterium: according to [5] $\text{D/H} = 2.8^{+0.8}_{-0.6} \times 10^{-5}$, while recent observations by [6] shows $\text{D/H} = (2.53 \pm 0.04) \times 10^{-5}$. The element ^{4}He is strongly constrained primordial value of X(^{4}He) = 0.228 ± 0.005 as obtained from seven metal-poor galaxies [7]. Other investigation by [8] indicate a range $0.232 < X(\,^4\text{He}) < 0.258$. Not only the difference is between observations and predictions but observing ^{7}Li in very metal poor stars does not confirm the known plateau [9]. This constitutes presently "the Lithium problem". The overproduction of ^{7}Li seems not to be resolved by the SBBN, for example on astrophysical or nuclear physics grounds, a non-standard scenario of the SBBN seems to be needed to resolve the ^{7}Li problem. In particular, modifying neutrinos properties (neutrinos degeneracy, oscillations, non-vanishing mass...) and adding the effect of a dark component [10-12].

2. Non-standard BBN

2.1. Non-vanishing neutrinos chemical potential

According to [13], neutrinos degeneracy affects elements production in two ways:
(i) It increases the energy density leading to speed up the expansion rate of the universe as appears in the Friedmann equation. The Hubble parameter is then given by:

$$\left(\frac{\dot{a}}{a}\right)^2 = H^2 = \frac{8\pi G}{3}\rho_{rad} = kT^4, \qquad k = \frac{\pi^2}{30}(g_b + \frac{7}{8}\times g_f), \tag{1}$$

where g's are the relativistic degrees of freedom, a is the scale parameter and ρ is the total energy density. The energy density of fermions and anti-fermions is given by [14]:

$$\rho_f + \rho_{\bar{f}} = \frac{7\pi^2}{120} gT^4 \left[1 + \frac{30\beta^2}{7\pi^2} + \frac{15\beta^4}{7\pi^4}\right], \tag{2}$$

where $\beta = \frac{\mu_f}{T}$ is the degeneracy parameter .In the SBBN, $\beta = 0$ (non-degenerate neutrinos). Clearly including β increases the energy density (or k) and this is the only way in which ν_μ and ν_τ modify the SBBN. As a consequence of a larger expansion rate, freeze-out temperature will be higher leading to higher neutrons mass fraction. This will result in increasing D and ^{4}He but ^{7}Li will decrease as shown in table 2.

(ii) A second effect concerns the electron neutrinos, which are involved in the weak interactions rates:

$$n + \nu_e \leftrightarrow p + e^- \ , \ n + e^+ \leftrightarrow p + \bar{\nu}_e \ , \ n \leftrightarrow p + e^+ + \bar{\nu}_e \tag{3}$$

A new equilibrium ratio is obtained (see [14]):

$$\frac{n}{p} = \exp[-(\beta_{\nu_e} + Q/T)] \tag{4}$$

Where $Q = m_n - m_p = 1.29$ MeV, T in MeV, and β_{ν_e} is the electron neutrino degeneracy parameter. The rate of the first reaction can be written as [14]:

$$\lambda_{n\nu} \approx 1.63 \left(\frac{T_\nu}{Q}\right)^3 \left(\frac{T_\nu}{Q} + 0.25\right)^2 s^{-1} \tag{5}$$

The other rates are related to it.

Taking $-0.5 < \beta_{\nu_e} < 0.5$, equation (5) becomes:

$$\lambda_{n\nu} \approx 1.63 \times \exp(+\beta_{\nu_e}) \left(\frac{T_\nu}{Q}\right)^3 \left(\frac{T_\nu}{Q} + 0.25\right)^2 s^{-1} \tag{6}$$

It is seen that the rate is enhanced exponentially when $\beta_{\nu_e} > 0$, thus it becomes faster and more neutrons are converted into protons, so that the freeze-out temperature is delayed (see table 1). As table 1 shows the effect on $^7\mathrm{Li}/H$ is significant. Using equation (6) we calculate the mass fraction of neutrons at freeze-out :

$$X_n^* = \int_0^\infty e^{\left(-5.42k^{-1/2}e^{\beta_{\nu_e}}\int_0^y (x+0.25)^2\left(1+e^{-\beta_{\nu_e}}\exp\left(-\frac{1}{x}\right)\right)dx\right)} \frac{dy}{2y^2(1+\cosh(\beta_{\nu_e}+1/y))} \tag{7}$$

And to get the freeze-out temperature T^* we use the modified $X_n^{eq} = \frac{1}{1+\exp(\beta_{\nu_e})\exp(Q/T^*)}$. (8)

The integral in equation (7) does not allow an explicit dependence of X_n^* on β's. For this reason we took different values of $-0.5 < \beta_{\nu_e} < 0.5$ and $0 < \beta_{\nu_\mu}, \beta_{\nu_\tau} < 2.5$ (or and 3.53<k<8.6) and we evaluated this integral in the above ranges, where some results are given in tables 1 and 2.

Table 1. Effect of adding (β_{ν_e}) on freeze-out temperature and light elements (numbers in brackets are power of ten in all tables and figures).

β_{ν_e}	X_n^*	T^*	X(^{4}He)	D/H	$^7\mathrm{Li}/H$
-0.2	0.1987	0.8110	2.987(−1)	2.832(−5)	5.224(−10)
0 or **(SBBN)**	0.1571	0.7695	2.464(−1)	2.509(−5)	4.681(−10)
0.2	0.1217	0.7279	2.030(−1)	2.291(−5)	4.073(−10)

Table 2. Effect of adding ($\beta_{\nu_\mu} = \beta_{\nu_\tau}$) on freeze-out temperature and light elements.

$\beta_{\nu_\mu} = \beta_{\nu_\tau}$	K	X_n^*	T^*	X(^{4}He)	D/H	^{7}Li/H
2.5	7.6511	0.1864	0.8773	3.133(−1)	4.882(−5)	2.710(−10)
1.0	4.0690	0.1624	0.7881	2.592(−1)	2.826(−5)	4.317(−10)
SBBN	3.53	0.1571	0.7695	2.464(−1)	2.509(−5)	4.681(−10)

IOP Conf. Series: Journal of Physics: Conf. Series **869** (2017) 012091 doi:10.1088/1742-6596/869/1/012091

Table 3. Effect of adding $\beta_{\nu_e} = 0.28$ and $\beta_{\nu_\mu} = \beta_{\nu_\tau} = 2$ together on light elements.

X(^{4}He)	D/H	^3He/H	^{7}Li/H
2.219(−1)	3.361(−5)	1.127(−5)	2.814(−10)

Adding β_{ν_e} or $\beta_{\nu_\mu}, \beta_{\nu_\tau}$ independently will have effect on ^{7}Li but it will violate observational constraints on D and ^{4}He. Therefore, adding $\beta_{\nu_e} = 0.28$ along with $\beta_{\nu_\mu} = \beta_{\nu_\tau} = 2$ together will lead to ^{7}Li depletion without violating observational constraints (see table 3).

2.2. *Adding dark energy density*

It is known that dark matter and dark energy play a decisive role in the evolution of the universe. Due to the fact that the identity of dark matter is not well known a unified adiabatically expanded fluid is adopted so that the dark energy and dark matter can be considered as two different aspects of the same component. To illustrate this model, temperature-dependent dark energy density is added to radiation density as follows [10,11]:

$$\rho_D(T) = \rho_D(T_0)\left(\frac{T}{T_0}\right)^{n_\rho}, \tag{9}$$

where $T_0 = 1.0$ Mev $=1.16 \times 10^{10}$ K, and n_ρ is a constant characterizing the power law. In case of $n_\rho = 4$, the dark component mimics a radiation density. The case $n_\rho = 3$ describes a matter density behavior, while $n_\rho = 6$ describes a scalar field. With these assumptions the Friedmann equation becomes:

$$\left(\frac{\dot{a}}{a}\right)^2 = H^2 = \frac{8\pi G}{3}(\rho_{rad} + \rho_D), \tag{10}$$

Where $\rho_{rad}(T)$ is as in equation (1). Including $\rho_D(T)$, two parameters will be introduced : $n_\rho = 3(w_D + 1)$, describing the behavior of the dark fluid ($w_D = \frac{P_D}{\rho_D}$ and P is the pressure of the dark component) and $k_\rho = \frac{\rho_D(T_0)}{\rho_{rad}(T_0)}$, which is the ratio of effective dark fluid density over the total radiation density at BBN time (taken to be $T_0 = 1$ Mev).

Then equation (9) becomes:

$$\rho_D(T) = k_\rho \rho_{rad}(T_0)\left(\frac{T}{T_0}\right)^{n_\rho}. \tag{11}$$

Knowing that the universe was radiation dominated during the time of BBN requires the constraints $n_\rho \geq 4.0$ and $K_\rho < 1.0$ [10]. Adding dark density alone will not be sufficient to resolve the Lithium problem. The point is that the dark energy density being temperature dependent will increase the expansion rate leading higher ^{4}He abundance and to higher production of ^{7}Li (see table 4). This is contrary to the case when adding β_{ν_μ} *and* β_{ν_τ} as described above where the abundance of ^{7}Li decreases because of higher abundance of Deuterium. These two different outcomes have different effects on the expansion rate and the freeze-out temperature and this illustrate how sensitive the Lithium production to these assumptions.

Table 4. Effect of adding $\rho_D(T)$ with $k_\rho = 0.5$ and $n_\rho = 6$.

X(^{4}He)	D/H	^3He/H	^{7}Li/H
2.920(−1)	2.816(−5)	1.016(−5)	5.082(−10)

2.3. *Adding Dark entropy*

In analogy to the dark energy component, a dark entropy component can be added as well. This is described by [4] and [12] as:

$$s_D(T) = k_s s_{rad}(T_0)\left(\frac{T}{T_0}\right)^{n_s}, \text{where } s_{rad}(T) = h_{eff}(T)\frac{2\pi^2}{45}T^3, \tag{12}$$

IOP Conf. Series: Journal of Physics: Conf. Series **869** (2017) 012091 doi:10.1088/1742-6596/869/1/012091

and $h_{eff}(T)$ are the effective degrees of freedom characterizing the contribution of particles to the entropy density. Then, the total entropy becomes:

$$s_{tot}(T) = s_{rad}(T) + s_D(T) \tag{13}$$

The direct effect of adding dark entropy is altering the time-temperature relation and clearly the light elements abundances. In this case the energy conservation equation will read:

$$\frac{ds_{tot}}{dt} = -3Hs_{tot} \quad \text{or} \quad \frac{d}{dt}(\rho_{tot}a^3) + P_{tot}\frac{d}{dt}(a^3) + T\frac{d}{dt}(s_D a^3) = 0, \tag{14}$$

where the third term corresponds to the dark entropy. The effect of adding dark entropy is increasing the temperature of the universe, altering the freeze-out of neutrons and the time when it takes place. Deuterium bottleneck (the highest D abundance where D/H$\sim 10^{-3}$ at $T \approx$0.07Mev) is shifted which leads to higher final D abundance. In addition, ^{7}Li abundance decreases to match the observational range which is shown in table 5 and figure 1 without violating observational constraints.

Table 5. Effect of adding $s_D(T)$ with $k_s = 5.5 \times 10^{-4}$ and $n_s = 5$.

X(^{4}He)	D/H	^3He/H	^{7}Li/H
2.306(−1)	3.489(−5)	1.154(−5)	2.781(−10)

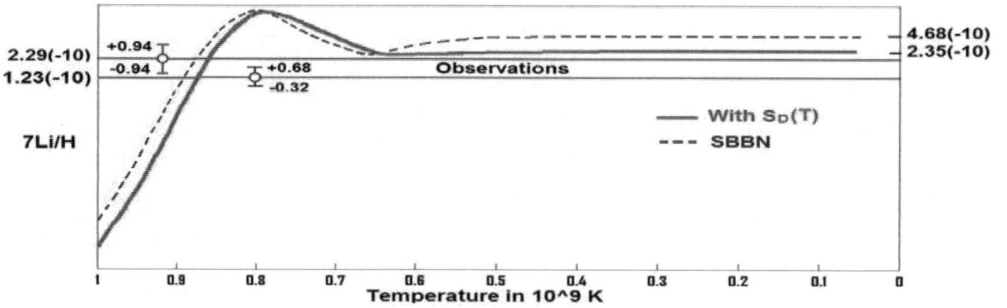

Figure 1. ^{7}Li as function of temperature in SBBN and when adding dark entropy ($k_s = 0.00075$, $n_s =$ 5). Notice that ^{7}Li decreases to match the range of observations.

References

[1] Cyburt R H *et al* 2009 *JCAP* **21** 0910
[2] Dunkley J *et al* 2009 *Astrophys. J. Suppl.* **180** 306
[3] Ryan S G *et al* 2000 *ApJ Lett.* **530** L57.
[4] Arbey A 2012 Computer Physics Communications 2012 **183** 1822.
[5] Noterdaeme P *et al* 2012 *A& A* **542** L33.
[6] Cooke R *et al* 2014 *ApJ* **781** 31.
[7] Songaila A *et al* 1994 *Nature* **368** 599.
[8] Fields D B and Olive K 2006 *Nuclear physics* **777** 208
[9] Spite F and Spite M 1982 Astron. Astroph. **115** 382.
[10] Arbey A and Mahmoudi F 2008 *Phys. Lett.* B **669** 46.
[11] Arbey A 2009 *AIP Conf. Proc.* **1241** 700
[12] Arbey A and Mahmoudi F 2010 *JHEP* **1005** 051
[13] Wagoner V *et al* 1967 *ApJ* **148** 3W.
[14] Mukhanov V 2005 *Physical foundations of cosmology*, Cambridge university press,

IOP Conf. Series: Journal of Physics: Conf. Series **869** (2017) 012092 doi:10.1088/1742-6596/869/1/012092

PPMXL photometric study of two newly discovered open clusters (SAI 19 and SAI 23)

W F Alotaibi

King Abdul Aziz University, Kingdom of Saudi Arabia-21442 Jeddah

E-mail : wassayf-@hotmail.com

Abstract. In our research, we studied the characteristics of two newly discovered open star clusters. Their structural (core radius, limiting radius, etc) and the astrophysical parameters (reddening, distance, age, etc) have been investigated using 2MASS database and PPMXL database extracted from VizieR web site for the studied clusters. The studied newly discovered clusters are SAI 19 and SAI 23 were not studied before, they have ages of 0.4 Gyr and 0.3 Gyr respectively. The luminosity and mass functions as well as the total masses have been determined.

1. Introduction

Stellar clusters can be classified into two environmental classes depending on their association with interstellar matter [1]. *Exposed clusters* are clusters with little or no interstellar matter within their boundaries. Almost all clusters found in standard open cluster catalogs fall into this category. *Embedded clusters* are clusters which are fully or partially embedded in interstellar gas and dust. They are completely invisible at optical wavelengths and best detected in the infrared. Koposov *et al* [2] and Glushkova *et al* [3] published new surveys through searching for new open star clusters. They have discovered 168 new clusters among them 26 are embedded clusters. We have selected two open star clusters; SAI 19 and SAI 23 from the new survey, especially the embedded clusters in the galactic disk. Table 1 lists the centres coordinates and the diameters of the two selected clusters as given in the survey of ref. [3]. The current work is the first study for clusters SAI 19 and SAI 23. This paper is organized as follows ; the clusters center and clusters structure are given in section 2, clusters membership is introduced in section 3, the color magnitude diagram and the obtained physical parameters are discussed in section 4, and the cluster's luminosity and mass functions are introduced in section 5. Finally, the conclusion is devoted to Section 6.

2. The Clusters Center and Clusters Structure

We determine the center of the two clusters; SAI 19 and SAI 23 by building histograms of star counts along α and δ using 2MASS extracted data. Then, the histograms were fitted by Gaussian function to obtain the location of the maximum number density in α and δ, which represents the cluster's center. The new values of centers for the two clusters are given in table 1 and presented in figure 1.

Radial Density profile (RDP) is defined as the number of stars per unit area at different values of radius from the cluster's center to outward [4]. It enables us to determine the cluster structural parameters such

IOP Conf. Series: Journal of Physics: Conf. Series **869** (2017) 012092 doi:10.1088/1742-6596/869/1/012092

as core radius, limiting radius ...etc. The RDP is constructed by counting stars inside concentric rings outward from the cluster center. The observed surface radial density profile was fitted using the King Model [5], in the form:

$$\rho(r) = f_{bg} \frac{f_0}{1+\left(\frac{r}{r_{core}}\right)^2}, \tag{1}$$

where r_{core}, f_0 and f_{bg} are the core radius, the central surface density and background surface density, respectively. The cluster limiting radius, $r_{\lim}$, was calculated by comparing $\rho(r)$ to the levels of the background density and it is defined as:

$$r_{\lim} = r_{core}\left(\frac{f_0}{\sigma_{bg}} - 1\right)^{1/2}, \tag{2}$$

where σ_{bg} is an uncertainty of f_{bg}.

Table 1. Equatorial, Galactic coordinates and the apparent diameters of the candidate clusters obtained from SAI open cluster catalog.

Cluster	α (h:m:s)	δ (°:′:″)	G. Long. (°)	G. Lat. (°)	D.′
SAI 19	02:26:31.9	+61:59:33	133.846	1.153	6.0
SAI 23	02:54:05.0	+60:39:52	137.404	1.286	2.0

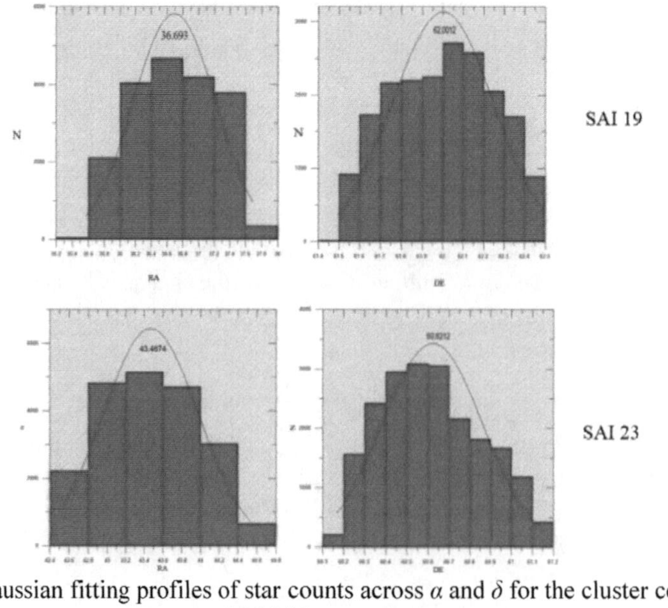

Figure 1. The Gaussian fitting profiles of star counts across α and δ for the cluster center for SAI 19 and SAI 23 open clusters.

Figure 2 shows the RDP from the new centers of the clusters. The core radius and limiting radius are shown for each cluster in table 2.

3. Clusters Membership

To remove the field stars (non-member stars) from the cluster region, we applied the kinematical method. First by building two histograms; one for the component of the proper motion in α and the other for the component of the proper motion in δ. The highest count of stars represents the cluster member stars, because they share nearly the same proper motion values, then by selecting a range for both pmRA and pmDE around the histogram peak, the stars within this range could be considered as members of the cluster as. Then cleaning the cluster data from the rest stars (field stars) to obtain cleaned CMDs for each cluster for further data analysis.

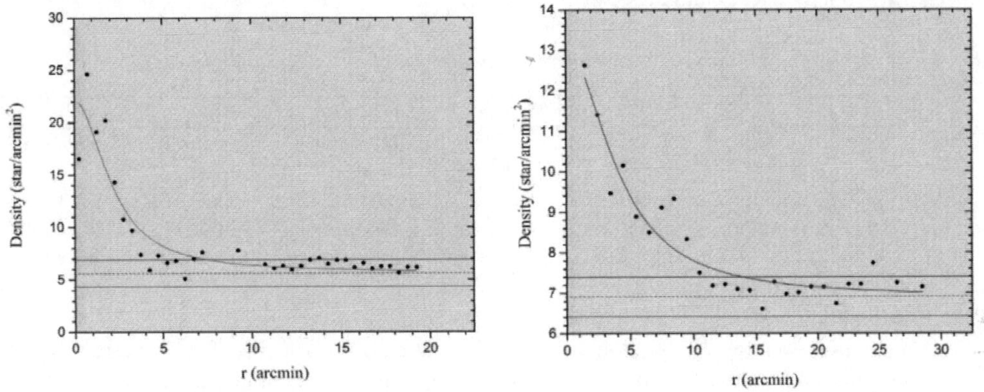

Figure 2. Radial density profile for SAI 19 (left) and SAI 23 in (right).

4. Clusters Physical Parameters

The cleaned CMDs are used to derive physical parameters (age, reddening and distance) as shown in table 2. Figure 3 presents the observed CMDs; J versus J-H and Ks versus J-Ks for our targets fitted with the suitable isochrones. These isochrones cover an age range from 4 Myr to 14 Gyr with step of 0.05 in the logarithm of age and for a metallicity of $Z = 0.017$ [6].

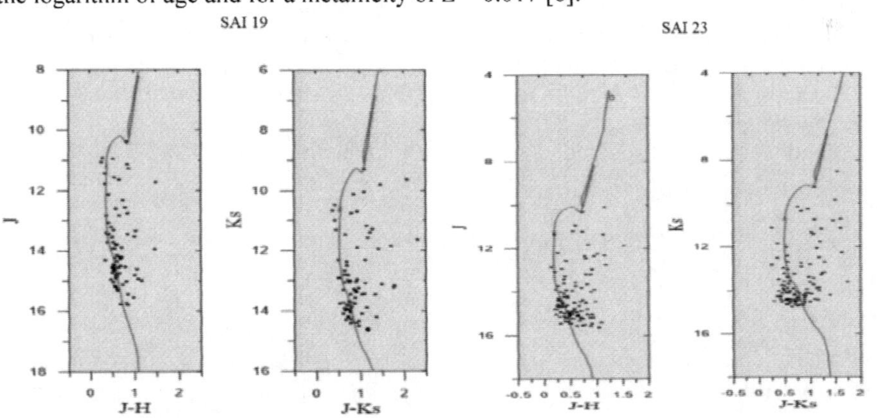

Figure 3. Isochrone fitted to the observed CMD of SAI 19 and SAI 23.

5. Clusters Luminosity and Mass Functions

We have used the cleaned cluster data after the field stars decontamination to construct the luminosity (LF) and mass functions (MF). We present the observed LFs in J band for the stars fainter than zero absolute magnitude, these LFs are given in absolute magnitude scales after applying the distance modulus derived in the previous section, at J band LF lies in the luminosity range (-2 to 4), with a maximum at round 3.25 mag for SAI 19, while J band lies in the luminosity range (-2.5 to 4), with a maximum at around 3 mag for SAI 23. This absolute magnitude are determined using a suitable bin size.

The LFs are transformed to the MFs using the mass-luminosity dependency according to the adopted isochrones. Scale [7] gave the distribution per mass interval by the relation:

$$\frac{dN}{dM} \propto M^{-\alpha}, \tag{3}$$

where α is the power index equals to 2.35 according to the original Salpeter value [8]. It is noted that the investigated MF slope ranging of the clusters under consideration are found to be - 2.59 and 2.42, which are found to be around the Salpeter's value.

Table 2. The astrophysical parameters of the candidate clusters.

Parameter	SAI 19	SAI 23	Parameter	SAI 19	SAI 23
Center (Ra. h: m: s) (Dec. °: ′: ″)	02: 26: 46.32 +62:8:13.2	02 :53 :52.18 +60:37:16.32	E(J-H) (mag.)	0.354± 0.02	0.199±0 .02
r_{lim} (arcmin.)	7.20 ± 1.1	14.32 ± 2.3	m-M (mag.)	10.89± 0.04	11.352±0.17
Core radius (arcmin.)	2.15±0.31	4.17 ± 0.62	Distance (pc.)	1509 ± 33	1863 ± 150
Log(age)	8.6 ± 0.05	8.5 ± 0.05	Membership	81 stars	108 stars

6. Conclusion

In this research, we present a photometric and kinematic study for two selected open clusters from the 26 embedded clusters discovered by Glushkova *et al* [3], through their automatic sky survey. For these embedded clusters and many other newly discovered clusters, the coordinates of their centers and diameters are only given in this survey. This is the first study that has been done for our target SAI 19 and SAI 23 to determine the structure and the physical parameters. In our study, we used the near-infrared JHKs data from 2MASS Point Source [9] and PPMXL [10] Catalogs.

Reference

[1] Lada C J and Lada E A 2003 *ARA&A* **41** 57
[2] Koposov S E, Glushkova E V and Zolotukhin I Y 2008 *A & A* **486(3)** 771
[3] Glushkova E V, Koposov S E, Zolotukhin I Y, Beletsky Y V, Vlasov A D and Leonova S I 2010 *Astron. Lett.* **36(2)** 75
[4] Haroon A A, Ismail H A and Alnagahy F Y 2014 *Astrophys Space Sci* **352** 665
[5] King I R 1966 *Astron. J.* **71** 64
[6] Girardi L, Bressan A, Bertelli G and Chiosi C 2000 *A&AS* **141** 371
[7] Scalo J M 1986 *Fundamentals of Cosmic Physics* **11** 1-278
[8] Salpeter E E 1955 *APJ* **121** 161
[9] Skrutskie M F *et al* 2006 *Astron. J.* **131** 1163
[10] Roeser S, Demleitner M and Schilbach E 2010 *Astron. J.* **139** 2440

IOP Conf. Series: Journal of Physics: Conf. Series **869** (2017) 012093 doi:10.1088/1742-6596/869/1/012093

Tidal disruption of the aged open cluster Berkeley 17

W P Chen[1], S Bhattacharya[2], I Mishra[3], K Vaidya[2] and B Lalchand[1]

[1]Graduate Institute of Astronomy, National Central University, Taoyuan, Taiwan
[2]Birla Institute of Technology and Science Pilani, Rajasthan, India
[3]Indian Institute of Technology, Guwahati, India

E-mail: wchen@astro.ncu.edu.tw

Abstract. Berkeley 17 is known to show a core-tail morphology. Pan-STARRS 1 photometry complete to $g_{P1} \approx 21.5$ mag reveals the effect of mass segregation between the core and the tail (and an antitail). Furthermore, there is an overall paucity of low-mass member stars --- likely as the consequence of tidal stripping by the foreground Perseus arm of this aged (10 Gyr) cluster. Our analysis indicates that about half of the blue straggler candidates suspected to be associated with Berkeley 17 may be field contaminations, and yet some may be confused with blue horizontal-branch stars usually found only in globular clusters. We propose Berkeley 17 to be a remnant super star cluster in the continuing process of disintegration.

1. Berkeley 17 as a Galactic star cluster

Located near the Galactic anticenter with a heliocentric distance 2700 pc [1, 2], Berkeley 17 ($\ell = 175.657°, b = -03.649°$) is among the oldest (10 Gyr) Galactic open clusters. In addition to the age, the cluster in many ways resembles globular clusters, with a subsolar metallicity ($[\mathrm{Fe/H}] \approx -0.33$) [3], an excessive number of blue stragglers [4], and a prominent blue horizontal branch [5], as illustrated in the near-infrared color-magnitude diagram in figure 1.

The stellar density distribution of Berkeley 17, as diagnosed by star-count of Pan-STARRS sources [6], shows an elongated shape, with a tail pointing to the Galactic plane [5,7]. An antitail is also visible.

Figure 2 depicts the Pan-STARRS color-magnitude diagrams for stellar population analysis for the core (about 5' in size) and the two tails (each about 2' in size). A reference field 1 deg away in longitude, having the same Galactic latitude as Berkeley 17, was used to remove possible field contaminations. The mass function thus derived for the core shows a flat or slow rising (up to incompleteness of $g \approx$ 21.5 mag, or member stellar mass 0.675 $M_\odot$) distribution with mass, whereas in the tails there is a paucity of relatively more massive members. This clearly manifests the effect of dynamical mass segregation, for which more massive stars tend to 'sink' to the center, while low-mass members occupy larger volumes.

Our analysis suggests some 200 to 300 member stars in the mass range 0.675 to 0.925 $M_\odot$, with an overall (core plus tails) lack of low-mass stars, in contrast to the ubiquitous mass function in the field or in young stellar systems for which low-mass stars outnumber massive stars. Vulnerable low-mass then-members of Berkeley 17 must have been evaporated, possibly by the Perseus arm, at a heliocentric distance 1.95 kpc [8] in the line of sight [5].

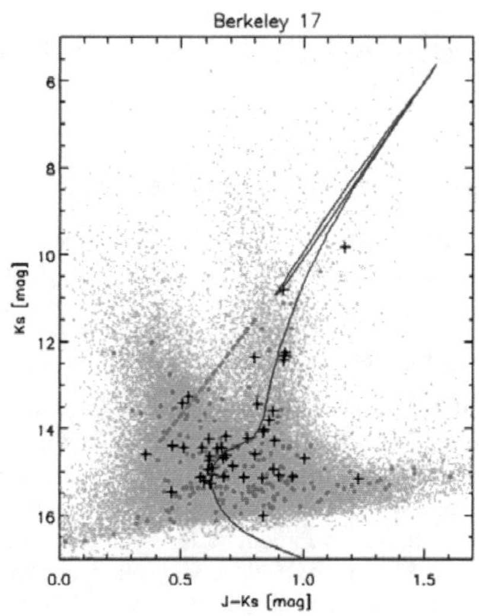

Figure 1. The 2MASS color-magnitude diagram of all the stars (in gray) within 2 deg diameter sky area toward Berkeley 17. Red circles represent the stars in the central 3', for which field contamination is relatively low, and the main-sequence turn-off is evidenced in comparison to the 10-Gyr PARSEC isochrone (in blue) at distance 2.7 kpc with $E(B-V)$=0.6. Black crosses mark stars with proper motions consistent with the cluster's mean values of $(\mu_\alpha \cos\delta, \mu_\delta)$ = (3.90, -2.31) marcses/year. The dashed line illustrates the horizontal branch extending to a bluer $(J-Ks)$ color as observed in the globular cluster NGC 3201 [9].

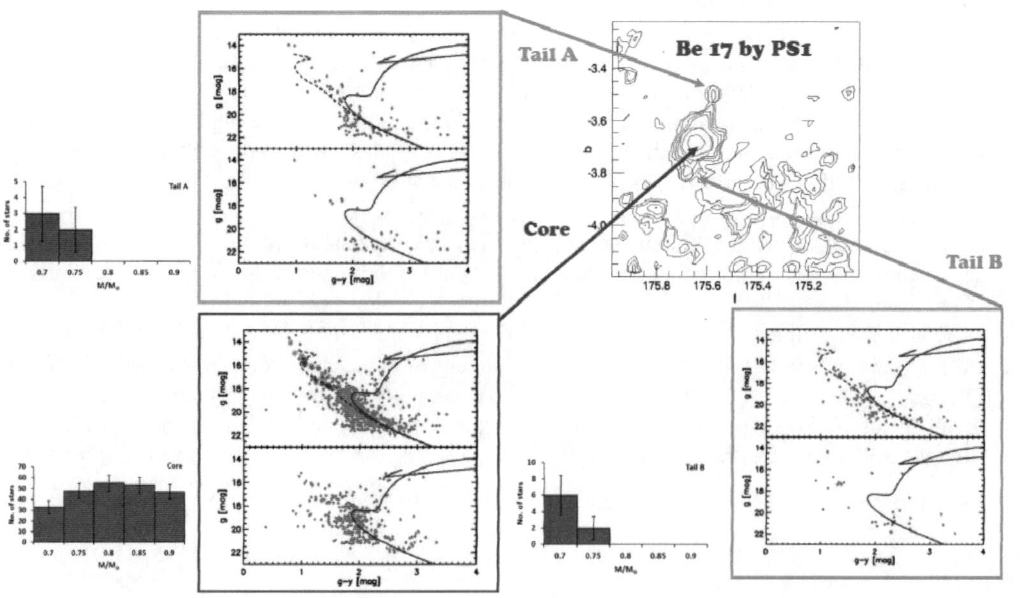

Figure 2. The Pan-STARRS g versus $(g-y)$ color-magnitude diagrams for the core and two tails of Berkeley 17, as labelled. In each case the observed (upper panel) and cleaned dataset with a reference field (bottom panel) are shown, with the histogram depicting the statistical member mass function.

2. Blue stragglers in Berkeley 17

Blue stragglers are stars in the color-magnitude diagram that, given the age of a cluster, should have been evolved off the main sequence according to stellar evolution theories. Blue stragglers are commonly found in old open clusters or globular clusters. It was suggested that Berkeley 17 was rich in blue stragglers [4], listed in table 1, along with their 2MASS (typical errors 0.02 mag) and Pan-STARRS (typical errors 0.002 mag) photometry, and URAT proper motions (typical errors 5.1 mas per year). Field decontamination analysis [5] indicates that half of these are likely field stars not related to the cluster, and a few could be blue horizontal-branch members (see figure 1), usually found only in globular clusters.

Table 1. Blue straggler candidates [4]

RA	DEC	J	H	Ks	g_{P1}	r_{P1}	i_{P1}	z_{P1}	y_{P1}	PM α	PM δ
[deg] J2000		[mag]								mas yr^{-1}	
80.09418	30.54449	13.643	13.361	13.210	15.566	15.040	14.798	14.677	14.552	3.8	-0.4
80.09688	30.56871	13.712	13.322	13.215	15.700	15.113	14.849	14.791	14.640	-2.1	1.9
80.09837	30.59620	15.139	14.861	14.472	17.602	16.938	16.607	16.412	16.248	12.3	4.2
80.09992	30.57075	13.234	12.913	12.803	15.178	14.646	14.399	14.283	14.194	-1.3	3.1
80.10205	30.57660	13.428	12.989	12.941	15.407	14.819	14.558	14.463	14.378	3.4	3.1
80.10609	30.54484	15.342	14.922	14.587	17.826	17.072	16.708	16.515	16.354	2.2	-1.7
80.10845	30.55068	14.809	14.318	14.220	17.143	16.409	16.085	15.935	15.803	6.1	-1.4
80.11208	30.57586	13.995	13.749	13.656	15.931	15.372	15.134	15.037	14.916	-1.9	0.1
80.11219	30.58856	14.949	14.630	14.592	17.080	16.462	16.197	16.012	15.865	2.2	-1.0
80.11382	30.52456	15.031	14.632	14.551	17.479	16.746	16.387	16.213	16.083	-3.4	-3.1
80.11759	30.58191	15.473	14.784	14.851	18.092	17.227	16.829	16.599	16.441	8.3	0.1
80.12391	30.57504	15.093	14.623	14.430	17.767	16.922	16.522	16.309	16.131	3.3	2.4
80.13012	30.53110	15.199	14.868	14.295	17.716	16.979	16.584	16.395	16.241	-8.3	-5.6
80.13029	30.52306	15.299	14.884	14.765	17.481	16.869	16.531	16.413	16.234	7.7	4.4
80.13438	30.56957	15.107	14.795	14.582	17.677	16.956	16.586	16.377	16.209	-1.4	-2.1
80.13631	30.54781	14.311	13.982	13.791	16.777	16.071	15.723	15.561	15.420	-2.9	-4.0
80.13709	30.56058	15.293	14.956	14.611	17.845	17.083	16.707	16.501	16.371	4.5	-2.9
80.15726	30.54103	13.614	13.296	13.148	16.084	15.517	15.185	15.193	14.924	-2.4	-10.8
80.15732	30.56170	15.343	15.161	14.732	17.771	17.065	16.687	16.488	16.331	8.6	-1.3
80.16198	30.53691	13.906	13.661	13.533	15.894	15.355	15.094	14.959	14.828	-2.1	-6.0
80.16384	30.57309	15.128	14.853	14.721	17.258	16.660	16.361	16.175	16.012	-3.0	-4.5
80.16466	30.53942	14.082	13.673	13.600	16.215	15.557	15.269	15.135	15.034	-3.7	-5.7
80.16670	30.57519	13.207	12.883	12.752	14.973	14.488	14.273	14.183	14.104	-5.0	-4.8
80.16988	30.59414	15.276	15.008	14.779	17.902	17.134	16.727	16.501	16.344	-9.9	-3.8
80.17392	30.58072	14.513	14.239	14.083	16.972	16.261	15.890	15.689	15.512	-3.0	-5.1
80.17400	30.57818	14.011	13.686	13.542	16.326	15.632	15.295	15.139	15.029	-1.4	-1.7
80.17703	30.58659	13.900	13.650	13.482	16.174	15.559	15.245	15.055	14.885	-3.0	-2.2
80.18008	30.55041	13.721	13.287	13.111	16.097	15.401	15.046	14.934	14.786	-5.9	-7.1
80.18151	30.53378	15.532	15.069	14.871	18.040	17.273	16.887	16.659	16.462	-1.7	-0.1
80.18355	30.59289	12.407	12.103	12.025	14.096	13.745	13.549	13.453	13.305	7.3	2.2
80.18609	30.55750	12.485	12.387	12.258	13.731	13.494	13.404	13.357	13.274	0.1	-7.9

3. Summary

Old open clusters such as Berkeley 17 provide a link to globular clusters to shed light on how star clusters form, age, and disintegrate to supply the field star population. Berkeley 17 shows a clear mass segregation between the cluster core and tails, with the low-mass then-members already escaped. Possibly the cluster is being tidally pulled, evidenced by its unusual radial velocity –84 kms^{-1}, suggestive of a large space U velocity [10], by the Perseus spiral arm in front of it. Still, the cluster manages to remain bound for its longevity with rich membership, likely owing to its location at the outer Galactic disk relatively in isolation from other perturbing sources (e.g., encounters with molecular clouds or star clusters, shear disruption by differential rotation) during its orbital course. It is plausible that Berkeley 17 is the remnant of a super star cluster, which as a class is considered as precursors of globular clusters [11]. Deeper photometry than Pan-STARRS data are required to identify the level of depletion of low-mass members in the cluster, and even the surrounding debris then-members.

The blue straggler population in Berkeley 17 should be investigated further. The scenario of merging binaries ("reviving" on the main sequence) is favoured in crowded environments such as in globular clusters. On the other hand, low metallicity is known to contribute at least partly to a horizontal branch extending to the blue. Spectroscopic classification of the blue straggler "candidates" presented in table 1 is a natural next step to clarify their nature as bona fide blue stragglers versus blue horizontal-branch stars or field stars.

Acknowledgments

Part of this work is financially supported by the grant MOST 103-2112-M-008-24-MY3. WPC is grateful to the organizing committee of FTAPS, and the hospitality of American University of Sharjah, UAE during the conference.

References

[1] Bragaglia A, Tosi M, Andreuzzi G and Marconi, G 2006 *Month. Not. Royal Astro. Soc.* **368** 1971
[2] Phelps R L 1997 *ApJ* **483** 826
[3] Friel E D *et al* 2002 *Astron. J.* **124** 2693
[4] Ahumada, J A and Lapasset E 2007 *A& A* **463** 789
[5] Bhattacharya S *et al* 2016 arXiv:1611.05593
[6] Chambers K C *et al* 2016, arXiv:1612.05560
[7] Chen W P, Chen C W and Shu C G 2004 *Astro. J.* **128** 2306
[8] Xu Y, Reid M J, Zheng X W and Menten K M 2006 *Science* **5757** 54
[9] Bono G *et al* 2010 *Astrophy. J. Lett.* **708** L74
[10] Hayes C R and Friel E D 2014 *Astron. J.* **147** 69
[11] Gallagher J S and Grebel E K 2002 IAU Symp. **207** *Extragalactic Star Clusters* ed Geisler D, Grebel E K and Minniti D (San Francisco, CA: ASP) 745

Planetary plasma and atmospheres explored by space missions in Japan: Hisaki, Akatsuki, and beyond

Y Kasaba[1], T Imamura[2], F Tsuchiya[1], N Terada[1], Y Miyoshi[3], Y Kasai[4], and Y Saito[5]

[1] Tohoku University, Japan
[2] University of Tokyo, Japan
[3] Nagoya University, Japan
[4] National Institute of Information and Communications Technology, Japan
[5] Japan Aerospace Exploration Agency, Japan
E-mail: kasaba@pat.gp.tohoku.ac.jp

Abstract. Planetary plasma and atmospheres have been challenged by space missions of Japanese science community from 1990s, with ISAS and JAXA. The first trial, a Martian orbiter Nozomi, was launched in July 1998. At the departure from Earth in Dec. 1998, she met an engine trouble but we struggled and found a narrow and long path connecting to the Dec 2003 arrival, which is the simultaneous arrival with ESA Mars Express. Unfortunately, we had an additional power trouble in Apr. 2002 associated with a solar flare event, and we gave up the trial at the gate of Mars in Dec. 2003. In parallel to the Kaguya Lunar orbiter in 2007-2009, a next trial to planets, the Akatsuki orbiter to Venus, was prepared. She departed from Earth in May 2010. However, she got an engine trouble at the arrival to Venus in Dec. 2010, and we again endured another long path, but this road was at last ended by a success of the orbit entry in Dec. 2015. We also created the UV/EUV space telescope, Hisaki, using the sensor and optics technologies extracted from Nozomi. It is going well after the launch in 2013 and actively looking planetary thin atmospheres collaborating with other space missions. This paper summarizes the Hisaki and Akatsuki missions which are now on orbit, with the next missions, Arase (ERG), BepiColombo, JUICE, and beyond.

1. Hisaki, a UV/EUV space telescope dedicated to planetary exospheres

The first success from planetary plasma and atmosphere by ISAS (Institute of Space and Astronautical Science) and JAXA (Japan Aerospace and Exploration Agency) was achieved by a spacecraft on the terrestrial orbit, Hisaki UV/EUV space telescope, launched in September 2013 [1]. Its Extreme Ultraviolet spectrograph, EXCEED, provides the spectral images (520–1,480 A) of tenuous gas and plasmas. We have used this telescope for the continuous observation campaigns of planetary magnetosphere, ionosphere, and exosphere and tried the long-term dataset for days to months, which are not easy even by the planetary orbiters and ground-based telescopes.

First targets were giant planets. Hisaki has a unique capability optimized to the long-term monitoring of H_2 aurora total flux in UV and the emission of satellite-origin gas and plasma clouds in EUV. Jupiter has a huge magnetosphere, driven by the plasma supply from the volcanic activities of Io with a typical rate of 1 ton/sec. Plasmas from Io plasma torus is transported outward and extracting huge energies from the Jovian fast rotation with strong magnetic field. Hisaki's 3-month observations per year were coordinated in 2013-2016 with Hubble Space Telescope (UV), X-ray space telescopes, and ground-based Visible-Infrared observatories. Those results are summarized in another paper in the proceedings [2]. In 2016-2017, Hisaki's priority is on the support observation of NASA Juno orbiter. For Saturn, water plume is supplied from Enceladus. We have examined the temporal and spatial variations of neutral oxygen produced associated with this plume [3].

For terrestrial planets, Hisaki tried to grasp the variation of oxygen and related molecules in ionospheres and exospheres, which are connected to the amount and process of atmospheric escape. Since Mars and Venus have weaker intrinsic magnetic dipole moments than Earth, their upper atmospheres are directly affected by solar wind collisions and can escape to space. Since solar activity could be more active in the early days, the knowledge of the solar wind effect on their upper atmospheres in current condition is the basic information for the estimation for the atmospheric evolution in those days, which can also be affected from the lower atmospheric conditions. For Venus, the variations of the dayglow brightness for OII 83.4 nm, OI 130.4 nm and OI 135.6 nm in the Venus' thermosphere was detected [4]. Those emissions vary similarly with the variations of the solar EUV radiation flux. And other periodicities with 1-13 days were also identified and assumed as the density oscillations of oxygen atoms or photoelectrons in the thermosphere modulated by gravity waves and/or planetary-scale waves propagating from the lower atmosphere (figure 1). Extension of these studies is planned with the cloud-deck turbulence observations by Akatsuki (see Sec. 2). Martian observation was also tried and the upper limit of the ionospheric and exospheric emissions will be determined, collaborated with the upper atmospheric observations by NASA MAVEN orbiter.

2. Akatsuki, a Venus climate orbiter

Akatsuki was designed as the 2nd planetary orbiter following Nozomi, to investigate the meteorological processes. The orbiter was launched in May 2010, and once reached Venus in December 2010. However, by the trouble of its orbital maneuver engine, its orbit insertion failed. As the result, Akatsuki was forced the long travel as Nozomi. But this 5-years tour was successfully finished in December 2015. Although the failed engine could not be used, Akatsuki went into the initial orbit on a westward, elongated equatorial orbit with an apoapsis altitude of ~440,000 km [5]. Now, the scientific operation is on the orbit with the apoapsis altitude of ~360,000 km, a periapsis altitude of 1,000–8,000 km, and an orbital period of 10.5 days, respectively. It is not the same one planned before the launch, the current orbit is still suitable to take continuous global images of the full Venusian disks in ultraviolet (UV), near infrared (NIR), and mid infrared (MIR), like terrestrial meteorological satellites. The observed multi-wavelength images contain the information of the top (mesosphere) and the base (troposphere) of thick cloud decks, with the temporal and spatial resolution of 10s km and 1-2 hours, and provides the capability of continuous wind vector tracking.

By the combination of five imaging instruments aboard Akatsuki, multiple height levels of the atmosphere can be sensed. The lower atmosphere and the surface on the nightside are investigated by the 1-μm micrometer Camera (IR1). The middle and lower cloud region is covered by IR1 in the dayside and by the 2-μm Camera (IR2) in the nightside. The dayside cloud top is observed by the Ultraviolet Imager (UVI) and IR2. The Longwave Infrared Camera (LIR) covering 10-μm observes the cloud top temperature on both dayside and nightside. The Lightning and Airglow Camera (LAC) also searches the lightning and airglows on the nightside. And, Radio Science (RS) support those imaging observations by determining the vertical temperature profile and its spatial and temporal variability. With the connection of ground-based support observations and numerical model studies, their large- and meso-scale features of multiple height ranges can provide the information of meridional circulation, mid-latitude jets, and their various wave activities, which provides a link to the vertical and horizontal couplings, the dynamical energy flow maintaining the super-rotation.

After the start of full observations in spring of 2017, Akatsuki started to reveal several unknowns in Venus atmosphere (figure 2). (1) LIR discovered large-scale bow-shaped features in the thermal map at the cloud top of Venus [6]. The feature was first observed on 7 December 2016 and lasted for 4 days. The features do not seem to rotate at the speed of the background atmospheric super-rotation

(~100 m/s), but seem to be stationary with respect to the ground surface. Comparison with a numerical modeling suggests that those features are manifestations of gravity waves from the ground level. (2) UVI imaged Venus at two wavelengths: 283 nm in the SO_2 absorption band and 365 nm in the

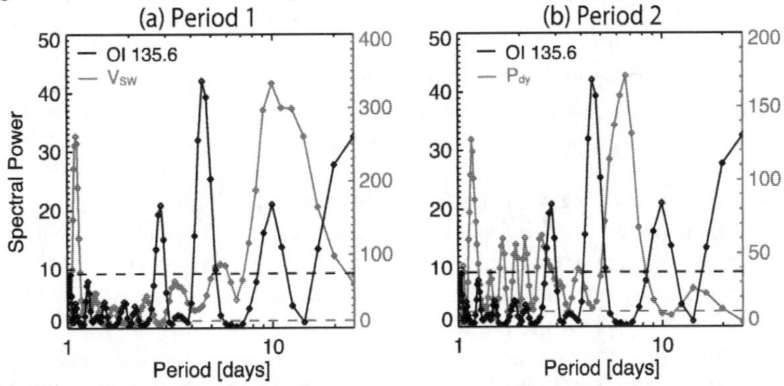

Figure 1. Period of Brightness variations of Venusian OI 135.6 nm observed by Hisaki. The period of solar wind velocity and the dynamic pressure observed by Venus Express is also shown.

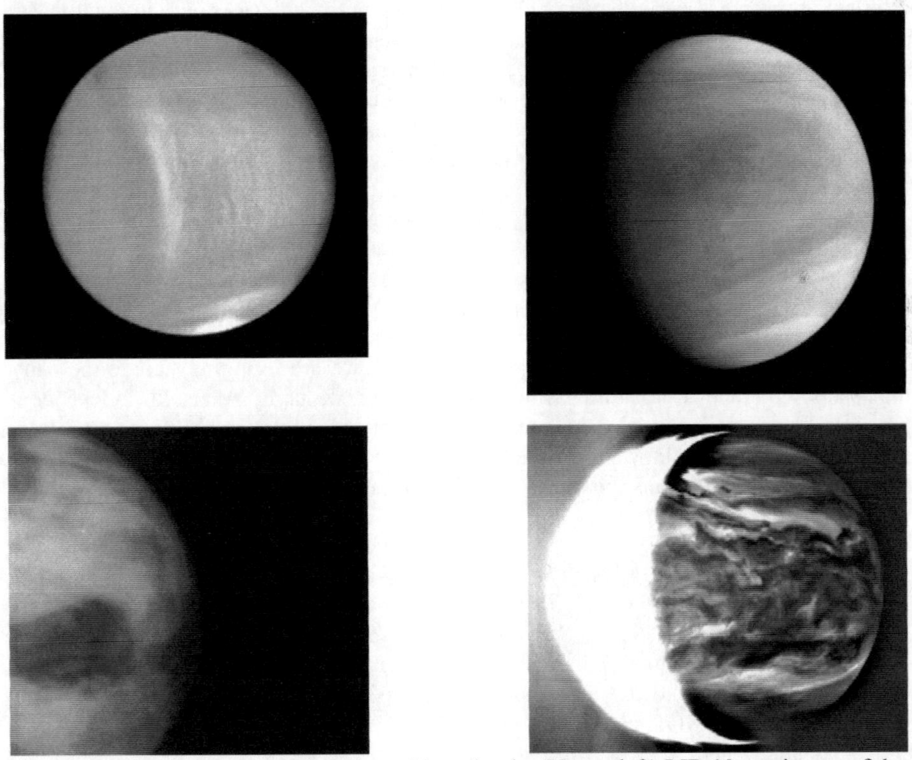

Figure 2. Venusian images taken by Akatsuki payloads. (Upper-left) LIR 10-μm image of the full disk. (Upper-right) UVI 283-nm SO_2 absorption band image of the dayside taken. (Lower-left) IR1 1.01-μm image of the night-side. (Lower-right) IR2 2.26-μm image of the nightside.

band of unknown absorber. Images at these wavelengths were not always well correlated with each other. Investigation of the chemistry and dynamics governing these species is ongoing by utilizing cloud-tracked wind data. (3) IR1 visualized the surface topography of Venus, as imaged at 1.01 μm on the nightside disk. Details of Aphrodite (elevation: 3 to 5 km) were successfully captured. IR1 also captured small-scale structures in the cloud layer on the dayside, which are clues to the cloud formation process. (4) IR2 acquired 2.26 μm images of night-side disk of Venus, in which waves and turbulences in various scales are visualized. On dayside, 2.02 μm CO_2-absorption band images showed the reflected sunlight image indicates the co-existence of super-rotating cloud features and topography-fixed features. (5) RS obtained vertical profiles of the atmospheric temperature with radio occultation technique. The profiles show significant temporal/spatial variability, the cause of which will be studied by combining RS and camera data. Initial results will be published as the first special issue at the end of 2017. We welcome interests and supports from international scientific community.

3. ERG, Collaborations with ESA (BepiColombo and JUICE), and beyond

As the next mission, we successfully launched the terrestrial inner magnetosphere mission, ERG, on 20 December 2016, and named as Arase ('a river raging with rough white water') [7,8]. In January 2017 we finished the initial operations including antenna extensions soon started the investigation of the formation and loss of the terrestrial radiation belt, as a new comer of space weather observatories with the international supports of ground-based observations and numerical modeling study teams.

We designed the international collaborations for further planets than Venus and Mars. Now we are waiting two missions with European scientists. The first is BepiColombo, ESA-JAXA joint mission to Mercury [9]. We provide the Mercury Magnetospheric Orbiter (MMO), attached on and transported with the European Mercury Planetary Orbiter (MPO), with hardware contributions to MGF (magnetic field), PWI (plasma waves), MPPE (charged and energetic neutral particles), MDM (dusts), MSASI (sodium exosphere imaging) aboard the MMO and Phebus (UV imaging) aboard the MPO. This project started in early 2000's, and the MMO spacecraft is already shipped from Japan (JAXA) to Netherlands (ESA). After the launch in 2018 and the long transfer including Venus flyby observations, we will observe the Mercury system from 2025 with two years as multi-spacecraft campaign. The next is the Jupiter ICy Moon Explorer (JUICE), targeting the Jovian system and icy moons with the final orbit around Ganymede. We will directly provide hardware contributions to three teams, GALA (laser altimeter), PEP (charged particles), and RPWI (radio and plasma waves). After the launch in 2022, we will investigate the largest planetary system and icy moons with sub-surface water worlds from 2030.

From 2016, JAXA also started a mission study of a Mars - Phobos sample return mission (the planned launch year: 2024). Using the visible and near infrared spectral imagers for the composition distribution of global and sampling points of Phobos, we are trying to include the capability of the atmospheric continuous monitoring which are done for Venus by Akatsuki. Independently, a piggy-bag-sized Mars mission is also planned in the team led by National Institute of Information and Communications Technology and University of Tokyo. Although those are not yet formal or approved status, the brief introductions and our background strategy for those missions will also be presented.

References

[1] Yoshikawa I *et al* 2014 *Space Sci. Rev.* **184** 237-258
[2] Tsuchiya F *et al. in this proceeding*
[3] Tadokoro H *et al* 2015 *Proc Symp. Planet. Sci* sps (ed. Tohoku Univ.) http://pparc.gp.tohoku.ac.jp/ collegium/proc/sps2015/sps2015_06_tadokoro.pdf.
[4] Masunaga K *et al* 2015 *J. Geophys. Res.* **120** 2037-2052
[5] Nakamura M *et al Earth Planet. Space* **68** 1
[6] Fukuhara M *et al* in press *Nature Geoscience*
[7] Miyoshi Y *et al* 2012 *AGU monograph* **199** 103-116
[8] Miyoshi Y *et al in this proceeding*
[9] Benkhoff J *et al* 2010 *Planet. Space Sci.* **58** 2-20

Geospace exploration project: Arase (ERG)

Y Miyoshi[1], Y Kasaba[2], I Shinohara[3], T Takashima[3], K Asamura[3], H Matsumoto[3], N Higashio[3], T Mitani[3], S Kasahara[3], S Yokota[3], S Wang[4], Y Kazama[4], Y Kasahara[5], S Yagitani[5], A Matsuoka[3], H Kojima[6], Y Katoh[2], K Shiokawa[1], K Seki[7], M Fujimoto[3], T Ono[2] and ERG project group

[1] Nagoya University, Nagoya, Aichi, Japan
[2] Tohoku University, Sendai, Miyagi, Japan
[3] Japan Aerospace Exploration Agency (JAXA), Japan
[4] Academia Sinica Institute of Astronomy and Astrophysics (ASIAA), Taipei, Taiwan
[5] Kanazawa University, Kanazawa, Ishikawa, Japan
[6] Kyoto University, Kyoto, Japan
[7] University of Tokyo, Tokyo, Japan,

E-mail: miyoshi@isee.nagoya-u.ac.jp

Abstract. The ERG (Exploration of energization and Radiation in Geospace) is Japanese geospace exploration project. The project focuses on relativistic electron acceleration mechanism of the outer belt and dynamics of space storms in the context of the cross-energy coupling via wave-particle interactions. The project consists of the satellite observation team, the ground-based network observation team, and integrated-data analysis/simulation team. The satellite was launched on December 20 2016 and has been nicknamed, "Arase". This paper describes overview of the project and future plan for observations.

1. The ERG project

The ERG (Exploration of energization and Radiation in Geospace) is Japanese geospace exploration project. The project focuses on relativistic electron acceleration mechanism of the outer belt and dynamics of space storms in the context of the cross-energy coupling via wave-particle interactions. The project consists of the satellite observation team, the ground-based network observation team, and integrated-data analysis/simulation team. Details of the mission concept and information have been described in Miyoshi et al. [2012].

2. The Arase (ERG) satellite

The comprehensive observations for plasma/particles, fields and waves near the magnetic equator are important for understanding the cross energy coupling for relativistic electron accelerations and dynamics of space storms. The Arase (ERG) satellite has comprehensive instruments for plasma/particles, and field/waves are installed as shown in figure 1.

The satellite was launched on December 20 2016 and has been nicknamed, "Arase". The Arase satellite is sun-aligned spin stabilized with 7.5rpm. The apogee altitude is about 6 Re and the perigee altitude is higher than ~300 km. The inclination angle will be 31deg. After the commissioning operations for a few months after the launch, the Arase satellite will start normal operations to explore the Geospace.

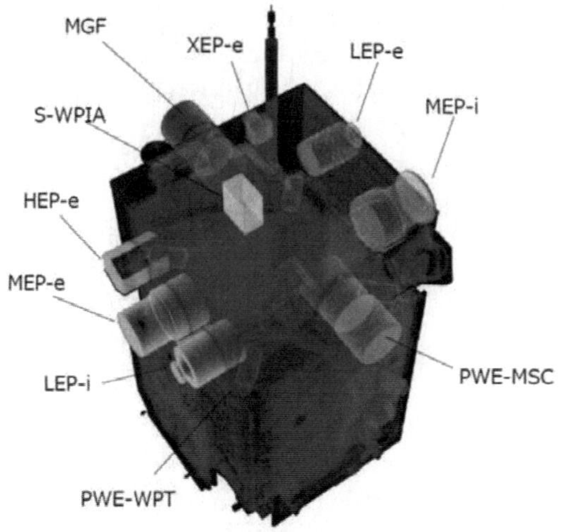

Figure 1. Appearance and Science Instruments of the Arase satellite.

2.1 Plasma/Particle Instruments

Four electron sensors (LEP-e, MEP-e, HEP-e, and XEP-e) and two ion sensors (LEP-i, and MEP-i). Electron sensors can measure electrons from 10 eV to 20 MeV, while ion sensors can measure ions from 12 eV/q to 180 keV/q with mass discrimination. The energy ranges of each detector are designed to overlap each other, which can provide seamless energy spectrum. About electron observations, both HEP-e and XEP-e instruments mainly observe relativistic electrons of the radiation belts, and these instruments are essential to derive the phase space density profile. On the other hand, LEP and MEP instruments observe hot electrons that are free energy source for plasma waves. Since anisotropies of the distribution function should be a free energy of plasma waves, observations of the distribution function is important to clarify how plasma waves generate inside the radiation belts. Measurements of particles at the energy range of tens keV is very difficult in the radiation belts. Newly developed technologies to remove background contamination can be applied in the instruments, and detail observations of tens keV electrons will be possible.

About ion observations, LEP-i and MEP-i instruments observe several ion species in the inner magnetosphere. Although there are same contamination problems as electron observations, especially, at tens keV energies, the new technology is realize to observe ions up to 180 keV/q in the radiation belts. These ion observation data will be used for study of evolution of ring current ions, and ion observations with mass discrimination are essential to study the composition of ring current particles that come from both solar wind and the ionosphere.

2.2 Electric Fields and Plasma Wave Observations

Plasma Wave and Electric Field (PWE) instrument observes electric fields at the frequency range from DC to 10 MHz as wells as the magnetic field at the frequency range from a few Hz to 20 kHz. The electric field is measured by two pairs of wire dipole antennas, and its length is about 30 m tip-to-tip. The high-frequency magnetic field is measured by the two orthogonal search coils.

There are various kinds of plasma waves in the inner magnetosphere. Whistler mode chorus waves [2] and the ion Bernstein mode waves will be important for non-adiabatic acceleration to generate relativistic electrons. Electromagnetic ion cyclotron (EMIC) waves that are generated from ring current ions will work for rapid pitch angle scattering of relativistic electrons. Whistler mode hiss waves inside plasmapause work for the pitch angle scattering of electrons. The PWE instrument can observe the frequency spectrum and wave-form of these plasma waves. The MHD pulsations with ~5 min periods are a driver for adiabatic acceleration by radial diffusion, which can be observed by the PWE instrument as well as the MGF instrument. Thermal plasma density that is important information for wave-particle interactions is determined from cutoff-frequency of the upper-hybrid resonance waves.

2.3 Magnetic Field Observations
Magnetic Field Experiment (MGF) instrument observes the ambient magnetic field as well as the MHD pulsations. The fluxgate sensor with the boom is used for measurements. Observations of ambient magnetic field are a key to know ambient plasma environment around the Arase satellite. The plasma distribution function and pitch angle distribution is obtained using the ambient magnetic field. The local cyclotron frequency is also determined from the MGF measurement. The MGF instrument observes MHD pulsations and EMIC waves as well as the PWE instrument. Since the ring current evolution produces distortions of the ambient magnetic field, and its distortion affects the particle distribution and trajectories in the inner magnetosphere, the accurate measurements of magnetic field deviation from the intrinsic magnetic field is important to evaluate the ring urrent effect. The MGF instrument can measure such deviations of magnetic fields during space storms.

2.4 Wave-particle Interaction Analyzers
In order to measure the wave-particle interactions, that is energy conversion process between plasma/particles and waves, the newly developed S-WPIA system will be installed in the Arase satellite. The vector cross-product of the particle velocity and the electric field velocity should be equal to the time derivative of the kinetic energy of particles; the positive means the acceleration of particles by waves, while the negative means the growth of waves. Therefore, the relative phase between the electric field and the velocity of particles determines the direction of the energy flow. Using the particle data from MEP-e/HEP and the wave form data from the PPE, the S-WPIA system can calculate the relative phase between the waves and the particles for each particle. This will be the first observation to identify directly the wave-particle interaction process in space, and we can observe the cross-energy coupling process via wave-particle interaction [3]. The images for observations of wave-particle interactions is shown in figure 2.

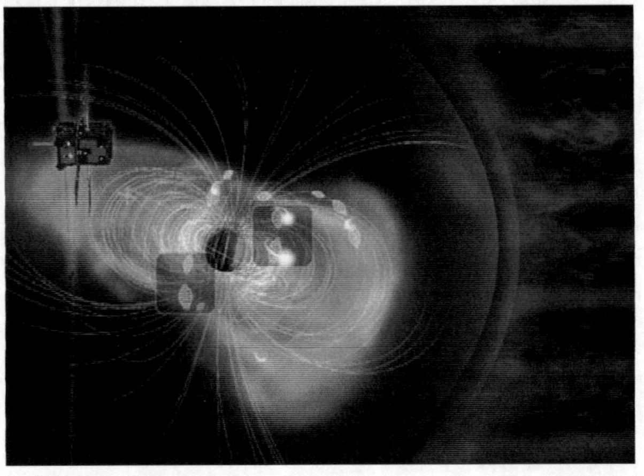

Figure 2. Artificial image of wave-particle interactions in the inner magnetosphere by the Arase satellite.

3. Ground-based network observations and integrated analysis

Observations from the ground will play a role to complement the Arase satellite observations. SuperDARN worldwide HF radar networks, magnetometer networks, optical image networks, riometer/VLF observation groups join the ERG project. These ground network observations provide global variation of electric field, magnetic field, current system, the plasma/particle distribution. Moreover, the riometer/aurora observations will show when and where the precipitation of energetic particles takes place by the wave-particle interactions. In this way, these ground observations are the remote-sensing tool to observe the global dynamics of the geospace.

In order to understand various kinds of data from satellite and ground observations, the integrated analysis using many data sets is essential to gain science output. Both global simulation and micro-process simulation such as wave-particle interactions are important to understand the physical process through quantitative comparisons with observations.

4. Science Data and Space Weather Data

The science data after Level-2 have been archived with NASA/CDF format in the ERG science center (http://ergsc.isee.nagoya-u.ac.jp). The science center will develop the integrated data analysis tool as a plug-in of the IDL/SPEDAS tool.

The Arase satellite provides not only the science data but also the space weather data (http://seesproxy.tksc.jaxa.jp/fw_e/dfw/SEES/English/Top/top_e.shtml). The Arase space weather data includes quasi-real time data for HEP-e, XEP-e and MGF. These data are not calibrated for the scientific purpose, but the data should be useful for monitoring the current space environment as well as inputs to drive the space weather forecast model.

5. International collaborations

Currently, a number of satellites have operated in geospace and there are many network observations at the ground. In fact, Van Allen Probes, THEMIS, MMS are operated to investigate inner magnetosphere and geospace. Russia has a plan to launch Resonance. And, several ground-based network observations provide a global-view of geospace from the ground. Simultaneous observations at different radial distance from the Earth and different local times are possible by the international fleet of satellites, which are highly desirable for the ERG project.

References

[1] Miyoshi Y *et al* 2012 *Geophys. Monogr. Ser.* **199** 103-116 (ed. D.Summers, I.R. Mann, D.N. Baker, and M Schulz) AGU Washington D.C. doi:10.1029/2012BK001304

[2] Kasahara Y, Miyoshi Y, Omura Y, Verkhoglyadova O P, Nagano I, Kimura I and Tsurutani B T 2009 *Geophys. Res. Lett.* **36** L01106 doi:10.1029/2008GL036454

[3] Katoh Y, Kitahara M, Kojima H, Omura Y, Kasahara S, Hirahara M, Miyoshi Y, Seki K, Asamura K, Takashima T and Ono T 2013 *Ann. Geophys.* **31** 503-512

Mössbauer spectroscopy of pyroxene in the light-dark structure of the Kapoeta meteorite: implications for thermal history of the Kapoeta parent body

Y A Abdu[1] and F C Hawthorne[2]

[1]Department of Applied Physics and Astronomy, University of Sharjah, P.O. Box 27272, Sharjah, United Arab Emirates.
[2] Department of Geological Sciences, University of Manitoba, Winnipeg, MB R3T 2N2, Canada.

E-mail: yabdu@sharjah.ac.ae

Abstract. We report an experimental study of the intracrystalline distribution of Fe^{2+} and Mg between the non-equivalent octahedral sites, M1 and M2, in orthopyroxene from the Kapoeta meteorite by single-crystal X-ray diffraction and Mössbauer spectroscopy. The Fe^{2+}-Mg ordering closure temperatures obtained by X-ray structure refinement for two orthopyroxene crystals from Kapoeta (357±30 °C and 411±18 °C) are very similar to those reported for diogenites, and indicate a slow cooling rate. The powder Mössbauer spectra of the light and dark regions in Kapoeta are identical, and their hyperfine parameters are characteristic of orthopyroxene. The closure temperature obtained by Mössbauer spectroscopy for the Kapoeta orthopyroxene (744 °C) is rather high, and indicates a fast cooling rate. The results are discussed in connection with the thermal history of the Kapoeta parent body.

1. Introduction

Pyroxene is one of the most common minerals in stony meteorites and on the surface of terrestrial planets and the moon. The intracrystalline Fe^{2+}-Mg exchange reaction between the M1 and M2 octahedral sites in pyroxenes records the latest thermal event of the host rock. The unshocked pyroxene slowly cools to ~ 500 °C and below. Fe^{2+} occurs at M2 site while Mg predominately occurs at M1 site. In crystals that have been rapidly cooled, a more disordered Fe^{2+}-Mg distribution is observed over the M1 and M2 sites. Kapoeta is an achondrite meteorite that belongs to the howardite group. Howardites, eucrites, and diogenites (HED) are believed to originate from igneous processes on the asteroid 4 Vesta. Kapoeta is a polymict regolith breccia with a light-dark structure, and the rock consists almost entirely of pyroxene [1].

In this paper, we report the distribution of Fe^{2+} between the M1 and M2 sites of pyroxene in the light and dark regions of the Kapoeta meteorite by Mössbauer spectroscopy and single-crystal X-ray diffraction (XRD). We compare our results with previous Mössbauer work on the whole-rock powdered sample [2] and single-crystal XRD data [3], in an attempt to better understand the thermal history of the Kapoeta parent body.

2. Experimental setup

The selection of single crystals of orthopyroxene suitable for X-ray diffraction was not easy, as the Kapoeta material is brecciated. Two homogeneous and inclusion-free orthopyroxene crystals, hereafter referred to as Xl1 and Xl2, were selected for X-ray diffraction. X-ray diffraction data were collected with a BRUKER P4 four-circle diffractometer equipped with monochromatic Mo Kα radiation and a 4K CCD area detector. X-ray diffraction intensities were collected up to 60° and unit-cell parameters were refined by using least squares. Mössbauer-spectroscopy measurements were done at room temperature (RT) using a ^{57}Co(Rh) point source. For preparing the powdered Mössbauer absorber, the sample was mixed with sugar and finely ground under acetone to avoid oxidation. The mixture was then loaded into a Pb ring (2 mm inner diameter) and covered by tape on both sides. The spectra were analyzed using a Voigt-based quadrupole-splitting-distribution (QSD) method.

3. Results, Discussion and Conclusion

The degree of Fe^{2+}-Mg ordering for each orthopyroxene crystal was determined from the intracrystalline distribution coefficient, K_D, given by:

$$K_D = [X_{Fe^{2+}(M1)} . X_{Mg(M2)}] \,/\, [X_{Fe^{2+}(M2)} . X_{Mg(M1)}] \quad (1)$$

where ($X_{Fe^{2+}(M1)}$, $X_{Mg(M1)}$) and ($X_{Fe^{2+}(M2)}$, $X_{Mg(M2)}$) are the M1 and M2 site occupancies, respectively. The closure temperature, T_C, of the Fe^{2+}-Mg ordering reaction was calculated from the relation [4]:

$$\ln K_D = -2557(\pm 49)/T(K) + 0.547(\pm 0.048) \quad (2)$$

The M1 and M2 site occupancies determined by X-ray structure refinement, along with K_D and T_C values, for the Kapoeta orthopyroxene crystals Xl1 and Xl2 are given in Table 1. The distribution of Fe^{2+} and Mg over the M1 and M2 sites gives T_C values of 411±18 °C and 357±30 °C for Xl1 and Xl2, respectively, in accord with those previously reported for orthopyroxene crystals from Kapoeta having similar compositions to our orthopyroxenes [3]. The low T_C values of the Kapoeta orthopyroxene crystals indicate a slow cooling rate. The composition of these pyroxenes and their T_C values are very similar to those reported for diogenites [5], and may suggest a diogenitic origin of these orthopyroxenes deep within the parent body of the HED meteorites. The studies of orthopyroxene single crystals are not representative of the bulk Kapoeta meteorite, as it is a polymict breccia, and hence the low Tc values obtained are not representative of bulk Kapoeta breccia. To obtain a closure temperature for the bulk Kapoeta orthopyroxene, we use Mössbauer spectroscopy, as discussed below.

Table 1. M1 and M2 site occupancies, intracrystalline distribution coefficient (K_D) and closure temperature (T_C) for the Kapoeta orthopyroxene (opx).

	Opx-Xl1 XRD	Opx-Xl2 XRD	Opx-bulk Mössbauer
$X_{Fe^{2+}(M1)}$	0.047(2)	0.018(3)	0.12(3)
$X_{Mg(M1)}$	0.953(2)	0.982(3)	0.88(3)
$X_{Fe^{2+}(M2)}$	0.548(2)	0.379(3)	0.50(5)
$X_{Mg(M2)}$	0.452(2)	0.621(3)	0.50(5)
Mg/(Mg+Fe^{2+})	0.70	0.80	0.69
K_D	0.041(2)	0.030(5)	0.14(4)
T_C (°C)	411 ± 18	357 ± 30	744 ± 119

IOP Conf. Series: Journal of Physics: Conf. Series **869** (2017) 012096 doi:10.1088/1742-6596/869/1/012096

The RT Mössbauer spectrum of the light region in the Kapoeta meteorite is shown in Fig. 1. It is fitted to a QSD model having one generalized QSD site for Fe^{2+} with two Gaussian components. The QSD profile for Fe^{2+} (Fig. 1) displays two well-resolved Gaussian components centered at QS ~ 2.6 and 2.1 mm/s. Such a QSD profile can be used to obtain Fe site occupancies [6], and we assign the two components to Fe^{2+} at the M1 (QS ~ 2.6 mm/s) and M2 (QS ~ 2.1 mm/s) sites in orthopyroxene. The spectrum of the dark region is almost identical to that of the light region, and both spectra have the same Mössbauer relative areas (within experimental error) for Fe^{2+} at the M1 and M2 sites (table 2).

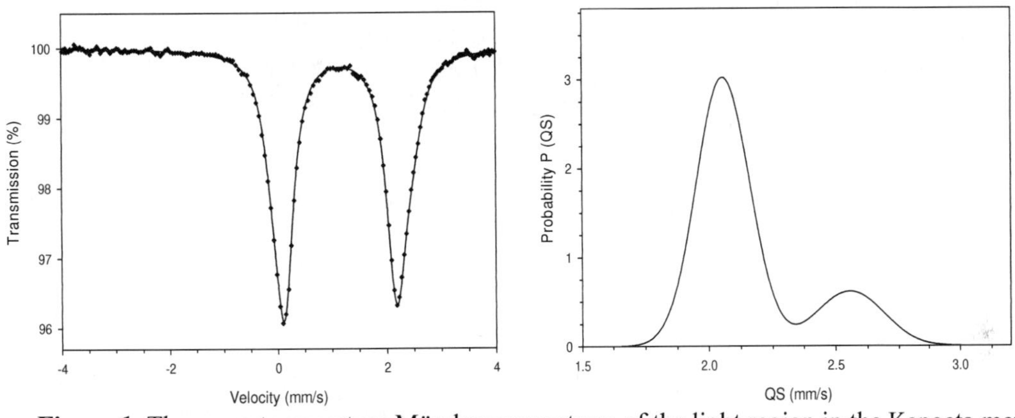

Figure 1. The room-temperature Mössbauer spectrum of the light region in the Kapoeta meteorite is shown along with the QSD profile for Fe^{2+}.

Table 2. Mössbauer hyperfine parameters for orthopyroxene (Opx) in the light and dark regions in the Kapoeta meteorite.

Sample	Fe-site	CS (mm/s)	QS (mm/s)	Γ (mm/s)	A (%)
Opx (Light region)	Fe^{2+}(M1)	1.18	2.56	0.29	20
	Fe^{2+}(M2)	1.15	2.06		80
Opx (Dark region)	Fe^{2+}(M1)	1.18	2.50	0.28	21
	Fe^{2+}(M2)	1.15	2.05		79

CS = center shift relative to α-Fe (±0.02 mm/s); QS = quadrupole splitting (±0.05 mm/s); Γ = linewidth of the elemental doublets of the QSD (±0.02 mm/s); A = relative area (±4%).

Using the Mössbauer spectroscopy the relative areas of Fe^{2+}(M1) and Fe^{2+}(M2) for orthopyroxene in the light region are observed and shown in Table 2. The site occupancies $X_{Fe^{2+}(M1)}$ and $X_{Fe^{2+}(M2)}$ are also observed and listed in Table 1 of 0.12 and 0.50, respectively. $X_{Mg(M1)}$ and $X_{Mg(M2)}$ are then calculated from 1- $X_{Fe^{2+}(M1)}$ and 1- $X_{Fe^{2+}(M2)}$, respectively, and the results are given in Table 1, along with K_D and T_C. A value of K_D equal to 0.14(4) for orthopyroxene (bulk sample) is obtained. This value is significantly smaller than that obtained from the Mössbauer spectrum of the whole-rock powdered sample (0.57) [2]. The difference between the two Mössbauer results could be attributed to the presence of carbonaceous-chondrite inclusions in the Kapoeta whole-rock sample and the different fitting procedures.

In contrast to the low T_C values determined by X-ray structure refinement, the T_C value obtained by Mössbauer spectroscopy for the Kapoeta orthopyroxene is high (744 °C, see Table 1), indicating that the rate of cooling is fast. This could be due to shock and reheating of the regolith material on or near the surface of the meteorite parent body. The interpretation of the single-crystal XRD and Mössbauer results is consistent with the nature of the Kapoeta rock (and howardites in general) being a polymict, eucrite+diogenite, breccia [7].

References

[1] Fredriksson K and Keil K 1963 *Geochim. Cosmochim. Acta* **27** 717
[2] Gismelseed A M *et al* 1996 *Conf. Proc.* **Vol. 50** 737, "ICAME-95" Ortalli I (Ed.)
[3] Domeneghetti M C et al 2000 *Meteoritics & Planetary Science* **35** 347
[4] Stimpfl M et al 1999 *Contrib. Mineral. Petrol.* **136** 297
[5] Zema M et al 1997 *Meteoritics & Planetary Science* **32** 855
[6] Abdu Y A and Hawthorne F C 2009 *Canadian Mineralogist* **47** 917
[7] Warren P M et al 2009 *Geochim. Cosmochim. Acta* **73** 5918